T0156278

Lecture Notes on Data Engineering and Communications Technologies

Volume 41

Series Editor

Fatos Xhafa, Technical University of Catalonia, Barcelona, Spain

The aim of the book series is to present cutting edge engineering approaches to data technologies and communications. It will publish latest advances on the engineering task of building and deploying distributed, scalable and reliable data infrastructures and communication systems.

The series will have a prominent applied focus on data technologies and communications with aim to promote the bridging from fundamental research on data science and networking to data engineering and communications that lead to industry products, business knowledge and standardisation.

** Indexing: The books of this series are submitted to ISI Proceedings, MetaPress, Springerlink and DBLP **

More information about this series at http://www.springer.com/series/15362

Kuo-Ming Chao · Lihong Jiang ·
Omar Khadeer Hussain ·
Shang-Pin Ma · Xiang Fei
Editors

Advances in E-Business Engineering for Ubiquitous Computing

Proceedings of the 16th International
Conference on e-Business Engineering
(ICEBE 2019)

 Springer

Editors
Kuo-Ming Chao
ECB, Faculty of Engineering
and Computing
Coventry University
Coventry, UK

Omar Khadeer Hussain
School of Business
University of New South Wales
Canberra, ACT, Australia

Xiang Fei
Faculty of Engineering and Computing
Coventry University
Coventry, UK

Lihong Jiang
School of Software
Shanghai Jiaotong University
Shanghai, China

Shang-Pin Ma
National Taiwan Ocean University
Keelung City, Taiwan

ISSN 2367-4512   ISSN 2367-4520   (electronic)
Lecture Notes on Data Engineering and Communications Technologies
ISBN 978-3-030-34985-1   ISBN 978-3-030-34986-8   (eBook)
https://doi.org/10.1007/978-3-030-34986-8

This Springer imprint is published by the registered company Springer Nature Switzerland AG
The registered company address is: Gewerbestrasse 11, 6330 Cham, Switzerland

Message from the ICEBE 2019 General Chairs

On behalf of the conference committee for ICEBE, it is our pleasure to welcome you to the 16th IEEE International Conference on e-Business Engineering (ICEBE 2019) and to Fudan University, China.

ICEBE is a prestigious conference which is initiated from 2003 by the IEEE Technical Committee on e-commerce. It is a high-quality international forum for researchers and practitioners from different areas of computer science and information systems to exchange their latest findings and experiences, as well as to help shape the future of IT-transformed consumers, enterprises, governments and markets. We look forward to many productive discussions during the conference and many memorable moments for the conference participants.

ICEBE 2019 is jointly organized by the Technical Committee on Business Informatics and Systems (TCBIS), which is formerly known as Technical Committee on Electronic Commerce (TCEC), China National Engineering Laboratory for E-Commerce Technologies, Tsinghua University, Fudan University, SAP Labs China and Coventry University. Strong support and valuable guidance from many organizations have made this year's conference possible. We are grateful for their generous support. The success of the conference also depended on all the chairs and members of the ICEBE 2019 committees. We would like to give our special thanks for their hard work and precious time to make this a success.

We all look forward to many excellent technical and social interactions during this year's conference. We encourage all of you to fully participate in the technical and social events. We wish you an enjoyable and impressive meeting in Shanghai and thank you for attending!

Jen-Yao Chung
Yinsheng Li
Hongming Cai
Kuo-Ming Chao
General Chairs

Ci-Wei Lan
Jingzhi Guo
Feng Tian
General Vice-chairs

Message from the ICEBE 2019 Program Chairs

Welcome to Shanghai, the hosting city of ICEBE 2019. Shanghai is not only China's financial capital but also a global financial center with the world's busiest container port. ICEBE 2019 features a wide spectrum of topics touching on many of the most important challenges faced by practitioners as well as researchers in e-business engineering. As the program chairs of the conference, we hope that you will enjoy the technical program that we have put together for you and will find your participation in this conference most fruitful and rewarding.

ICEBE 2019 has a total of nine tracks, including the new industry track addressing the application of emerging applications in the digital enterprise. Due to fast developing e-business applications, the number and the quality of the submissions keep raising. We received 86 submissions this year. To ensure quality and fairness, each paper was reviewed by two to five reviewers whose recommendations formed the basis of the decisions. The final decisions were then made after a careful process of scrutinizing and weighing the reviewers' comments and recommendations by the track chairs and program chairs. Twelve papers were accepted as long papers, which represent an acceptance rate of 13.9%. Additional 35 papers were accepted as regular papers.. The accepted papers represent 31 countries and regions around the world, with the top spots occupied by China, Taiwan, UK and Australia. Congratulations to all authors!

We are indebted to all the track chairs for their support, dedication and hard work throughout the entire process of paper submission, review and selection:

- Adriana Giret, Universidad Politecnica de Valencia, Spain
- Grigoris Antoniou, University of Huddersfield, UK
- Raymond Y. K. Lau, City University of Hong Kong, Hong Kong
- Chunping Li, Tsinghua University, China
- Fu-Ming Huang, Soochow University, Taiwan
- Morteza Saberi, UNSW Canberra, Australia
- Boyi Xu, Shanghai Jiao Tong University, China
- Diaiping Hu, Shanghai Jiao Tong University, China
- Hongwei Wang, Zhejiang University, China

- Chang-Tsun Li, Deakin University, Australia
- Tao Zhang, Northwestern Polytechnical University, China
- Hsi Min Chen, Feng Chia University, Taiwan
- Jiang Jinlei, Tsinghua University
- Yu-Sheng Su, National Taiwan Ocean University, Taiwan
- Shin-Jie Lee, National Cheng Kung University, Taiwan
- Wen-Tin Lee, National Kaohsiung Normal University, Taiwan
- Abderrahmane Leshob, University of Quebec at Montreal, Canada
- Muhammed Younas, Oxford Brookes University, UK
- Jiaqi Yan, Nanjing University, China
- Khaled Salah, Khalifa University, UAE
- Lei Xu, SAP, China
- Zhiyuan Fang, Sun Yat-sen University, China

We cannot express our gratitude to the members of the Program Committee and other external reviewers. Without their timely and quality reviews, we would not have been able to arrive at our final program, which is so important for the success of ICEBE 2019. We are also deeply indebted to Ci-Wei Lan who has played a strong part in the conference organization. Finally, we would like to thank all the authors who have submitted their latest work to the conference. We would not have been able to put together a strong and interesting program without their support and contribution.

We are looking forward to a thought-provoking and informative conference. We appreciate your active engagement.

Omar Hussain
Lihong Jiang
Program Chairs

Shang-Pin Ma
Program Vice-chair

Organization

ICEBE 2019 Organizing Committee

General Chairs

Jen-Yao Chung	Quanta Cloud Technology, USA
Yinsheng Li	Fudan University, China
Hongming Cai	Shanghai Jiao Tong University, China
Kuo-Ming Chao	Coventry University, UK

General Vice-chairs

Ci-Wei Lan	IBM China Software Development Lab, Taiwan
Jingzhi Guo	University of Macau, China
Feng Tian	Xi'an Jiaotong University, China

Program Chairs

Omar Hussain	University of New South Wales, Australia
Lihong Jiang	Shanghai Jiao Tong University, China

Program Vice-chair

Shang-Pin Ma	National Taiwan Ocean University, Taiwan

Advisory Committee

Ying Huang	Lenovo, China
Ray Farmer	Coventry University, UK
Liangzhao Zeng	Apple Inc., USA
Hui Lei	IBM, USA

Kwei-Jay Lin   University of California, Irvine, USA
Zhiyuan Fang   Sun Yat-sen University, China

Local Organizing Committee Chair

Yinsheng Li   Fudan University, China

Agent for e-Business Track Chairs

Adriana Giret   Universidad Politecnica de Valencia, Spain
Grigoris Antoniou   University of Huddersfield, UK

Big Data for e-Business Track Chairs

Raymond Y. K. Lau   City University of Hong Kong, Hong Kong
Chunping Li   Tsinghua University, China
Fu-Ming Huang   Soochow University, Taiwan
Morteza Saberi   UNSW Canberra, Australia

Internet of Things (IoT) Track Chairs

Boyi Xu   Shanghai Jiao Tong University, China
Diaiping Hu   Shanghai Jiao Tong University, China

Mobile and Autonomous Computing Track Chair

Hongwei Wang   Zhejiang University, China

Security, Privacy, Trust, and Credit Track Chairs

Chang-Tsun Li   Deakin University, Australia
Tao Zhang   Northwestern Polytechnical University, China

Service-Oriented and Cloud Track Chairs

Hsi Min Chen   Feng Chia University, Taiwan
Jiang Jinlei   Tsinghua University
Yu-Sheng Su   National Taiwan Ocean University, Taiwan

Software Engineering for e-Business Track Chairs

Shin-Jie Lee   National Cheng Kung University, Taiwan
Wen-Tin Lee   National Kaohsiung Normal University, Taiwan
Abderrahmane Leshob   University of Quebec at Montreal, Canada

E-Commerce Trading Technologies Including Blockchain Track Chairs

Muhammed Younas	Oxford Brookes University, UK
Jiaqi Yan	Nanjing University, China
Khaled Salah	Khalifa University, UAE

Publication Chair

Xiang Fei	Coventry University, UK

ICEBE 2019 Program Committee

Agent for e-Business Track

George Baryannis	University of Huddersfield, UK
Tianhua Chen	University of Huddersfield, UK
Victor Sanchez-Anguix	Coventry University, UK
Andreas Symeonidis	Aristotle University of Thessaloniki, Greece
Ilias Tachmazidis	University of Huddersfield, UK

Big Data for e-Business Track

Alireza Abbasi	UNSW Canberra, Australia
Amin Beheshti	Macquarie University, Australia
Ripon Chakrabortty	University of New South Wales, Australia
Feng Chong	Beijing Institute of Technology, China
Hamidreza Izadbaksh	Kharazmi University, Iran
Mustafa Jahangoshai	Urmia University of Technology, Iran
Hokyin Jean Lai	Hong Kong Baptist University, Hong Kong
Qing Li	Arizona State University, USA
Xiao-Dong Li	Hohai University, China
Yanghui Rao	Sun Yat-sen University, China
Dumitru Roman	SINTEF, Norway
Mehran Samavati	The University of Sydney, Australia
Yain-Whar Si	University of Macau, Macau
Yu-Sheng Su	National Taiwan Ocean University, Taiwan
Xiaohui Tao	University of Southern Queensland, Australia
Marzie Zarinbal	Iranian Research Institute for Information Science and Technology (IRANDOC), Iran
Xiong Zhang	Beijing Jiaotong University, China

Internet of Things (IoT) Track

Chi-Hua Chen   Fuzhou University, China
Tao Dai   Donghua University, China
Hwb Hu   School of Computer, Wuhan University, China
Guoqiang Li   Shanghai Jiao Tong University, China
Ying Li   East China University of Science
                        and Technology, China
Ching-Lung Lin   Minghsin University of Science and Technology,
                        Taiwan
Chunhui Piao   Shijiazhuang Tiedao University, China
Wei Wang   University of Skövde, Sweden
Cai Xiantao   Wuhan University, China
Cheng Xie   Yunnan University, China
Shin-Jer Yang   Soochow University, Taiwan
Lijuan Zheng   Shijiazhuang Tiedao University,
                        School of Information Science
                        and Technology, China

Mobile and Autonomous Computing Track

Aylmer Johnson   University of Cambridge, UK
Xiaodong Liu   Edinburgh Napier University, UK
Yusheng Liu   Zhejiang University, China
Gongzhuang Peng   University of Science and Technology Beijing,
                        China
Miying Yang   University of Exeter, UK
Heming Zhang   Tsinghua University, China

Security, Privacy, Trust, and Credit Track

Irene Amerini   University of Florence, Italy
Ahmed Bouridane   Northumbria University, UK
Lennon Chang   Monash University, Australia
Jiankun Hu   University of New South Wales, Australia
Amani Ibrahim   Deakin University, Australia
Richard Jiang   Lancaster University, UK
Hae Yong Kim   University of São Paulo, Brazil
Yue Li   Nankai University, China
Xufeng Lin   Deakin University, Australia
Massimo Tistarelli   University of Sassari, Italy
Theo Tryfonas   University of Bristol, UK
Weiqi Yan   Auckland University of Technology,
                        New Zealand

Service-Oriented and Cloud Track

Che-Cheng Chang	Feng Chia University, Taiwan
Hai Dong	RMIT University, Australia
Yucong Duan	Hainan University, China
Marcelo Fantinato	University of São Paulo, Brazil
Yinjin Fu	PLA University of Science and Technology, China
Fazhi He	Wuhan University, China
Donghui Lin	Kyoto University, Japan
Jianxun Liu	Hunan University of Science and Technology, China
Xiaoyi Lu	The Ohio State University, USA
Shang-Pin Ma	National Taiwan Ocean University, Taiwan
Chang-Ai Sun	University of Science and Technology Beijing, China
Ruizhi Sun	China Agricultural University, China
Paul Townend	University of Leeds, UK
Guiling Wang	North China University of Technology, China
Hailong Yang	Beihang University, China
Yun Yang	Swinburne University of Technology, Australia

Software Engineering for e-Business Track

Nik Bessis	Edge Hill University, UK
Anis Boubaker	École de Technologie Supérieure (ETS), Canada
Hongming Cai	Shanghai Jiao Tong University, China
Yong-Yi Fanjiang	Fu Jen Catholic University, Taiwan
Pierre Hadaya	UQAM, Canada
Kuo-Hsun Hsu	National Taichung University of Education, Taiwan
Jong Yih Kuo	National Taipei University of Technology, Taiwan
Chien-Hung Liu	National Taipei University of Technology, Taiwan
Shang-Pin Ma	National Taiwan Ocean University, Taiwan
Laurent Renard	UQAM, Canada
Feng Tian	Xi'an Jiaotong University, China
Hsiao-Ping Tsai	National Chung Hsing University, Taiwan

E-Commerce Trading Technologies and Blockchain Track

Adnan Akhunzada	RISE Research Institutes of Sweden, Sweden
Junaid Arshad	University of West London, UK
Muhammad Ajmal Azad	University of Derby, UK

Pengfei Chen	Sun Yat-sen University, China
Fan Jing Meng	IBM, China
Marek Ogiela	AGH University of Science and Technology, Poland
Aneta Poniszewska-Maranda	Institute of Information Technology, Lodz University of Technology, Poland
Muhammad Habib Rehman	FAST NU, Pakistan

Industry Track

Zhenjia Hu	SAP Labs China, China
Liu Jia	SAP Labs China, China
Yongyuan Shen	SAP Labs China, China
Leon Xiong	SAP Labs China, China
Leiyi Yao	SAP Labs China, China
Keguo Zhou	SAP Labs China, China

Message from the IoS 2019 Workshop Co-chairs

Welcome to the Second International Workshop on Internet of Services and Applications (IoS 2019) in Shanghai, China.

IoS is being held second time in conjunction with ICEBE 2019. The IoS workshop mission is to provide an international forum for the IoS researchers and practitioners around the world to present their research results and exchange their ideas and experience.

It is a pleasure to be part of an endeavor that brings in the emerging use of IoS and challenges associated with it, while outlining the way forward. I hope that IoS 2019 Workshop will be a success in bringing together IoS researchers, foster thought-provoking discussion and set future research directions. The workshop will place emphasis on more thorough discussions and project future trends in IoS applications and their associated technologies.

Finally, I would like to thank program committee member for their excellent teamwork.

Nazaraf Shah
Chi-Hua Chen
IoS 2019 Workshop Co-chairs

IoS 2019 Workshop Co-chairs and Program Committee

Workshop Co-chairs

Nazaraf Shah	Coventry University, UK
Chi-Hua Chen	Fuzhou University, China

Program Committee

Ci-Wei Lan	IBM Taiwan
Chi-Hua Chen (Co-chair)	Fuzhou University, China
Jen-Hsinag Chen	Shih Chien University, Taiwan
Farookh Khadeer Hussain	University of Technology Sydney, Australia
Ahmad Al-Daraiseh	American University of Madaba, Jordan
Zahid Usman	Rolls-Royce, UK
Muhammad Usman	Quaid-i-Azam University, Pakistan
Sergiu-Dan Stan	Technical University of Cluj-Napoca, Romania
Tomasz Zlamaniec	Ocado Technology, UK
Rong Zhang	East China Normal University, China
Guodong Long	University of Technology Sydney, Australia
Xiufeng Liu	Technical University of Denmark, Denmark
Kashif Saleem	King Saud University, Saudi Arabia
Santhosh John	Middle East College, Oman
Seyed Mousavi	Coventry University, UK
Khoula Alharthy	Middle East College, Muscat, Oman

Keynote Abstracts

SAP: An Engine for Intelligent Enterprise Transformation Under Experience Economy

Bill Xu
Director of Strategic Planning and Operation of SAP Labs China
Head of SAP Labs China Internal Innovation & Incubation
Head of SAP Labs China Digital School
Master of Software Engineering, Shanghai Jiao Tong University
Professional Project Manager and Certified PMP
Division Director of Division Q ToastMasters International (2015–2016)

Abstract. Now, we are approaching the fourth industrial revolution that enterprises face the challenges of using big data and intelligent technologies (AI, analytics, IoT, blockchain) to help businesses run more smartly and efficiently. Meanwhile, experience is a new battleground and is becoming a key factor for enterprises to win and retain their customers. SAP, which is a leader in traditional ERP markets, is also leading this intelligent enterprise transformation in the experience economy. This speech will talk about how SAP transforms itself in the new industrial era and deliver the intelligent enterprises (with some examples on Industry 4.0, digital supply chain, analytic cloud). It will also explain SAP's new strategy in intelligent enterprises and how SAP combines operational data together with experience data to help enterprises win in an experience economy.

Biography

Bill Xu joined SAP BusinessObjects in 2007, is Director of Strategic Planning and Operation of SAP Labs China, Head of SAP Labs China Internal Innovation & Incubation, and Head of SAP Labs China Digital School.

Prior to this position, Bill Xu, as Global Program Manager, has been responsible for global business intelligence product development and delivery. He has rich experiences in leading international teams across countries like China, Canada, France, India with multi-culture background.

Bill Xu has taken different roles in Toastmaster International such as the Area I2 Director from 2014 to 2015 and the Division I Director from 2015 to 2016.

He graduated as a bachelor of computer science at Yancheng Institute of Technology, finished his master study at Shanghai Jiao Tong University in software engineering and continues his study at Fudan University on MiniEMA program.

Event Sequence Learning and Its Applications

Junchi Yan
Independent Research Professor (PhD Advisor) in the
Department of Computer Science and Engineering,
Shanghai Jiao Tong University
Adjunct Professor in the School of Data Science,
Fudan University

Abstract. In this talk, I will first introduce the challenge for continuous time domain event sequence learning and prediction, especially in the comparison of traditional time series modeling. Specifically, I will present our recent works in parametric temporal point process-based approaches as well as the more recent neural network-based point process models. Some applications in preventative maintenance, health care and social media analysis are also briefly discussed. The talk ends up with an outlook for the potential extension of current works.

Biography

Dr. Junchi Yan is currently Independent Research Professor (PhD Advisor) in the Department of Computer Science and Engineering, Shanghai Jiao Tong University. He is also affiliated with the Artificial Intelligence Institute of SJTU and Adjunct Professor in the School of Data Science, Fudan University. Before that, he was Research Staff Member in IBM Research, China, where he started his career since April 2011. He obtained his Ph.D. in the Department of Electronic Engineering from Shanghai Jiao Tong University, China. His work on graph matching received the ACM China Doctoral Dissertation Nomination Award and China Computer Federation Doctoral Dissertation Award. His research interests are machine learning, data mining and computer vision. He serves as Associate Editor for IEEE Access, (Managing) Guest Editor for IEEE Transactions on Neural Network and Learning Systems, Pattern Recognition Letters, Pattern Recognition, and Vice Secretary of China CSIG-BVD Technical Committee, and on the executive board of ACM China Multimedia Chapter. He has published 50+ peer-reviewed papers in top venues in AI and has filed 20+ US patents. He won the Distinguished Young Scientist of Scientific Chinese for the year 2018.

Contents

Domain Knowledge Synthesis for e-Business Management

Application of Big Data Analytics for Facilitation of e-Business

Analytics as a Service for e-Business

Data and Big Data for e-Business

Semantic Document Classification Based on Semantic Similarity Computation and Correlation Analysis

Shuo Yang[1(✉)], Ran Wei[2], and Jingzhi Guo[3]

[1] School of Computer Science and Cyber Engineering, Guangzhou University, Guangzhou, China
yangshuo@gzhu.edu.cn
[2] Department of Computer Science, University of California, Irvine, CA, USA
[3] Faculty of Technology and Science, University of Macau, Macau, China

Abstract. Document (text) classification is a common method in e-business, facilitating users in tasks such as document collection, analysis, categorization and storage. However, few previous methods consider the classification tasks from the perspective of semantic analysis. This paper proposes two novel semantic document classification strategies to resolve two types of semantic problems: (1) polysemy problem, by using a novel semantic similarity computing strategy (SSC) and (2) synonym problem, by proposing a novel strong correlation analysis method (SCM). Experiments show that the proposed strategies improve the performance of document classification compared with that of traditional approaches.

Keywords: Semantic document · Document classification · Semantic similarity · Semantic embedding · Correlation analysis · Machine learning

1 Introduction

Automatic document classification is applied in numerous electronic business (e-business) scenarios [1,16]. For example, a medium-sized company may receive quite a few emails daily without accurate and concrete information such as recipient's name or department, which have to be read by an assigned agent so that the destinations can be determined. Thus, it is possible that an automatic document classification system can reduce human workload to a great extent.

More generally, given the rapid growth of web digital documents, it is often beyond one's ability to categorize information by reading thoroughly the pool of documents. Accurate and automatic text classification techniques are hence needed to classify the incoming text documents into different categories such as news, contracts, reports, etc. Users can hence estimate the content and determine the priorities of each document, maintaining more organized working schedule.

© Springer Nature Switzerland AG 2020
K.-M. Chao et al. (Eds.): ICEBE 2019, LNDECT 41, pp. 3–18, 2020.
https://doi.org/10.1007/978-3-030-34986-8_1

A typical method of automatic text classification is that given a training set of documents with known categorical labels and word dependency information, calculate the list of possibilities for each test document on the each label assigned. Certainly, the label with the highest likelihood corresponds to the predicted category that a test document belongs to. Classical machine learning (ML) algorithms such as Bayesian classifier, decision Tree, K-nearest neighbor, support vector machine and neural network were often applied in text classification [11]. In recent years deep learning algorithms are also introduced in these tasks. One representative trial was the application of convolutional neural network (CNN), a powerful network in computer vision [12]. Recurrent neural network, which can capture information that has been calculated so far, was later introduced and became a popular method to handle sequence-formed information, with satisfactory classification performance [24].

However, most strategies mentioned above seldom view the classification problem from the perspective of semantic analysis. For example, the traditional Bayesian-based text classification method constructs a classification model based on the frequencies of some feature words in corpus. Unfortunately, it does not consider polysemous words (a word which holds different meanings depending on the context) and synonymous words (different words which hold a similar meaning) for semantic analysis during the classification procedure. For example, the Chinese word "Xiaomi" can mean either an agricultural product or a high-tech company; hence documents including "Xiaomi" possibly be classified as "agriculture" or "technology" when using the traditional Bayesian method. Similar problems also exist in the classification of English documents. For example, English documents containing the word "program" may not only represent computer code programs and be classified as "computer", but also represent a scheduled radio or television show and be classified as "entertainment".

On the other hand, synonymous words can also cause mis-classification of documents. For example, the word "people" is synonymous with "mass" and "mob" and they may occur in documents with various topics (e.g., architecture, culture and history). Therefore, choosing these words as features of the classification model may cause classification errors. These situations also exist in document classification tasks of word-embedding-based deep learning methods. For example, during feature extraction procedure the word dependence is calculated based on the statistical analysis on the posterior probability of a word following another one. However, a single embedding cannot represent multiple meanings, while similar embeddings may refer to different topic types.

More precise descriptions of the two problems:

(1) **Problem of polysemy**: some words have multiple meanings, which may lead to mis-classification of documents;
(2) **Problem of synonym**: different words with similar meanings are often used in different scenarios, but when they appear in an article at the same time, it may lead to mis-classification of documents;

In the later sections of this paper, the authors will try to resolve these two research problems.

Khan et al. [11] suggested that semantic analysis could help enhance the performance of classification. In practice, semantic analysis is generally implemented by the introduction of ontology that represents terms and concepts in domain-wise manner, and the domains are pre-defined by expert knowledge bases [11]. Although a few attempts have been made, such as using ontological knowledge [6] and WordNet for word sense disambiguation (WSD) [13,21], so far limited progresses have been achieved. This is mainly due to domain constraint of ontology or the ambiguity across different natural languages, which may lead to polysemy and synonymy issues [19] and finally result in uncertainty of document classification [7].

In this research, we report a novel semantic embedding and similarity computing approach to implement semantic document categorization. The *first* strategy aims to solve polysemy problem by using a novel semantic similarity computing method (SSC) so that the most context-fitting meaning of a word in one sentence can be determined by referring to the meaning of similar sentences expressing this word in a common dictionary. In this paper, *CoDic* [8,22] and *Hownet* [5] are used as common dictionaries for meaning determination and term expansion. With their help, words with ambiguity will be removed from the feature list, enabling more distinctive features to be selected. The *second* strategy aims to solve the synonym problem by adopting a strong correlation analysis method (SCM), where synonyms unrelated to the classification task are deleted. Otherwise, select the specific meaning of one word in the synonym group from the common dictionary and replace others in the same group.

2   Related Work

Automated document classification, also called categorization of document, has a history that can date back to the beginning of the 1960s. The incredible increase in online documents in the last decades intensified and renewed the interests in automated document classification and data mining. In the beginning, document classification focused on heuristic methods, that is, solving the task by applying a group of rules based on expert knowledge. However, this method was proved to be inefficient, so in recent years more focuses are turned to automatic learning and clustering approaches. These approaches can be divided into three categories based the characteristics of their learning phases:

(1) *Supervised document classification*: this method guides the whole learning process of a classifier model by providing complete training dataset that contains document content and category labels at the same time. The process of supervision is like training students using exercises with "correct" answers.
(2) *Semi-supervised document classification*: a method with a mixture of supervised and unsupervised document classification. Part of documents have category labels while the others do not.
(3) *Unsupervised document classification*: this method is executed without priori knowledge of the document categories. The process of unsupervised learning

is like that of students doing final examination which they do not have standard answers for reference.

However, regardless of whichever learning methods, many of them require the conversion of unstructured text to digital numbers in the data pre-processing stage. The most traditional (and intuitional) algorithm is one-hot representation, which uses N-dimension binary vector to represent vocabulary with each dimension stands for one word [11]. However, this strategy easily incurs the curse of dimensionality for representation of long texts. This is because a big vocabulary generates high-dimension, but extremely sparse vectors for long documents. Therefore, dimensionality reduction operation which removes redundant and irrelevant features is needed [2]. This demand is satisfied by the methodology called feature extraction/selection. The goal of feature extraction is the division of a sentence into meaningful clusters and meanwhile removing insignificant components as much as possible. Typical tasks at the pre-processing stage include tokenization, filtering, lemmatization and stemming [20]. After that, feature selection aims to select useful features of a word for further analysis. Compared with one-hot representation that generates high-dimensional, sparse vectors, an improved solution called TF-IDF produces more refined results. In this frequency-based algorithm, the "importance" of a word is represented by the product of term frequency (how frequent the word shows up in a document) and inverse document frequency (log-inverse of the frequency that documents containing such word in the overall document base) [14,20]. These two algorithms, however, clearly suffer from limitations as a result of neglecting the grammar and word relations in documents. More recently, distributed representation that illustrates dependencies between words are more widely used, as it reflects the relationships of words in one document [15]. Currently, the most widely used strategy to learn the vectorized words is to maximize the corpus likelihood (prediction-based), with the word2vec toolbox being one of the most popular tools. Implementation of this algorithm is dependent on the training of representation neural network with words in the form of binary vectors generated by one-hot representation. The weights of the network keep being updated until convergence, which generates a vector that lists the possibility of each word could follow the input word in a document [11,15].

3   Semantic Document Classification

This section proposes two novel strategies to resolve the research problems mentioned above.

3.1   Strategy to Resolve Polysemy Problem: SSC

The first strategy aims to solve polysemy problem by using a novel semantic similarity computing method. As previously mentioned, the most context-fitting meaning of a word can be determined by referring to the semantics of related sentences in a common dictionary (e.g., CoDic for English and Hownet for Chinese).

In our method, we implement the semantic similarity computing method (*SSC*) for the comparison of similarity between two sentences. The SSC splits a text document into sentences. For each word (w) in a sentence (s), all of its concepts from the dictionary are extracted based on its Part-of-speech (*PoS*) tag in the sentence. Then, semantically compare each concept of w with s and return the concept with the highest similarity score. Words without determinative meaning will be removed from the list of features, and hereby more distinctive terms are more likely to be left and selected as features. The pseudocode of the *SSC* algorithm is shown in Table 1.

Table 1. Semantic similarity computing (SSC)

Algorithm: *semantic similarity computing (SSC)*
Input: *target sentence (ts); a set of test sentences (ss)*
Output: *the most similar sentence (s in ss) to ts with its maximum similar score (max)*

```
def sentence_similarity(sentence1, sentence2)
# Tokenize and tag
   sentence1 = pos_tag(word_tokenize(sentence1))
   sentence2 = pos_tag(word_tokenize(sentence2))
# Get the synsets for the tagged words
   synsets1 = [tagged_to_synset(*tagged_word) for tagged_word in sentence1]
   synsets2 = [tagged_to_synset(*tagged_word) for tagged_word in sentence2]
# Filter out the Nones
   synsets1 = [synset1 for synset1 in synsets1 if synset1]
   synsets2 = [synset2 for synset2 in synsets2 if synset2]
   score, count = 0.0, 0
# For each word in the first sentence
   for synset1 in synsets1
   # Get the similarity score of the most similar word in the second sentence
      best_score = max([synset1.path_similarity(synset2) for synset2 in synsets2])

      # Check that whether the similarity could have been computed
      if best_score is not None
         score += best_score
         count += 1
   # Average the values
   score /= count
   return score # end of sentence_similarity function

# __main__
max = 0.0
most_similar_sentence = None

for s in ss
   value1 = sentence_similarity(s, ts)
   value2 = sentence_similarity(ts, s)
   avg_similarity = (value1 + value2) / 2
   if avg_similarity >maximum
      most_similar_sentence = s
      max = avg_similarity
print("The most similar sentence is {}, with score {}".format(most_similar_sentence, max))
```

The workflow of the *SSC* is quite simple. According to Table 1, it is clear that the first step is the segmentation of each sentence into words (*word_tokenize*) and tokenize them (*pos_tag*) with their parts of speech. Then, we get the synonym set (*synset*) for each tagged word in the sentence according to their PoS (*tagged_to_synset*). After that, we remove the none component in each synset. Next, for each synset in the first sentence (*sent1*), we compute the similarity score of the most similar word (*path_similarity*) in the second sentence (*sent2*). After computing the similarity score of all synsets of sent1 with that of sent2, an average similarity value between them can be returned. By using this method, we can acquire the similarity values between all test sentences (*ss*) and the target sentence (*ts*). In the end, the test sentence with the highest similarity value can be chosen as the most semantically similar sentence.

Cosine Similarity (CS) is an often used method to compute the similarity score between two vectors (e.g., m for $sentence_1$, n for $sentence_2$) by measuring the cosine of angle θ between them.

$$CS = \cos(\theta) = \frac{m * n}{\|m\| * \|n\|} = \frac{\sum\limits_{i=1}^{N} m_i n_i}{\sqrt{\sum\limits_{i=1}^{N} m_i^2} \sqrt{\sum\limits_{i=1}^{N} n_i^2}} \tag{1}$$

Therefore, for CS, the most important step is to convert sentences into vectors. A common way is to use the model of bag of words with TF (term frequency) or TF-IDF (term frequency-inverse document frequency). Another method is to utilize Word2Vec or self-trained word embedding to implement the mappings from words to vectors.

3.2   Strategy to Resolve Synonym Problem: SCM

There may be many synonyms in a large text, but not all of them are suitable text features. As is known to all, selecting effective text features can reduce the dimension of feature space, enhance the generalization ability of the model and reduce overfitting, so as to improve the effect and efficiency of classification and clustering [3]. Therefore, effective feature selection is particularly important. In this paper, we refine the synonym problem into a sub-problem: how to determine the degree of the relevance between a feature and the classification task and then remove the feature words in the synonym group that are less or not relevant to the classification task.

In this paper, a novel correlation analysis algorithm, named SCM, is proposed to obtain effective feature sets. The idea of the SCM contains two important considerations:

(1) The feature words with strong category discrimination ability are extracted by using the category discrimination method (CDM), and then the correlation between other feature words and categories is measured by the feature correlation analysis (FCA). That is, the selected feature is guaranteed to be the most relevant to the category first, and then the degree of correlation between other features and selected features is calculated.

(2) When a feature showing a strong correlation with the selected feature is found, the SCM will not include it into the feature candidate set even if the feature has a strong correlation with the category. Because compared with existing feature candidate set, the new undetermined features cannot provide additional category-related information. The mathematical foundation of this idea is that linear dependent vectors cannot construct the base of a vector space but orthogonal vectors can.

This paper adopts TF-IDF (Term frequency-Inverse Document Frequency) as the implementation of CDM. By applying TF-IDF to the synonym group in undetermined features, we can get a feature candidate set composed of a number of features with strong category discrimination ability. In TF-IDF, the importance of a word is represented by the product of the word frequency (i.e., the frequency with which the word appears in the document) and the inverse document frequency (i.e., dividing the total number of documents by the number of documents containing the term, and then taking the logarithm of that quotient). The formulas of TF-IDF are as follows.

$$tf_{i,j} = \frac{n_{i,j}}{\sum_k n_{k,j}} \tag{2}$$

$$idf_i = \lg \frac{|D|}{|\{D_j : t_i \in d_j\}| + 1} \tag{3}$$

$$tf\text{-}idf_{i,j} = tf_{i,j} \times idf_i \tag{4}$$

where (2) refers to the importance of a term t_i in a particular document d_j. The molecule $n_{i,j}$ is the number of occurrences of t_i in d_j, and the denominator is the sum of the number of occurrences of all words in d_j. Formula (3) is a measurement of the general importance of a word in all documents. Its molecule represents the total number of documents in the corpus. The denominator represents the number of documents containing the word t_i. Formula (4) is the product of "term (word) frequency (TF)" and "inverse document frequency (IDF)". The more important a word is to a certain category of texts, the higher its tf-idf value will be, and vice versa. Therefore, TF-IDF tends to filter out common words and retain important words to certain category of texts.

The SCM proceeds to calculate how strongly all features (in each synonym group) are related to category (C) in the feature candidate set. The formulas are as follows,

$$H(X) = \sum_{i=0}^{n} (p_i * \lg \frac{1}{p_i}) \tag{5}$$

$$H(X|Y) = \sum_j p(Y_j) \sum_i p(X_i|Y_j) \lg \frac{1}{p(X_i|Y_j)} \tag{6}$$

$$I(X|Y) = H(X) - H(X|Y) \tag{7}$$

$$Corr(X,Y) = \frac{I(X|Y) + I(Y|X)}{H(X) + H(Y)} \tag{8}$$

where X is an n-dimensional random variable and Y is a certain of class (or category). Formula (5) represents the entropy of X, that is the uncertainty of X. Formula (6) means the uncertainty of X given the occurrence of Y. Formula (7) represents the information gain between $H(X)$ and $H(X|Y)$. Formula (8) is used to measure the degree of correlation between a feature (X) and a category (Y).

According to the degree of correlation, the features in each synonym group are arranged in a descending order respectively, and then the ordered feature sequences are sent back into the feature candidate set. Select the first feature in the sequence, that is, the feature with the strongest correlation with the category (C), and remove it from the feature candidate set and put it into the feature result set.

In order to eliminate redundant features, it is necessary to calculate the degree of mutual independence between any two features (within a synonym group). Thus, this section proposes a novel feature correlation analysis method, called FCA, to exclude unnecessary features in synonym groups of the feature candidate set. The idea of the FCA is simple: if a remaining feature in the candidate set is a strong category-correlated feature, and its mutual independence with the selected feature is greater than or equal to a threshold *alpha*, it indicates that the candidate feature is independent of the selected feature, and it needs to be included in the feature result set. Otherwise, the feature is considered as redundant and should be deleted. Repeat this process until the feature candidate set is empty. The formulas are as follows:

$$IDP(X_i, X_j) = \frac{I(X_i; Y|X_j) + I(X_j; Y|X_i)}{2H(Y)} \tag{9}$$

$$I(X; Y|Z) = lg\frac{p(X|YZ)}{p(X|Z)} \tag{10}$$

where (9) is used to measure the degree of mutual independence between feature X_i and feature X_j when the category (Y) is known. Formula (10) describe the mutual information between feature X and feature Y in the case of given condition Z.

4 Experiments

This section designs experiments for a comparison between classical document classification algorithms and our improved ones.

4.1 Datasets Description

To test the reliability and robustness of our strategy, we use:

Dataset 1: a movie review dataset from Rotten Tomatoes [17,25]. This dataset contains 10662 samples of review sentences, with 50% positive comments and the remaining negative ones. The size of the vocabulary of the dataset is 18758. Since the dataset does not come with an official train/test split, we simply extract 10% of shuffled data as evaluation (dev) set to control the complexity of the model. In the next research stage, we will use 10-fold cross-validation on the dataset.

Dataset 2: 56821 Chinese news dataset. It is available in the PaddlePaddle[1], an open source platform launched by Baidu for deep learning applications. It contains 10 categories: international (4354), culture (5110), entertainment (6043), sports (4818), finance (7432), automobile (7469), education (8066), technology (6017), stock (3654) and real estate (3858).

4.2 Experiment on Neural Network (NN)

In this experiment, the baseline CNN is taken as an example to compare the performance of classical NN and the improved one with our proposed strategy in document classification. We set the same hyper-parameters to make a comparison between CNN and our method (Sem_{CNN}) (see Table 2). From Table 2, it is known that both of the two trained models are evaluated on the *dev* dataset every 100 global steps and then they are stored in checkpoints before the training process starting again. After multiple training epochs, the models stored in the checkpoint can be recovered and used for testing on a new dataset. Detailed neural network structure can be found in the open source code[2]. The experimental procedure is described as follows.

Table 2. Hyper-parameters used in CNN and Sem_{CNN}

Parameters	Values
Percentage of splitting a dataset for training, testing and validating, respectively	0.8/ 0.1/ 0.1
Dimensionality of character embedding	128
Filter sizes	3,4,5
Number of filters per filter size	128
Dropout keep probability	0.5
L2 regularization lambda	0.01
Batch Size	64
Number of training epochs	1/ 5/ 10/ 50/ 100
Evaluate model on evaluation (dev) dataset after these steps	100

---

[1] PaddlePaddle: http://www.paddlepaddle.org/.
[2] Partial source code of the experiment can be found at https://github.com/yangshuodelove/DocEng19/.

(1) Each document in the corpus will be firstly transformed into our semantic document (i.e., documents with semantics embedding) [23] by extending each polysemous word and category-correlated synonymous word with its context-fitting concepts from the common dictionary (i.e., CoDic for English and Hownet for Chinese) with the help of the SSC and the SCM strategies, which aims for accurate semantic interpretation and term expansion.   CoDic is a semantic collaboration dictionary constructed under our CONEX project [8,22,23]. In CoDic, each concept is identified by a unique internal identifier (iid). The reason of this design is to guarantee semantic consistency and interoperability of documents while transferring across heterogeneous contexts. For example, from Figs. 1 and 2, it is clear that in CoDic, the word "program" with the meaning of "a scheduled radio or television show" is uniquely labeled by an iid "0x5107df021015", while its another meaning "a set of coded instructions for insertion into a machine ..." has another unique iid "0x5107df02101c". Currently, CoDic is implemented in XML, where each concept is represented as an entry with a unique iid (see Fig. 3). It is convenient to extract all different meanings of any given word for later semantic analysis by using existed packages (e.g., *xml.etree.cElementTree* for Python and *javax.xml.parsers* for Java). Hownet as a common dictionary to handle Chinese documents is used similarly.

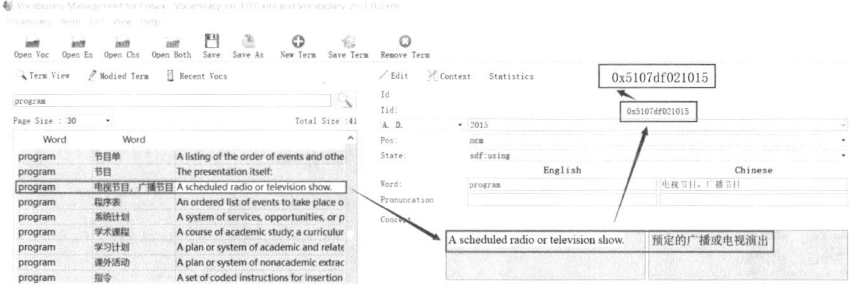

Fig. 1. Word "program" with the meaning "a scheduled radio or television show" in CoDic.

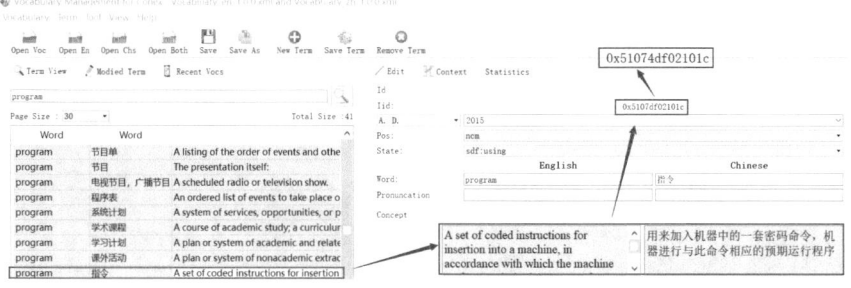

Fig. 2. Word "program" with the meaning "a set of coded instructions for insertion into a machine" in CoDic.

(2) Build a Sem_{CNN} network. The first layer embeds words and their extracted accurate concepts into low-dimensional vectors. The second layer performs convolutions over the semantic-embedded document tensors using different sized filters (e.g., $filter\_size = [3, 4, 5]$). Different sized filters will create different shaped feature maps (i.e., tensors). Third, max-pooling is used to merge the results of the convolution layer into a long feature vector. Next, dropout regularization is added in the result of max-pooling to trade-off between the complexity of the model being trained and the generalization of testing on evaluation dataset. The last layer is to classify the result using a Softmax strategy.

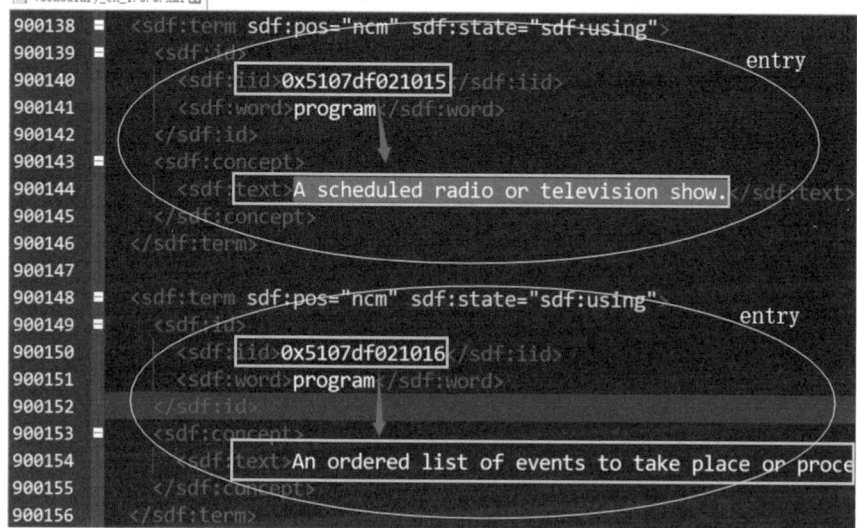

Fig. 3. CoDic in XML.

(3) Calculate loss and accuracy. The general loss function for classification problems is the cross-entropy loss which takes the prediction and the real value as input. Accuracy is another useful metric being tracked during training and testing processes. It can be used to prevent model overfitting during the model training by interrupting the training process at the turning point where the classification accuracy on the evaluation dataset starts decreasing regardless of the continuously declining error on the training dataset. The parameters taken at this critical point are then used as the model training results.

(4) Record the summaries/checkpoints during training and evaluation. After an object declaration of CNN/Sem_{CNN} class, batches of data are generated and fed into it to train a reliable classification model. While the loss and accuracy are recorded to keep track of their evolvement over iterations, some

important parameters (e.g., the embedding for each word, the weights in the convolution layers) are also saved for later usage (e.g., testing on new datasets).

(5) Test the classification model. Data for testing are loaded and their true labels are extracted for computing the performance of prediction. Then, the classification model is restored from the checkpoints, executing on the test dataset and producing a prediction for each semantic document. After that, the prediction results are compared with the true labels to obtain the testing accuracy of the classification model.

4.3   Experiment on Machine Learning (ML) Approaches

The procedures of training classification models using classical machine learning algorithms with the proposed strategy are listed as follows, while the details can be also found in our open source code.

(1) Transform words into vectors based on inputted texts (Note: Chinese document needs to execute word segmentation beforehand.). Collect all words used in texts, perform a frequency distribution and then find out effective features suitable for document classification by using the proposed strategies (SSC and SCM). After that, each text will be converted to a long word vector, where True (or 1) means a word (or a feature) exists while False (or 0) means absence.

(2) Execute multiple classical machine learning approaches (e.g., Naïve Bayes, NB) based on the word vectors from Step (1). In this experiment, three variants of NB classifier are used. They are Original NB, multinomial NB and Bernoulli NB classifier. All of them take word features and corresponding category labels as input to train classification models. It is of note that sometimes the classifier should be modified based on realistic cases. For example, in order to avoid the probability being close to zero and underflow problem in NB, it is better to initialize the frequency of each word to one and take natural log of the product in the computation of posterior probability, respectively.

(3) Save the trained classifiers for later usage. This is because the training process might be time-consuming, which depends on numerous factors such as dataset size and the computation complexity during model training. Thus, it is impractical to train classification models each time when you need them.

(4) Boost multiple classifiers to create a voting system that is taken as a baseline for comparison. To do this, we build a typical classifier (i.e., *VoteClasssifier*) with multiple basic classical machine learning classification algorithms (i.e., taking multiple basic classifier objects as input when initialized), each of which gets one vote. In *VoteClasssifier*, the *classify* method is created by iterating through each basic machine learning classifier object to classify based on the same input features. This experiment chooses the most popular metrics (e.g., accuracy) among these classifiers. The classification can be regarded as a vote. After iterating all the classifier objects, it returns the most popular vote.

4.4   Experiment Result and Analysis

In the actual testing process, we need to maintain a common synonymous word dictionary and a common polysemous word dictionary. The reason we need to maintain these two dictionaries is that the computation workload to judge polysemy and synonyms in a long text are very heavy. For example, if there are n words in a text and each word has m different meanings, then the computational complexity of determining polysemous words is $O(n*m)$, and the computational complexity of determining synonyms is $O(n*(n-1))$, so that the total computational complexity is $O(n*(m+n-1)) \geqslant O(n^2)$. Therefore, maintaining these two dictionaries can reduce computational complexity and reduce the pre-processing time of text classification.

Table 3 shows the experimental comparison between classical machine learning algorithms and their improved counterparts on Dataset 1. In this experiment, classical machine learning algorithms include Original Naïve Bayes (NB), Multinomial Naïve Bayes (MNB), Bernoulli Naïve Bayes (BNB), Logistic Regression (LR), support vector machine (SVM) with stochastic gradient descent (SGD), Linear SVC (SVC) and Nu-Support Vector Classification (NSVC).

Table 3. Comparison of classical machine learning algorithms and our improved ones on Dataset 1

Accuracy (%)		Accuracy (%)	
NB	66.265	Improved NB	78.464
MNB	65.813	Improved MNB	79.518
BNB	66.716	Improved BNB	79.819
LR	67.169	Improved LR	76.506
SGD	65.813	Improved SGD	74.096
SVC	66.716	Improved SVC	73.946
NSVC	60.09	Improved NSVC	76.355
VoteClassifier	65.663	Improved VoteClassifier	74.398

From Table 3, it is clear to see that our improved algorithms outperform the classical machine learning algorithms in the accuracy of model prediction on the evaluation dataset. It is of note that three-variant NB algorithms and LR perform better than three-variant SVM algorithms, in both of the classical ones and improved ones. The VoteClassifier plays a role of baseline for the comparison between different algorithms. Tables 4 and 5 show that Sem_{CNN} performs better than CNN in terms of accuracy and loss in different numbers of epochs. With the increase of the epoch, both of them increase in the accuracy of evaluation and decrease in the loss continuously (**before reaching overfitting**).

Table 4. Comparison of Sem_{CNN} and traditional CNN on Dataset 1.

Number of epochs	Accuracy		Loss	
	Sem_{CNN}	CNN	Sem_{CNN}	CNN
Epoch = 1	0.586	0.568	0.818	0.876
Epoch = 5	0.713	0.676	0.567	0.59
Epoch = 10	0.744	0.722	0.519	0.62
Epoch = 50	0.841	0.724	0.621	0.742
Epoch = 100	0.902	0.739	0.961	0.999

Table 5. Comparison of Sem_{CNN} and traditional CNN on Dataset 2.

Number of epochs	Accuracy		Loss	
	Sem_{CNN}	CNN	Sem_{CNN}	CNN
Epoch = 1	0.861	0.828	0.473	0.560
Epoch = 5	0.956	0.923	0.211	0.295
Epoch = 10	0.990	0.953	0.095	0.212
Epoch = 20	0.990	0.966	0.0919	0.170

5   Conclusion

This paper introduces a novel semantic document classification approach. It mainly has two improvements: (1) solving the polysemy problem by using a novel semantic similarity computing method (SSC). The SSC implements semantic analysis by executing semantic similarity computation and semantic embedding with the help of the common dictionary. In this paper, we use CoDic for English texts and Hownet for Chinese texts. (2) solving the synonym problem by proposing a novel strong correlation analysis method (SCM). The SCM consists of the CDM strategy for the selection of feature candidate set and the FCA strategy for the determination of the final feature set. Experiments show that our strategy can improve the performance of semantic document classification compared with that of traditional ones.

We will continue going further after this research. More multiple deep learning models (e.g., DualTextCNN, DualBiLSTM, DualBiLSTMCNN or BiLSTMAttention) will be tested for semantic text similarity on document datasets with different natural languages. We would also try to compare this strategy with state-of-the-art embedding methods such as FastText [10], BERT [4] and ULMFit [9] and ELMo [18].

Acknowledgment. This research is supported by both the National Natural Science Foundation of China (grant no.: 61802079) and the Guangzhou University Grant (no.: 2900603143).

References

1. Altınel, B., Ganiz, M.C.: Semantic text classification: a survey of past and recent advances. Inf. Proces. Manag. **54**(6), 1129–1153 (2018)
2. Cerda, P., Varoquaux, G., Kégl, B.: Similarity encoding for learning with dirty categorical variables. Mach. Learn. **107**(8–10), 1477–1494 (2018)
3. Chandrashekar, G., Sahin, F.: A survey on feature selection methods. Comput. Electr. Eng. **40**(1), 16–28 (2014)
4. Devlin, J., Chang, M.W., Lee, K., Toutanova, K.: Bert: pre-training of deep bidirectional transformers for language understanding. arXiv preprint arXiv:1810.04805 (2018)
5. Dong, Z., Dong, Q., Hao, C.: HowNet and the computation of meaning (2006)
6. Fang, J., Guo, L., Wang, X., Yang, N.: Ontology-based automatic classification and ranking for web documents. In: Fourth International Conference on Fuzzy Systems and Knowledge Discovery (FSKD 2007), vol. 3, pp. 627–631. IEEE (2007)
7. Gambhir, M., Gupta, V.: Recent automatic text summarization techniques: a survey. Artif. Intell. Rev. **47**(1), 1–66 (2017)
8. Guo, J., Da Xu, L., Xiao, G., Gong, Z.: Improving multilingual semantic interoperation in cross-organizational enterprise systems through concept disambiguation. IEEE Trans. Industr. Inf. **8**(3), 647–658 (2012)
9. Howard, J., Ruder, S.: Universal language model fine-tuning for text classification. arXiv preprint arXiv:1801.06146 (2018)
10. Joulin, A., Grave, E., Bojanowski, P., Mikolov, T.: Bag of tricks for efficient text classification. arXiv preprint arXiv:1607.01759 (2016)
11. Khan, A., Baharudin, B., Lee, L.H., Khan, K.: A review of machine learning algorithms for text-documents classification. J. Adv. Inf. Technol. **1**(1), 4–20 (2010)
12. Kim, Y.: Convolutional neural networks for sentence classification. arXiv preprint arXiv:1408.5882 (2014)
13. Liu, Y., Scheuermann, P., Li, X., Zhu, X.: Using WordNet to disambiguate word senses for text classification. In: International Conference on Computational Science, pp. 781–789. Springer (2007)
14. Manning, C.D., Raghavan, P., Schütze, H.: Scoring, term weighting and the vector space model. In: Introduction to Information Retrieval, vol. 100, pp. 2–4 (2008)
15. Mikolov, T., Sutskever, I., Chen, K., Corrado, G.S., Dean, J.: Distributed representations of words and phrases and their compositionality. In: Advances in Neural Information Processing Systems, pp. 3111–3119 (2013)
16. Mirończuk, M.M., Protasiewicz, J.: A recent overview of the state-of-the-art elements of text classification. Expert Syst. Appl. **106**, 36–54 (2018)
17. Pang, B., Lee, L.: Seeing stars: exploiting class relationships for sentiment categorization with respect to rating scales. In: Proceedings of the 43rd Annual Meeting on Association for Computational Linguistics, pp. 115–124. Association for Computational Linguistics (2005)
18. Peters, M.E., Neumann, M., Iyyer, M., Gardner, M., Clark, C., Lee, K., Zettlemoyer, L.: Deep contextualized word representations. arXiv preprint arXiv:1802.05365 (2018)
19. Thangaraj, M., Sivakami, M.: Text classification techniques: a literature review. Interdisc. J. Inf. Knowl. Manag. **13** (2018)
20. Wang, Y., Wang, X.J.: A new approach to feature selection in text classification. In: 2005 International Conference on Machine Learning and Cybernetics, vol. 6, pp. 3814–3819. IEEE (2005)

21. Wawer, A., Mykowiecka, A.: Supervised and unsupervised word sense disambiguation on word embedding vectors of unambigous synonyms. In: Proceedings of the 1st Workshop on Sense, Concept and Entity Representations and Their Applications, pp. 120–125 (2017)
22. Xiao, G., Guo, J., Gong, Z., Li, R.: Semantic input method of chinese word senses for semantic document exchange in e-business. J. Ind. Inf. Integr. **3**, 31–36 (2016)
23. Yang, S., Wei, R., Shigarov, A.: Semantic interoperability for electronic business through a novel cross-context semantic document exchange approach. In: Proceedings of the ACM Symposium on Document Engineering 2018, p. 28. ACM (2018)
24. Young, T., Hazarika, D., Poria, S., Cambria, E.: Recent trends in deep learning based natural language processing. IEEE Comput. Intell. Mag. **13**(3), 55–75 (2018)
25. Zhang, Y., Wallace, B.: A sensitivity analysis of (and practitioners' guide to) convolutional neural networks for sentence classification. arXiv preprint arXiv:1510.03820 (2015)

An Equity-Based Incentive Mechanism for Decentralized Virtual World Content Storage

Bingqing Shen, Jingzhi Guo$^{(\boxtimes)}$, and Weiming Tan

Faculty of Science and Technology, University of Macau, Taipa, Macao,
Special Administrative Region of China
{daniel.shen, wade.tan}@connect.um.edu.mo,
jzguo@um.edu.mo

Abstract. Virtual worlds have become the arena for many entertainment, social, and business activities and provided a platform for user content generation. To protect user innovation, persistency is an important property. Unfortunately, existing virtual worlds, owned by some entities, are not immune from death due to entity failure. To provide a persistent virtual world, a decentralized architecture is explored, which is constructed on user contributed devices. However, there are many challenges to realize a decentralized virtual world. One important issue is user cooperation in reliable content storage. The devices contributed by users may not be reliable for maintaining all user contents, but users do not have the incentive to provide reliable devices for others. This paper addresses the issue by two steps. First, an indicator is provided to users, called replica group reliability which is based on the proposed replicability index. Based on the indicator, users can learn the reliability of their content storage. Then, a new user incentive mechanism, called equity-based node allocation strategy, is proposed to promote user cooperation to collectively maintain reliable content storage. A decentralized algorithm implementing the strategy is designed and the evaluation results show its effectiveness and efficiency.

Keywords: Virtual world · Persistency · Replication · Incentive · Cooperation · Reliability

1 Introduction

Virtual worlds, including MMOGs and Metaverses [1], are computer-generated environments constructed by numerous networked computing devices and populated by many avatars of users and non-player characters (NPCs). The growth of virtual worlds has received significant attentions in the last decade both from industry [2] and academic [3] communities. Virtual worlds allow users to create their own contents. Users can utilize their expertise and creativity to create virtual products for virtual customers [4]. Thus, persistency is an important feature, which demands that a virtual world, together with all the generated contents, exists forever regardless of any changes on virtual world owners or users. Currently, nearly all existing virtual worlds are created and owned by certain entities (often commercial companies), and user-generated

© Springer Nature Switzerland AG 2020
K.-M. Chao et al. (Eds.): ICEBE 2019, LNDECT 41, pp. 19–32, 2020.
https://doi.org/10.1007/978-3-030-34986-8_2

contents are stored on the servers of these entities. When the entity owning a virtual world dies, bankrupts, or withdraws its operations, the affected virtual world will collapse together with the loss of its user-generated contents. The result is that user-generated contents become non-persistent (See the inactive virtual world list in http://opensimulator.org/wiki/Grid_List) and virtual world users suffer great loss on their virtual assets.

To protect virtual world users and their virtual assets, the persistency feature of virtual worlds must be maintained. In the previous research [5], a self-organization system, called Virtual Net, has been introduced to prevent the possible collapse of virtual world and maintain the persistency feature. The central idea of Virtual Net is that nobody owns the virtual world but everybody collectively creates a self-organized virtual world so that it will not fail due to the departure of any entity. In this approach, each virtual world user contributes a part of his/her computing resources such as a certain amount of CPU time, memory, storage and bandwidth of his/her computing device(s). Devices that have contributed computing resources can be smart phones, personal computers, virtual appliances, etc. The contributed part of each device is virtualized into one or multiple nodes which install and run the virtual world program. Users of Virtual Net can store their contents or deploy some applications on the nodes without a central management. Thus, Virtual Net aims at providing virtual world solution based on peer-to-peer (P2P) computing [6].

Nevertheless, it is a great challenge to realize the self-organization approach because of the dynamics and unreliability of user contributed resources. One outstanding issue is content availability. The user contents, including user-generated virtual objects, object states, and virtual world programs, might not be available when the nodes they are stored on go offline. To address the issue, a new redundancy model, called logical computer, is proposed in [5]. A logical computer includes one or multiple nodes for content storage and computing in a distributed manner. To combat with node failure, every node of a logical computer is replicated to many nodes (called replicas or a replica group) on different devices. In a replica group, when some of the nodes fail, new replicas will be created by the surviving replicas.

Creating new replicas requires data transfer from surviving replicas to new nodes. But, the size of user content could be large. Compared to small-size content, it takes longer time to create new replicas for large content, and it will be more likely that no sufficient number of new replicas are created before existing replicas fail, if nodes are not reliable. For example, a replica group with five replicas can only create three new replicas in one replication, and these three replicas can only create one new replica in the next replication. With the reduction of replica amount, eventually, the replica group may fail and the content is unavailable until it is re-deployed to a new replica group. Thus, maintaining the storage of larger content requires nodes to be more reliable to create sufficient replicas in a replication. However, users do not have the incentive to maintain the high reliability of their contributed devices. This is the user cooperation problem in reliable content storage, which includes the following challenges: (1) how to determine the reliability of a replica group, and (2) how to promote user cooperation for providing reliable nodes as needed.

To resolve the user cooperation problem in content storage, this paper firstly proposes a new metric, called replicability, to determine if the current reliability of a replica group is sufficiently high to maintain the required number of replicas. Based on the metric, users may have different node reliability requirements. For example, users with larger size content may want their content to be migrated to more reliable nodes to improve content availability. Then, a novel user incentive mechanism, called equity-based node allocation (EBNA) is proposed. It leverages equity theory to improve users' cooperation even when they have different requirements. Equity theory [7, 8] reveals that users have incentive to change an inequitable allocation result. By linking device reliability to user input and outcome, EBNA motivate users who need higher reliable nodes to maintain the same reliability of their devices. In return, all users with different node reliability requirements will be allocated their needed nodes and they can benefit from high content availability. In summary, this paper makes the following contributions.

1. A new replicability index is proposed for gauging the reliability of a replica group.
2. A novel equity-based node allocation strategy is proposed for incentivizing users to collectively maintain high reliability of their content storage.
3. A new decentralized node matchup algorithm is devised to implement the EBNA strategy.

The remainder of this paper is organized as follows. Section 2 discusses the related work. Section 3 proposes the replicability index for measuring replica group reliability. Section 4 introduces the EBNA strategy. Section 5 presents the algorithm design, which implements the proposed strategy. In Sect. 6, experiments are conducted on evaluating the correctness and the performance of the algorithm. Finally, a conclusion is made to summarize the paper.

2   Related Work

The research of the proposed EBNA approach is related to the existing work on P2P incentive mechanism and the authors' prior work of the logical computer redundancy model [5]. In P2P applications, a problem of free-riding has been found [9] such that nearly 70% users share no files and only 1% peers serve almost 50% file requests [10], which may lead to the tragedy of common [11]. To resolve the free-riding problem, three user incentive approaches have been developed, which are bartering, reputation, and payment [12].

The bartering approach proposes that the computing resources that a peer can receive should be directly proportional to the computing resource that a peer currently contributes to other peers. Well-known examples are BitTorrent protocol for file sharing [13] and Cooperative Internet Backup Scheme [14] for mutual file storage. Differently, reputation-based approaches (either direct or indirect) resolve the free-riding problem by recording the resource sharing histories of other users as their reputations in each peer node [15]. Once a peer requests the resource on another peer, the requested peer will check the reputation record and decide the amount of resource that the requester can acquire [12]. Similar to the reputation-based approach, the payment-based approach also relies on partner's historical cooperative behavior to determine resource sharing. The difference is that one peer's long-term reputation is quantitatively measured as "money" and hence the history statistics on reputation is unnecessarily maintained by any peer. For example, [16] employs a decentralized micropayment system [17] to "forge" currencies for the payment of data storage and data verification.

The existing bartering approaches allows some nodes to be unaccommodated for their noncooperation, while the reputation approach and the payment approach rely on long-term information accumulation. Thus, they are not applicable in our problem. The proposed EBNA approach in this paper can be regarded as a revised bartering approach based on equity theory, in which incentives are given to users for their contributions that can barter future returns.

3 Replica Group Reliability

This section introduce the measure of determining replica group reliability. Firstly, a new index, called replicability is introduced, which can determine the expected number of new replicas created in a replication. Based on replicability, replica group reliability is discussed.

3.1 Replicability

Figure 1 shows the logical computer redundancy (LCR) model. In LCR, A replica group maintains a minimal number (n) of replicas to achieve a required content availability and an extra number (e) of replicas to reduce replication overhead [5]. Thus, the size of a replica group is up to $n + e$. When the group size is less then n, new replica(s) will be created by the surviving replicas. The residual life of the surviving replicas limits the number of new replicas which can be created in a replication. Replicability (w) measures the expected amount of data that can be transferred before all the surviving replicas fail. The expected replicability can be estimated by

$$w = BW_{up} \cdot \frac{(n-1)^2}{n+E+1} \cdot MTTF \qquad (1)$$

where BW_{up} is the network bandwidth of a replica assigned for replica creation, E is the maximal number of e which is a system configuration, and $MTTF$ is the node mean time-to-failure, describing node reliability. The detailed derivation process of replicability can be found in [5]. Since BW_{up}, n, and E are constants and selected in configuration, w is only changed by $MTTF$. Thus, (1) can be simplified to

$$w = \alpha \cdot MTTF, \text{where } \alpha = BW_{up} \cdot \frac{(n-1)^2}{n+E+1} \qquad (2)$$

With replicability, the expected number of replicas created in a replication can be calculated by dividing w with the content size.

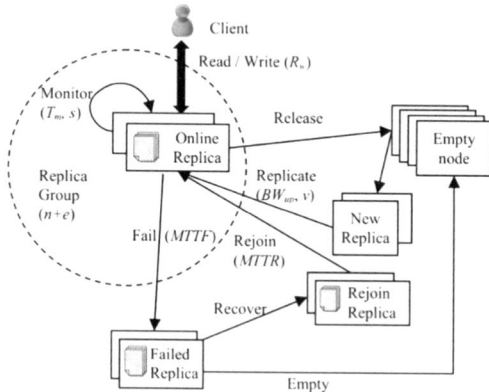

Fig. 1. Logical computer redundancy (LCR) model [5].

3.2 Determining Group Reliability

Replicability alone does not determine the group reliability. For a given replicability, if the content size is larger, then fewer replicas can be created, resulting in lower group reliability. Thus, group reliability is determined by replicability (w) and content size (v), as below listed.

1. $w \geq n \cdot v + \delta$: A replica group is highly reliable.
2. $n \cdot v \leq w \leq n \cdot v + \delta$: A replica group is weakly reliable.
3. $w < n \cdot v$: A replica group is unreliable.

Since w only describes the mean value of replicability, the actual value could be changed by the actual residual life length of the surviving replica(s). Thus, the actual amount of data can be transferred in a replication is likely to be lower than expected. δ is introduced as a confidence variable for the estimation of replicability. If w is greater than $n \cdot v + \delta$, then it is highly likely that w is greater than $n \cdot v$, implying the sufficient replicability to create n replicas in a replication. In contrast, if w does not reach $n \cdot v + \delta$, but is still higher than $n \cdot v$, then the likelihood that n replicas can be maintained is moderate. Lastly, if w even cannot reach $n \cdot v$, the replica group is unlikely to maintain n replicas after one or several replications.

Group reliability provides an indicator for users to be aware of the availability of their content storage. If it is low, it is likely that the required number of replicas (n) cannot be maintained, increasing the probability of content unavailability. Thus, the users possessing larger files need more replicability to maintain reliable content storage, which needs the cooperation of other users to provide more reliable nodes.

4 Equity-Based Node Allocation

Equity theory, firstly developed by Adams [7, 18], explores the psychological relation between the *outcome/input* ratio of a person. A person can receive psychological influence from the comparison of one's current *outcome/input* ratio with others', and he/she will be satisfied only if the comparison result is equitable to them [8, 18–21]. If the person is in an inequitable relation, he/she will change the input to match the outcome [22]. Therefore, an incentive mechanism can be designed to improve user cooperation by eliminating inequality.

In the context of content storage in Virtual Net, given the content size v and the minimal replica group size n, group reliability is determined by replicability. Thus, the replicability defined in Eq. (2) can be linked to user outcome, denoted by w_r. On the other hand, users can control the $MTTF$ of their contributed nodes (denoted by $MTTF_c$), which is related to the replicability of other replica groups. Thus, the $MTTF$ of user contributed nodes can be linked to user input by introducing a special replicability w_c, defined by

$$w_c = A \cdot MTTF_c \tag{3}$$

where $MTTF_c$ denotes the $MTTF$ of user contributed device(s). In summary, user outcome is w_r, called received replicability, and user input is w_c, called contributed replicability.

Node allocation is a function that allocates the suitable nodes to different replica groups. Based on equity theory, an allocation is equitable if it conforms to the following definitions.

Definition 1 (Equitable Allocation). For any two users, p and q, A node allocation leads to equity if the allocation result satisfies $w_r^p/w_c^p = w_r^q/w_c^q$, where w_r^p/w_c^p and w_r^q/w_c^q are the respective *outcome/input* ratio of p and q.

Users may increase or decrease their content size, resulting in the change of replicability needs. Thus, a node allocation strategy should also accommodate the temporal dynamics of the *outcome/inputratio*, and meanwhile it can maintain the equity property. Such allocation is called improvable allocation, as defined below.

Definition 2 (Improvable Allocation). For any two users, p and q, let $w_r^p|_{t1}$, $w_c^p|_{t1}$, $w_r^q|_{t1}$, $w_c^q|_{t1}$ be their respective input and outcome at time t. For any two different times $t1$ and $t2$, if both $\frac{w_r^p|_{t1}}{w_c^p|_{t1}} = \frac{w_r^q|_{t1}}{w_c^q|_{t1}}$ and $\frac{w_r^p|_{t2}}{w_c^p|_{t2}} = \frac{w_r^q|_{t2}}{w_c^q|_{t2}}$ hold, then the node allocation is an improvable allocation.

In practice, unfortunately, it is unlikely that w_r and w_c can always be perfectly matched due to the finite number of nodes. Thus, a weak improvable allocation is proposed to minimize the inequity among users, as defined below.

Definition 3 (Weak Improvable Allocation Strategy). For any two users, p and q, let $w_r^p|_{t1}$, $w_c^p|_{t1}$, $w_r^q|_{t1}$, $w_c^q|_{t1}$ be their respective input and the outcome at time t. For any two different times $t1$ and $t2$, if an allocation can minimize both $\left| \frac{w_r^p|_{t1}}{w_c^p|_{t1}} - \frac{w_r^q|_{t1}}{w_c^q|_{t1}} \right|$ and $\left| \frac{w_r^p|_{t2}}{w_c^p|_{t2}} - \frac{w_r^q|_{t2}}{w_c^q|_{t2}} \right|$, then it is a weak improvable node allocation.

Definition 3 implies that, to achieve weak equity, the difference of *outcome/input* ratio among users should be minimized all the time, which can be formulated to the following optimization problem.

$$Min \left| \frac{w_r^i|_t}{w_c^i|_t} - \frac{w_r^j|_t}{w_c^j|_t} \right|, \text{ for any } t \qquad (4)$$

To resolve Problem (4), a practical EBNA strategy can be described with the following steps.

4. Calculate the w_c by Eq. (3) for each user device.
5. Sort the user contributed devices by their w_c.
6. For each user, allocate him/her one node from the devices with the w_c closest to his/her contributed devices.
7. Repeat Step 3 until n nodes are allocated to the user.

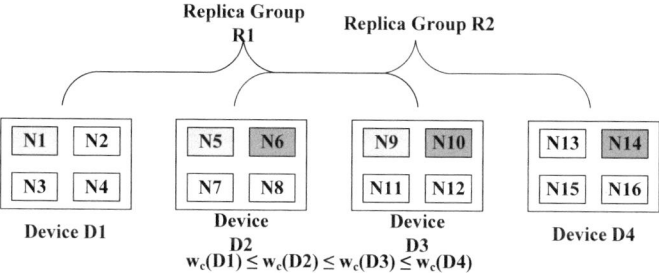

Fig. 2. An illustration of node allocation on different devices.

The proposed EBNA strategy is executed in loop. If the w_c of a device is found to be changed, Step 1–4 will be repeated until all users are settled. For a user, his/her content is replicated to the nodes with the w_c closest to his/her contributed device(s). As illustrated in Fig. 2, the user of Device $D2$ has a replica group $R1$. With $n = 3$, node $N1$, $N5$, and $N9$ belong to $R1$.

It can be expected that if a user increases the size of his/her content, the current replicability may not provide sufficient group reliability, indicated by $w_c - (n \cdot v + i)$. To avoid content availability reduction, the user may increase the reliability of his/her contributed device to w_c'. Then, the system will allocate him/her the new nodes with the w_c closest to w_c'. Through monitoring, if the new group reliability indicator $w_r - (n \cdot v + \delta)$ is greater than 0, the user may stop improving the reliability of his/her device. w_c can be increased for a user by maintaining a device online for longer time, installing a monitoring program, or replacing the device with a more reliable one, which is out of the scope of this paper.

5 Decentralized Node Matchup

Since replicability is only changed by *MTTF*, devices can be sorted in their observed *MTTF*. Without a central service for sorting devices by *MTTF*, implementing the practical EBNA strategy proposed in the previous section becomes a challenging problem. We propose a decentralized node matchup (DNM) algorithm to efficiently solve Problem (4). In DNM, each device is connected to some remote devices, called neighbors. By sorting the *MTTF* of neighbors, a device chain can be constructed such that the device with the lowest *MTTF* is located at one end and the device with the highest *MTTF* is located at the other end. If a device changes its *MTTF*, it will leave the current position in the chain and move to the correct position to maintain the *MTTF* monotonicity of the chain. In case of long chain, remote links are maintained to accelerate position lookup. Moreover, chain split caused by device departure is handled to maintain the global consistency.

5.1   Device Chain Structure

The DNM algorithm adopts a chain structure linking all devices together. They are sorted by *MTTF*, as shown in Fig. 3. The device of greatest *MTTF* is called chain head; the device with the least *MTTF* is called chain tail. A device chain has two directions. The direction from head to tail is called downward direction; the reverse direction is called upward direction.

In a device chain, each device maintains one direct neighbors at each directions. The direct neighbor with smaller *MTTF* is called lower direct neighbor; the one with larger *MTTF* is called upper direct neighbor. Each device also maintains k links to the near devices in the chain. Half of them link to the near devices with smaller *MTTF* (if exist), called lower neighbors; another half of them link to the near devices with larger *MTTF* (if exist), called upper neighbors. k is chosen by configuration. As shown in Fig. 3, k is configured to 4. Devices $D3$ has direct lower neighbor $D2$, direct upper neighbor $D4$, lower neighbors $D2$, $D1$, and upper neighbors $D4$, $D5$.

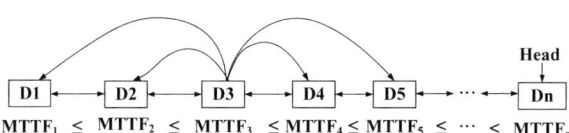

Fig. 3.   Device chain structure.

5.2   Remote Links

With the increase of devices, a device chain could be too long to quickly move a device to the correct position, in case of *MTTF* change. Thus, remote links are introduced to enable long hop. Specifically, each device maintains l remote devices at each direction, called remote neighbors. The distance d is configured such that two adjacent remote neighbors are separated by $d - 1$ device(s) in the middle. A device to the first remote neighbor also has distance d. As illustrated in Fig. 4, l is configured to 3 and d is configured to 2. Thus, Device $D3$ maintains 3 upper remote links to Device $D5$, $D7$, … and 1 lower remote link to $D1$

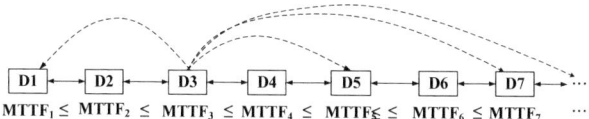

Fig. 4. Remote links.

Algorithm 1 Position Lookup

Notations:

$MTTF_d$: MTTF of the device that is the algorithm invoker;

$MTTF_U(i)$: MTTF of the ith upper remote neighbor;

$MTTF_L(i)$: MTTF of the ith lower remote neighbor;

$N_U(i)$, $N_L(i)$: i^{th} upper remote neighbor, i^{th} lower remote neighbor;

$P(N)$: position of remote neighbor N.

---

1. **IN THE CASE** of upward lookup,
2.   **FOR EACH** $i \in [1, l]$
3.   **IF** $MTTF_d < MTTF_U(1)$
4.   **Return**
5.   **ELSE IF** $MTTF_d > MTTF_U(l)$
6.   Delegate the position lookup to $N_U(l)$
7.   **ELSE**
8.   **IF** $MTTF_U(i) < MTTF_d < MTTF_U(i+1)$
9.   Return $P(i)$
10.   **END IF**
11.   **END IF**
12.   **END FOR**
13. **IN THE CASE** of downward lookup,
14.   **FOR EACH** $i \in [1, l]$
15.   **IF** $MTTF_d > MTTF_L(1)$
16.   **Return**
17.   **ELSE IF** $MTTF_d > MTTF_L(l)$
18.   Delegate the position lookup to $N_L(l)$
19.   **ELSE**
20.   **IF** $MTTF_L(i+1) < MTTF_d < MTTF_L(i)$
21.   Return $P(i)$
22.   **END IF**
23.   **END IF**
24.   **END FOR**
25. **END**

5.3 Position Change

Periodically, each device in the chain contacts with its upper direct neighbor to compare their *MTTF*s. If the *MTTF* of the device is greater than the upper direct neighbor, it will firstly invoke the position lookup algorithm (Algorithm 1) to look for a remote position. If the algorithm returns the current position, then the device will swap the position with the upper direct neighbor. Otherwise, the device is moved to the remote position. Once a device changes the position, it will learn new neighbors through the new upper direct neighbor. A position change process without remote hop is similar to bubble sort [23]. (Their performance will be compared in the next section.)

5.4   Chain Merge

If a device leaves Virtual Net, the chain will be broken into two. DNM needs to merge multiple chains into one for global consistency. In the algorithm, chain discovery and chain merge is achieved by chain heads. Specifically, a global chain head is maintained through an out-of-band approach, for example, a well-known URL with a dynamic domain name service [24]. Then, each chain head periodically checks and compares the *MTTF* with the global chain head. If the *MTTF* is higher than the global chain head, the chain head will take over the global position. Otherwise, the chain head, together with the rest devices in the chain, joins the chain of the global chain head. Consequently, two chains merge into one. Chain merge is illustrated in Fig. 5. It is possible that both chains grow with new device joins. Thus, devices in Chain 2 might be sent to different positions.

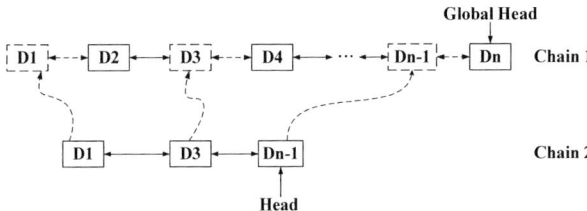

Fig. 5. Chain merge.

6   Evaluations

In this section, the DNM algorithm is evaluated from its correctness and performance. The correctness is evaluated from two aspects. First, experiments validate whether the algorithm can sort a device chain in the correct order. Based on the order, the overall improvement of equity is observed, which is calculated by the variance of the user *outcome/input* ratio (*V*) with the following function:

$$V = \frac{1}{N-1} \cdot \sum_{i=1}^{N} \left(\frac{w_r^p}{w_c^p} - \overline{\left(\frac{w_r^p}{w_c^p} \right)} \right)^2 \qquad (5)$$

Second, individual user's replicability change is simulated by changing the device *MTTF*. The experiment observes the *outcome/input* ratio change along with the position change of a user device in the device chain to validate the proposed node allocation strategy. The performance of the algorithm is evaluated from efficiency and reliability.

6.1   Equity Improvement

First, the algorithm is evaluated in the overall equity improvement. The overall equity is measured with the variance in Eq. (5). The initial variance is randomly generated by randomly connecting the devices into a chain. The final variance is obtained after applying the DNM algorithm. The equity improvement is then calculated by comparing the two variances. Moreover, the simulation is run with different network scales, ranging from 10 to 10^5. The experiment result (Table 1) shows that the algorithm can decrease the variance, implying equity improvement. Particularly, the variance decreases more along with the increase of network scale, because a larger network has a larger chance to contain devices with similar *MTTF*s.

Table 1. Equity improvement.

Network size	Initial variance	Final Variance	Improvement
10	0.589	0.26	55.94%
100	59.70	1.80	96.99%
1,000	58.28	0.022	99.96%
10,000	70.73	1.03×10^{-4}	99.999854%
100,000	74.08	3.72×10^{-7}	99.999999%

The second experiment begins by changing the *MTTF* of a randomly selected device, which is then sent to the correct position with the proposed algorithm. The changed MTTF is reflected by the contributed replicability (w_c). The experiment also records the initial received replicability (w_r) before running the algorithm and the final received replicability (w_r') after running the algorithm. The differences of the contributed replicability and the receive replicability are calculated by $|w_c - w_r|$. The experiment result (Table 2) shows that the received replicability approaches to the contributed replicability after running the algorithm. This implies that the proposed algorithm can match users' input to his/her outcome. Like the last experiment, moreover, a larger network has a larger chance to decrease the replicability difference.

Table 2. Replicability improvement.

Network size	Contributed replicability (KB)	Initial received replicability (KB)	Final received replicability (KB)
10	4.17×10^7	2.64×10^7	5.39×10^7
100	4.17×10^7	1.22×10^7	4.22×10^7
1,000	4.17×10^7	5133760.16	4.1650×10^7
10,000	4.17×10^7	5064004.06	4.1693×10^7
100,000	4.17×10^7	5081727.64	4.1697×10^7

6.2   Efficiency and Reliability

The performance of the DMN algorithm is evaluated in this section by studying its efficiency and reliability. Algorithm efficiency studies how fast an equity can be reached. If device sort is slow, *MTTF* change may increase the *outcome/input* ratio variance, adding inequity. Algorithm reliability studies the effect of device join and departure to the change of overall equity.

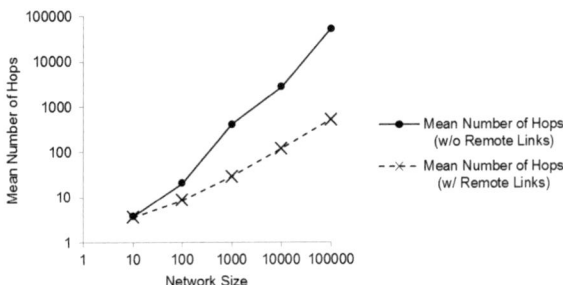

Fig. 6. *Outcome/Input* ratio variance change trend with device departures and device joins.

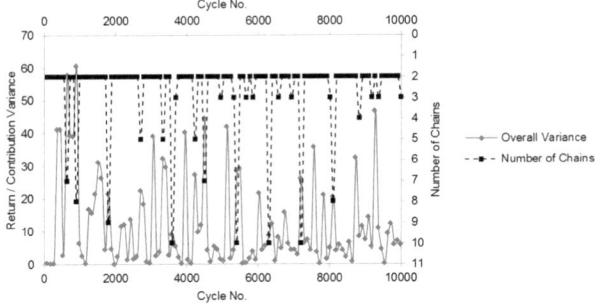

Fig. 7. Mean number of hops in device join.

Algorithm efficiency is measured by the communication overhead, i.e., the number of hops from one device to another before sending a device to the correct position. The proposed algorithm is compared to bubble sort. Figure 6 shows that DMN needs less number of hops for devices to reach their correct positions. Particularly, when the network size grows to 10^5 devices, DMN needs only 1/100 hops of bubble sort, indicating much higher efficiency. Moreover, the experiment result shows that the number of hops increases in $O(n^{1/2})$ in the proposed algorithm. Compared to bubble sort which has a linear increase of communication overhead, DNM is more scalable.

To evaluate algorithm reliability, device departure and device join are introduced in the simulation. Figure 7 shows the trend of the *outcome/input* ratio variance when both device departure and device join are applied. It can be found that the fluctuation of variance is roughly aligned with the change of the number of chains. Greater number of chains indicates larger number of device departure, creating more near neighbor loss and increasing the variance. The trend in Fig. 7 also shows the variance reduction along with the decrease of the number of chains, implying the effectiveness of the chain merge function.

6.3   Discussion

The experiment results shows that the proposed DNM algorithm can improve the overall equity, by matching users' contributed to their outcome. It also shows that if a user changes the reliability of his/her device (reflected by the device *MTFF*), his/her received replicability will also be changed proportionally, affecting the reliability of the replica group. By observing the group reliability indicator, if a user needs more replicability to storage a larger content, he/she has to firstly provide more replicability to the network. The result is that users have the incentive to improve the reliability of their contributed devices. In return, their content stored on Virtual Net will become highly available.

7   Conclusions

This paper addresses the important problem of user cooperation in decentralized virtual world content storage. If users do not have the incentive to maintain high reliability of their contributed devices, virtual world content may become unavailable due to device failure, affecting user experience. In the paper, the replica group reliability is measured with the proposed replicability index. Based on the measure, users can learn their needs of node reliability. To motivate users to cooperatively provide high group reliability to each other, the proposed EBNA strategy firstly relates node reliability to users' input and outcome by leveraging equity theory. Then, it provides a node allocation strategy to solve the user cooperation problem. Lastly, the EBNA strategy is implemented by the DNM algorithm in a decentralized setting. Experiment results show that the proposed algorithm is effective, efficient, and reliable.

Future work includes the study of node allocation in the case that some devices can only be virtualized to a small number of devices. In this case, the DNM algorithm cannot guarantee that each logical computer can contain the same number of replicas. This will put algorithm design in a more realistic system model, which is important to Virtual Net implementation.

Acknowledgement. This research is partially supported by the University of Macau Research Grant No. MYRG2017-00091-FST.

References

1. Dionisio, J.D.N., Gilbert, R.: 3D virtual worlds and the Metaverse: current status and future possibilities. ACM Comput. Surv. (CSUR) 45(3), 34 (2013)
2. Tredinnick, L.: Virtual realities in the business world. Bus. Inf. Rev. 35(1), 39–42 (2018)

3. Poppe, E., Brown, R., Recker, J., Johnson, D., Vanderfeesten, I.: Design and evaluation of virtual environments mechanisms to support remote collaboration on complex process diagrams. Inf. Syst. **66**, 59–81 (2017)
4. Zhou, M., Leenders, M.A., Cong, L.M.: Ownership in the virtual world and the implications for long-term user innovation success. Technovation (2018)
5. Shen, B., Guo, J., Li, L.X.: Cost optimization in persistent virtual world design. Inf. Technol. Manage. **19**(3), 155–169 (2018)
6. Lua, E.K., Crowcroft, J., Pias, M., Sharma, R., Lim, S.: A survey and comparison of peer-to-peer overlay network schemes. IEEE Commun. Surv. Tutorials **7**(1–4), 72–93 (2005)
7. Adams, J.S.: Towards an understanding of inequity. J. Abnorm. Soc. Psychol. **67**(5), 422 (1963)
8. Austin, W., Walster, E.: Reactions to confirmations and disconfirmations of expectancies of equity and inequity. J. Pers. Soc. Psychol. **30**(2), 208 (1974)
9. Krishnan, R., Smith, M.D., Telang, R.: The economics of peer-to-peer networks. Available at SSRN 504062 (2003)
10. Adar, E., Huberman, B.A.: Free riding on Gnutella. First Monday **5**(10) (2000)
11. Hardin, G.: The tragedy of the commons. J. Nat. Resour. Policy Res. **1**(3), 243–253 (2009)
12. Oualha, N., Roudier, Y.: Peer-to-Peer Storage: Security and Protocols. Nova Science Publishers Inc., New York (2010)
13. Liu, W., Peng, D., Lin, C., Chen, Z., Song, J.: Enhancing tit-for-tat for incentive in BitTorrent networks. Peer-to-peer Network. Appl. **3**(1), 27–35 (2010)
14. Lillibridge, M., Elnikety, S., Birrell, A., Burrows, M., Isard, M.: A cooperative internet backup scheme. In: Proceedings of the Annual Conference on USENIX Annual Technical Conference, p. 3. USENIX Association, June 2003
15. Shen, H., Lin, Y., Li, Z.: Refining reputation to truly select high-QoS servers in peer-to-peer networks. IEEE Trans. Parallel Distrib. Syst. **24**(12), 2439–2450 (2012)
16. Oualha, N., Roudier, Y.: Securing P2P storage with a self-organizing payment scheme. In: Data Privacy Management and Autonomous Spontaneous Security, pp. 155–169. Springer, Heidelberg (2010)
17. Vishnumurthy, V., Chandrakumar, S., Sirer, E.G.: Karma: a secure economic framework for peer-to-peer resource sharing. In: Workshop on Economics of Peer-to-peer Systems, vol. 35, no. 6, June 2003
18. Adams, J.S.: Inequity in social exchange. In: Advances in Experimental Social Psychology, vol. 2, pp. 267–299. Academic Press (1965)
19. Messe, L.A., Dawson, J.E., Lane, I.M.: Equity as a mediator of the effect of reward level on behavior in the Prisoner's Dilemma game. J. Pers. Soc. Psychol. **26**(1), 60 (1973)
20. Radinsky, T.L.: Equity and inequity as a source of reward and punishment. Psychon. Sci. **15**(6), 293–295 (1969)
21. Wicker, A.W., Bushweiler, G.: Perceived fairness and pleasantness of social exchange situations: two factorial studies of inequity. J. Pers. Soc. Psychol. **15**(1), 63 (1970)
22. Schmitt, D.R., Marwell, G.: Withdrawal and reward reallocation as responses to inequity. J. Exp. Soc. Psychol. **8**(3), 207–221 (1972)
23. Donald, E.K.: The art of computer programming. Sorting Searching **3**, 426–458 (1999)
24. Knoll, M., Wacker, A., Schiele, G., Weis, T.: Bootstrapping in peer-to-peer systems. In: 2008 14th IEEE International Conference on Parallel and Distributed Systems, pp. 271–278. IEEE, December 2008

A Group-Centric Physical Education Supporting System for Intelligent Campus Application

Lei Wang, Hongyan Yu, Yang Cao, Xiaofan Zhou, Lihong Jiang$^{(\boxtimes)}$, and Hongming Cai

Shanghai Jiao Tong University, Shanghai, China
{wangleiCORE, yuhongyan, caoyangss, zhouxiaofan, jianglh, hmcai}@sjtu.edu.com

Abstract. The concept of intelligent campus has been put forward for many years, aiming to provide high quality services dynamically and proactively for students, teachers and other employees in university in the era of Internet of Things. There exists a phenomenon that the health status of college students is continuously declining, which has aroused public concerns. This phenomenon could be improved via offering intelligent management services to college physical instructors by taking advantage of advanced Internet of Things (IoT) technologies. In this paper, a framework for group-centric physical education supporting system based on wristbands and cloud platform is proposed. The specific methods are divided into three parts based on event streams: IoT nodes networking and configuration before class, real-time health status monitoring and emergency detection in class, and students' performance and course quality evaluation after class, which provide an overall insight into students' fitness and exercise condition for the physical teachers. In addition, a prototype system is implemented for teaching test and to prove the feasibility of the proposed methodology, and the comparison of our proposed methods with other systems is also discussed.

Keywords: Internet of Things (IoT) · Wristband · Physical education · Intelligent campus

1 Introduction

The investigation shows that suboptimal health status (SHS), which is considered to be an intermediate status between disease and health, occurs in 55.9% in Chinese students [1]. The continuous declining of Chinese college students' health status [2] shows in an all-round way including their strength, speed, explosive power, endurance, etc. This phenomenon has attracted the attention of the education department. College teachers also pay much attention on the situation. When students' health status become worse, teachers tend not to assign excessive or strenuous exercise to them, but this could be inevitable in some physical examination like 800-m race or during daily training. But if not well-trained, students are more easily prone to get hurt in the exercise. This put a dilemma in front of college physical education teachers. They need a new approach that

© Springer Nature Switzerland AG 2020
K.-M. Chao et al. (Eds.): ICEBE 2019, LNDECT 41, pp. 33–47, 2020.
https://doi.org/10.1007/978-3-030-34986-8_3

can help them monitor the health status of students in real time when they are doing some strenuous exercise in case of an emergency happening and can support course design that suits the special requirements of each case or class by offering analysis on students' physical fitness and teaching quality.

Apart from the physical education needs, in some similar conditions, like in a gymnasium, where the coach needs to monitor the trainer's exercise intensity in real-time and adjust training plan after evaluation, it also shows similar requirements.

At present, the concept of "intelligent" has swept the world, evolving from "smart", and integrates smartness and intelligence. And the improvement of wireless protocols and necessary infrastructures, the development of cloud services and the lower cost of hardware [3] has started a new era in intelligent field, like intelligent campus, intelligent transportation and intelligent business. Assisted by the Internet of Things (IoT), mobile networks and other advanced technologies, the health data of users are becoming available. However, there exist some problems when giving analysis on these data, as listed follows:

- Objective problems: Due to the discontinuous nature of the data collection [4] and inevitable transmission mistakes, the problem of data sparsity and data missing is common and it adds difficulties to monitor and analyze users' health condition. There must be a method to approximate and simulate the real data and improves the fault tolerance of the system.
- Subjective requirements: In many sports activities like basketball and volleyball, training in groups is a common phenomenon. Appropriate grouping based on users' fitness status will improve the efficacy of exercise and reduce the probability of emergency.

Therefore, we proposed a group-centric framework for supporting physical education, which integrates smart devices and cloud platform to provide effective and accurate services to teachers in intelligent campus environment. The group centric idea is shown in three aspects: physical fitness grouping, gateway and devices grouping and exercise intensity grouping. We will introduce it in detail in methodology part.

Specifically, our main contributions are summarized as follows:

1. A complete framework for supporting physical education in intelligent campus is proposed, which is group-centric and mainly includes three parts: preparation and student physical fitness grouping before class, real-time health monitoring and emergency prediction in class, and students' performance and course quality evaluation after class. The specific methods in each part are also discussed in detail.
2. The application scenario is introduced. A prototype system is implemented to show the feasibility of the methodology we presented, and comparisons with other systems are discussed.

The rest of this paper is organized as follows. Section 2 discusses some related works about intelligent campus and IoT nodes networking technologies. Section 3 illustrates the framework we proposed for the group-centric physical education supporting system at length, including networking and preparation before class, real time monitoring and emergency prediction in class, and exercise intensity grouping of students and course assessment after class. Section 4 introduces the application scenario and the prototype system. Section 5 concludes the work in this paper and discusses about the direction of future study.

2 Related Works

In this section, we divide our related works into two parts: intelligent campus including its definition, key technologies and application researches, and IoT nodes networking methods.

2.1 Intelligent Campus

The intelligent campus is depicted as "a central digital nervous system that dictates the end-to-end learning lifecycle of a knowledge ecosystem" [5]. The concept of "intelligent campus", also known as "iCampus", is initially explored in an MIT program named MIT-Microsoft Alliance Program, which fostered barrier-less collaborations and exciting new technologies developments aimed at revolutionizing the practice of higher education [6]. Intelligent campus is to utilize the resources efficiently and deliver high quality services thereby getting the operational cost reduced to make the ecosystem of university more interactive and creative [7]. Based on 29 articles, the paper [8] concluded that the basic idea of intelligent campus is an effort to integrate intelligent technologies by the university to provide service including learning activities, social interaction, office management and energy saving in a dynamic, user-oriented and proactive way in the environment of Internet of Things and cloud platform, etc.

The model of intelligent campus is comprised of six domains to emphasize the characteristics, and they are: iLearning, iManagement, iGovernance, iSocial, iHealth and iGreen [5]. In various studies, there are seven key technologies that support smart campus: radio-frequency identification (RFID), Internet of Things (IoT), cloud computing, 3D visualization technology like augmented reality, sensor technology, mobile technology like NFC, QR code and GPS and web service [8].

There are many efforts put in the applications for intelligent campus based on IoT platform. [9] presents an IoT-based platform deployed in the Moncloa Campus of International Excellence, which offers two pilot services, namely people flow monitoring based on Wi-Fi tracking and environmental monitoring, aiming to enable experimenting with smart city services and fine-tuning them. Xin Dong et al. designed and implemented a mobile platform, which focuses on three aspects in smart campus: the information of social circles based on interests mining, the provision of educational guidance based on emotional analysis and campus trading aiming to optimize the allocation of campus resources [10]. [11] uses the agent-oriented approach which is specifically based on the agent-based cooperating smart object methodology and related middleware to develop an IoT-based multi-agent system, namely a Smart University Campus. There are difficulties in integrating wide range of existing technologies and protocols to offer all the aforementioned services in campus ecosystem and many of the work are still in research.

2.2 IoT Nodes Networking

To provide systems that interact with the environment and anticipate or predict the actions of residents, four elements have to be interacted: IoT nodes that collect data, IoT nodes that embed actuators, the cloud platform and the interconnected IoT gateways that exchange messages with the IoT nodes and with the cloud [3].

The main driving technologies of IoT include ZigBee, BLE (Bluetooth Low Energy), Wi-Fi, RFID, 4G, 5G, etc. The IoT devices that are embedded with sensors usually use short range radio frequency (RF) communications to get access to other nodes, where the gateway plays an important role in bridging between various sensor networks and mobile communication networks or Internet and managing the devices of sensor networks.

There are many designs in intelligent systems using this kind of structure to network the IoT nodes. An IoT gateway system based on ZigBee and GPRS protocols after well consideration for typical IoT application scenario and requirements from telecom operators is proposed in [12]. [3] presents ZiWi, a distributed fog computing Home Automation System (HAS) that allows seamless communications among ZigBee and Wi-Fi devices. Amir-Mohammad Rahmani et al. exploit the strategic position of gateway to offer high-level services like local storage, real-time local data processing, embedded data mining, etc., and cope with many challenges such as energy efficiency, scalability and other issues in ubiquitous healthcare system [13].

3 Methodology

3.1 A Framework for Group-Centric Physical Education Supporting System for Intelligent Campus Application

The structure of the framework is proposed based on IoT nodes and cloud platform, as illustrated in Fig. 1. The IoT nodes are responsible for sensing and data acquisition. A gateway that can exchange message between IoT nodes and the cloud server is needed. The cloud platform supports data processing and provides alarming service to end users through gateway via wireless network. The other three parts are divided according to event streams. Before class, some preparation work including IoT infrastructure configuration and preprocessing of the input information of system is disposed, meanwhile preliminary physical fitness grouping of students is analyzed. In class, emergency prediction and alarming service is provided by modeling on the student's fitness group and his real-time health data. After class, students' performance and course quality evaluation is analyzed and displayed in visual mode through a web-based dashboard.

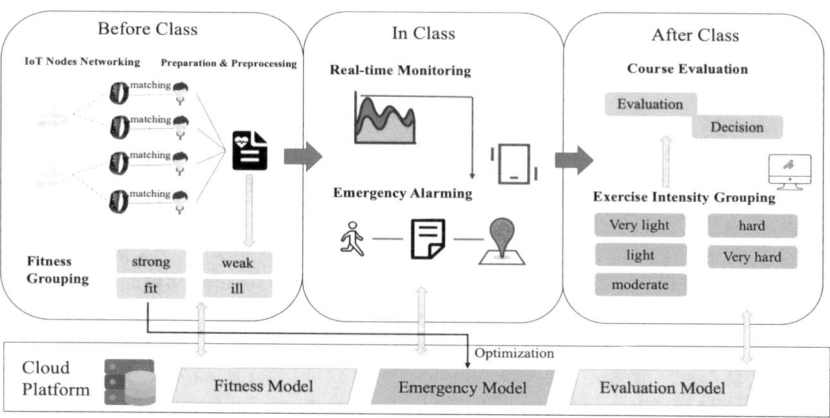

Fig. 1. Framework for group-centric physical education supporting system

3.2   Before Class: IoT Nodes Networking and Preprocessing

In this part, the preparation for hardware and preprocessing of some necessary input information of the system are discussed.

Wristband, as one of the optimal choice as sensor nodes, is a convenient and suitable smart device to monitor students' health condition when they are having PE (Physical Education) lesson for its miniaturization and portability. Common low power LAN protocols include ZigBee, Wi-Fi, Bluetooth etc., and compared with low power WAN protocols like LoRa and NB-IoT, these technologies are more mature in industry area and have many successful application cases. The unavoidable drawbacks of the former protocols are the limited coverage range and limited maximum number of connections, like Bluetooth Low Energy (BLE), which has gained a dominant position in communication with wristbands but one Bluetooth module can only connect to seven modules at most at the same time. Therefore, gateways are needed to bridge the sensor devices and the cloud platform, where the sensor nodes use short-range RF to communicate with the gateways and the gateways communicate with the application server with long-range communication channels. The mobile device keeps broadcasting via BLE till establishing stable connections to the gateway at its vicinity. The gateway and its connected devices or sensor nodes form a group, which imply the context information including location and ambient environment of these nodes. Apart from the method of constructing a wireless local area network, the newly published networking topology design could be leveraged, e.g., the Bluetooth mesh that is compatible with BLE 4.0 and above version and its flooding-based solution that allows for end-to-end data transmission overcomes the coverage and path diversity limitations [14]. After networking of these sensor nodes and gateways, the connection between the gateways and the cloud servers that have considerable computing capabilities should be configured.

While network topology is constructed, the necessary input data of the system including information about teachers, students, courses, devices and the relationships between them is preprocessed. Before class, a list that distributes devices to different students according to their course time is generated. The NFC (Near Field Communication) module in wristbands could facilitate to remove the roll call and assist rapid identification.

The third preparation work before class, physical fitness grouping, aims to analyze and group students based on their health condition, which can assist in emergency modeling and course design that caters for students' physical ability to achieve effective training. The fitness model FM are defined as follows:

FMI = function <age, sex, BMI, BF>

It uses four basic physiological indexes. The BMI is short for body mass index and BF is short for body fat. Z-score normalization is used to normalize BMI and BF because these indexes are supposed to follow Gaussian distribution. Min-max normalization is used to process the age. After preprocessing, cluster method like K-means, DBSCAN (Density-Based Spatial Clustering of Applications with Noise) is used to group these students. What's more, the medical history of students need to be considered separately. After the health persona is analyzed and we may get the four fitness groups: strong, fit, weak and ill. The computing of fitness group has an impact on the risk model in emergency prediction since different groups are variable in their physical functions.

3.3   In Class: Real Time Monitoring and Emergency Alarming

When students are doing intensive exercise in class, the most important thing is to know if there is or there will be an emergency in case of more serious consequences. Therefore, their health condition must be monitored by instructor in real time and an emergency model that could predict the risk of their physical status need to be constructed.

There are four main characteristics of exercise prescription that have been widely adopted which are the intensity, duration, frequency and mode of exercise [15]. Exercise intensity has been given more attention in the literature due to its relative efficacy in improving cardiorespiratory fitness [16]. In 1957, Karvonen et al. [17] developed an equation to gauge intensity, which was termed the heart rate reserve (HRR) method and was based on the difference between the resting heart rate (HR_{rest}) and the estimated maximum heart rate (HR_{max}). The equation is shown in (1).

$$HRR = HR_{max} - HR_{rest} \tag{1}$$

This method takes training workload into account comparing with the maximum heart rate (MHR) method, which only uses HR_{max} and does not consider the resting heart rate.

The heart rate intensity HRI(t) at time t is defined as (2).

$$HRI(t) = \frac{HR(t) - HR_{rest}}{HRR} \tag{2}$$

In (2), HR(t) means the heart rate at time t. And according to the American College of Sports Medicine (ACSM) guidelines [18], classification of physical activity intensity based on HRI is defined as follows in Table 1.

Table 1. Classification of physical activity intensity

Physical activity intensity	HRI (%)
Very light	<20
Light	20–<40
Moderate	40–<60
Vigorous (hard)	60–<85
Vigorous (Very hard)	85–<100
Maximal	100

The parameters we monitor are comprised of their own physical index, i.e., heart rate, which is a key factor of exercise prescription and is a relative precise index we could acquire, and ambient environment factors, including temperature and humidity that will have an outside impact on people's physical function and exercise performance as well. The heart rate is measured and sent to cloud server one pulse per second, and the environment condition can be acquired by a set of sensors that record and transmit these parameters. In some conditions, linear interpolation is used to deal with message missing during the transmission. While in extreme conditions when the measuring equipment happens to be in failure, this group-centric mechanism could be leveraged to simulate and approximate the student's real-time health data based on the data simultaneously collected in the same fitness group and gateway group.

The emergency model is constructed considering aforementioned two factors and students fitness group. The emergency model EM is defined as follows:

EM = function <fitness group, HRI, environment condition, timestamp>

- Fitness group is classified based on students' four physiological indexes and the calculation method has been introduced.
- HRI means the heart rate intensity.
- The influence of ambient environment condition is considered, consisting of temperature, humidity and air quality.
- Timestamp describes the current time.

The parameters depict all the correlated input of emergency detection module. To generate the binary output that signifies whether there is an emergency with the input streams of EM, machine learning method, e.g., LSTM (Long Short-Term Memory) neural networks, a RNN (Recurrent Neural Network) model and suitable for processing serial data, is used to build the model. At the beginning, amounts of data should be used to pre-train the model for continuous iteration of network weights.

3.4 After Class: Student Performance Analysis and Course Evaluation

The two modules after class are student performance analysis and course quality evaluation, and the results are displayed in visual mode in a web-based dashboard for teacher's reference.

The student's performance is assessed by their exercise intensity in class and they are divided into groups for teachers' better management. According to the method proposed in the former part, we divide students into five groups after calculating the time distribution of HRI(t) on interval [0,1] in one class time. Specifically, we calculate the area enclosed by the curve in intervals $[0, 0.2)$, $[0.2, 0.4)$, $[0.4, 0.6)$, $[0.6, 0.85)$ and $[0.85, 1]$ respectively and pick the largest. If the largest one hits intervals $[0, 0.2)$, $[0.2, 0.4)$, $[0.4, 0.6)$, $[0.6, 0.85)$ or $[0.85, 1]$, by referring to Table 1, accordingly the exercise intensity groups are: very light, light, moderate, vigorous(hard) and vigorous (very hard). From the example shown in Fig. 2, we could see that the student's performance group is the fourth one: vigorous (hard). Furthermore, the students could be divided into more refined groups based on exercise intensity in different fitness group calculated before.

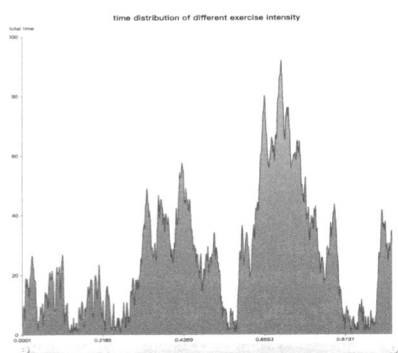

Fig. 2. Time distribution of different heart rate intensity

The course quality is usually measured by the students' performance in class. Therefore, based on the former grouping result of students, we will quantify the exercise intensity and the corresponding number of students that achieves this level to assess the courses. The evaluation criteria are adjusted by the instructor based on his professional knowledge.

The course evaluation model CE is defined as follows:

CE = function <exercise intensity, percentage, duration, professional knowledge>

- Exercise intensity describes the intensity of physical activity and it has five categories as listed in Table 1.
- Percentage means the percentage of students in a course that achieve the accordingly exercise intensity.
- Duration describes how long this intensity is lasted.

- Professional knowledge is added to be a standard when evaluating the course quality. Apparently, different courses have different levels of requirements for exercise intensity, e.g., the yoga class is less intensive than the basketball class.

The course could be divided into groups like exercise-intensive and exercise-causal based on the instructors' professional knowledge. By comparing the quantified physical activity intensity of students with the professional knowledge that gives indexes of each course, teachers could have an intuitive concept of the effect of sports activity design, i.e., the gap between their goal and actual effect. Moreover, the results give them insights to make more scientific decisions when making course design.

4 Case Study

4.1 Case Study

The framework we proposed is a composition of intelligent campus ecosystem, which aims to support physical education in an all-around way, including monitoring in class and evaluation after class. It provides services for teachers and students in a user-oriented and proactive way like prescription and alarming of emergencies and decision-making support.

Moreover, this kind of early warning mechanism can not only be used in physical education, but in some sports related campaign such as in a marathon race where there is a high chance that accidents happen. In addition, the fitness grouping method can be applied in a gymnasium for a gym coach to analyze the customers' health condition and make targeted plans for them in group granularity. It also could be used for health monitoring in hospital, in home, etc. With some modifications, this framework can apply to these similar scenarios.

But in this paper, we focus on the intelligent physical education scenario and a prototype system will be introduced next.

4.2 System Architecture

In this case, we implemented a prototype system for the group-centric physical education supporting system. The system architecture is shown in Fig. 3, and it is divided into four layers: perception layer, data layer, service layer and application layer.

In perception layer, due to widespread use of biosensors or smart devices like bracelets and smart phones, we can have access to users' health data and ambient environment data by using interfaces offered by the manufactures.

In data layer, databases are used to store the data collected in perception layer. The wristband raw data collected per second is stored in the NoSQL database for requirements of fast writing. In SQL database, master data that describes basic attributes about students, teachers, courses, classes and devices and transaction data that is produced during the analysis procedure are stored. The data in NoSQL database are scheduled to clean to save capacity.

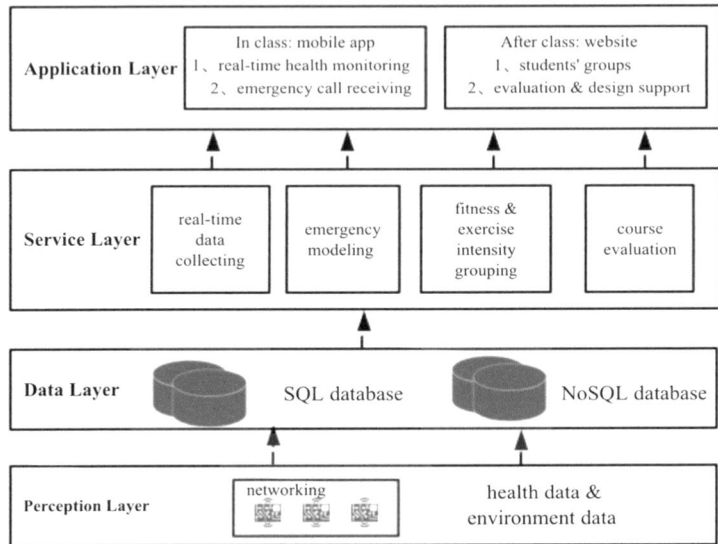

Fig. 3. System architecture

In service layer, it consists of four modules. Assisted by IoT nodes in perception layer that constructs a location-based LAN to gather students' health data, real-time data collecting is implemented. The emergency detection and emergency alarming module builds a prediction model to prescribe whether a student is at risk or not. And after class, exercise intensity of each student is calculated to evaluate their course performance. Also, teaching indexes are incorporated to evaluate the quality of class and then assist in course design decision making.

In application layer, the system is divided into two parts. One is apps in mobile devices for real-time monitoring and emergency notification in class, and the other is a visual web-based dashboard for after-class evaluation. The two applications are all implemented for instructor's use.

4.3   Prototype System

In this part, the prototype system is implemented based on our proposed methodology. It is supposed to be comprised of the three modules: the networking and configuration before class, real-time monitoring and emergency prediction in class, and course evaluation after class. And the laboratory dataset is used.

Networking and Configuration. For the purpose of supervising students' health status in real-time, after comparison among different devices, we choose to use smart wristbands for its portability and ease of use. But the wristbands in the market usually only provide such service that measures the user's physical indexes through the embedded biosensors and then shows results on the manufacturer's app. However, in our application scenario, all users' data should be collected and sent to one centric server to be analyzed. Therefore, wristband that supports custom coding to acquire health data for further processing is chosen.

Since most of the wristbands use Bluetooth, a short-range RF protocol, to communicate, a gateway that is able to exchange messages between wristbands via Bluetooth signal and the cloud platform via the Internet is needed. The gateway we chose could cover a circular area with 200 m in diameter theoretically, which is satisfactory to most physical activities in campus. More gateways can be deployed to cover larger areas. When the terminal enters the coverage of the gateway, its signal will be broadcast to gateway, and then the message will be transmitted to cloud server after establishing stable connection to proximate gateway around. Similarly, the gateway can transmit the emergency warning message from the cloud platform to wristbands and to mobile app.

The wristband is set to collect the user's physical indexes and exchange with the gateway pulse per second for real-time monitoring. The gateway is configured to connect to the server by filling in server's IP address and listening port where the program that deals with the business logic runs. The communication protocol between the server and the gateway is TCP protocol and we use Netty framework, a Java based open source framework supported by Jboss, to implement it. The server receives the byte array and translates it into hexString, the hexString is translated into byte array and then sent to the gateway.

Till now, we have constructed the network between the IoT nodes and the cloud platform. Apart from this, preliminary process for the system is needed, including inputting information of students, teachers, courses, devices, etc., matching students and devices and clustering students based on their physical fitness condition. The data of students we simulated is clustered into four fitness groups by DBSCAN: strong, fit, weak and ill.

Real-Time Monitoring and Emergency Prediction. After the wristbands are distributed in class, the user's health data is collected by the gateway per second for real-time processing. The teacher could monitor the location and the heart rate change of students on the phone. If a student is prescribed to be in danger, the phone will receive the message from the server and vibrate to notify the teacher. In addition, the wristband can remind the wearer to pay attention to their own health status once the heart rate is out of the set scope. The warning is shown in Fig. 4. In the monitoring page of teacher's app, as displayed in Fig. 5, the different colors of students signify their different physical condition groups.

Fig. 4. Wristband alarming

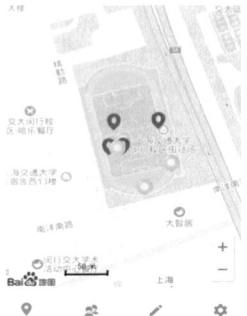

Fig. 5. Teacher's mobile app

Course Evaluation. The evaluation module after class consists of two parts: students' performance evaluation and course assessment. Figure 6 shows the basic information of a student including his fitness group, exercise intensity distribution in a selected class and the comparison with his group and with the whole class are displayed.

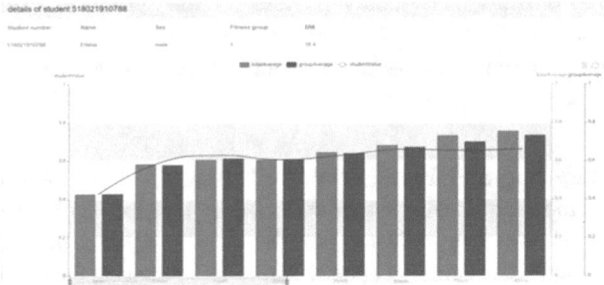

Fig. 6. Health condition and exercise intensity of a student

Figure 7 displays the heart rate distribution of different groups of students in one class, and in addition, sliding window is used to analyze the changes and generate text reports. Teachers could get the course evaluation result after setting the assessment criteria of the class based on their professional knowledge and check the detailed information of those that achieve the exercise requirements. These evaluations are intuitive and could guide the teacher in making decision about course design.

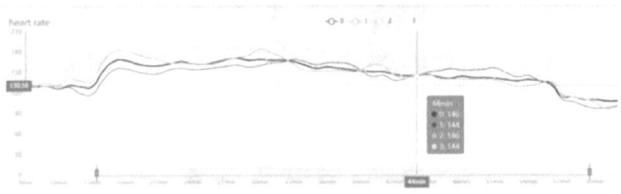

Fig. 7. Heart rate distribution of different groups

4.4   Discussion

Our proposed methods aim to integrate smart IoT devices into college physical education and a prototype system which supports in-class emergency prediction and alarming and after-class course evaluation quantified in visual mode is implemented. The other two systems we compare with is [19] and [20], among which, [19] uses a wearable wristband to measure worker's %HRR-based physical demand to avoid excessive demands, and [20] proposes a patient-centric eHealth ecosystem using fog-driven IoT architecture and the case study of early warning system is analyzed for our comparison.

The three systems are evaluated in the following metrics in Table 2. Though it is difficult to make a fair comparison since not all the solutions are designed for the same application, but we can see that they all use heart rate (HR) to do health-related monitoring and analysis, and compared with centric cloud computing, distributed and decentralized fog computing paradigm could reduce the service latency. However, this gap of our proposed method could be narrowed by deploying our server at the vicinity of the IoT nodes within the campus network. And the devices of our system are more convenient and cost-saving comparing with [20]. But the security and privacy issues in our methods need further careful consideration.

Table 2. Comparisons between the systems

	[19]	[20]	Our system
Functionality	Measuring continuous physical demand of workers	Early warning service	Emergency monitoring and course evaluation
Computing paradigm	Not discussed	Fog computing	Cloud computing
Protocol	Not discussed	BLE and Wi-Fi	BLE and Wi-Fi
Device	Wristband	Smart eyeglasses	Wristband, sensors, etc.
Operation modality	Continuous measuring	Continuous monitoring	Continuous collecting (1 pulse per second)
Response time level	Average	Fast	Average
Security and privacy	Believing the collected HR data is a minor risk	Watermark added	Not discussed

5   Conclusion

This paper presents a group-centric physical education supporting system based on wristbands and cloud platform for smart campus application in the era of Internet of Things. The methods we proposed focus on three parts: before class, in class and after class. To provide these services, the IoT infrastructure is firstly deployed before class by building a wireless local area network within the university using wristbands, gateways and cloud servers. Moreover, students' physical fitness is grouped according to their basic physiological indexes for emergency prediction. In class, the data stream of students is collected and input into the LSTM model to give prediction of the emergency. After class, the teachers could view the results of course evaluation in a web-based dashboard intuitively, which provides guidance for course design. What's more, a prototype system is implemented to prove the feasibility of our proposed methods and comparison with other similar systems is discussed. The significant contribute is that our work provides an overall insight into students' exercise condition for college physical teachers and thereby assist them effectively in managing students and making targeted teaching plans to allow preemptive action to reduce serious consequences.

However, in the critical emergency prediction part, the requirements for faster alarming feedback may be affected by network congestion. The future work will mainly focus on transferring the emergency prediction service to the edge of network, which may enable more critical real-time response compared with the centralized cloud servers. What's more, our proposed system needs practical test in real-world environment.

Acknowledgements. This research is supported by National Social Science Foundation of China(NSSFC) under Grant No. 18BTY082.

References

1. Bi, J., Huang, Y., Xiao, Y., et al.: Association of lifestyle factors and suboptimal health status: a cross-sectional study of Chinese students. BMJ Open **4**(6), e005156 (2014)
2. State Physical Culture Administration: Results of the national health survey of students in 2014. Chin. J. Sch. Health **36**(12), 4 (2015)
3. Froiz-Míguez, I., Fernández-Caramés, T., Fraga-Lamas, P., Castedo, L.: Design, implementation and practical evaluation of an IoT home automation system for fog computing applications based on MQTT and ZigBee-WiFi sensor nodes. Sensors **18**(8), 2660 (2018)
4. Zhang, Y.: GroRec: a group-centric intelligent recommender system integrating social, mobile and big data technologies. IEEE Trans. Serv. Comput. **9**(5), 786–795 (2016)
5. Ng, J.W., Azarmi, N., Leida, M., Saffre, F., Afzal, A., Yoo, P.D.: The Intelligent Campus (iCampus): end-to-end learning lifecycle of a knowledge ecosystem. In: 2010 Sixth International Conference on Intelligent Environments, pp. 332–337. IEEE, July 2010
6. The iCampus project under MIT-Microsoft Alliance. http://icampus.mit.edu
7. Alghamdi, A., Shetty, S.: Survey toward a smart campus using the Internet of Things. In: IEEE 4th International Conference on Future Internet of Things and Cloud (FiCloud), pp. 235–239. IEEE (2016)
8. Muhamad, W., Kurniawan, N.B., Yazid, S.: Smart campus features, technologies, and applications: a systematic literature review. In: 2017 International Conference on Information Technology Systems and Innovation (ICITSI), pp. 384–391. IEEE, October 2017
9. Alvarez-Campana, M., López, G., Vázquez, E., et al.: Smart CEI Moncloa: an IoT-based platform for people flow and environmental monitoring on a Smart University Campus. Sensors **17**(12), 2856 (2017)
10. Dong, X., Kong, X., Zhang, F., et al.: OnCampus: a mobile platform towards a smart campus. SpringerPlus **5**(1), 974 (2016)

11. Fortino, G., Russo, W., Savaglio, C., et al.: Agent-oriented cooperative smart objects: from IoT system design to implementation. IEEE Trans. Syst. Man Cybern. Syst. **48**(11), 1939–1956 (2017)

12. Zhu, Q., Wang, R., Chen, Q., et al.: IoT gateway: bridging wireless sensor networks into Internet of Things. In: 2010 IEEE/IFIP International Conference on Embedded and Ubiquitous Computing, pp. 347–352. IEEE (2010)

13. Rahmani, A.M., Thanigaivelan, N.K., Gia, T.N., et al.: Smart e-health gateway: bringing intelligence to Internet-of-Things based ubiquitous healthcare systems. In: 12th Annual IEEE Consumer Communications and Networking Conference (CCNC), pp. 826–834. IEEE (2015)

14. Darroudi, S., Gomez, C.: Bluetooth low energy mesh networks: a survey. Sensors **17**(7), 1467 (2017)

15. American College of Sports Medicine: ACSM's Guidelines for Exercise Testing and Prescription, 7th edn. Lippincott Williams & Wilkins, Baltimore (2006)

16. Swain, D.P., Franklin, B.A.: VO2 reserve and the minimal intensity for improving cardiorespiratory fitness. Med. Sci. Sports Exerc. **34**(1), 152–157 (2002)

17. Karvonen, M.J., Kental, E., Mustala, O.: The effects of on heart rate a longitudinal study. Ann. Med. Exp. Fenn. **35**(3), 307–315 (1957)

18. American College of Sports Medicine: ACSM's Guidelines for Exercise Testing and Prescription. Lippincott Williams & Wilkins, Baltimore (2013)

19. Hwang, S., Lee, S.H.: Wristband-type wearable health devices to measure construction workers' physical demands. Autom. Constr. **83**, 330–340 (2017)

20. Farahani, B., Firouzi, F., Chang, V., et al.: Towards fog-driven IoT eHealth: promises and challenges of IoT in medicine and healthcare. Future Gener. Comput. Syst. **78**, 659–676 (2018)

Data Driven Techniques for e-Business Management

Speech Evaluation Based on Deep Learning Audio Caption

Liu Zhang, Hanyi Zhang, Jin Guo, Detao Ji, Qing Liu, and Cheng Xie[✉]

Yunnan University, Kunming, Yunnan, China
xiecheng@ynu.edu.cn

Abstract. Speech evaluation is an essential process of language learning. Traditionally, speech evaluation is done by experts evaluate voice and pronunciation from testers, which lack of efficiency and standards. In this paper, we propose a novel approach, based on deep learning and audio caption, to evaluate speeches instead of linguistic experts. First, the proposed approach extracts audio features from the speech. Then, the relationships between audio features expert evaluations are learned by deep learning. At last, an LSTM model is applied to predict expert evaluations. The experiment is done in a real-world dataset collected by our collaborative company. The result shows the proposed approach achieves excellent performance and has high potentials in the application.

Keywords: Speech evaluation · Deep learning · Audio caption

1 Introduction

Speech evaluation is an important language learning method. Currently, many language training institutions in the market use speech evaluation to evaluate the language ability of trainees. This makes it necessary for language training institutions to hire a large number of linguists to serve speech evaluation. However, due to the different evaluation criteria of different linguists and the insufficient number of linguists, speech evaluation on the market is expensive and not objective enough or accurate enough. Technically reducing the labor cost of speech evaluation and improving the accuracy of evaluation has become a significant demand in the current market.

With the rapid development of deep learning and speech recognition technology, some studies have been able to effectively perform speech recognition, can accurately identify different people's speech, and can also score each person's pronunciation [1,2]. However, in addition to identification and scoring, these methods currently do not simulate linguists' evaluation of Speech, which is difficult to meet the need of various language training institutions in the market.

Aiming at the above problems, this paper proposes an audio caption method based on speech recognition and natural language generation technology. The

Supported by organization ICEBE.

method takes the individual's speech as input and outputs it like an expert's comment. In the training process, the method trains the features of the speech with the comments of the linguist. In the prediction process, the method will output language expert comments according to the characteristics of the speech according to the grammar of the natural language. The method was tested in a real data set, and the experimental results show that the method can effectively generate corresponding expert comments based on speech. The accuracy and efficiency of the method are both acceptable to the market. In general, this article has the following three main contributions:

- This paper technically reduces the costs of speech evaluation and effectively solves the practical problem that speech evaluation is not objective enough and accurate.
- The audio caption method proposed in this paper can not only perform speech recognition and scoring, but also simulate language experts to evaluate various languages.
- This article uses real data sets to conduct experiments, which can meet market requirements in terms of accuracy and efficiency.

In Sect. 1, a brief introduction to speech evaluation and the proposed audio caption method is presented. In Sect. 2, the overall background of deep learning, the current research status of speech recognition, and the development of natural language is introduced. In Sect. 3, related techniques for the intelligent speech evaluation model are described in detail and the overall framework and model details of the audio intelligent evaluation model are described. Section 4 evaluates based on the evaluation indicators of the audio intelligent evaluation model.

2   Related Work

In recent years, deep learning is changing the various fields of traditional industries [3]. As deep learning becomes commoditized, people's needs become a creative application in different fields. For example, the applications in the medical field are mainly reflected in auxiliary diagnosis, rehabilitation intelligent equipment, medical record and medical image understanding, surgical robots, etc. [4]; the applications in the financial field mainly include intelligent investment, investment decision, intelligent customer service, precision marketing, risk control, anti-fraud, intelligent claims, etc. [6]; applications in the field of education mainly include one-on-one intelligent online counseling, job intelligence correction, digital intelligent publishing, etc. [5]. At present, deep learning has made the most outstanding progress in image and speech recognition, involving image recognition, speech recognition, natural language processing and other technologies [7–9]. The intelligent evaluation technology used in this paper also relies on deep learning and adopts a multi-model composite method to realize the language evaluation system.

In the field of speech recognition, the results have been fruitful in recent years. Shi et al. proposed to replace the traditional projection matrix with a higher-order projection layer [10]. The experimental results show that the higher-order

LSTM-CTC model can bring 3% compared with the traditional LSTM-CTC end-to-end speech model. The decline in the 10% relative word error rate. Xiong Wang et al. proposed the use of confrontational examples to improve the performance of Keyword Spotting (KWS) [11]. Experiments were carried out on the wake-up data set collected on the intelligent voice. The experimental results showed that the threshold was set. In the case of 1.0 false wake-ups per hour, the proposed method achieved a reduction of 44.7% of the false rejection rate. Cloud has explored the end-to-end speech recognition network based on the technology. The proposed method is based on the improvement of the original CNN-RNN-CTC network [12]. This method is based on the Deep Speech 2 CNN-RNN-CTC model proposed by Baidu. Focus on improving the RNN part of the original network. In addition, in the field of speech recognition, the idea of cascading structure was first proposed, which improved the accuracy of these difficult samples. The experimental results showed that the WER of the Librispeech test-clean test reached 3.41%, saving 25% on training time. With the rapid development of deep learning, the rapid development of computing power, the rapid expansion of data volume, and the deep learning of large-scale applications in the field of speech recognition have made breakthroughs. This paper combines the research results of the previous speech in the field of speech recognition, and carries out a series of preprocessing and feature extraction on the input speech, which improves the accuracy of speech recognition.

Natural language generation is the focus of today's machine learning research. Google launched the project of its automatic summary module in 2016. Textsum [13] this module also uses RNN to encode the original text, and uses another RNN digest, which also uses cluster search in the final stage of digest generation (beam-search) Strategies to improve summary accuracy. Britz et al. [14] conducted some experiments and analysis on the sequence mapping model. The results show that the cluster search has a great influence on the quality of the abstract. In 2017, Facebook's AI Lab published its latest model [15], which uses a convolutional neural network as an encoder to add word position information to the word vector, using a Gated Linear Unit (GLU). As a gate structure, and refreshing the record on the automatic summary dataset, this does not mean that the CNN-based sequence mapping model must be better than RNN [16,17]. Although CNN is efficient, but there are many parameters, and CNN can't be sensitive to the sequence of words like RNN. It is necessary to introduce the position information of words into the word vector to simulate the timing characteristics of RNN. It can be seen that RNN has its natural advantages in dealing with serialized information. Advantage. In this paper, the model of cyclic neural network is used to realize the process from feature to natural language generation.

In summary, speech evaluation based on deep learning audio caption studied in this paper can be fully tested for various languages. In this paper, the audio caption method inputs various language audios, and the output is a specific evaluation of various languages, as shown in Fig. 1. In this paper, the MFCC feature extraction technology [18] and the LSTMP model [19] are combined to extract the audio features; the word vector [20]and the memory length model

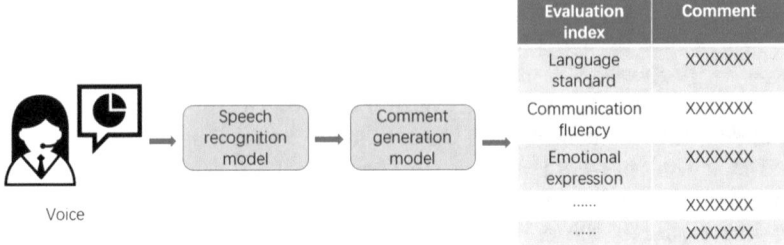

Evaluation index	Comment
Language standard	XXXXXXX
Communication fluency	XXXXXXX
Emotional expression	XXXXXXX
......	XXXXXXX
......	XXXXXXX

Voice

Fig. 1. Language intelligence evaluation model process.

[21] are combined to evaluate the audio [22, 23]. According to relevant research, it has been found that this audio caption method has not appeared in the field of deep learning before this article. On the contrary, deep learning has a lot of correlation research on image loading subtitles in the field of image recognition [24]. Aiming at the shortage of models for deep learning in speech evaluation, this paper combines the relevant knowledge in the field of image recognition, and successfully designs an audio caption method with accurate speech feature extraction and accurate guidance.

3   The Method

3.1   Preliminaries

Audio Feature Extraction: At present, there are many audio feature extraction methods. The most common feature parameters are pitch period, formant, Linear Predictive Cepstral Coding (LPCC), and Mel-frequency Cepstrum Coefficients (MFCC). Among them, MFCC is a feature widely used in automatic speech and speaker recognition. MFCC audio feature extraction algorithm is based on human auditory characteristics and can be used to display the characteristics of a piece of speech energy, bass frequency, etc., so that a speech can be well expressed through a digital matrix. In this paper, the MFCC feature extraction method is applied to extract each time step feature, and the relationship between the frequency and the Mel frequency can be approximated as shown in (1). The preprocessing stage of the speech signal includes processes such as emphasis, framing, windowing, endpoint detection, and de-noising processing. After pre-processing, the data is distributed from the time domain to the frequency domain by the fast Fourier transform to obtain the energy distribution of the spectrum. The square of the modulus obtains the spectral line energy, and then it is sent to the Mel filter bank to calculate the energy of the Mel filter, taking the logarithmic energy of each filter output. Finally, the DCT transform is performed according to (2) to obtain the MFCC characteristics parameter.

$$Mel(f) = 2595lg(1 + \frac{f}{700}) \tag{1}$$

$$C_n = \sum_{m=1}^{m} S(m)cos[\frac{\pi n(m - 0.5)}{M}] \tag{2}$$

The MFCC feature is composed entirely of N-dimensional MFCC feature parameters. The MFCC feature parameter of each time step is taken as a basic feature, and this feature can be used as a voice input feature after further processing. When processing the obtained MFCC feature parameters of a single time step, this paper adopts the MFCC feature parameters of multiple speech segments as the MFCC feature of the intermediate speech segment. This processing method overcomes the traditional MFCC feature because the speech segment is too short. The problem caused by the inability to fully express the syllables in the discourse. This paper will combine to obtain a more representative MFCC feature as an input to the final speech feature.

Speech Processing: In a deep neural network, a cyclic neural network uses a cyclic structure with time delay in the time dimension to make the network have a memory structure. Long short term memory is a deformed structure of a cyclic neural network. On the basis of the ordinary RNN, the memory unit is added to each neural unit in the hidden layer, so that the memory information in the time series is controllable, and the forgetting gate, the input gate, the candidate gate, and the output gate are passed each time passing between the hidden layer units. It can control the memory and forgetting of the previous information and the current information, so that the RNN network has long-term memory function. The LSTM model uses three gate structures to store and control information. For example, at time t, there are three inputs to the LSTM: the cell unit state c_{t-1} at the previous moment, the output value h_{t-1} of the LSTM at the previous moment, and the input value x_t of the current time network, LSTM there are two outputs: the cell unit state c_t at the current time and the output value h_t of the current time LSTM. Where x, h, t are vectors. The forgetting gate f_t is used to control how much the cell state c_{t-1} at the previous moment is retained to the current time c_t, the input gate i_t controls how much the input x_t of the network is saved to the cell state c_t, and the output gate o_t is the control unit. The effect of state c_t on the current output value h_t of the LSTM. The calculation formulas of the three gates at time t are as shown in (3), (4), and (5). Where W_i, W_f, W_o are the weight matrix of the corresponding gate, and are the splicing of two matrices, corresponding to h_{t-1} and c_t, b_i, b_f, b_o are corresponding offset matrices, $\sigma(\cdot)$ Activate the function for sigmoid. The current input unit state is as shown in (6), and the current unit state at time t is as shown in (7). Where ○ represents the product by point, and the above calculation is combined. The output of the final LSTM is as shown in (8).

$$i_t = \sigma(W_t \cdot [h_{t-1}, x_t] + b_i) \tag{3}$$

$$f_t = \sigma(W_t \cdot [h_{t-1}, x_t] + b_f) \tag{4}$$

$$O_t = \sigma(W_0 \cdot [h_{t-1}, x_t] + b_o) \tag{5}$$

$$\tilde{c}_t = tanh(W_c \cdot [h_{t-1}, x_t] + b_c) \qquad (6)$$

$$c_t = f_t \circ c_{t-1} + i_t \circ \tilde{c}_t \qquad (7)$$

$$h_t = o_t \circ tanh(c_t) \qquad (8)$$

In the speech processing phase, the LSTM model used in this paper is not a traditional LSTM architecture, but an improved LSTMP architecture [25]. The LSTMP layer adds a linear recursive projection layer to the original LSTM layer. The linear recursive projection layer acts like a fully connected layer, which compresses the output vector to reduce the dimensionality of the high-latitude information and reduce the dimensions of the neuron unit, thereby reducing the number of parameters in the associated parameter matrix. As a result, the model converges quickly and makes the model superior to deep feedforward neural networks with more orders of magnitude. LSTMP has proven to be very impressive in a variety of speech recognition problems.

Natural Language Generation: In natural language generation, this paper refers to an image captioning architecture based on deep loops [24]. This architecture combines the latest advances in computer vision and machine translation to generate natural language for describing images. The model has verified the fluency of the sentences of the model description image and the accuracy of the model giving a specific image description on several data sets. This model also takes advantage of long-short term memory models in cyclic artificial neural networks, and LSTM plays a key role in this architecture. This architecture requires training the LSTM model to predict what the words are after each word and what the corresponding image is for each sentence. In this process, an LSTM memory is created, so that all LSTMs share the same parameter, and when the word t−1 is outputting, the word at time t can be obtained. All duplicate connections are converted to image features and corresponding sentences for the feedforward connection in the expanded version. It is worth noting in this model that words are represented as a one-hot vector, and the dimension is equivalent to the size of a dictionary. Among them, S_0 and S_N are defined as the words at the beginning and end of each sentence, and a complete sentence is formed when the LSTM issues a stop word. The input images and words are mapped to the same space, and the images are entered only once, in order to inform the LSTM to output sentences related to the image. The experimental results show that the robustness of the image captioning model based on the deep loop and the quantitative evaluation have obtained good results.

3.2   The Overview of the Method

With the extensive application of deep learning related techniques in speech feature extraction and natural language generation, speech evaluation based on deep learning audio caption designed in this paper uses two independent cyclic artificial neural networks. The task processing flow of this model is shown in Fig. 2, which is divided into the following three steps:

– Audio feature extraction phase. The input audio is subjected to cepstral analysis on the Mel spectrum to extract features of each time step. The process also includes short-term FFT calculation for each frame of the spectrogram, cepstral analysis of the speech spectrogram, Mel frequency analysis, and Mel frequency cepstral coefficient calculation to obtain MFCC characteristic parameters. On this basis, the MFCC feature parameters of multiple time steps are combined to obtain the MFCC feature.
– Voice processing stage. Entering the MFCC feature of the audio feature extraction stage. The extracted MFCC features are further analyzed by the three-layer LSTMP, and the recognition rate of the audio features is improved, and finally a more effective audio feature vector is obtained. At this stage, the paper cites the MSE loss function to score audio classifications and improve the accuracy of audio classification.
– Natural language generation phase. The feature vector of the audio was obtained in the previous stage. At this stage, the comments are first segmented, and then the word vectorization is performed. The resulting word vector is used as an input to the natural language model. The LSTM in the natural language generation model has a memory function, and the trained LSTM model maps to the relevant word vector to form a corresponding comment after the milk transmission audio.

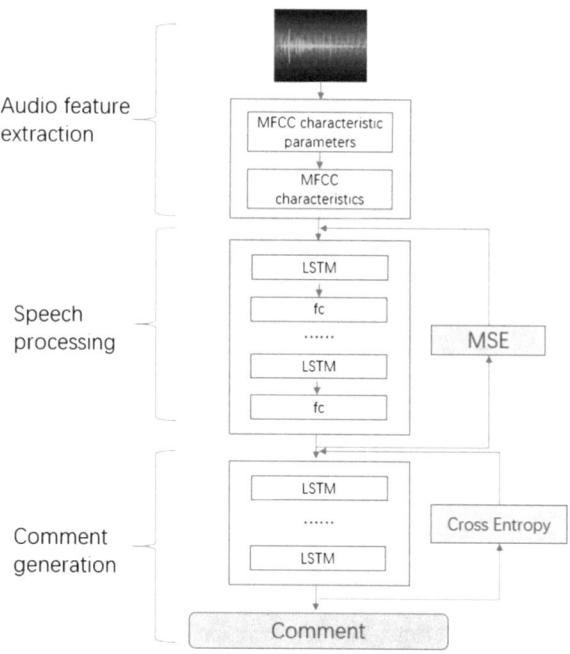

Fig. 2. Language intelligence evaluation model process.

3.3   Audio Feature Extraction

In the audio intelligent evaluation model, the audio features are first extracted. In other words, the recognizable components of the audio signal need to be extracted and the other useless information is removed, such as background noise, emotions and the like. The model design of the audio feature extraction stage introduces a feature that is widely used in automatic speech and speaker recognition, namely MFCC. The MFCC feature extraction is performed on the audio file of the model input format of .MP3, and the MFCC feature parameters are obtained. In order to reduce the amount of calculation and improve the accuracy of audio features, this paper cites the MFCC feature parameters of 18 time steps to be combined to obtain more representative MFCC features. The value here should be noted. If there is insufficient depreciation in the merge time step, the insufficient part is complemented by 0. The audio feature extraction process is shown in Fig. 3 below.

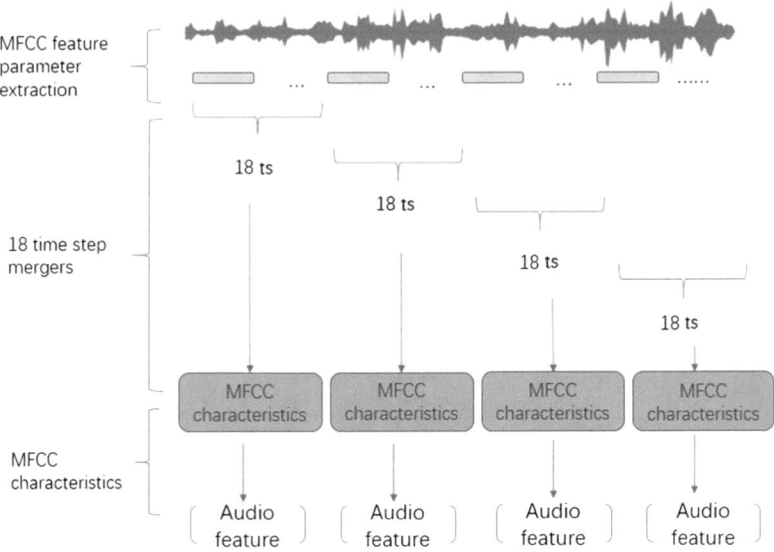

Fig. 3. Audio feature extraction process.

3.4   Speech Processing Model

In the speech processing stage, the recognition effect of MFCC under noise is drastically reduced. This paper adopts the batch gradient descent algorithm in deep neural network adaptive technology, which compresses the most important information of speaker characteristics in low-dimensional fixed length. In order to reduce the parameter optimization of the matrix, the three-layer LSTMP is used to extract the audio features. The LSTMP used in this paper is built by the improved LSTM algorithm, and finally the score is obtained after the fully connected layer. Thus, the output characteristics of the LSTM model represent a mapping between speech and its score, with which comments are automatically generated. The model designs a loss function of the language ability score, citing the maximum entropy MSE, N data are divided into r groups, and the sample variance of the i-th group is s_i^2, then the overall MSE is as shown in (9), the loss function As shown in (10).

$$MSE = \frac{\sum_{i=1}^{r}(n_i - 1)s_i^2}{N - r} \tag{9}$$

$$loss = \frac{1}{MSE}(pred - label)^2 \tag{10}$$

In deep neural networks, features are defined by the pre-L-1 layer and are ultimately learned jointly by the maximum entropy model based on the training data. This not only eliminates the cumbersome and erroneous process of artificial feature construction, but also has the potential to extract invariant and discriminative features through many layers of nonlinear transformations, which are almost impossible to construct manually. Here, the two-classification method is applied to make the model output the degree of speech in the language standard (x = 0 means that the language is not standard; x = 1 means the language standard), the degree of emotional fullness (y = 0 means that the feeling is dull; y = 1 means that the feeling is rich. And the degree of fluency (z = 0 indicates that the expression is not smooth; z = 1 indicates that the expression is smooth).

In deep neural networks, the closer the hidden layer is the input layer, the lower the layer, and the closer the output layer is, the higher layer. Lower-level features typically capture local patterns, while these local patterns are very sensitive to changes in input characteristics. However, higher-level features are more abstract and more invariant to changes in input features because they are built on top of low-level features. Therefore, the model extracts the output of the penultimate layer to represent the audio features as an audio feature vector with each feature dimension being (1, 64). The voice processing process is shown in Fig. 4.

3.5   Comment Generation Model

In the comment generation model, in order to output the high-accuracy and fluent comments, this paper builds the basic natural language generation model and draws on the solution of the image caption problem. This method obtains the feature training natural language generation model through the pre-trained model. In the image captioning architecture, the long-short-term memory model in the cyclic artificial neural network is utilized. The long-short term memory model has a memory function. Its memory means that in a sequence, memory propagates in different time steps. The long-short term memory model solves the problem that in the traditional RNN, when the training time is long, the

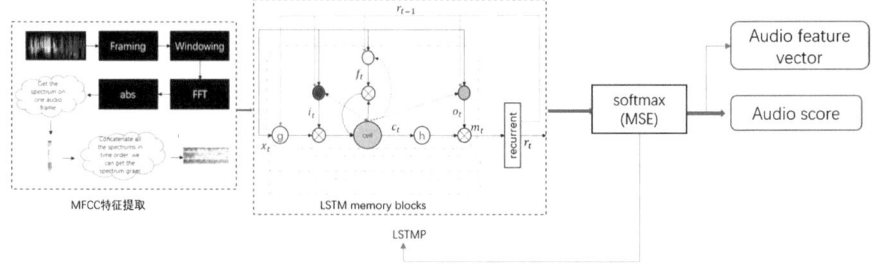

Fig. 4. Speech processing.

residuals that need to be returned will exponentially decrease, and the network weight update will be slow.

The comment generation model receives the output characteristics of the language scoring model LSTM layer and substitutes it into the LSTM natural language generation model. Before that, we need to process the corresponding comments of the corresponding audio, perform word vectorization processing, generate a digital vector of comments, and finally form a comment vocabulary. The model is based on the accumulation of multi-layer traditional LSTM units. The input audio features correspond to the digital vector of the generated comments. Finally, according to the corresponding vocabulary, the natural language composition comments corresponding to each number are found and the comments are output.

The model uses word level to generate texts, input enough comment datas for training, and LSTM has a memory function that can predict the next word. In the model, S_0 and S_N are defined as the words $< start >$ and $< end >$ at the beginning and end of each sentence. When the LSTM issues the $< end >$ word, a complete sentence is formed. For example, the source sequence contains [$< start >$], 'Language', 'standard', 'emotional', 'full', 'expression', 'smooth'] and the target sequence is one containing ['Language', 'standard', 'emotional', 'full', 'express', 'smooth', '$< end >$']. Using these sources and targets sequences and eigenvectors, the LSTM decoder is trained into a language model subject to eigenvectors. The LSTM language generation model is shown in Fig. 5 below. Finally, the vector corresponds to the comment and the comment corresponding to the audio is output.

4   Experiment

4.1   Data Setup

This model requires a certain amount of audio for training. It is difficult to find all kinds of language audio needed in this article on real social media. Therefore, we cooperate with external related companies to manually collect audio for model training and different degrees of testing. Because we need more data sets, all

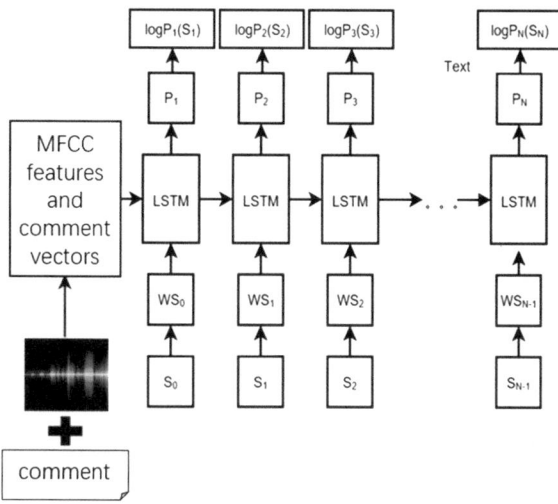

Fig. 5. LSTM language generation model.

kinds of languages can not be enumerated, here we train and test models only use slang audio. Taking into account the uncertainties in the voice recording process, the environment is uncertain, etc., we use high-fidelity mobile audio recording equipment to collect. The data set statistics for the training and test audio of this article are shown in Table 1 below.

In order to train the speech recognition model and the comment generation model, this paper collects the audio of different people's proverbs. At the same time, we will submit the collected audio data to professional linguists for evaluation. After analysis, experts will score the audio standard level, emotional fullness and fluency, and give the three aspects according to the above three aspects. When comments are outputted. First, the collected audio data sets and corresponding comment sets are input into the intelligent voice evaluation model for training. The collected test data sets are then imported into the trained model, and the model automatically generates comments and evaluations. Finally, testing the model.

Table 1. Audio file dataset statistics

	Train	Test
Mandarin	159	26
Dialect	21	4
Emotionally rich	57	7
Emotionally dull	123	23
Smooth expression	103	16
The expression is not smooth	77	14

4.2 Evaluation Metric

Speech evaluation based on deep learning audio caption consists of a speech recognition model and a comment generation model. The evaluation index of the speech recognition model is the auto-generated speech feature evaluation. Firstly, the model will be scored from three aspects, namely the standard degree of language, the degree of emotional fullness and the degree of fluency. Model performance is evaluated based on recall rate, accuracy and accuracy. The evaluation index of the comment generation model is a manual indicator. The manual selection index is given by the external personnel after the evaluation of the degree of conformity.

$$Recall = \frac{TP}{TP + FN} \tag{11}$$

$$Precision = \frac{TP}{TP + FP} \tag{12}$$

$$Accuracy = \frac{TP + TN}{TP + TN + FP + FN} \tag{13}$$

The formula for recall, precision and accuracy is as shown in (11), (12), (13) above. Where TP means predicting positive class as positive class number; TN means predicting negative class as negative class number; FP means that positive class prediction is negative class number misstatement; FN means that the number of negative classes predicted is the number of positive classes missed.

For the whole model, this paper adopts two aspects of time complexity and response time scalability in real-world applications, and estimates the response effect under real-life conditions. The time complexity will be compared and analyzed through the average response time of multiple tests. The scalability assessment is evaluated as the number of training data increases.

4.3 Experiment Result

In this paper, the model response time is tested several times to get the average response time. The results are shown in Table 2. Five audio numbers are randomly selected from the obtained data. It can be seen from the table that the response time of the evaluation model is in the range of 300 to 600 ms; the response time of the generated model is about 20 ms; and the total duration is about 500 ms. In the scalability analysis, we give a line graph of the model training as the number of training sets increases with time complexity, as shown in Fig. 6. As the number of samples in the test set grows linearly, the response time growth trend is basically linear. Explain that the overall operation of the model can be adapted to the real application environment.

The collected various language audios are input into the intelligent evaluation model of this article for training. The trained model is used as the benchmark model. On the test set, use the trained benchmark model for testing. In the language processing model, the scores of the standard level, emotional fullness and fluency of the language are obtained through testing, and Recall, Precision and

Table 2. Time complexity of the algorithm

Audio number	19	23	29	22	8
Evaluation model (ms)	513	343	455	399	497
Generate model (ms)	20	20	22	19	22
Total duration (ms)	534	364	478	419	520

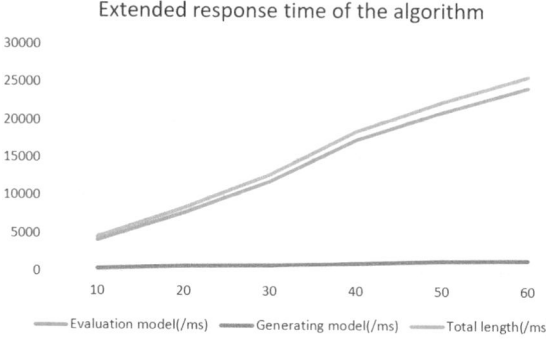

Fig. 6. Extended response time line graph of the algorithm.

Accuracy are displayed. Among them, the degree of language standard Recall is 1.0, Precision is 0.919, Accuracy is 0.922; the degree of Recall is 1.0, Precision is 0.805, Accuracy is 0.861; the smoothness of Recall is 0.982, Precision is 0.848, and Accuracy is 0.939. The scoring results show that the language processing model can be accurately evaluated in terms of audio feature extraction and speech scoring. The audio test set is input through the trained model in the comment generation model, and the comment corresponding to the audio is output. The output audio test set comment result is then sent to the outside commentator of the project, and the accuracy result is given against the audio and the comment. The final result is shown in Table 3. According to Table 3, we can see that the model gives an average accuracy of 85.9% for comments and manual comments.

Table 3. Comment accuracy

Commentator number	1	2	3	4	5
Accuracy	0.900	0.866	0.833	0.833	0.866
Average accuracy	0.859				

4.4   Comparison

A Baseline in Speech Evaluation is compared with our Audio Capturing method to show the superiority of our Audio Capturing method by comparing the Accuracy evaluation indicators in the language processing model with the average accuracy of the comment generation model.

The basic structure of Baseline also uses two independent cyclic artificial neural networks. In the audio feature extraction part, the MFCC is applied to extract the audio features. In the speech processing stage, only one LSTM model is used for speech processing. In the natural language generation stage, the trained LSTM model is applied.

The experimental results of our Audio Capturing method have been obtained in Sect. 4.4. Through this comparison experiment, we can further illustrate the advantages of our proposed method. The comparison experimental results are shown in Table 4 below. By comparing the average accuracy and accuracy in the Accuracy evaluation index generation model, we can see that our proposed Speech Evaluation effectively solves the problem that the Speech Evaluation is not objective and accurate, and the evaluation of various languages will be more objective.

Table 4. Comparison results of baseline and ours

Evaluation metric	Aspects	Baseline	Ours
Accuracy	Standard degree	0.233	0.982
	Emotional expression	0.767	0.848
	Smoothness	0.33	0.939
Comment accuracy	Average accuracy	0.601	0.859

4.5   Discussion

Throughout the development of the audio caption method, we found that the generated comment fluency has a certain gap with the manual evaluation, and the comment generation model needs further correction and optimization. There is insufficient attention to some sensitive parts of speech and some rare phonetic words, and further optimization of the model is required, for example, to add an attention model. However, the method proposed in this paper can be trained under a small training set, and it can also have better effects. And our model is small and easy to load on other devices.

5   Conclusion

This paper develops speech evaluation based on deep learning audio caption for various language assessments. The system is divided into two modules: speech recognition and comment generation, which respectively extract the speech features and give an objective evaluation for the speech features. The audio caption method in this paper evaluates the speech feature extraction and audio correspondence evaluation. The results show that the audio caption method can output objective and smooth comments for different voice features. The comments have certain reference value and can help enterprises to use each. Intelligent training in language literacy.

References

1. Heigold, G., Moreno, I., Bengio, S., et al.: End-to-end text-dependent speaker verification. In: IEEE International Conference on Acoustics. IEEE (2016)
2. Sadjadi, S.O., Ganapathy, S., Pelecanos, J.W.: The IBM 2016 speaker recognition system (2016)
3. LeCun, Y., Bengio, Y., Hinton, G.: Deep learning. Nature **521**(7553), 436–444 (2015)
4. Esteva, A., Kuprel, B., Novoa, R.A., et al.: Dermatologist-level classification of skin cancer with deep neural networks. Nature **542**(7639), 115–118 (2017)
5. Bulgarov, F.A., Nielsen, R., et al.: Proposition entailment in educational applications using deep neural networks. In: AAAI vol. 32, no. 18, pp. 5045–5052 (2018)
6. Kolanovic, M., et al.: Big data and AI strategies: machine learning and alternative data approach to investing. Am. Glob. Quant. Deriv. Strategy (2017)
7. Zhang, J., Zong, C.: Deep Learning: Fundamentals, Theory and Applications. Springer, Cham (2019). 111
8. Chen, L.-C., Zhu, Y., et al.: Computer Vision – ECCV 2018. In: European Conference on Computer Vision, vol. 833, Germany (2018)
9. Goldberg, Y.: A primer on neural network models for natural language processing. Comput. Sci. (2015)
10. Shi, Y., Hwang, M.Y., Lei, X.: End-to-end speech recognition using A high rank LSTM-CTC based model (2019)
11. Shan, C., Zhang, J., Wang, Y., et al.: Attention-based end-to-end models for small-footprint keyword spotting (2018)
12. Zhou, X., Li, J., Zhou, X.: Cascaded CNN-resBiLSTM-CTC: an end-to-end acoustic model for speech recognition (2018)
13. Abadi, M., Barham, P., Chen, J., et al.: TensorFlow: a system for large-scale machine learning (2016)
14. Britz, D., Goldie, A., Luong, M.T., et al.: Massive exploration of neural machine translation architectures (2017)
15. Gehring, J., Auli, M., Grangier, D., et al.: Convolutional sequence to sequence learning (2017)
16. Hochreiter, S.: Recurrent neural net learning and vanishing gradient. Int. J. Uncertainty Fuzziness Knowl. Based Syst. **6**(2), 107–116 (1998)
17. Chien, J.T., Lu, T.W.: Deep recurrent regularization neural network for speech recognition. In: IEEE International Conference on Acoustics. IEEE (2015)
18. Wang, W., Deng, H.W.: Speaker recognition system using MFCC features and vector quantization. Chin. J. Sci. Instrum. **27**(S), 2253–2255 (2006)
19. Sak, H., Senior, A., Beaufays, F.: Long short-term memory based recurrent neural network architectures for large vocabulary speech recognition [EB/OL]. https://arxiv.org/abs/1402.1128v1 5 February 2014

20. Peng, Y., Niu, W.: Feature word selection based on word vector. Comput. Technol. Sci. **28**(6), 7–11 (2018)
21. Hochreiter, S., Schmidhuber, J.: Long short-term momery. Neural Comput. **9**(8), 1735–1780 (1997)
22. Kulkarni, G., Premraj, V., Ordonez, V., et al.: Baby talk: understanding and generating simple image descriptions. IEEE Trans. Pattern Anal. Mach. Intell. **35**(12), 2891–2903 (2013)
23. Hong-Bin, Z., Dong-Hong, J., Lan, Y., et al.: Product image sentence annotation based on gradient kernel feature and N-gram model. Comput. Sci. **43**(5), 269–273, 287 (2016)
24. Vinyals, O., Toshev, A., Bengio, S., et al.: Show and tell: lessons learned from the 2015 MSCOCO image captioning challenge. IEEE Trans. Pattern Anal. Mach. Intell. **39**(4), 652–663 (2016)
25. Prabhavalkar, R., Alsharif, O., Bruguier, A., McGraw, I.: On the compression of recurrent neural networks with an application to LVCSR acoustic modeling for embedded speech recognition [EB/OL]. https://arxiv.org/abs/1603.08042 25 March 2016

Cross-language Keyword Analysis of Digital Transformation for Business

Ziboud Van Veldhoven[1]([⊠]) (ID), Rongjia Song[1,2] (ID),
and Jan Vanthienen[1] (ID)

[1] KU Leuven, Leuven, Belgium
{ziboud.vanveldhoven, rongjia.song,
jan.vanthienen}@kuleuven.be
[2] Bejing Jiaotong University, Beijing, China

Abstract. Digital transformation enjoys a growing interest from both practitioners and researchers. It is an encompassing term to describe the wide scope of the changes that are happening due to the increased use of digital technologies and comprises terms such as e-business and industry 4.0. Although the change is prominent worldwide, it is still not well defined and one might wonder if the concept is researched in the same way across the globe. For this reason, we conduct a thorough network analysis of the keyword co-occurrences in both the English and Chinese literature. We construct, compare, and analyze two conceptual networks to advance our understanding of digital transformation further and to investigate if there exists a difference between the English and Chinese literature. Furthermore, we utilize the obtained keyword networks to validate a recently proposed digital transformation framework. The results indicate that the digital transformation has a broad scope, is ill-defined, and is closely related to digitalization and industry 4.0. In English, digital transformation discusses many technologies, industries, and topics whereas the Chinese core journals mostly discuss the publishing and media industries. Finally, we present a research agenda that calls for more research in the role of society, more research in the interactions between society and business, and the usage of the right terminology.

Keywords: Digital transformation · Keyword analysis · Bibliometric study · Network analysis · Scientometrics

1 Introduction

The world has changed significantly in the past 20 years with the rise of digital technologies and their widespread usage. Not only businesses but every aspect of society is changing [1]. Originally, digital technologies were only used for specific purposes such as calculating, making invoices and enhancing business processes [2]. However, digital technologies proved to be useful for much more. Slowly, more and more aspects of business and society got digitalized and will continue to do so [3]. Nowadays, digital technologies have completely changed the business and societal landscape [4]. It revamped the way we communicate with each other, the way we consume products and services, our work-life balance, our values and beliefs [5], and introduced new business models that were not possible before [6]. For example, the rise of Peers Inc. collaborations [7] such as Airbnb, Uber, Alibaba, OFO, and eBay was made possible thanks to the new possibilities a digitalized society brings forward, which companies can exploit to introduce novel business models.

This change process has frequently been called digital transformation (DT) and enjoys an increasing interest by both business [8] and researchers [9, 10]. Despite the increased interest by academia to study this change process, DT is not well defined [11–13] and little is known about its exact scope and impact [14, 15]. The research field is hard to grasp, considering the broad scope and the high amount of publications every year. Furthermore, research has indicated that there exists no comprehensive framework that summarizes the changes DT brings forward [12]. Additionally, there exists a noteworthy lack of cross-country or cross-language literature studies.

The goal of this paper is to extend our knowledge of DT by mapping the current literature using a keyword co-occurrence network. This method of visualizing the literature can be useful because it provides additional insights about the current literature and can reveal communities of related research, different scopes, research priorities, and the width of the research field [16]. We investigate not only the English literature but also the Chinese literature and examine whether there exists a difference between the two flows of literature in terms of focus points and scopes. The keyword networks are then compared with the recently proposed DT framework [17] and used to provide further support for it. Lastly, we suggest a research agenda for future research.

Several studies have been conducted over the years with similar goals. Recently, [18] investigated the industry 4.0 field, which is closely related to DT [19, 20], using a bibliometric analysis of relevant papers published between 1990 and 2018. Another study [20] provides an overview of the current DT literature with a systematic literature review and bibliometric analysis. They report that DT literature got a small amount of attention around 2004, but the real research started in 2011 and the especially boomed since 2014. They continue by stating that DT is not well defined, but most include technology, organizations, and society. Furthermore, they present the top keywords in the DT literature as DT, digitalization, management, Internet of Things (IoT), and internet. Other research mainly focuses on literature reviews and less on bibliometric analyses, such as [10]. Although some effort has been made, a thorough study of the keywords in the DT literature is not yet conducted which motivates us to do so. Furthermore, this paper is the first cross-language literature study on DT and may be welcomed as a worthy endeavor for researchers to expand their notion on this interesting phenomenon.

The rest of this paper is structured as follows. The next section discusses the DT research and its meaning as defined in the literature, and discourses the DT framework [17]. Next, an overview of the methodology is presented in Sect. 3, followed by our results in Sect. 4. Section 5 contains the discussion and research agenda. Finally, section six gives a conclusion.

2 Background

2.1 Digital Transformation

In order to do a detailed bibliometric study, it is important to know how to define the search terms. In past research, DT has been defined in many ways, which makes it difficult to construct a search query. Three definition scopes can be identified. As an example of the first scope, [1] define DT as 'changes that the digital technology causes or influences in all aspects of human life'. In the second scope, researchers focus on business improvements, such as 'use of technology to radically improve the performance or reach of enterprises' [21] or 'the use of new technologies to enable major business improvements' [8] and 'Extended use of advanced IT, such as analytics, mobile computing, social media, or smart embedded devices, and the improved use of traditional technologies, such as enterprise resource planning (ERP), to enable major business improvements' [22]. Definitions of the third scope discuss the bigger changes, as opposed to improvements, such as 'the combined effects of several digital innovations bringing about novel actors, structures, practices, values and beliefs that change, threaten, replace or complement existing rules of the game within organizations, ecosystems, industries or fields' [5]. In the Chinese literature, DT is defined as 'the fundamental change or transformation of business and industry rather than merely applying digital technologies' (translated from [23]) or as 'the choice of changing together with the customer demands' (translated from [24]). It is clear that DT has a wide scope and is hard to define [3].

Previous research has suggested merging the various definitions into one, such as 'an evolutionary process that leverages digital capabilities and technologies to enable business models, operational processes and customer experiences to create value' [25] or 'a process that aims to improve an entity by triggering significant changes to its properties through combinations of information, computing, communication, and connectivity technologies' [10]. Despite these efforts, several aspects of DT are not mentioned in these definitions. First, the definitions do not stipulate the reason why DT is happening, which is an important aspect of a definition. Secondly, it is known that DT does not only affect businesses but also society. Yet, the link with society is not touched upon, which indicates that the definitions are not comprehensive with the literature.

2.2 Digital Transformation Framework

For the reasons above, we define DT as 'the continuously increasing interaction between digital technologies, business, and society, which has transformational effects and increases the change process's velocity, scope, and impact' [17]. This definition explicitly mentions the increasing interaction between digital technologies, business, and society as the reason why DT is happening. Furthermore, it makes a clear link with not only business and technology but also with society, which captures both people and communities. In previous research, it was shown that the definition is comprehensive with the existing literature while being exhaustive [17].

In order to better understand the exact scope and impact of DT, we refer to the DT framework which is depicted in Fig. 1. The DT framework consists of three axes that symbolize the three transformations that are happening in business, in technology, and in society. The changes that are happening in each of these dimensions are mentioned chronologically and summarized into five main bullet points. As digital technologies influence every aspect of business and society, their impact propels through the framework to affect the other dimensions retrospectively. For example, when smartwatches were released, they influenced customers as some of them purchased and used them. This, in turn, impacted businesses because these customers wanted their smartwatch to be useful when interacting with the businesses. Hence, the business developed new services to integrate with the changing customers' habits, which in turn increased the number of customers adopting smartwatches. As such, DT can be understood as 'the increasing interaction between digital technologies, business, and society' [17] which will steer the network to balance out disparities. As a result, the change process the world is going through increases in velocity, speed, and impact.

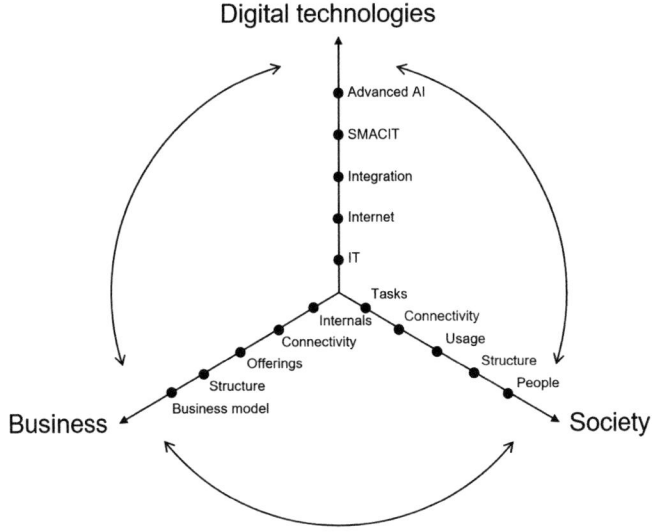

Fig. 1. The digital transformation framework (adapted from [17])

2.3 Geographical Importance

It has been suggested that DT depends on the geographical context [26]. In different parts of the world, DT might follow a different journey. One of the reasons why is that Western countries were more likely to have iteratively adopted innovative technologies at their initial stages. As a result, these countries nowadays deal with numerous legacy infrastructures, technologies, dated operational backbones, and resistance from people who do not want to change their behavior again. On the other hand, emerging countries are more likely to have skipped the first iterations of novel technologies that steer DT, which makes them less prone to legacy infrastructure and dated technologies problems.

For example, the payment system in Europe went from cash to card, to tap to pay (NFC payments), and now in the process to mobile payments. In China, the payment system transformed straight from cash to mobile. For this reason, many Western countries are stuck with various technologies that they must continue to support and provide. Secondly, the DT opportunities inherently depend on the local context (laws, labor market, and so forth), the local customers, and the local culture. This could translate into different DT approaches for different countries and their companies. Therefore, a cross-country literature study is an interesting endeavor. For this reason, we expect several differences between the Chinese literature and the Western literature.

3   Methodology

We conducted a systematic literature retrieval on two different databases, Scopus and Chinese National Knowledge Infrastructure (CNKI), for the English and Chinese literature respectively. We use our notion of DT, as described above, to construct the search query and to validate whether the papers retrieved are valid and precise. We iteratively evaluated various queries and manually checked whether the articles that were retrieved corresponded to our description of DT. Unfortunately, searching for DT or frequently-used related terms such as digitalization [3] often retrieves papers which are irrelevant to our interest. There exists a tradeoff between a strict query which will retrieve the right but fewer papers and a wider query which will retrieve more papers of which there will be more irrelevant papers. In line with the research goal, we chose a stricter query so that the manual translation of Chinese literature will be feasible. We decided to investigate the last 10 years (2008–2018), as the literature before 2008 is limited [20]. To limit the potential biases that an uncompleted year may bring forward, we omitted the year 2019. The outcome of the iteratively testing of queries resulted in the following queries in Scopus and CNKI, of which we provide the translation:

- AUTHKEY ("digital transformation") AND LANGUAGE (english) AND PUBYEAR > 2007 AND DOCTYPE (ar OR cp) AND (EXCLUDE (PUBYEAR, 2019))
- AUTHKEY ("digital transformation" OR "digitalization transformation" OR "digital revolution" OR "digitalization revolution") AND LANGUAGE (Chinese) AND PUBYEAR > 2007 AND DOCTYPE (ar) AND (EXCLUDE (PUBYEAR, 2019))

DT was translated into four different Chinese search terms to be as complete as possible. This is because of the interchangeable use between 'digital' and 'digitalization', and between 'transformation' and 'revolution', together with the Chinese rule of noun combination. In contrast, most of the English literature agrees on one spelling, i.e. DT.

The queries retrieved all journal articles (ar) or conference papers (cp) that were published between 2008 and 2018, and that had DT in the keywords. For the Chinese literature, only journal papers that are indexed in the databases of PKU list of core journals, CSSCI (i.e. NJU list of core journal) and CSCD (i.e. Chinese Science Citation Database), are included to ensure the research quality and to limit the paper quantity to feasible levels of analysis. The queries respectively retrieved 348 and 596 papers (date of retrieval 03.07.2019). We extracted the bibliometric data of all papers the queries returned, including the year of publication and the author keywords. In Scopus, this process was done automatically which we manually checked for errors. Four errors were identified, which brings the total number of papers down to 344. For the Chinese literature, we manually created a CSV file with the authors, year, and keywords based on the file that we retrieved from CNKI. The keywords were translated manually into English by the second author, who is a native Chinese speaker. To reduce the discrepancy between the English and Chinese keyword spelling, a thesaurus was constructed in which the authors agreed upon common keywords and their spelling.

Next, we cleaned the keywords in several steps. First, we fixed the spelling errors that occurred in several papers. Secondly, we analyzed the synonyms found in the keywords and merged them to have more consistent results. For example, both IoT and the Internet of Things were found in the keywords, which were standardized into IoT. Thirdly, we reduced many different spelling styles for the same word. For example, the keywords contained plural and singular variants (e.g. business models and business model), British and American English (e.g. digitalization and digitalisation), hyphens (e-commerce and ecommerce), acronym variants (SMACIT and SMAC-IT), capitalizations and so on. This required manually checking the keywords and mutually agreeing on which keywords to standardize. Finally, some generic keywords, which do not provide any useful information, were omitted for the analysis such as 'study', or keywords that relate to the conference or journal where the paper was published.

After the cleaning, we used VOSviewer software tool [27] on both datasets to create keyword co-occurrence networks that consist of nodes (keywords) and links (co-occurrence) to visualize the literature. The weight of each link corresponds to the number of papers that have the keywords together, whereas the size of the nodes corresponds to the total number of occurrences. A minimum of three occurrences for a keyword to be included was selected. Lastly, we looked at the total volume of published papers over time and discussed our findings.

4   Results

4.1   English Literature

Figure 2 shows the English keyword co-occurrence network. As it can be observed, the DT literature is broad and consists of numerous linked keywords, related terminologies, and several subgroups with a specific research focus. Several general remarks can be made. First, the key focus of the English DT literature finds itself in the business and deals with digitalization, digitization, industry 4.0, digital economy, and IoT. Albeit the fact that DT covers a wider scope than the business changes, it is logical that the research focuses on the business side of things. It is indeed of major importance that the information system (IS) and economics research offers guidelines and aid to businesses to help with their DT journey. The term industry 4.0 is often related with DT [19] as it represents the fourth industrial revolution in which the whole sphere of industrial production is transformed to smartly connect the entire manufacturing process for performance improvements [18]. To reach this 'fourth stage', businesses must digitalize their processes, mutual communications, logistics, and manufacturing plants. One of the major technology drivers behind this change process is IoT. It is proposed that industry 4.0 is the goal of many traditional manufacturing firms, whereas DT is the general change process the entire world goes through without having an end goal. Furthermore, it is important to note that digitization and digitalization are often used together with DT even though they are different, demonstrating the vagueness surrounding the terminology. Digitization is the process of converting analog sources to digital [28] whereas digitalization is the sociotechnical phenomena of adopting and using more digital technologies [28]. Given this definition, it is unlikely that many IS researchers investigate digitization as it relates to very specific, and technological approaches which were more prominent in the 90s. A probable explanation is that digitization is often mistakenly interchanged with digitalization. In our view, DT is inherently different from both digitization and digitalization, as it is not limited to a certain change but deals with the long-term process and evolution the world is going through. Additionally, it goes one step further as digitalization; DT is not only about adopting digital technologies in everyday life but also transforming everyday life with them. As such, digitalization can be seen as an evolution of digitizing, and both of them can be seen as part of the DT process [10].

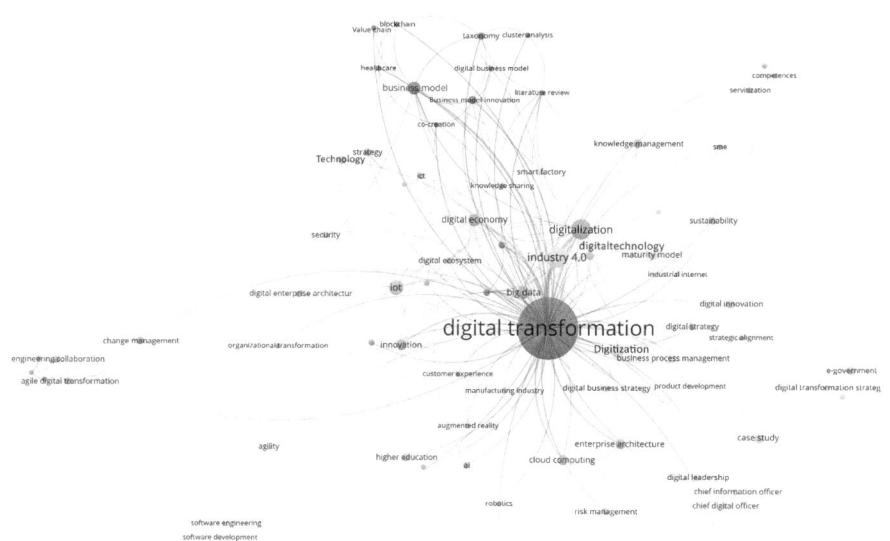

Fig. 2. The English keyword co-occurrence network

Secondly, there exist several sub-communities with strong co-occurrence connections between specific sets of keywords. These communities represent specific research focuses and collaborations. From the top of Fig. 2 in clockwise order, it is possible to identify the following research domains: (1) business models, (2) knowledge management and sustainability, (3) digital innovation, strategy, and e-government, (4) leadership, (5) enterprise architecture and technology (artificial intelligence (AI), cloud computing, robotics), (6) customer experience and innovation, (7) change management, engineering, and agility, and (8) IoT, security, and technology. Although these communities do not represent the entire academic work, it gives us a clear idea about their current focuses. Furthermore, it shows interesting relationships in-between the communities. Most of the academic efforts are closely entangled with each other, whereas change management, agility, and collaboration are more isolated.

When we compare the DT framework with the keyword analysis, we notice that most of the research talks about the business transformation dimension (business model, knowledge management, digitalization, industry 4.0, digital economy, innovation, and digitization). The technology dimension is mentioned less (cloud computing, IoT, robotics, technology, ICT, and internet), and society transformation is only mentioned sporadically (e-government, co-creation, and customer experience). Furthermore, little research seems to be talking about the impact, the struggles, or the steering force of DT. It is important to note that the keywords do not show any discrepancies with the framework. Nevertheless, a valuable addition to the technology dimension might be robotics when the framework is used in the context of industry 4.0.

4.2 Chinese Literature

Figure 3 presents the keyword co-occurrence network of the DT literature published in Chinese. The key focus of the Chinese literature is around the publishing (publishing, journal, newspaper, digital publishing, and press) and media (media convergence, traditional media, and media) industry. This is a more specific focus compared to the broader scope found in the English literature. This result is somewhat counterintuitive, as DT affects all sectors, yet only these two are presented in the keyword network.

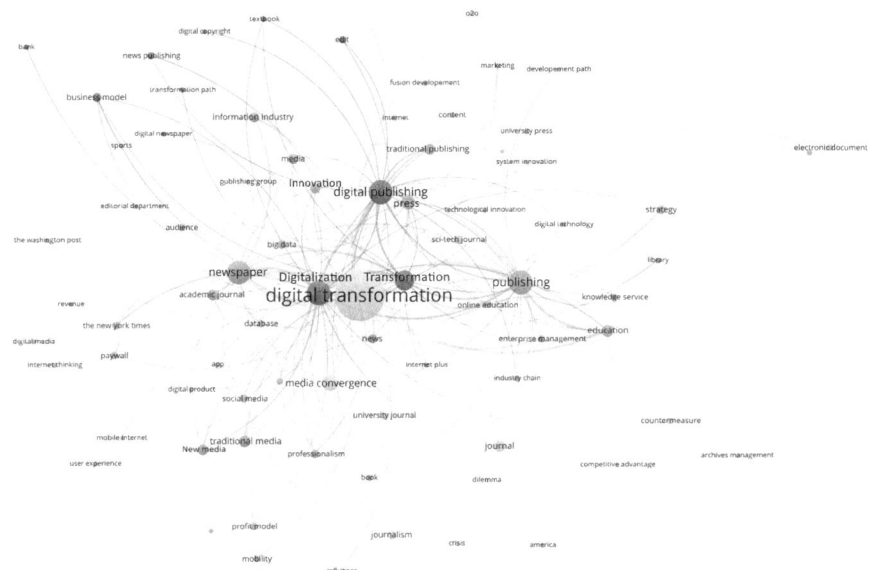

Fig. 3. The Chinese literature keyword co-occurrence network

We can distinguish several sub-communities, starting from the top in clockwise order: (1) publishing (traditional publishing, digital publishing, and press), (2) strategy, (3) education and publishing, (4) journalism, (5) media (traditional media, books, and new media), (5) newspapers, and (6) business models. It is important to note that these sub-communities are less distinct than in the English literature as nearly all the keywords are clearly linked with the publishing or media sector.

The biggest difference between the two networks is that publishing, news, and media have a much bigger research interest in the Chinese literature than in the Western literature. Another difference is that research about business models, technologies that steer DT, or different sectors that are being impacted by DT is more popular in the Western literature. There are several possible explanations for this difference: first, the online database may be biased towards research in publishing and media. Secondly, publishing and media industries have likely been impacted the most in the past 10 years, thus have enjoyed the most interest by researchers. This explains why there are many more papers on these sectors than on sectors that have not yet been impacted this severely by DT. Finally, we hypothesize that the DT research in social sciences is still in its early phase except for the research in the publishing industry that is already matured. Therefore, core social sciences journals (CSCD, CSSCI, and PKU) mostly consists of matured research fields such as DT in publishing. If this is true, we should expect a different result when we include all journals in the online database CNKI.

We evaluated the last hypothesis by redoing the above analysis but this time without journal limitations when retrieving data from CNKI. To see what other topics are researched in the DT literature, we excluded papers discussing the publishing sector from further analysis. We limited the papers to those issued in 2019 to decrease the translation work necessary for this research to feasible levels, which resulted in a total of 243 papers (date of retrieval 04.07.2019). We constructed the CSV file with the year and keywords that were translated manually by the authors. The result of the keyword co-occurrence network of this dataset is shown in Fig. 4.

By lifting the restriction of indexed papers of core journals, a considerable different result is obtained that more closely resembles the English literature. Here, DT research is focused on digital economics, digitalization, manufacturing, e-government, novel technologies, telecom, and banking. Several technologies are also mentioned, such as AI, 5g, industrial internet, and blockchain. The number of sectors increased considerably and now includes banking, telecom, e-government, smart cities, retail, and oil & gas.

Looking at the different communities, we can identify starting from the top in clockwise order: (1) AI and cloud computing, (2) industrial internet and the internet platform, (3) e-government, one belt one road project, digital china, smart city and manufacturing, (4) mobile operators and 5g, (5) banking and customer experience, and (6) big data, IoT, and digitalization. An interesting observation is that Chinese literature discusses more government projects, such as one belt one road and digital china, than the Western literature.

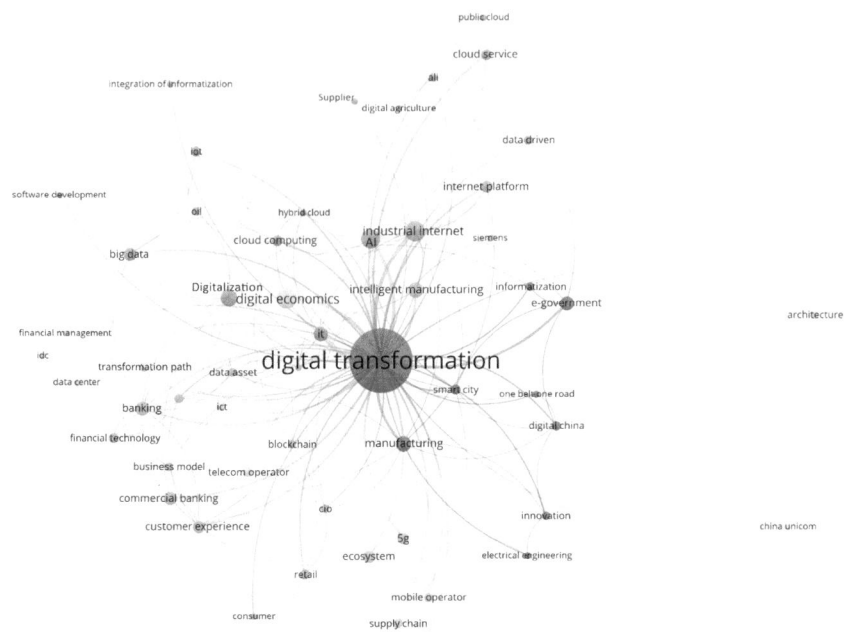

Fig. 4. The 2019 Chinese keyword co-occurrence network (without journal limitation)

4.3   Evolution Over Time

We also present the paper volumes over time in Fig. 5 according to the two original queries we utilized. Both the Chinese and English literature, as well as the total amount, is shown in the graph. This figure shows that DT research is booming since 2015 in the English literature, whereas the term DT is thriving in the Chinese literature since 2010. While it is unlikely that DT research was not present in the English literature between 2008 and 2015, a possible explanation is that previously most authors referred to this phenomenon as digitalization, IT-enabled transformation, or digital innovation. Furthermore, part of the growth of DT research in English can be attributed to the merging of several similar terms such as digitalization, IT-enabled transformation, digitizing, and business model innovation into DT. Lastly, part of the recent interest is because ICT is the main business driver nowadays for performance improvements, sustainability, and profitability. Digital technologies are so important that they are the reason more than 50% of the fortune 500 companies have disappeared since the year 2000 [29]. This spikes the interest in DT and thus the number of papers published. This trend will likely continue in the following years.

Paper volume over time

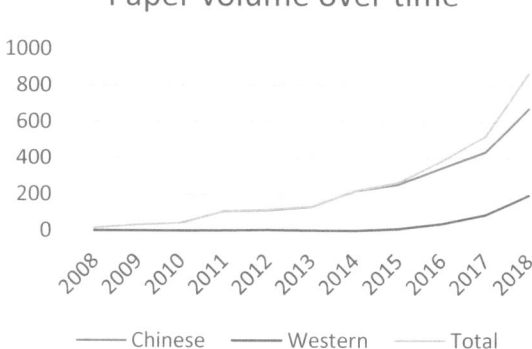

Chinese ——— Western ——— Total

Fig. 5. Paper volume over time

5 Discussion

The Chinese and English literature of DT between 2008 and 2018 were mapped using keyword co-occurrence graphs. The graphs depict the broad scope of DT. Many sectors, such as education, healthcare, manufacturing, publishing, media, telecom, finance, and e-governance are being studied. These sectors are often associated with novel technologies such as blockchain, cloud computing, AI, IoT, software, robotics, 5G, and others. Finally, the graphs show that many other keywords are used in relation to DT, such as change management, innovation, industry 4.0, business models, strategy, government projects, and leadership. Several of these keywords are often used in conjunction with DT as if they were synonyms, such as digitalization, digitizing, business transformation, and digital economy. This result demonstrates that DT misses a clear definition and might not be well understood.

Furthermore, we looked at the evolution over time of the keywords by creating graphs limited to the literature between 2011 and 2016 and compared them to Figs. 2 and 4. This clearly showed that DT research is expanding to new sectors, research aspects, and terminology. On the one hand, this indicates that DT research is still far from mature. On the other hand, this broadening of the research also fades the meaning of DT.

There exist significant differences between the two languages, with the English literature having a broad scope and many research hubs whereas the Chinese core journals were focused around the publishing and media industry. If we excluded the journal restrictions and investigated the keyword co-occurrence of the entire Chinese literature in 2019, we get a graph that more closely resembles the English literature. One important reason for these differences is the semantic deviation between the two languages. In Chinese, DT is more related to the direct transformation triggered by digital technologies, such as the increasing digitizing of business processes and the movement from paper to digital devices. This traditional meaning of DT is more applicable in the publishing and media industry, and less applicable in other sectors where the change is also in the business model. However, in English DT mainly focuses on the deeper transformation such as the change of business models driven by digital technologies. Another reason is the potential bias in journals indexed by databases of CSCD, CSSCI, and PKU. These journals are considered core social sciences journals which prefer matured research such as the academic work around publishing and media. Nevertheless, more sectors are researched in other journals, as depicted in Fig. 4, but its research is less matured and thus has less chance to be included in the core journals. It is probable that in the near future these differences might change, as social sciences play an important role in DT research for all sectors.

One of the interesting remarks is that Chinese literature has terminology and interest for its government projects, such as 'digital china' and 'one belt one road', while this was not observed in the English literature. This research is encouraged, as the government plays a key role in the DT process [5]. It can decide what practices are appropriate for digital disruptors, the policy legitimation and can influence the standard and network effects through its purchasing power [5].

Unexpectedly, the graphs depict a lack of mentions to society transformation even though society has been mentioned frequently as part of DT [1, 4, 5, 20]. Even though it is likely that a considerable amount of research exists that investigates the changes in society due to the increased use of digital technologies, the analysis shows that the research streams are not yet linked. The term DT has the power and academic attention to link these two important research streams together. This connection is important as it is impossible to look at businesses alone in the complex interactive process of DT. This paper calls for the use of the term DT in social sciences for papers discussing the changes as explained in this paper not only in business and technologies but also in society.

The DT framework consolidates the numerous terms, definitions, and views on DT. It is shown to be exhaustive and comprehensive with the existing notions of DT in the literature [17]. The keywords co-occurrence networks do not contradict the DT framework as they are covered by it. While the framework does not mention several technologies, such as robotics and 5g, explicitly, the technology dimension can easily be adapted to sector specifics, such as manufacturing, to include these. Nevertheless, the DT interactions of the framework are not present in the keywords networks.

We propose a research agenda that consists of several attention points. First, research must make more effort into unifying their understanding of DT and using the right terminology. This is vital for the further development of the DT research field. The DT framework [17] can serve as a foundation for this problem. Secondly, more research is needed in the society aspect of DT. While it is known that society plays a key role in DT, the exact relation it has with businesses and technology is not well understood. Thirdly, we explained the steering force of DT as the increasing interaction between technology, society, and business. Considerably more research is needed to investigate exactly how these interactions operate. Finally, more business sectors should be studied as all sectors will be impacted by DT.

This study has several limitations. First, it is impossible to retrieve all papers that deal with DT. While we tried to limit this selection bias by iteratively testing and evaluating the queries, other queries could bring forward different results. Secondly, human translations are prone to errors. To mitigate this problem, the keywords were checked manually for errors and a mutual thesaurus was constructed in which the authors agreed upon common spelling and naming conventions. Nevertheless, other translations might have been obtained with different translators. Third, the scope of this study was limited due to manual labor involved in such research. Hence, future studies could deploy broader queries and utilize multiple databases to produce a more detailed summary of the DT literature or to investigate the literature in other languages.

6   Conclusion

In this paper, we gave an overview of DT and its definition and investigated the DT literature's keywords both in the English and Chinese literature. We deployed a systematic literature review on two databases, Scopus and CNKI, and retrieved their bibliometric data to construct keyword co-occurrence networks. The networks reveal that DT has a broad scope, DT is not well defined, and there exists a significant difference between the English and the Chinese literature. While the English literature had many different aspects, such as business models, industry 4.0, e-government and leadership, the Chinese literature was mostly focused around publishing and media sectors. Consequently, we analyzed the Chinese literature on DT that was published in 2019 without restricting the query to the core journals and found out that the other sectors, such as the ones found in the English literature, were indeed present. This result indicates that the papers in the core journals are focused on publishing and media industry but non-core journals include many sectors. The results of this paper suggest that more research is needed to unify the meaning of DT, to investigate how society impacts the DT, and to study more sectors as DT is inherently sector dependent.

References

1. Stolterman, E., Fors, A.C.: Information Technology and the Good Life. Springer, Boston (2004)
2. Brynjolfsson, E., Hitt, L.M.: Beyond computation: information. J. Econ. Perspect. **14**(4), 23–48 (2000)
3. Parviainen, P., Tihinen, M., Kääriäinen, J., Teppola, S.: Tackling the digitalization challenge: how to benefit from digitalization in practice. Int. J. Inf. Syst. Proj. Manage. **5**(1), 63–77 (2017)
4. Ebert, C., Duarte, C.H.C.: Digital transformation. IEEE Softw. **35**(4), 16–21 (2018)
5. Hinings, B., Gegenhuber, T., Greenwood, R.: Digital innovation and transformation: An institutional perspective. Inf. Organ. **28**(1), 52–61 (2018)
6. Porter, M.E., Heppelmann, J.E.: How smart, connected products are transforming competition. Harv. Bus. Rev. **92**, 1–23 (2014)
7. Chase, R.: We need to expand the definition of disruptive innovation. Harv. Bus. Rev. **7**, 1–4 (2016)
8. Fitzgerald, M., Kruschwitz, N., Bonnet, D., Welch, M.: Embracing digital technology: a new strategic imperative. MIT Sloan Manage. Rev. **55**, 1–12 (2013)

9. Bharadwaj, A., El Sawy, O.A., Pavlou, P.A., Venkatraman, N.: Digital business strategy: toward a next generation of insights. MIS Q. **37**(2), 471–482 (2013)
10. Vial, G.: Understanding digital transformation: a review and a research agenda. J. Strateg. Inf. Syst. **28**, 1–27 (2019)
11. Haffke, I., Kalgovas, B., Benlian, A.: The role of the CIO and the CDO in an organization's digital transformation. In: ICIS 2016 Proceedings, vol. 1, January 2017, pp. 1–20 (2016)
12. Henriette, E., Feki, M., Boughzala, I.: The shape of digital transformation: a systematic literature review. In: MCIS 2015, pp. 431–443 (2015)
13. Schallmo, D.R.A., Williams, C.A.: Digital Transformation Now! Guiding the Successful Digitalization of Your Business Model, vol. 35, no. 4 (2018)
14. Kane, G.C.: Digital maturity, not digital transformation. MIT Sloan Manage. Rev. (2017). http://sloanreview.mit.edu/article/digital-maturity-not-digital-transformation/
15. Matt, C., Hess, T., Benlian, A.: Digital transformation strategies. Bus. Inf. Syst. Eng. **57**(5), 339–343 (2015)
16. Lozano, S., Calzada-Infante, L., Adenso-Díaz, B., García, S.: Complex network analysis of keywords co-occurrence in the recent efficiency analysis literature. Scientometrics **120**(2), 1–21 (2019)
17. Van Veldhoven, Z., Vanthienen, J.: Designing a comprehensive understanding of digital transformation and its impact. In: 32nd Bled eConference: Humanizing Technology for a Sustainable Society, pp. 745–763 (2019)
18. Janik, A., Ryszko, A.: Mapping the field of Industry 4.0 based on bibliometric analysis. In: IBIMA Conference, pp. 1–15 (2018)
19. Deloitte: Industry 4.0. Challenges and solutions for the digital transformation and use of exponential technologies (2015)
20. Reis, J., Amorim, M., Melão, N., Matos, P.: Digital transformation: a literature review and guidelines for future digital transformation. In: World Conference on Information Systems and Technologies, pp. 411–421, March 2018
21. Westerman, G., Calméjane, C., Bonnet, D., Ferraris, P., McAfee, A.: Digital Transformation: A Road-Map for Billion-Dollar Organizations (2011)
22. Chanias, S., Myers, M.D., Hess, T.: Digital transformation strategy making in pre-digital organizations: the case of a financial services provider. J. Strateg. Inf. Syst. **28**(1), 17–33 (2019)
23. Liu, P., He, X.: The digital transformation of traditional industries. People's Trib. **26**, 87–89 (2018)
24. Wang, J.: Digital transformation of commercial banks. China Financ. **22**, 48–50 (2018)
25. Morakanyane, R., Grace, A.A., O'Reilly, P.: Conceptualizing digital transformation in business organizations: a systematic review of literature. In: Bled eConference: Digital Transformation - from Connecting Things to Transforming Our Lives, pp. 427–443 (2017)
26. Ulez'ko, A., Demidov, P., Tolstykh, A.: The effects of the digital transformation. In: International Scientific and Practical Conference "Digitization of Agriculture - Development Strategy (ISPC 2019), vol. 167, pp. 125–129 (2019)
27. van Eck, N.J., Waltman, L.: Software survey: VOSviewer, a computer program for bibliometric mapping. Scientometrics **84**(2), 523–538 (2010)
28. Legner, C., et al.: Digitalization: opportunity and challenge for the business and information systems engineering community. Bus. Inf. Syst. Eng. **59**(4), 301–308 (2017)
29. von Leipzig, T., et al.: Initialising customer-oriented digital transformation in enterprises. Procedia Manuf. **8**, 517–524 (2017)

API Prober – A Tool for Analyzing Web API Features and Clustering Web APIs

Shang-Pin Ma$^{(\boxtimes)}$, Ming-Jen Hsu, Hsiao-Jung Chen,
and Yu-Sheng Su

National Taiwan Ocean University, Keelung, Taiwan
albert@ntou.edu.tw

Abstract. Nowadays, Web services attract more and more attentions. Many companies expose their data or services by publishing Web APIs (Application Programming Interface) to let users create innovative services or applications. To ease the use of various and complex APIs, multiple API directory services or API search engines, such as Mashape, API Harmony, and ProgrammableWeb, are emerging in recent years. However, most API systems are only able to help developers to understand Web APIs. Furthermore, these systems do neither provide usage examples for users, nor help users understand the "closeness" between APIs. Therefore, we propose a system, referred to as API Prober, to address the above issues by constructing an API "dictionary". There are multiple main features of API Prober. First, API Prober transforms OAS (OpenAPI Specification 2.0) into the graph structure in Neo4J database and annotates the semantic concepts on each graph node by using LDA (Latent Dirichlet Allocation) and WordNet. Second, by parsing source codes in the GitHub, API Prober is able to retrieve code examples that utilize APIs. Third, API Prober performs API classification through cluster analysis for OAS documents. Finally, the experimental results show that API Prober can appropriately produce service clusters.

Keywords: Web API analysis · Semantic annotation · GitHub · Cluster analysis

1 Introduction

Nowadays, Web APIs (Application Programming Interface) attract more and more attention. Many companies, such as Google, Facebook, Netflix, and Microsoft, expose their data or services by publishing Web APIs to let users create innovative services or applications. Meanwhile, REST (Representational state transfer) [1–3] is a software architecture design style, and the services designed in this style are called RESTful services. REST has been widely recognized as the recommended way to provide Web APIs. For example, there are more than 20,000 RESTful Web APIs published in ProgrammableWeb [4], a well-known API directory system, and the number continues to grow. Therefore, when a user faces numerous Web APIs with complex functionality, how to provide a mean to let users quickly understand the functionality of services and find related services has become a critical issue. To address the above issue, multiple

© Springer Nature Switzerland AG 2020
K.-M. Chao et al. (Eds.): ICEBE 2019, LNDECT 41, pp. 81–96, 2020.
https://doi.org/10.1007/978-3-030-34986-8_6

API directory services or API search engines, such as APIs.io [5], Mashape [6], and APIs.guru [7], are emerging in recent years. However, most systems only help users to get the basic information of the API or provide relevant articles; it is definitely insufficient.

Therefore, in this paper, we propose a novel API directory system, referred to as API Prober, to help users to understand the characteristics and client code examples of RESTful Web APIs based on the OAS (OpenAPI Specification 2.0) [8], the most widely used API description specifications, and related services. API Prober searches for the actual usage examples on GitHub. API Prober structurally analyzes and semantically annotates the elements of each service to help users to retrieve and filter services based on the characteristics of REST and common Web service design practices [9]. Notably, OAS documents were collected from APIs.guru, an API directory system on which numerous RESTful services with corresponding OAS documents are published. Note that "Web Service", "Web API", "RESTful Service", "RESTful Web Service", and "RESTful Web API" indicate the same meaning in this paper.

The remainder of this paper is organized as follows. Section 2 outlines background work related to this study and conducts surveys on the current API directory systems. Section 3 describes the design and implementation of API Prober. Section 4 shows the experimental results to illustrate the benefits of this study. Section 5 summarizes the characteristics of this study and future work.

2 Background and Related Work

2.1 Background Work

The OpenAPI Specification (OAS, originally called Swagger) is a document specification for describing, generating, and consuming RESTful Web services, which allows both humans and computers to discover and understand the capabilities of a service. As a result, a number of extended open source projects have emerged, such as automatic generation of back-end service framework programs, automated test programs, and front-end graphical interface frameworks. OAS can handle the needs of most RESTful services. Our approach is devised based on OAS.

The graph database is a kind of NoSQL database, which can store strongly related data in a high performance and high development efficiency than the related database. Neo4j [10] is an open-sourced graph database which is written in Java. It supports rapid development of graph powered systems that take advantage of the rich connection between data. The architecture is designed for the management, storage, and traversal of nodes and relationships. In Neo4j, the total amount of data stored in a database will not affect operations runtime. We apply the graph database to store the structure of API documents.

Cluster analysis (or called clustering briefly) [11, 12], a common technique in statistical data analysis, performs the tasks of grouping objects. It is used in many fields, including machine learning, image analysis, information retrieval, biometric analysis, and data compression. Notably, clustering is commonly used in text analysis, especially search engines, to organize multiple documents into categories, thereby improving the ability of search to look for related documents and let users to browse related information more conveniently. API Prober also provides clusters of Web APIs to let users to locate similar or related services more easily.

2.2 Related Work

Mashape [6] collects various APIs and displays APIs with the market-oriented style, like finding applications on Google Play or the App store. Mashape provides a personal key for each account as an identifier. After that, the user can consume all the APIs in Mashape through the key. Notably, Mashape provides a fixed code example for each API and allows online tests. APIs.io [5] is an experimental API website. APIs.io can use APIs.json to describe API operation information and find APIs on the Internet. Users can access the API page through query results. APIs.guru [4] is a website that displays APIs based on OAS. It not only contains the basic information of the API, but also adds more detailed information through OAS. ProgrammableWeb [4] published a lot of news and information about the Web API. So far, ProgrammableWeb has collected more than 20,000 RESTful APIs. In ProgrammableWeb, only basic API information is provided, while other API detailed information is linked to external web pages. For the client code examples, developers can publish code templates or related articles. It should be noticed that new or non-mainstream APIs do not have enough resources. When a user is providing a service to ProgrammableWeb, the user must manually mark the type of service. In addition to providing detailed API content, API Harmony [13] also analyzes features of RESTful services and retrieves relevant supplemental information, such as code examples on GitHub and articles on Stack Overflow.

By comparing with existing systems, only API Prober is able to extract important features of REST and common service design practices, find client code examples, and provide API clusters to help users understand how to use Web APIs.

3 API Prober: Approach Descriptions

Figure 1 shows the architecture of the proposed API Prober system. API Prober is mainly divided into two blocks, API Prober Runtime and API Prober Analyzer. The main function of API Prober Analyzer is data collection and analysis. Firstly, the features of the OAS files collected from APIs.guru are analyzed by Service Feature Analyzer. We refer to [9] to identify 9 service features for common Web service design and REST characteristics. This helps users to pick up well-designed services and understand how to utilize the retrieved services. Then, the services' descriptions and their supported service features are converted into the node set and the relationship set in the graph database. Next, API Usage Explorer searches for examples that consume services by crawling GitHub. Code examples could provide the users good references to build new applications or services. The main function of API Prober Runtime is responsible for the presentation of the front-end interface (Front-end UI) for the users. When a user enters the service content page, the Service Web Controller converts the user request into commands to the Service Manager to perform the tasks, such as searching services, filtering services by automatically annotated tags, viewing service information, and providing service client code examples.

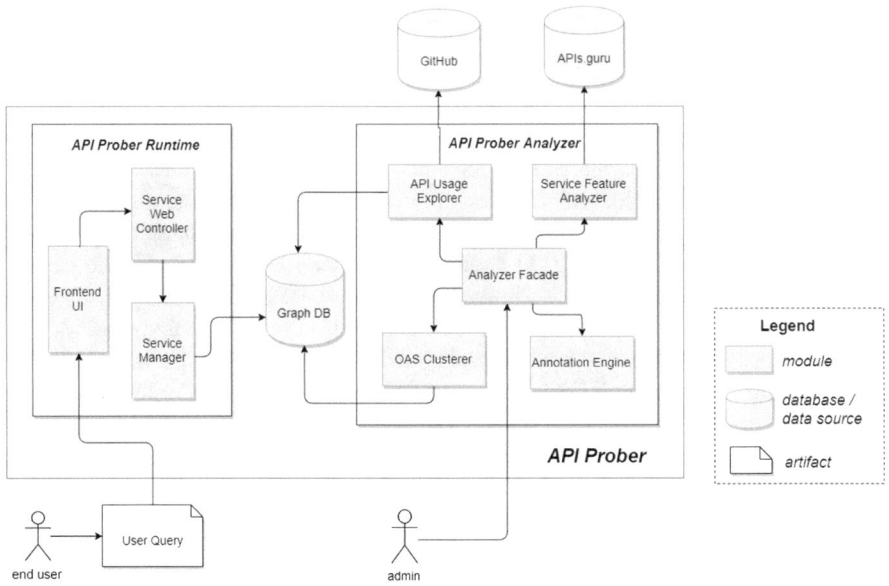

Fig. 1. API Prober: system architecture.

In next subsections, we fully discuss how to convert OAS documents of RESTful services into nodes and relationships in Neo4J (graph database), and analyze the services based on the identified service features. Next, we describe how to use GitHub to collect client code examples for services and conduct clustering for OAS documents.

3.1 Transforming OAS into Nodes

Based on our past research results [14], we divide the OAS into six parts to ease further processing. We store the corresponding nodes in the Neo4J graph database: (1) a resource node includes the title, logo, providerName, description, host, and basePath of the service; (2) a path node includes only path name; (3) an operation node includes action and description of the path; (4) a parameter node includes the name, description, and the media_type; (5) a status code node includes the status code and description of each operation; and (6) a response node includes the name, description and media_type of each status code response.

After building nodes, we establish the relationships between the nodes. A Resource node links one or more path nodes represents multiple service paths. A path node links to one or more operation nodes to represent different operations in the service path. An Operation node links to one or more parameter nodes and status code nodes to represent multiple input parameters and multiple status code. A status code node links to one or more response nodes to represent the output parameters of each status code (shown in Table 1).

Table 1. Relationships between nodes.

Node	Relationship type (direction: left to right)	Node
Resource	Endpoint	Path
Path	Action	Operation
Operation	Input	Parameter
Operation	Have	StatusCode
StatusCode	Output	Response

The above rules are used to show the Nodes and Relationships that the OAS maps to Neo4J. An example of OAS can be found in the case of the Google Gmail API (https://www.googleapis.com/discovery/v1/apis/gmail/v1/rest). Because the example of Gmail API is too complex, in Fig. 2, we provide a simplified example to show the Neo4J graph structure for an OAS document. Next, we use the information stored by the node to annotate the concepts of the four nodes. The properties extracted and used for sematic concept annotation are listed below.

- Resource: x-tags and description.
- Operation: description.
- Parameter: description and parameter name.
- Response: description and response name.

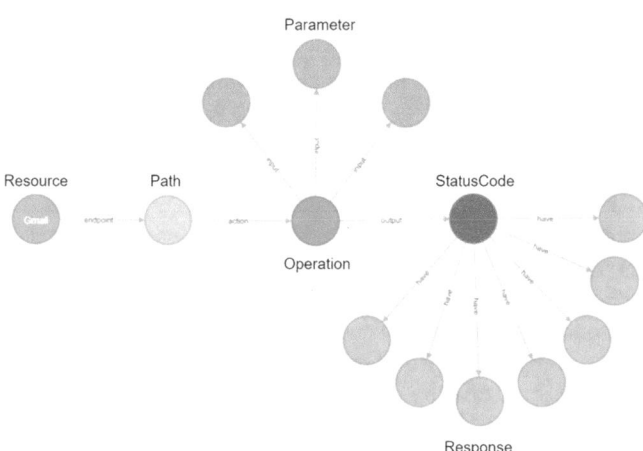

Fig. 2. Example of OAS nodes in Neo4j.

After retrieving the information, based on the previous research results [14], we divided the annotation process into the following three steps:

- The first step of annotation is tokenization and stemming. The IR tool, Lucene Analysis API[1], is used to chop up the service descriptions into tokens and remove stop words. For stop words, we also collect some service provider names and terms that are commonly used for RESTful services to avoid affecting the accuracy of service matching. For stemming, we used the well-known stemming algorithm, Porter Stemming Algorithm [15], to strip suffixes from a derived word.
- The second step is to use the LDA algorithm to extract topic words of OAS documents. The LDA topic model [16] is an unsupervised learning method to statistically discover topics for documents. API Prober applies LDA and annotates OAS documents by retrieved LDA topics so that API Prober can perform service queries, parameter pairing, and service grouping based on the identified words.
- The third step of annotation is term expansion. It is common that a query term issued by the user and a term included in the service descriptions are different terms but represent the same meaning. Based on our previous method [14], we use the Edge Counting Method [17] to calculate the similarity between each term of synset (synonyms, hypernyms, hyponyms) and the original term. Only the terms whose similarities are larger than the given threshold, 0.9, are put into the set of expanded terms. Thus, API Prober performs different kinds of term expansion based on parts of speech for different nodes:
  - Nodes of operations: expansion for verb terms
  - Nodes of resources, parameters, and responses: expansion for noun terms

Finally, we put the collection of the original words analyzed in step 2 and the expanded words analyzed in step 3 into the corresponding nodes.

3.2   Analysis for RESTful Service Features

As mentioned, REST design style has become the standard for most Web services [9, 18]. The main principle of REST is designing services based on the resources. Any information that can be named should be a resource: a document or image, a temporal service, a collection of other resources, a non-virtual object (e.g. a person), and so on. REST uses a resource identifier to identify the particular resource involved in an interaction between components. But the REST style does not pose strict regulations; there are a lot of decisions that the developers must decide when providing a public service API; it may produce high-quality APIs or lead to poorly designed APIs. For example, one of the common bad smells is using a single HTTP verb to search and delete resources. This kind of design causes that the service client needs to carefully confirm the usage of each endpoint and tend to misuse the service. Therefore, we identify multiple RESTful service characteristics and the common practices of Web service design to analyze the content of OAS files. Based on analyzed services with

---

[1] https://lucene.apache.org.

features, users can retrieve services with the desired service features and avoid inappropriate use of services by inspecting the features of a target service.

Based on the nodes transformed from OAS, the service feature analyzer automatically analyzes the information of nodes. We divide it into two different levels, the service level and the endpoint level. The service level is defined as the overall feature of the OAS file, and the endpoint level is service path features, as shown in Fig. 3.

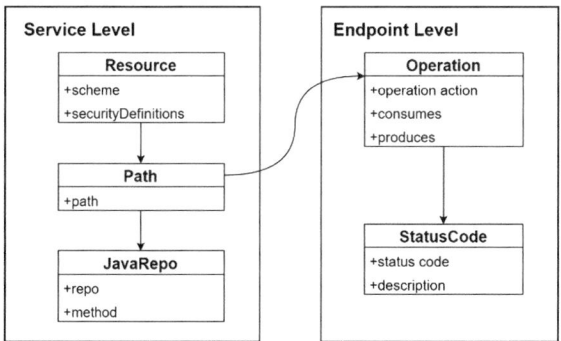

Fig. 3. Nodes in two levels for OAS.

In the service level, the resource node (for the target RESTful service), path nodes, and java repository nodes are identified and produced. The content of each java repository node is an example about the corresponding service path. In the endpoint level, the operation nodes and status code nodes are identified and produced.

Based on [9], we adopt 9 REST Web service design principles commonly used by service developers to help users filter and detect services are listed below.

1. HTTPS support: The main purpose of HTTPS development is to provide identity authentication for the web server and protect the privacy and integrity of the exchanged data to protect the user's personal data.
2. User authentication: Some Web services allow users to invoke freely without user verification, which may lead to improper use, such as frequent and long-term use of Web API. It may cause the web server to be overloaded and even unavailable for normal users.
3. At most 20 operations: This service feature is only the statistics and observation about the number of Web service operations. Twenty operations are usually an indicator to show that the service is fully functional and able to be used in various conditions.
4. REST-style URIs: This feature is the basic principle of the REST architecture. Resource addressability is defined by the service path using a URI as a unique address, and each resource can be accessed through the URI. For example, the URI for a single resource, https://example.com/resources/{19}, indicates the unique Web resource of id "19".

5. HTTP status code use: The HTTP status code is a 3-digit code that represents the response status of the Web server transport protocol. The service provider needs to declare supported status codes and set an appropriate status code for each service request.
6. Explain error messages: This feature indicates if a service declares the descriptions of the error status for each operation. This feature can help users to understand the service responses and allow service providers to test or maintain the service more easily.
7. Example API conversations: When users want to know how to use the Web service, in addition to viewing the files provided by the service, they will try to find examples of the service for reference. Users view service files for more detailed service content. But for most users, a practical example will speed up the use and understanding of the service.
8. Output format JSON: JSON (JavaScript Object Notation) is a lightweight data exchange language. According to the statistics of 500 Web services in [9], up to 91% is JSON as the output format. It shows that JSON has become the output standard of most services. Service providers can design services while their services can focus on mainstream output formats, thereby avoiding wasted resources by supporting many different output formats.
9. Input format JSON: Similar to output format JSON, the service path takes JSON as the input format. With JSON data format, the communication between services and services is more convenient.

After analyzing the above nine features through the Service feature analyzer, API Prober stores the service level service feature into the resource node, and the endpoint level service feature are stored in the operation node.

3.3   Extraction of Service Client Codes

In this section, we describe the process of extracting real-world service client examples by the API Usage Explorer (AUE) module. GitHub is a platform for version control via Git and provides hosting services for software source code. In addition to allowing individuals and organizations to create and access code, it also provides the social network services that allows users to track other users, organizations, and software libraries. It has become the world's largest open source code community. In API Prober, AUE bases each RESTful service to find relevant code through GitHub. According to [19], there are three ways to collect codes in GitHub repositories: GitHub Archive, GHTorrent, and GitHub Search API provided by GitHub. Since GitHub Archive and GHTorren can only provide old archived data, we selected the GitHub Search API as the code collection method in this research.

At the beginning of the extraction process, AUE retrieves the scheme, host, basePath from a resource node and the path name from a path node in the graph database, and then combines the information into a target service path, for example, https://api.github.com/emojis, as shown in Table 2. Readers could see examples and related files here:

https://raw.githubusercontent.com/APIs-guru/unofficial_openapi_specs/master/github.com/v3/swagger.yaml

Next, AUE uses the combined service path as the query parameter and calls the GitHub Search API to search for possible service client codes. The service path of the search is as follows:

https://api.github.com/search/code?q=https://api.github.com/emojis+language:java

- q: service path information to be queried
- language: the target programming language

The code extraction process is divided into five steps. The first step is to retrieve the first 100 search results returned by the GitHub Search API. Based on the scores given by GitHub, AUE also sorts the search results in the descending order. The second step is to tokenize the target service path so that we can match the path and the retrieved code files. The third step is to perform tokenization on the text_matches, i.e. the matching fragments, of each retrieved code file. The fourth step is to compare the service path with the text_matches to determine if all the tokens in the service path are also included in the fragment, thereby filtering out some inappropriate results and avoiding the retrieval of incorrect examples. The fifth step is to collect the qualified results and stores them in the Neo4J database.

Table 2. Example of service path combination.

Node	Information
Resource	$.info.scheme:https
	$.info.host:api.github.com
	$.info.basePath:/
Path	$.paths.{path_name}:/emojis

3.4   OAS Clustering

To achieve better service recommendation, it is required to pre-classify services to help users to browse relevant services more easily. Based on the techniques of cluster analysis, we divide the OAS documents into multiple clusters. To implement the OAS clustering, several preparatory tasks are conducted: (1) devising methods to calculate the similarity scores between OAS documents; (2) determining parameters involved in the clustering; and (3) using evaluation methods for clustering to find out the best combination of parameters.

Document Concept Score (DCS)
The goal of the design for the Document Concept Score (DCS) is to calculate the similarity between the target OAS file and a candidate OAS file. We consider the information of the Resource node and information of the Operation node for the target OAS. The calculation of DCS is based on VSM (Vector Space Model). API Prober converts the original words of Resources and Operations into V_{ow}, and the set of WordNet-extended words into V_{ww}, and aggregates the above sets as $Resource = \{V_{ow}, V_{ww}\}$ and $Operation = \{V_{ow}, V_{ww}\}$. Note that only the original words are considered for the target OAS, whereas both the original and the extended words are taken into account for the candidate OAS. Then we use the Term Count Model of VSM to

calculate the Resource and Operation scores between files. The calculation method is shown in Eq. 1.

$$sim^{RS}(T,C) = sim\left(V_{ow}^T, V_{ow}^C\right) + W_{WN} * sim\left(V_{ow}^T, V_{ww}^C\right) \tag{1}$$

T and C are the target OAS file and the candidate OAS file respectively. $sim^{RS}(T,C)$ is the VSM score of the Resource, and W_{WN} represents the weight of WordNet (the setting of W_{WN} is discussed later). $sim^{OP}(T,C)$ is also calculated in the same way as the above Eq. 1.

Based on the VSM scores of Resources and Operations, we can obtain the final DCS, as shown in Eq. 2.

$$DCS = W_{RS} * sim^{RS}(T,C) + W_{OP} * sim^{OP}(T,C) \tag{2}$$

Where W_{RS} is the weight of $sim^{RS}(T,C)$ in DCS; W_{OP} is the weight of $sim^{OP}(T,C)$ in DCS.

Parameter Setting for Clustering

We identified the following four parameters that mostly affect the results of clustering, and tried to find the best combination of these parameters.

- The number of LDA (Latent Dirichlet Allocation) Topics: As mentioned, API Prober applies LDA and annotates OAS documents by retrieved LDA topics. How to set an optimal number of topics is a critical issue when applying LDA. Thus, the number of LDA topics is the first parameter that we need to determine in API Prober.
- The weights of the Resource part and the Operation part in DCS: As mentioned, API Prober analyzes the Resource part and the Operation part of the OAS file to calculate the proposed DCS score (Eq. 2). The Resource part describes the purpose and motivation of the whole service, while the Operation part specifies the functionality of a service endpoint. In API Prober, both W_{RS} and W_{op} are ranging from 0 to 1 and their sum is 1. The values of W_{RS} and W_{op} are the second critical parameter required to assign for the OAS clustering.
- WordNetScore weights: As mentioned, we applied WordNet to expand the topic words produced by LDA to include more similar words. The word expansion is an important method to increase the possibility of matching services and enhance the matching precision. In API Prober, the weight W_{WN} (shown in Eq. 1), also ranging from 0 to 1, is the third parameter that needs to be set.
- Applied clustering method: In addition to the above three parameters, we also need to determine which clustering method should be applied. Note that Single Linkage [20] is not included to be a candidate method in this study. During clustering, the Single Linkage method finds out the most similar nodes in two different clusters and merges these two clusters. This procedure is conducted iteratively. Obviously, the major disadvantage of this method is producing unreasonably large clusters rather than producing size-balanced clusters. Other well-known clustering methods, including Complete Linkage, Average Linkage, Centroid Linkage, and Weighted Linkage, are regarded as alternatives in this research.

Finally, we combine various values or alternatives for the above four parameters, evaluate the clustering results for each combination based on the clustering evaluation method, and determine the best parameter combination. The evaluation methods and results are fully discussed in the next subsection.

Evaluation Methods and Results for Parameter Setting
For the clustering evaluations, there are two main tracks: external and internal evaluations. Internal evaluation directly calculates the similarities between clusters and the similarities within each cluster to determine the efficacy of clustering. Good clustering should have low similarities between clusters and high similarities within each cluster; Similar to the internal evaluation, external evaluation is based on the label data given by external evaluators.

This study applied Adjusted Rand Index (ARI) [21], a widely-used external evaluation method. ARI is based on Rand Index (RI), a simple method for assessing the similarity between two clustering results. Its calculation formula is shown in Eq. 3. The RI value ranges from 0 to 1. A higher RI score means that the expected clustering results and the actual clustering results is more similar.

$$RI = \frac{a+b}{\binom{n}{2}} \tag{3}$$

It is assumed that A is the actual (manual) clustering results, B is the clustering results obtained by a clustering model; a represents the logarithm of the elements of the same cluster in both A and B; b represents the logarithm of the elements of different clusters in A and B; $\binom{n}{2}$ represents the number of unordered pairs of n elements in the clustering result. Although RI is good indicator to show the efficacy of clustering, for the results of random distribution, RI does not ensure that its value is close to zero. The Adjusted Rand Index (ARI) addresses the above issue and ensures that its results for the random distribution is close to zero. The ARI ranges from -1 to 1, as shown in Eq. 4.

$$ARI = \frac{RI - Expected\_RI}{\max(RI) - Expected\_RI} \tag{4}$$

Where $\max(RI) = \frac{1}{2} \left[\sum_i \binom{n_i}{2} + \sum_j \binom{n_j}{2} \right]$ represents the maximum value of R1,

i.e. the actual results that are exactly matched with the expected results; $Expected\_RI =$

$\left[\sum_i \binom{n_i}{2} \sum_j \binom{n_j}{2} \right] / \binom{n}{2}$ indicates the expected value for the case of random distribution.

To find out the best combination of parameters, we randomly selected 100 OAS documents and invite three well-trained software engineers to discuss and determine clusters manually for these documents to construct the benchmark. The clustering results can be accessed by https://github.com/SOSELab401/API-Prober.

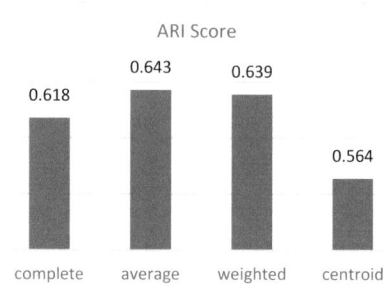

Fig. 4. ARI scores for clustering methods.

Next, we used four well-known clustering methods, including Complete Linkage, Average Linkage, Weighted Linkage, and Centroid Linkage, to produce clusters separately by integrating various values of other parameters and various thresholds of clustering, and calculated the highest ARI scores for each clustering method. Finally, the highest ARI values of all four methods as shown in Fig. 4. The results show that the Centroid Linkage method is relatively unsuitable to produce clusters of OAS documents. For other three methods, the Average Linkage method yields better ARI. Thus, we decided to apply the Average Linkage method and its setting of parameters for the highest ARI to be final combination of parameters. The final setting of parameters is shown in Table 3.

Table 3. Final combination of parameters.

Parameter name	Value/used method
LDA topic	4
Resource weight	0.6
Operation weight	0.4
WordNet weight	0.9
Clustering method	Average linkage method
Clustering threshold	0.8

4   Experiments and Discussions

This section outlines the experiments used to verify the efficacy of the proposed clustering approach for Web APIs (Web Services) based on their OAS documents.

4.1   Evaluation Indicators

We adjusted the calculation method of the ordinary Top-K Precision based on the characteristics of clustering. Two different cases were considered in the experiments: positive relevance and negative relevance. Positive relevance indicates that two or more relevant services should be clustered into the same cluster, while negative relevance means that two or more unrelated services should clustered into different clusters. Two new indicators, Positive Precision and Negative Precision, are devised as shown in Eqs. 7 and 8.

$$PP^k(\text{OAS}) = \frac{|SDC_{relevant}^k|}{|SDC^k|} \tag{7}$$

$$NP^k(\text{OAS}) = \frac{|SDC_{irrelevant}^k|}{|SDC^k|} \tag{8}$$

Where SDC stands for the Service Document Cluster, SDC^k represents the closest k OAS documents for a target OAS (determined by DCS) in a cluster; $SDC_{relevant}^k$ is the number of relevant OAS documents (determined by the evaluators) in SDC^k; Similarly, $SDC_{irrelevant}^k$ is the number of irrelevant OAS documents (also determined by the evaluators) in SDC^k. Finally, the Adjusted Precision (called AP briefly) is calculated by summing and averaging the results of PP^k and NP^k, as shown in Eq. 9 (n represents the number of clusters).

$$AP(n) = \frac{\sum_{i=1}^{n} PP^K + NP^K}{n} \tag{9}$$

4.2   Experimental Setup

We choose the category system of APIs.guru as our comparison target since no existing clustering methods/tools for OAS are available. Based on the evaluation indicators, we compared the clustering efficacy for APIs.Guru and API Prober by two ways: retrieving test cases based on the clusters of API Prober and retrieving test cases based on the categories of APIs.Guru. The details of these two experiments:

1. The total number of OAS documents is 1253.
2. For the first experiment, we randomly selected 20 clusters in API Prober, randomly selected a target OAS in each cluster, and find its closest 3 OAS documents. Based on these retrieved OAS documents, we calculated the AP scores for APIs.Guru and API Prober.
3. For the second experiment, we randomly selected 20 categories in APIs.Guru, randomly selected four OAS documents (one for the target OAS and three for the relevant OAS) in each category. Based on these retrieved OAS documents, we also calculated the AP scores for APIs.Guru and API Prober.

4.3   Experimental Results

The evaluation result of the first experiment is as shown in Fig. 5. The experiment result shows that both the APIs.Guru and API Prober yield AP of more than 0.75 and API Prober is obviously better than APIs.Guru. The reason why API Prober is not able to achieve higher precision is due to the quality of OAS documents. Although OAS has become a standard for Web API, document granularities likely vary in different organizations. For example, the service developer may roughly write the overall service description or the descriptions for service operations; it causes difficulties to perform accurate clustering.

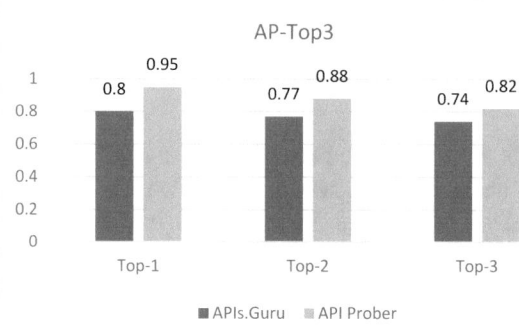

Fig. 5. AP (Adjusted Precision) comparison of experiment 1.

The result of experiment 2 is shown in Fig. 6. It can be clearly observed that the proposed clustering method is significantly better than the APIs.Guru category system. The main reason is that the APIs.Guru category system classifies OAS documents very roughly. For example, many Web APIs provide cloud-related functionalities, almost all these APIs are classified as "Cloud" in APIs.Guru; it causes inappropriate and imprecise categorization.

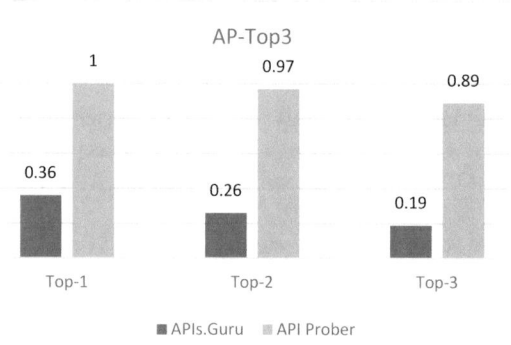

Fig. 6. AP (Adjusted Precision) comparison of experiment 2.

5   Conclusion

In this paper, we propose a system, referred to as API Prober, to analyze RESTful service descriptions, extract service client code examples, and produce service clusters to let users know how to use a target RESTful service quickly and find related services. The main features of API Prober are four-fold: (1) transforming OAS documents into the graph structure in Neo4J database; (2) analyzing the service features of common Web service design practices and REST characteristics; (3) retrieving practical examples in GitHub; and (4) devising a clustering scheme to group RESTful services.

Our future plans include: (1) extracting examples in various programming languages; and (2) developing a recommendation mechanism based on the service clusters to ease web service integration.

Acknowledgment. This research was sponsored by Ministry of Science and Technology in Taiwan under the grant MOST 108-2221-E-019-026-MY3.

References

1. Gat, I., Succi, G.: A Survey of the API Economy. Cut. Consort (2013)
2. Fielding, R.T., Taylor, R.N.: Principled design of the modern Web architecture. ACM Trans. Internet Technol. (TOIT) **2**(2), 115–150 (2002)
3. Amundsen, M.: RESTful Web Clients: Enabling Reuse Through Hypermedia. O'Reilly Media, Inc, Sebastopol (2017)
4. ProgrammableWeb. http://www.programmableweb.com/
5. APIs.io. http://apis.io/
6. Mashape. https://www.mashape.com/
7. APIs.guru. https://apis.guru/
8. OpenAPI Specification (OAS). https://swagger.io/docs/specification/
9. Neumann, A., Laranjeiro, N., Bernardino, J.: An analysis of public REST web service APIs. IEEE Trans. Serv. Comput. **2018**, 1 (2018)
10. Webber, J.: A programmatic introduction to neo4j. In: Proceedings of the 3rd Annual Conference on Systems, Programming, and Applications: Software for Humanity. ACM (2012)
11. Agrawal, R., Phatak, M.: A novel algorithm for automatic document clustering. In: 2013 3rd IEEE International Advance Computing Conference (IACC) (2013)
12. Reddy, V.S., Kinnicutt, P., Lee, R.: Text document clustering: the application of cluster analysis to textual document. In: 2016 International Conference on Computational Science and Computational Intelligence (CSCI) (2016)
13. Wittern, E., et al.: API harmony: graph-based search and selection of APIs in the cloud. IBM J. Res. Dev. **60**(2–3), 12:1–12:11 (2016)
14. Ma, S., et al.: Real-world RESTful service composition: a transformation-annotation-discovery approach. In: 2017 IEEE 10th Conference on Service-Oriented Computing and Applications (SOCA) (2017)
15. Porter, M.: The Porter Stemming Algorithm. http://www.tartarus.org/~martin/PorterStemmer/
16. Blei, D.M., Ng, A.Y., Jordan, M.I.: Latent dirichlet allocation. J. Mach. Learn. Res. **3**, 993–1022 (2003)

17. Li, Y., Bandar, Z.A., Mclean, D.: An approach for measuring semantic similarity between words using multiple information sources. IEEE Trans. Knowl. Data Eng. **15**(4), 871–882 (2003)
18. Haupt, F., et al.: A framework for the structural analysis of REST APIs. In: 2017 IEEE International Conference on Software Architecture (ICSA) (2017)
19. Cosentino, V., Izquierdo, J.L.C., Cabot, J.: Findings from GitHub: methods, datasets and limitations. In: 2016 IEEE/ACM 13th Working Conference on Mining Software Repositories (MSR). IEEE (2016)
20. Aggarwal, C., Zhai, C.: A Survey of Text Clustering Algorithms (2012)
21. Vinh, N.X., Epps, J., Bailey, J.: Information theoretic measures for clusterings comparison: Variants, properties, normalization and correction for chance. J. Mach. Learn. Res. **11**, 2837–2854 (2010)

Service-Oriented and Cloud

Design a Distributed Fog Computing Scheme to Enhance Processing Performance in Real-Time IoT Applications

Shin-Jer Yang[✉] and Wan-Lin Lu

Soochow University, Taipei, Taiwan
sjyang@csim.scu.edu.tw, gra231403@gmail.com

Abstract. The technology evolution and application popularity in cloud computing has driven the rapid development of the Internet of Things (IoT) services and applications. When these real-time events in IoT applications are transmitted to the cloud for processing, the load on cloud computing becomes heavier, which can result in processing delays and exceeding the processing deadline. To improve upon and solve problems derived from many and more real-time events in the cloud computing process, some have proposed a fog computing concept. Previous studies on fog computing are mainly focused on accurately defining the fog computing concept and its possible applications. The nodes in fog computing are closer to edge devices, which can process such these real-time events in this layer. With this fog process, processing efficiency can be significantly increased and delays can also be solved in cloud computing.

This paper design a set of fog computing framework called D-FOG scheme that can improve the better processing performance of real-time IoT applications than conventional cloud computing processing architecture. The experimental results of KPIs indicate that in Expired event rate can be lower by 28%, 34% and 28.2%; Event success rate can be increased of 21.2%, 26.9% and 20.48%, and also Average processing time can be reduced by 436.87 ms, 558.72 ms and 1320.45 ms in 100, 200 and 500 real-time events, respectively. Consequently, the proposed D-FOG scheme is more effective and efficient in real-time events processing than conventional cloud computing processing architecture.

Keywords: Cloud computing · Fog computing · D-FOG · Real-Time IoT applications

1 Introduction

The maturation of cloud computing technology is driving rapid development of Internet of Things (IOT) services and applications. Cloud computing uses a concentrated processing architecture and virtual technology to produce a cloud environment that can provide large quantities of calculation. This enables users to process a large amount of IoT events without the need to install high-performance computing equipment on the user-end. However, there are also disadvantages. One disadvantage in cloud computing is that to process each IoT event, the event must be uploaded to the cloud before it can be processed. Many factors in this process can increase the uploading time, which can

© Springer Nature Switzerland AG 2020
K.-M. Chao et al. (Eds.): ICEBE 2019, LNDECT 41, pp. 99–112, 2020.
https://doi.org/10.1007/978-3-030-34986-8_7

cause the entire processing time to become longer. IoT has many real-time applications, including product production monitoring in manufacturing and road monitoring in traffic and transportation, etc. These services and applications have many events that require real-time processing. When a large number of events are uploaded to the cloud for processing at the same time, it can result in delayed processing and exceeding the processing deadline.

To solve delay problems derived from a large number of IoT application events in real-time cloud computing and processing, CISCO proposed an early fog computing concept in 2014. Fog computing technology uses a distributed processing framework to disperse calculation, transmission, control, and storage services on the user's equipment and system or on nearby systems. Examples are mobile phones and gateways, etc. Fog computing can be used to process urgent real-time events on edge devices near the user. This not only can reduce delays, but can also shorten process time and ensure that the calculation is completed within the deadline, which makes real-time event processing more efficient and lightens the load on cloud computing processing.

Although the fog computing concept was proposed in 2014, the OpenFog Consortium that specializes in the study of fog computing was not established until the end of 2015. The Consortium officially published its white paper on fog computing architecture in 2017 and fog computing gradually began to attract people's attention. Currently, most studies are still focused on the precise definition of the fog computing concept. However, the purpose of this paper is to propose a fog computing scheme that can increase the efficiency of a real-time IoT application and solve processing delays. This scheme can also effectively shorten processing time to ensure that real-time IoT events can be processed within the deadline. Thus, the main purposes of this paper are as follows:

- To solve processing delays and deadline problems that result from a large number of real-time IoT application events in cloud processing, we propose a distributed fog framework into the conventional cloud computing processing architecture called the distributed fog computing scheme (D-FOG) for existing real-time IoT applications, which can effectively improve processing efficiency.
- In this paper, we will propose three types of key performance indicators (KPIs): Expired event rate, Event success rate, and Average processing time. Also, we will set up experimental environments for simulations.
- The simulations conducted in this paper used the proposed D-FOG scheme to simulate a real-time IoT applications. The D-FOG can be used to effectively improve the processing efficiency of real-time IoT applications.

The rest of this paper is organized as follows. Section 1 describes the background of the D-FOG and the research scope and purpose. Section 2 surveys all literature reviews about cloud computing, IoT and fog computing. In Sect. 3, this paper examines the operational flows of the D-FOG research processes and design of the D-FOG algorithm. Section 4 covers the simulated experiments setup and analyzes the results. Finally, we draw a conclusion and illustrate and analyze the simulation results, also we indicate the further research direction in Sect. 5.

2   Related Work

2.1   Cloud Computing

The American National Institute of Standards and Technology (NIST) defines cloud computing as "a type of computing resource access and sharing mode that can be adjusted at any time according to the user's needs." This type of computing can use minimal management work or service supplier interaction to achieve rapid configuration and release. Cloud computing defined by NIST is composed of five essential characteristics, three service models, and four deployment types [1, 2], as shown in Fig. 1. The descriptions of cloud computing are as follows.

The following are the five main essential characteristics:

1. On-demand Self-service: The user can use the cloud service according to their own needs, and does not require using the cloud supplier to make settings. Users can use the webpage to adjust their own settings.
2. Broad Network Access: Use the Internet connection to make service available everywhere so that the user can use the service just by connecting to the Internet.
3. Resource Pooling: Cloud service providers can use a multi-rental model for users and allocate resources according to the user's requirements.
4. Rapid Elasticity: Users can rapidly adjust the resource scale according to needs.
5. Measured Service: Resources on the cloud must be able to be measured. This means that the cloud provider can implement resource planning, control access, and tabulate fees.

There are three main types of services models:

1. *Software as a Service (SaaS):* Users can directly implement programs provided on the cloud architecture and do not need to install any software system or worry about updating problems. Users can just pay according to their own use status.
2. *Platform as a Service (PaaS):* The cloud provider provides a development platform for users so users can use this platform's development tools to install their own programs on the cloud for development and testing.
3. *Infrastructure as a Service (IaaS):* The cloud provider provides infrastructure facilities such as storage equipment and network. Users can flexibly control the required development environment in this environment. Users can freely choose the operating system or programs, and do not need to understand base level cloud architecture. They just directly use the services provided by the cloud provider.

There are four main deployment types:

1. *Private Cloud:* Used by a single organization. This prevents security problems caused by different users. Generally, this cloud is managed by the organization itself or managed by a third-party.
2. *Community Cloud:* A community cloud is mutually shared by multiple organizations. Generally, these organizations have similar objectives. Members in the organization can use the community cloud service to share important topics and use this to complete specific tasks or cooperate with policies.

3. *Public Cloud:* Compared to a private cloud, a public cloud is provided for use by the public. It can be managed and operated by businesses or academic or government organizations.
4. *Hybrid Cloud:* A hybrid cloud refers to a cloud formed by two or more of the aforementioned cloud types. Although different types of clouds are independent and separate, the two can be connected through a professional technology to transfer data and programs.

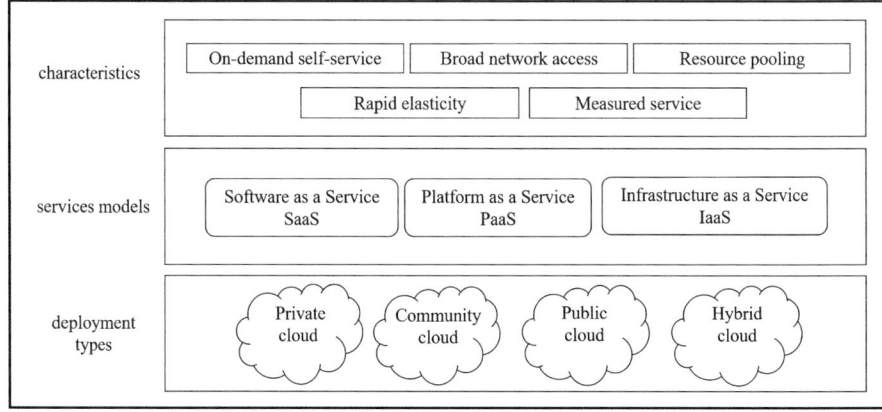

Fig. 1. NIST's definition of the cloud computing

2.2   Internet of Things (IoT) and Related Applications

The simplest definition of IoT is to connect all things through the Internet. The concept of IoT was actually proposed a long time ago, but IoT was not officially defined until the 2005 International Telecommunication Union's (ITU) report, The Internet of Things. The report officially proposed the concept of IoT and states that IoT is based on revolutionary success in mobility and the Internet [3, 4].

The European Telecommunications Standards Institute (ETSI) divides IoT into three conceptual layers, the application layer, the network layer, and the perception layer [7], as shown in Fig. 2.

1. *Application Layer:* The application layer conducts analysis and processing of network layer data according to the user's needs to provide smart services.
2. *Network Layer:* The network layer is responsible for transferring data received from the perception layer to the application layer for processing.
3. *Sensor Layer:* The perception layer uses four key technologies, radio frequency identification (RFID), near field communication (NFC), wireless sensor technologies, and embedded technology to collect data. The data is compiled and transmitted to the network layer.

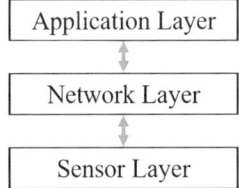

Fig. 2. IoT conceptual architecture

Currently, IoT has different applications and services in different fields. The following describes a few commonly seen IoT applications, including:

1. *Smart Grid:* A smart grid uses information technology to monitor and control electricity supply, as well as analyze the collected data. This helps to adjust power production and output and saves energy [5].
2. *Smart Traffic:* Smart traffic uses video monitors on the roadside to monitor status and adjust light signals to reduce congestion.
3. *Smart Medical Care:* This uses various wearable devices to monitor physiological data, which can prevent certain diseases from occurring or notify family or a doctor to implement necessary procedures when the values are abnormal. This can make the best use of the crucial rescue treatment time.

2.3   Fog Computing and Related Applications

The white paper published by CISCO in 2015 states that "fog computing adds an additional layer between edge devices and cloud computing in the original cloud computing architecture. In this layer there is a lot of equipment that can be called a fog node. In reality, any equipment with calculation, storage, and Internet connection capability can be used as a fog node, including network interface cards (NIC), access points, switches, routers, embedded servers, video monitors, and industrial controllers. Using these fog nodes to process and analyze IoT data can significantly reduce the maximum allowable delay. The following is the fog computing framework, as shown in Fig. 3.

Fog computing and cloud computing actually have a complimentary relationship. This does not mean that you do not need cloud computing after having fog computing. It is just that the addition of fog computing can take some burden off of the cloud computing processing. The CISCO report also specially mentioned how fog computing and cloud computing allocate each other's tasks to achieve efficient computation processing. For example, emergency real-time events are directly allocated to fog computing for processing while less urgent events are uploaded to cloud computing for processing [6].

Fog computing can be utilized in many real-time IoT applications such as smart grid, smart factory, industrial automation, smart traffic management, smart medical care, and real-time analysis application systems, which are separately introduced below [8–12]:

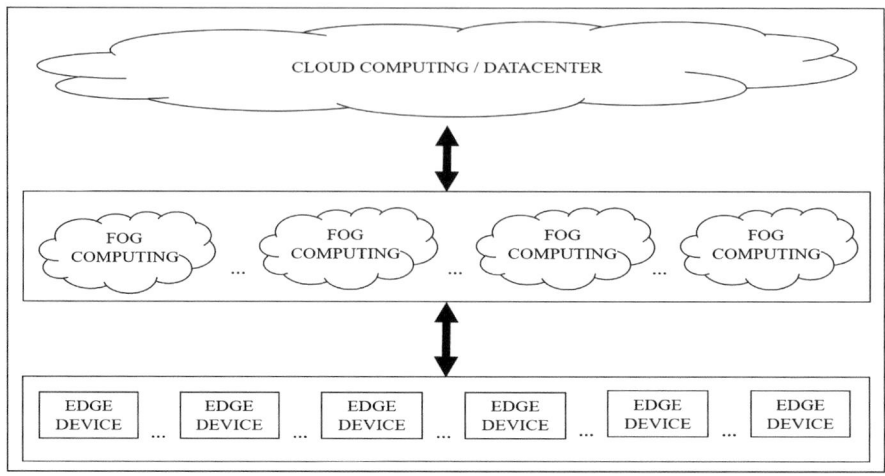

Fig. 3. Fog computing framework

1. *Smart Grid:* This uses smart meters to monitor real-time electricity supply in areas and uses fog computing platform for analysis. If there are any special changes, the grid will respond and use the fastest speed to stabilize the grid.
2. *Smart Factory and Industrial Automation:* This uses factory environment sensors to collect temperature or gas content values inside factories. The information is transmitted to fog nodes for real-time data analysis. If an abnormal value is discovered, the system can automatically notify factory personnel in real-time for emergency processing. This increases the factory's operational safety. Production line sensors can also be used to detect the line's work status and transmit the data to fog nodes for analysis. This can be used to check in real-time whether the products on the automated production line conform to product specifications. If an abnormality is discovered, the fog nodes will transmit a correction command to the automated production line and adjust the production line work in real-time. Combining production line monitoring and fog nodes' real-time data analysis can increase the factory's production efficiency.
3. *Smart Traffic Management:* This uses roadside deployment of fog nodes to collect vehicle traffic information and conduct real-time data analysis to respond to traffic in real-time. This can be used to automatically adjust traffic lights and alleviate traffic congestion. If a car accident happens, the system can alert rescue units in real-time, suggest to other drivers to use alternative routes, and achieve real-time sharing of traffic information.
4. *Smart Medical Care:* This uses wearable devices to monitor physiological values and upload the monitored data to fog nodes for real-time data analysis. When there is an abnormal physiological value, the system can send a help signal in real-time to the family or directly to a rescue unit for emergency processing to make the best use of the crucial treatment period.

5. *Real-time Analysis Application:* Weather monitoring is a real-time analysis application. Various types of sensors are used to collect temperature, wind speed, and rainfall quantity data, which are uploaded to the fog nodes for real-time data analysis. This provides the public with accurate weather data and can also send out real-time severe weather warnings so that the public can have more time to make disaster prevention preparations.

3   Operations Issue and Algorithm Design in D-FOG

The D-FOG scheme proposed in this paper introduces a distributed fog computing framework into the conventional cloud computing processing architecture to effectively process real-time events in IoT applications. Each event's priority sequence is used to conduct the process queuing. The D-FOG operating procedures are as described below and shown in Fig. 4.

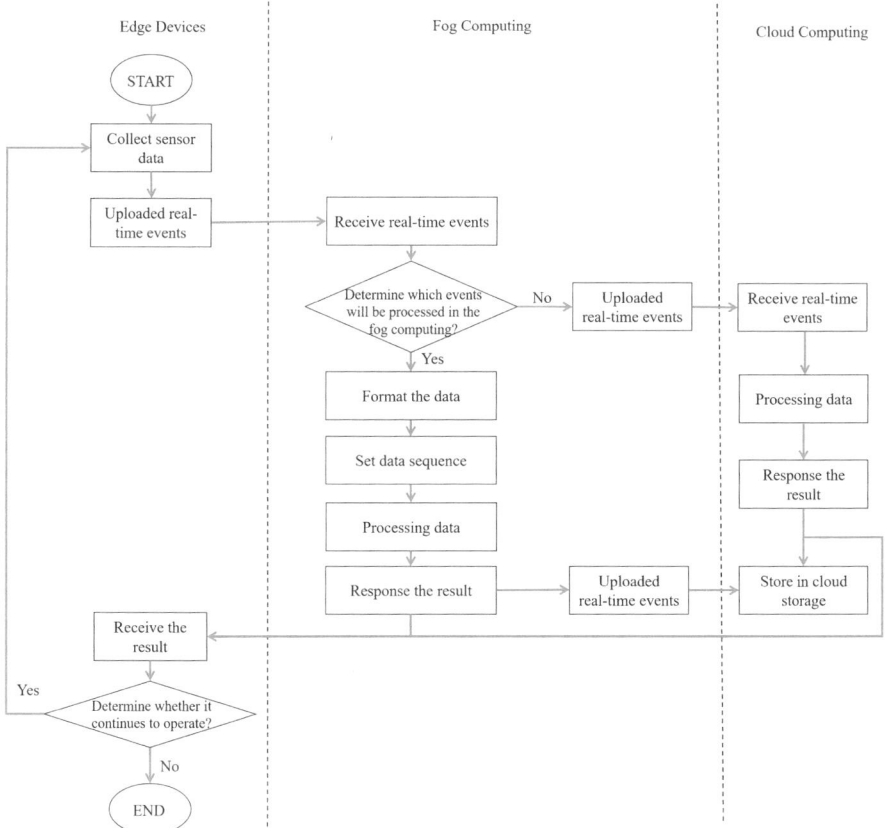

Fig. 4. D-FOG operating procedure

1. The fog nodes will receive events produced by edge devices.
2. The D-FOG scheme will determine which events will be processed in the fog computing layer based on port, and first format the data. Events that do not require real-time processing will be uploaded to cloud computing for processing.
3. The D-FOG scheme will check each event's priority level, data quantity, and processing time. Higher priority events that require real-time processing are immediately placed into the processing queue.
4. If too many events in the queue, call trust fog nodes to assist in processing the real-time events. In the same time, the fog node will keep working.
5. After the events are processed, the results are immediately sent back to the edge devices. After the event data that is to be stored is formatted, they will be uploaded to the cloud storage in lots.
6. After all the processing procedures have been completed, the fog nodes will release memory and wait for new events to be uploaded.

Based on the operational flow of the D-FOG as shown in Fig. 4, the pseudo code of the D-FOG algorithm can be designed as follows:

```
Algorithm D-FOG(){
Input:
RTED = []
FRTE = []
Seqlist = []
Output:
To complete real time IoT events and response messages.
Method:
BEGIN{
Get RTED Based on port;   // Receive the real-time events
FRTE=DATAFOMAT(RTED);     // Format the data into a format
that the system can process
SEQ(FRTE);                //Set event processing sequence
```

```
if(Too many events in the queue)then
  PROCESSINGHELP();       // Call trust fog nodes to assist
in processing the real-time events
  PROCESSINGDATA();       // Processing the real-time events
else
  PROCESSINGDATA();       // Processing the real-time events
end if
UPLOADDATA();             // Upload data to cloud storage
RELEASE-MEM();            // Release memory
}END
Procedre DATAFOMAT(data){
  if(data format is incorrect)then
    Format the data;
  end if
  return data
}END DATAFOMAT
Procedure SEQ(data){
  if(data is higher priority events)then
    Add to Seqlist;
  end if
}END SEQ
Procedure PROCESSINGHELP() {
  Call trust fog nodes and send the event;
  return response
}END PROCESSINGHELP
Procedure PROCESSINGDATA()
{
  Processing the event;
  if (event data is unusual)then
    return "error";
  else
    return "cloud"
  end if
}END PROCESSINGDATA
Procedure UPLOADDATA(){
  Upload data to cloud storage;
}END UPLOADDATA
Procedure RELEASE-MEM(){
  Release memory;
}END RELEASE-MEM
End D-FOG.
```

4   Simulations Environment Setup and Results Analysis

This section describes the simulations environment setup including software and hardware settings, list and illustrate KPIs and their calculation formulas. Also, we make the analysis of each KPI and summarize the simulations results.

4.1   Experiments Configuration Setup

For this experiment, we can utilize the VirtualBox to set up a fog computing simulated environment and also use Web Services to set up a cloud computing virtual environment on the Windows. Also, we take the environmental monitoring of smart factory as a practical case for simulations. The data types of real time events include temperature and humidity. The simulations data is generated by the random function in the sensor emulation of the smart factory system. The simulated environment is divided into three main parts. The first part is the edge devices, which are responsible for collecting data and transmitting the events regularly to the fog computing in the second part. The fog selects the real-time event that it can process and the remaining events are uploaded to the third part: cloud computing, for processing. The simulated environment configuration architecture is shown in Fig. 5. The related hardware/software configuration specifications are as shown in Table 1.

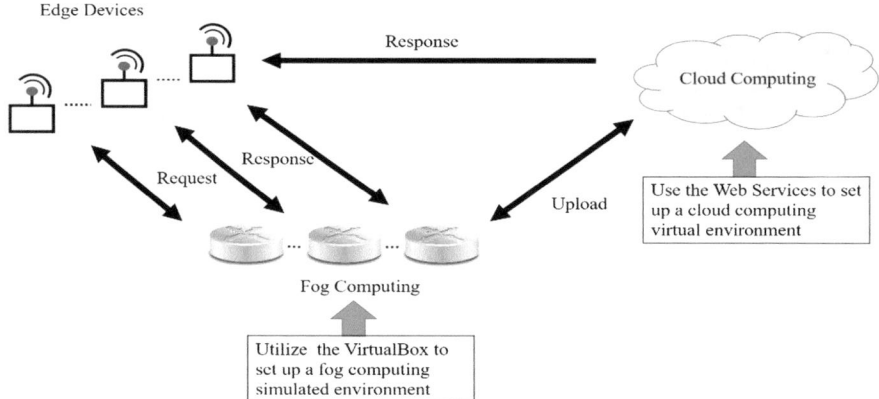

Fig. 5. Simulated environment configuration architecture

Table 1. Simulations environment specifications

HW/SW	Physical machine	VM
O.S.	Windows 10	Ubuntu
CPU	Intel® Core™ i7-4790 @ 3.60 GHz	1 Cores with 2.6 GHz
Memory	8 GB	2 GB
Disk	250 GB	50 GB

4.2   Descriptions of KPIs

In this paper, we propose three key performance indicators (KPIs) in terms of Expired event rate, Event success rate, and Average processing time which are used to analyze the simulation results to be compare between the D-FOG and conventional cloud schemes. The descriptions of the KPIs are shown in Table 2.

Table 2. The descriptions of three keys performance indicators

Key Performance Indicators (KPIs)	Purposes
Expired Event Rate (EER, Unit: %)	It represents that the percentage of processing time of each real-time event is exceeded the deadline time (EE) for all real-time event (RTE), as shown in Formulas (1)
Event Success Rate (ESR, Unit: %)	This KPI to confirm that the percentage of real-time event is successfully completed within the deadline (CE), as shown in Formulas (2)
Average Processing Time (APT, Unit: ms)	This KPI is to measure the average processing time (PT) of each real-time event, as shown in Formulas (3)

The correctly identified for Expired event rate, Event success rate, and Average processing time are shown in Formulas (1), (2), and (3).

$$EER(\%) \ = \ (EE/RTE\ ) \times 100 \tag{1}$$

$$ESR(\%) \ = \ (CE/RTE\ ) \times 100 \tag{2}$$

$$APT(ms) \ = \ PT/RTE \tag{3}$$

4.3   Results Analysis

We performed five time experiments averaging the simulation results under D-FOG and conventional cloud processing architectures (i.e. Cloud). Also, the tests were carried out using 100, 200, and 500 real-time events, respectively. The simulation results in terms of KPIs: Expired event rate, Event success rate, and Average processing time are as shown in Figs. 6, 7 and 8. Hence, the total simulation results are illustrated as shown in Table 3.

The experimental results indicate that when using the D-FOG scheme in the case of 100 events, the Expired event rate can be reduced by 28%, the Event success rate can be increased by 21.2%, and the Average processing time can be reduced by 436.87 ms to achieve 74.71% improvement ratios. In the case of 200 events, the Expired event rate can be reduced by 34%, the Event success rate can be increased by 26.9%, and the Average processing time can be reduced by 558.72 ms to achieve 75.86% improvement ratios. In the case of 500 events, the Expired event rate can be reduced by 28.2%, the Event success rate can be increased by 20.48%, and the Average processing time

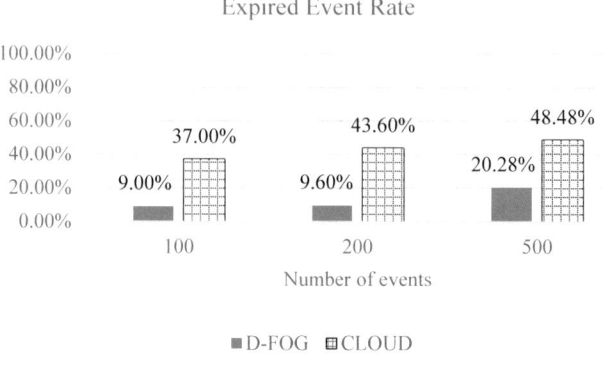

Fig. 6. Expired event rate

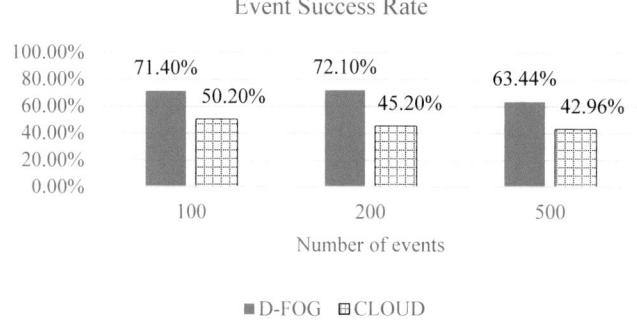

Fig. 7. Event success rate

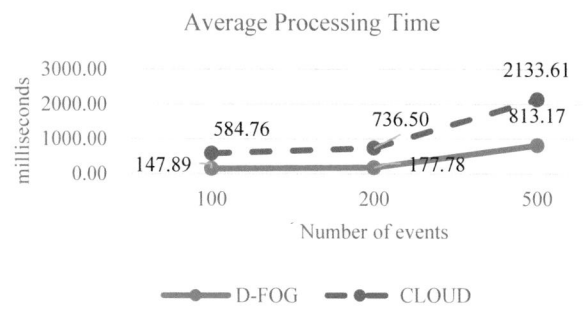

Fig. 8. Average processing time

Table 3. The summarized experimental results

D-FOG

KPIs	Events		
	100	200	500
Expired event rate	9.00%	9.60%	20.28%
Event success rate	71.40%	72.10%	63.44%
Average processing time (ms)	147.89	177.78	813.17

CLOUD

KPIs	Events		
	100	200	500
Expired event rate	37.00%	43.60%	48.48%
Event success rate	50.20%	45.20%	42.96%
Average processing time (ms)	584.76	736.50	2133.61

can be reduced by 1320.45 ms to achieve 61.89% improvement ratios. The comparative simulations are illustrated in Table 4. Hence, this proves that use of the D-FOG scheme can effectively improve the better processing efficiency of real-time IoT applications that conventional cloud architecture.

Table 4. Comparative improving ratios for simulations

Improving ratios

KPIs	Events		
	100	200	500
Expired event rate	28.00%	34.00%	28.20%
Event success rate	21.20%	26.90%	20.48%
Average processing time	74.71% (−436.87 ms)	75.86% (−558.72 ms)	61.89% (−1320.45 ms)

5 Conclusion

As the quantity of real-time IoT applications increase, the number of events that require real-time processing will increase. When a large number of events are simultaneously uploaded to cloud computing for processing, it can cause processing delays and make the processing time pass the processing deadline. To solve this problem, some people have proposed the fog computing concept. Current studies on fog computing are focused on defining its concept and application. In this paper, we propose a novel scheme that actually utilizes the fog computing framework, which we call the D-FOG.

The D-FOG scheme proposed in this paper uses fog computing as the processing scheme framework. In this paper, we set up the experimental environments for simulations. Also, we proposed three types of key performance indicators (KPIs): Expired event rate, event success rate, and average processing time. The experimental results of KPIs indicate that D-FOG in Expired event rate can be reduced by 28%, 34%, and 28.2%; Event success rate can be increased by 21.2%, 26.9%, and 20.48%; and Average processing time can be reduced by 436.87 ms, 558.72 ms, and 1320.45 ms in 100, 200, and 500 events, respectively. In summary, the results of this D-FOG system are compared to that of conventional cloud computing processing to prove that the D-FOG scheme we have proposed is more efficient and has better results than a conventional cloud computing processing architecture. Thus, the D-FOG scheme can be applied to all real-time IoT applications and solve processing delays caused by processing of real-time IoT application events, thereby providing a more efficient distributed processing architecture for all real-time IoT applications. Consequently, the D-FOG scheme not only can solve processing delays in real-time IoT application event processing, but can also provide superior service quality to achieve better processing efficiency and effective result for real-time events.

References

1. The NIST Definition of Cloud Computing. https://nvlpubs.nist.gov/nistpubs/Legacy/SP/nistspecialpublication800-145.pdf. Accessed 30 Apr 2019
2. Cloud computing. https://en.wikipedia.org/wiki/Cloud_computing. Accessed 30 Apr 2019
3. The Internet of Things. https://www.itu.int/net/wsis/tunis/newsroom/stats/The-Internet-of-Things-2005.pdf. Accessed 30 Apr 2019
4. Internet of Things. https://en.wikipedia.org/wiki/Internet_of_things. Accessed 30 Apr 2019
5. Smart grid. https://en.wikipedia.org/wiki/Smart_grid. Accessed 30 Apr 2019
6. Fog Computing and the Internet of Things: Extend the Cloud to Where the Things Are. https://www.cisco.com/c/dam/en_us/solutions/trends/iot/docs/computing-overview.pdf. Accessed 30 Apr 2019
7. Kotha, H.D., Gupta, V.M.: IoT application, a survey. Int. J. Eng. Technol. 7(2.7), 891–896 (2018)
8. Hu, P., Dhelim, S., Ning, H., Qiu, T.: Survey on fog computing: architecture, key technologies, applications and open issues. J. Netw. Comput. Appl. 98, 27–42 (2017)
9. Yi, S., Hao, Z., Qin, Z., Li, Q.: Fog computing: platform and applications. In: 2015 Third IEEE Workshop on Hot Topics in Web Systems and Technologies (HotWeb) on Proceedings, Washington, D.C, pp. 73–78 (2015)
10. Bonomi, F., Milito, R., Zhu, J., Addepalli, S.: Fog computing and its role in the internet of things. In: The 1st Edition of the MCC Workshop on Mobile Cloud Computing on Proceedings, Helsinki, Finland, pp. 13–16 (2012)
11. Atlam, H., Walters, R., Wills, G.: Fog computing and the internet of things: a review. Big Data Cogn. Comput. 2(2), 1–18 (2018)
12. de Brito, M.S., Hoque, S., Steinke, R., Willner, A.: Application of the fog computing paradigm to smart factories and cyber-physical systems. Trans. Emerg. Telecommun. Technol. 29(4), 1–14 (2017)

Design Issues of a Hybrid Wrapping Attack Protecting Scheme in Cloud Computing Environment

Shin-Jer Yang[✉] and Yu-Hsuan Huang

Department of Computer Science and Information Management,
Soochow University, Taipei, Taiwan
sjyang@csim.scu.edu.tw, stepyou2010@gmail.com

Abstract. In the cloud era, cloud security and user privacy have become important issues. Since cloud users often use web browsers to request services from the cloud service providers, when a signed message request is sent to the service receiver from the service provider, attackers can use wrapping attacks to tamper with the SOAP message transferred through the internet in order to avoid legal verifications and access web services without being detected to implement wrapping attacks. This paper is to integrate and improve Node Counting mechanism to propose a hybrid wrapping attack protection scheme called HWRAP.

This HWRAP is a continuation on the Node Counting method and the determining conditions were improved and divided into three modules interception, detection and logging; it performs analysis to the incoming SOAP requests. Not only does it compare the number of times the child node appears, it also performs detection to the elements on the path between the root node to the final node in order to enhance the verification procedure on the detection module. The simulation results of KPIs indicate that the Detection rate can be increased by 2%, 3.8%, 3.7%, and 2.8%; the Accuracy rate can be increased by 8%, 7%, 10.5%, and 9.5% in 50, 100, 200, and 500 packet requests, respectively. Consequently, the HWRAP scheme has a higher detection rate and better accuracy rate than Node Counting for detecting wrapping attacks. The final results show that the proposed HWRAP can identify attackers more accurately and enhance the security and quality of cloud services.

Keywords: Cloud computing · Wrapping attacks · HWRAP · Node counting · SOAP

1 Introduction

1.1 A Subsection Sample

In the cloud computing era, compared to the early stages of computer development where software was installed on personal computers, more and more software and information service can be acquired through the internet. However, cloud computing did not come out from nowhere; cloud computing exists today due to decades of development of the information industry. Its foundation came from the evolution of key

© Springer Nature Switzerland AG 2020
K.-M. Chao et al. (Eds.): ICEBE 2019, LNDECT 41, pp. 113–127, 2020.
https://doi.org/10.1007/978-3-030-34986-8_8

technologies including virtualization, grid operation, distributed computing, the internet, SOA, and Web 2.0 etc. Cloud computing is a type of service that virtualizes IT resources including networks, storage, calculation or software/hardware and platform, optimizes resource utilization, and has quantifiable billing service; it is a service platform that is distributed through the network so that users can use its service at any time. As cloud computing matures, more and more enterprises transfer their services and data to cloud for operation in order to reduce cost, improve internal business processes, and increase customer demands to increase the value of the enterprise. As cloud technology becomes more popular, not only do enterprises use cloud to increase competitive advantages, the general public also benefit from the advantages brought by the wide range of cloud applications. For example, uploading photos to Instagram, learning what friends are doing in real-time, or the various services provided by Google: Gmail, Google Calendar and Google Docs etc., have made our lives more rich and convenient. In the cloud era, cloud security and user privacy have also become important issues.

As the use of cloud services continues to increase over the internet, users using Web browsers to request services from cloud service providers send sensitive content such as important data that can easily attract intrusions from malicious parties and hackers. When the web server verifies the signed request, wrapping attacks can be completed during the SOAP message transmission process between legitimate users and the Web server. By copying the user account and password during login, hackers embed a forged element (wrapper) into the message structure, and switch the content of the original message in the wrapper by replacing it with malicious code, and then send the message to the server. Since the original content is still valid, the server will be tricked to authorize the tampered message. Therefore, hackers will be able to gain unauthorized access to protected resources, and execute malicious operations.

In studies related to wrapping attacks, appropriate ID verification can be used to encrypt the entire message, and establish trust relationships with any intermediate servers in order to prevent the occurrence of wrapping attacks. A protection method (Node Counting) against wrapping attacks was proposed recently and is divided into three modules—interception, detection and logging—to perform analysis to the incoming SOAP requests. Node counting evaluates whether the request is a wrapping attack by calculating the number of times the child node appears in the SOAP Header element, but when the attack behavior has slightly changed or if new attacks appear, the detection rate might be reduced. In other words, if attackers try to change the message structure, they can easily bypass this protection mechanism. Therefore, this paper corrects the flaws of the Node Counting method described above and proposes a method that can clearly point out each element on the path between the root node of the SOAP Header file to the final node in order to increase the protection capability. It is called the Hybrid Wrapping Attacks Protecting (HWRAP) scheme. Hence, the main purposes of this paper can be listed as follows:

- Under cloud computing environments, this paper is an extension to the wrapping attack prevention mechanism proposed by Gupta (2016) and proposes HWRAP to improve the verification process of the original Node Counting mechanism.

- The SOAP requests sent from the transmitter will be intercepted, their attribute values will be detected, and the absolute path of the signed request will be taken into consideration to determine whether they are malicious requests and deny the requests in order to increase the detection rate.
- With the HWRAP proposed in this paper, the Burp Suite tool is used to simulate actual environments of cloud computing and will be analyzed and compared with the prevention mechanism proposed in Gupta (2016) to show that HWRAP can determine wrapping attacks more effectively.
- We list the two key performance indicators: detection rate and accuracy rate to analyze whether the proposed HWRAP can reduce and slow down the occurrences of wrapping attacks in cloud services.

The rest of this paper is organized as follows. Section 1 describes the background and motivations for the studies on wrapping attack detection mechanisms and the scope and purpose of this paper. Section 2 presents cloud computing and security, and then describes the types and countermeasures of wrapping attacks. Section 3 examines the operational flow of the HWRAP processes and designs the HWRAP algorithm. Section 4 covers the simulated experiments setup and analyzes the results. Finally, we draw a conclusion, main contribution, and future research directions of this paper.

2 Related Work

2.1 Cloud Computing and Security

The National Institute of Standards and Technology (NIST) defines cloud computing as "Cloud computing is a model for enabling ubiquitous, convenient, on-demand network access to a shared pool of configurable computing resources (e.g., networks, servers, storage, applications, and services) that can be rapidly provisioned and released with minimal management effort or service provider interaction." [1]. In this definition by NIST, cloud model includes five basic characteristics: On-demand self-service, Broad network access, Resource pooling, Rapid elasticity, and measured service, as well as four deployment modes: Private cloud, Public cloud, Community cloud, and Hybrid cloud. Private cloud is used by single organizations and is usually cloud computing architectures established for enterprises or organizations themselves to prevent having different users and causing security concerns. Compared to private clouds, public clouds provide computing services to the general public; most of the common computing services are public clouds. For example Evernote, Dropbox, and Google Docs for individuals, or Microsoft Azure, etc. for businesses. Community clouds are established by multiple organizations; the benefits and purposes of these organizations are usually very similar and use services to have groups with common demands, such as the education clouds and medical clouds, that are currently popularly used. And as the name suggests, hybrid cloud refers to cloud architectures composed of two or more cloud types.

Service modes of cloud computing mainly include three types SaaS, PaaS and IaaS. SaaS allows users to not have to download software to personal computers and access services through the network without using up hardware resources; for example, Google Map and Gmail provided by Google. PaaS provides software developers with complete cloud environment development platforms. Developers can deploy their application programs to the cloud environment to perform development and testing without having to worry about maintaining the infrastructure behind the platform, reducing the past cost of constructing the infrastructure on their own, and managing development tools, for example, Microsoft Azure and Google App Engine. IaaS provides users with fundamental computation facilities such as servers, network and storage devices. Users can flexibly rent the operation system, storage space, and application programs, etc., based on their business needs, but do not have to control the underlying cloud architecture, for example, EC2 from Amazon.

Once the cloud computation era comes, data security for cloud computing is an issue that has to be faced whether for individuals or enterprises. In 2013, the Cloud Security Alliance (CSA) proposed nine major threats that cloud computing will face [2] and that includes the following: data breaches, data loss, account hijacking, insecure APIs, denial of service, malicious insiders, abuse and nefarious use, insufficient due diligence, and shared technology issues.

Many questions exist on whether cloud is secure enough. As users frequently use web browsers to request services from cloud service providers, sensitive messages exist during the service communication transmission process, and these important data and contents easily attract intrusions from malicious parties and hackers. Considering malicious intruders, there may be many types of attacks. For example, wrapping attacks can be completed by copying the user's account and password during the login stage to intercept the SOAP message exchange between the Web browser and server.

2.2 XML Signature

XML signature is a W3C recommended standard for defining the XML syntax of digital signatures. In terms of function, the XML signature is similar to PKCS#7 in many ways, but the XML signature has better expandability, and was adjusted for signing XML files. The XML signature is used in many web technologies such as SOAP and SAML [3].

XML signatures can be used to sign any type of resource and is most commonly used for XML files. However, any resource that can be accessed through a URL may be signed. If XML signature is used to sign resources other than XML files that includes the signature, it is called a detached signature; if the XML signature is used to sign a certain part of XML files that includes it, it is called an enveloped signature; if the XML signature includes data that was signed, it is called an enveloping signature [3].

An XML signature includes a signature element; its basic structure is as shown in Fig. 1:

```
<Signature>
    <SignedInfo>
        <SignatureMethod/>
        <CanonicalizationMethod/>
        <Reference URI>
            <Transforms>
            <DigestMethod>
            <DigestValue>
        </Reference>
    </SignedInfo>
    <SignatureValue/>
    <KeyInfo/>
    <Object/>
</Signature>
```

Fig. 1. XML signature structure

2.3   SOAP Message

SOAP (Simple Object Access Protocol) is a message transmission protocol specification based on XML; it is used to exchange structured messages in web services of computer networks. SOAP relies on application layer protocols such as HTTP or SMTP and other communication protocols. It allows programs running on different operating systems to perform data exchange and transmission using XML message format [6].

2.4   Types of Wrapping Attacks

A wrapping attack is an attack triggered by the use of malicious intermediate nodes; it is executed during the SOAP message transmission process between legal users and the cloud server. By copying the user account and password during login, hackers embed a forged element (wrapper) into the message structure, and switch the content of the original message in the wrapper by replacing it with malicious codes, and then send the message to the server. Since the original content is still valid, the server will be tricked into authorizing the tampered message. Therefore, hackers will be able to gain unauthorized access to protected resources and execute malicious operations.

McIntosh and Austel proposed four types of wrapping attacks for the first time in 2005 [7]:

Simple Ancestry Context Attack: In simple ancestry context attacks, the requested SOAP body is signed with the signature, and this signature is located in the security header of the request. The message receiver checks whether the signature is correct and legalizes trust in the signature credential. Finally, the receiver substitutes the "id" of the SOAP body in the signature to the ID reference to find out whether the element needed is actually signed. A classic simple ancestry contact attack uses the malicious SOAP body to replace the original SOAP body, and the original SOAP is placed in the <Wrapper> element of the SOAP Header [7].

Optional Element Context Attack: In optional element context attacks, signed data is included in the SOAP header and is random. Compared to a simple context attack, the main purpose of this attack is not the location of the signed data in the SOAP header; in fact, the signed data is optional.

Sibling Value Context Attack: In sibling value context attacks, the security header includes a signed element; this actually is an alternative sibling value for <Signature>. A common model of this attack can be the <Timestamp> element; the <Times-tamp> element and <Signature> are both direct descendants of the SOAP security header. The difference between this attack and the attacks previously mentioned is in the data signed—in this attack, it is a sibling value of <Signature>. The main purpose of this attack is to ignore the sibling value of the signature element [7].

Sibling Order Context Attack: This type of attack deals with the protection of sibling elements; their semantics are related to their order relative to each other and are not affected by rearrangement by an adversary [7].

2.5   Wrapping Attack Detection Countermeasure

In the real world, since cloud users usually request services from cloud service providers through web browsers, wrapping attacks usually cause damage to the cloud systems as well. Amazon discovered that their EC2 was susceptible to this type of attack in 2008 [8], and research showed that EC2 had vulnerabilities in its SOAP message security verification mechanism. Hackers were able to intercept and modify signed SOAP requests from legal users and perform unauthorized actions using the victim's account in the cloud. In Somorovsky [8], an analysis called "Black box" was used to perform research to the public cloud platform on Amazon EC2.

Wrapping Attack Defense Mechanisms. Currently there aren't many studies related to XML signature wrapping attacks. XML signature wrapping attacks utilize loopholes generated while processing XML signatures; attackers use forged elements to replace original elements during the message transmission process, or reposition the original position of the original element in the SOAP message in order to inject malicious data into the requests. XML wrapping attacks were first proposed by McIntosh and Austel in 2005 [7], and preliminary discussions were made on how to protect against wrapping attacks. Setting security policy specifications for both the sender and receiver prevents the signature from referring to elements in non-agreed collection of members, and therefore, reject messages that include unexpected signed elements. Although this method is satisfactory under certain circumstances, it is not applicable for all situations.
　　Wrapping attack defense mechanisms in recent years include the following:

FastXPath: FastXPath demands that every element on the path between the root directory of the file to the signing subtree must be clearly specified so that there is no flexibility to move signed contents to any other positions of the file [9]. In addition, since the FastXPath expression can include any ID attributes, its safety supports both ID-based and structure-based XML access methods. Since these two methods both point to the same element, there will be no deviation between the access method of the signature verification function and the access method of the application program logic. However, when access rules are applied to different program logics, the threat from wrapping attacks cannot be completely avoided.

Schema Hardening: Schema Hardening allows a collection of valid XML files from the description of the specification itself that are identical to the original schema, but it still needs to be enhanced to strictly prohibit any other contents not included in the schema. It involves the technology to delete all schema expansion points that appeared in the original schema file from the specification itself [10]. To be exact, schema investigates every <xs:any> declaration that appears intensively to verify whether they are used for general SOAP messages and either reconstructs them or deletes them. However, the enhanced schema application also brings significant performance drawbacks.

SESoap: The SESoap method signs the entire SOAP method; it applies the digital signature structure on the entire SOAP package element in order to ensure the security of the entire file [11]. Therefore, attackers will not be able to change the positions of elements or add/delete elements randomly into the original file.

Node Counting: Currently, the Node Counting is the new proposed wrapping attack defense mechanism, its system design is divided into three main modules: interception, detection, and logging [12]. The first interception module is responsible for intercepting incoming digitally - signed SOAP requests and forwarding them to the detection module. Then, the detection module applies algorithms to analyze the request attributes and calculate the frequency of child node appearances to detect whether a wrapping attack exists. If it was detected, the request will be rejected and a log will be generated by the logging module; if a wrapping attack was not detected, the request will be forwarded to the target receiver and another request successfully forwarded log will be generated.

3   Operation Issues and Algorithm Design in HWRAP

When the cloud server verifies the signed request, by copying the user account and password during login, hackers can embed forged elements into the original message structure, and replace the contents of the original message with malicious code, and then send the message to the server so that wrapping attacks can be completed during the SOAP message conversion process between the legitimate user and the cloud server. This paper is an extension on the Node Counting method proposed by Gupta [12] and improves its detection flaws to propose a new hybrid wrapping attack protecting mechanism called HWRAP (Hybrid Wrapping Attack Protecting).

The detection of a wrapping attack was performed for the three modules individually on the Burp Suite Professional simulation tool. The first is the interception module; on the virtual machine, a digitally signed SOAP request is sent by the sender and forwarded to the detection module; then, the second detection module uses algorithms to thoroughly analyze the request attributes for filtering. First, the XPath expression is used to check whether <Signature> is the child element of <Header>, and also checks the <Body> child element in <Header>; then, a test is performed to the URI value of the signed request. Not only is the number of times child nodes appeared in the SOAP header compared, it also detects the elements on the path from the root node of the SOAP header to the final node in order to improve the defense performance against wrapping attacks. If an illegal request was detected by the detection module, the request will be denied and logs will be generated by third-stage logging modules; if a legal request was detected, the request will be forwarded to the target recipient, and the logging module will log the successfully forwarded request. Detailed operation processes of the HWRAP detection mechanism is as shown in Fig. 2.

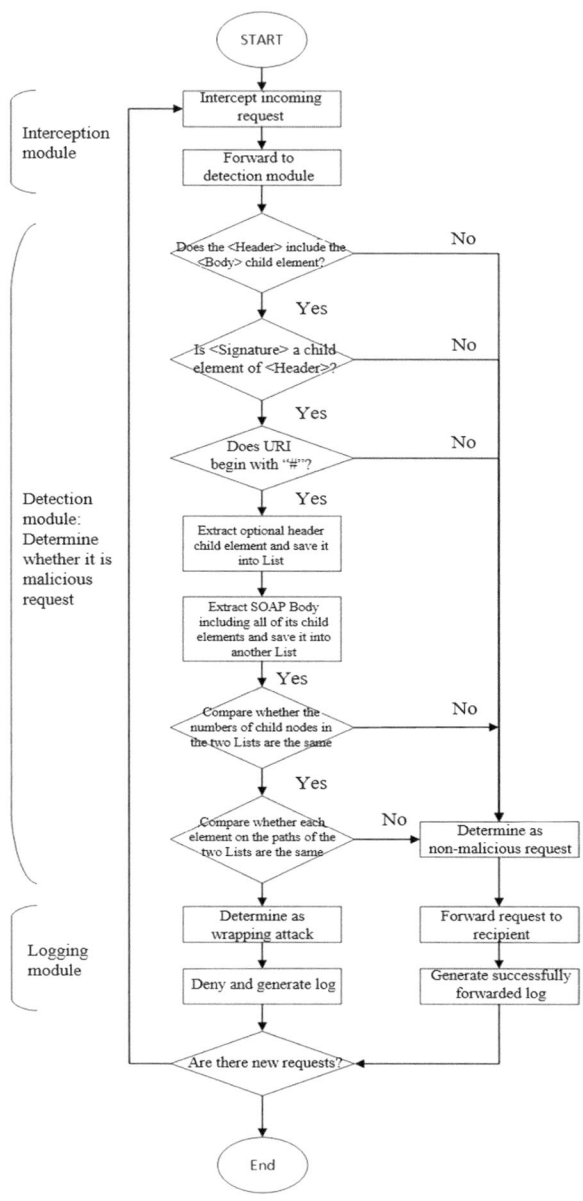

Fig. 2. The operational flow of HWRAP

Based on the operational flow of the HWRAP as depicted in Fig. 2, the pseudo code of the HWRAP can be designed as follows.

```
Algorithm HWRAP {
Input :
String[] HWRAP;
bool ifHWRAP = true;
bool ifAttacker;
bool ifFakeRequest;
Int[] List1;
Int[] List2;
String[] ListPath1;
String[] ListPath2;
Output:
To complete HWRAP Method.
Method:
BEGIN {
While(ifRequestInput) {
Load_HWRAP();
HWRAP_Body();
HWRAP_Sign();
HWRAP_Begin();
List_Count();
List_Path();
Input_Check();
ifAttacker = Check_HWRAP();
if(ifAttacker = true(){
```

```
List_Illegal_Request();
}else{
List_Legal_Request();
}
ifRequestInput = HWRAP_Check();
}
} END
Procedure Load_HWRAP() {
String HWRAP = getHWRAP();
HWRAP = HWRAP.Split(' ');
} END Load_HWRAP
Procedure HWRAP_Body() {
If(head.contains(body)){
return true;
}else{
return false;
}
} END HWRAP_Body
Procedure HWRAP_Sign() {
If(head.contains(signature){
return true;
}else{
return false;
}
} END HWRAP_Sign
Procedure HWRAP_Begin() {
If(str.startsWith("#"){
return true;
}else{
return false;
}
} END HWRAP_Begin
Procedure List_Count() {
if(List1.length == List2.length){
return true;
}else{
return false;
}
} END List_Count
Procedure List_Path() {
if(ListPath1.equals(ListPath2)){
return true;
}else{
return false;
}
```

```
} END List_Path
Procedure Input_Check() {
If(AnyInput()){
return true;
}else{
return false;
}
} END Input_Check
}END HWRAP
```

4   Simulations Setup and Results Analysis

This section describes the simulation environment setup including software and hardware settings, and list and illustrate KPIs and their calculation formulas. Also, we make the analysis of each KPI and summarize the simulation results.

4.1   Simulations Setup

This study will use virtual machines and physical machines to simulate the test environment of the cloud; the virtual machine software Oracle VM Virtual Box version 6.0.4 will be installed on a physical computer; the Linux OS software Ubuntu version 18.04 will be installed on the virtual machine and the JDK version is 1.9.0. This study will use the first virtual machine and set it as the main server of the cloud environment architecture, and then set up another virtual machine that also has cloud computing platform constructed and can execute the HWRAP simulator. First, start the SoapUI tool to send legal and illegal SOAP requests. Then, intercept the incoming request in Burp Suite Professional v1.7.37 and forward it to the detection module on SoapUI. Next, analyze the request attribute through the scanner module to detect whether there is a wrapping attack. Finally, show the result in the Log module. The simulation environment computer configuration is as shown in Table 1. Also, this experiment simulates the transmission situation of the request sent by wrapping attackers in the cloud environment, and further performs analysis to the SOAP request in combination with the HWRAP detection mechanism to confirm whether the SOAP request is a malicious request, and finally sets restrictions to the wrapping attacker as depicted in Fig. 3.

Table 1. Hardware and software specifications

HW/SW	Physical machine	VM
O.S.	Windows 10	Ubuntu 18.04
CPU	Intel® Core™ i5-7200U CPU @ 2.50 GHz 2.7 GHz	2 Cores, 2.5 GHz
Memory	8 GB	2 GB
Disk	224 GB	40 GB
Simulation Tool	x	Burp Suite Pro and SoapUI

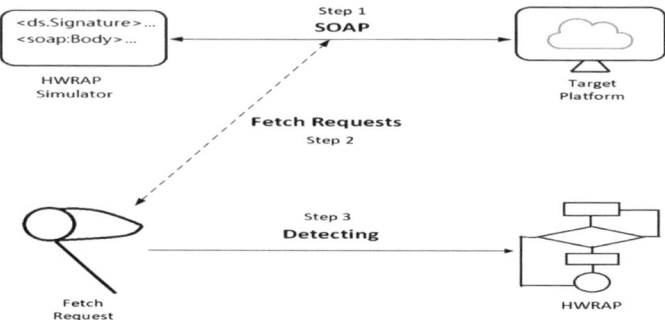

Fig. 3. Simulation environment architecture

4.2   Evaluation Matrix

HWRAP and Node Counting defense mechanisms are compared and analyzed through the simulation experiment, and two key performance indicators (KPIs) of detection rate and accuracy rate were proposed to analyze the simulation results, as shown in Table 2.

$$DR(\%) = DAR / TAR \times 100 \tag{1}$$

$$AR(\%) = CR / N \times 100 \tag{2}$$

Table 2. Two key performance indicators

Key Performance Indicators (KPIs)	Purposes
Detection rate (DR, Unit: %)	The DR calculates the ratio of detected abnormal requests (DAR) to the total number of abnormal requests (TAR) as shown in Formulas (1) and compare DR between the HWRAP and the Node Counting methods
Accuracy rate (AR, Unit: %)	The AR computes the ratio of correct requests (CR) to the total requests number (N) as shown in Formulas (2), then evaluating the accuracy of the HWRAP and Node Counting methods

4.3   Results Analysis

This experiment is performed on the Burp Suite Professional and SoapUI simulation tool. First, send a request from the SoapUI simulation tool, and intercept the incoming requests from the sender in the Proxy module on Burp Suite and forward them to the Scanner module. Then, thoroughly analyze the requests in the Scanner module for filtering and monitoring to identify the illegal incoming requests on SoapUI tool. If a request is detected as illegal, it is rejected. Finally, the analysis and statistics are performed in the Log module to generate the experimental result.

The experimental result shows that in the case of 50, 100, 200, and 500 requests, the HWRAP method has detection rates of 98%, 98.8%, 99.2%, and 98.5%, as shown in Fig. 4; the accuracy rates are 98%, 98%, 98.5%, and 99.2% respectively, as shown in Fig. 5. The summarized experimental results are shown in Table 3. Compared with the Node Counting method proposed by Gupta et al. [10], the experimental results of the KPIs indicate that the Detection rate can be increased by 2%, 3.8%, 3.7%, and 2.8%; the Accuracy rate can be increased by 8%, 7%, 10.5%, and 9.5% in 50, 100, 200, and 500 packet requests, respectively, as shown in Table 4. The HWRAP method has improvements in both the detection rate and the accuracy rate. Consequently, the HWRAP scheme proposed by this paper can provide better and more secure service quality.

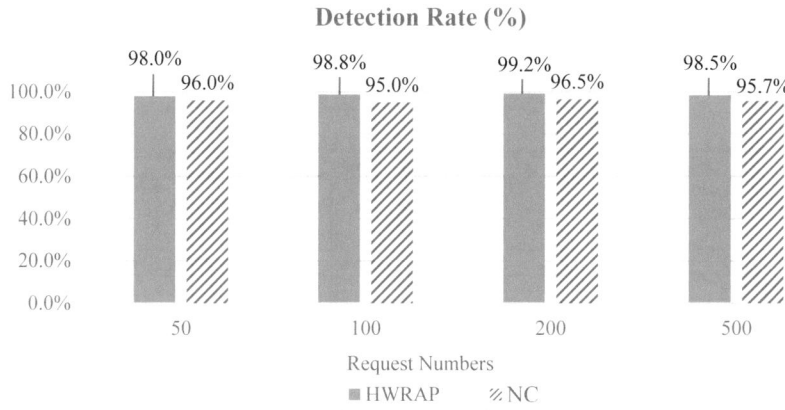

Fig. 4. Detection rate

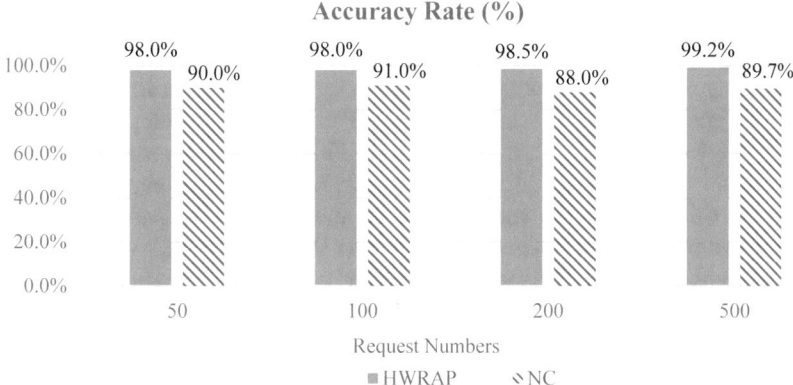

Fig. 5. Accuracy rate

Table 3. Summarized simulation results

HWRAP				
Request Numbers KPIs	50	100	200	500
Detection Rate	98%	98.8%	99.2%	98.5%
Accuracy Rate	98%	98%	98.5%	99.2%
Node Counting				
Request Numbers KPIs	50	100	200	500
Detection Rate	96%	95%	96.5%	95.7%
Accuracy Rate	90%	91%	88%	89.7%

Table 4. Comparative improving ratios for simulations

Improving Ratios				
Request Numbers KPIs	50	100	200	500
Detection rate	2%	3.8%	3.7%	2.8%
Accuracy rate	8%	7%	10.5%	9.5%

5   Conclusion

As users increasingly use web browsers to request services from cloud service providers, sensitive content and important data can easily attract intrusions from malicious parties and hackers. Attackers can use wrapping attacks to inject malicious elements into legal message structures and avoid legal verifications and access Web services without being detected to implement wrapping attacks. Hence, this paper designed new Hybrid Wrapping Attacks (HWRAP) to improve the verification process of the original Node Counting mechanism. Also, we utilized the Burp Suite Professional v1.7.37 tool to perform simulations for the incoming SOAP requests and comparison analysis between HWRAP and Node Counting schemes. First, the SOAP request sent by the sender, which will be intercepted; as compared from Node Counting, not only is the frequency of child node appearances compared, elements on the path from the root node to the end node were also detected to discover illegal incoming requests and deny the requests, enhancing the verification process of the detection module.

We compared the simulation results in the HWRAP scheme to the Node Counting method in the case of 50, 100, 200, and 500 SOAP packet requests. As a result, the detection rate of HWRAP increased by about 2%, 3.8%, 3.7%, and 2.8%, the accuracy rate is increased by 8%, 7%, 10.5%, and 9.5%, respectively. Finally, the proposed HWRAP can determine wrapping attackers more accurately and improve the security and quality of cloud services. In the future, we will perform more simulations to consider the other KPI of APT (Average Processing Time) to obtain more effective and accurate results.

References

1. Mell, P., Grance, T.: The NIST Definition of Cloud Computing. National Institute of Standards and Technology (2011)
2. Top Threats Working Group: The Notorious Nine: Cloud Computing Top Threats in 2013. Cloud Security Alliance (2013)
3. XML Signature. https://zh.wikipedia.org/wiki/XML_Signature
4. XML Signature Wrapping. http://www.ws-attacks.org/XML_Signature_Wrapping
5. XML Signature Syntax and Processing Version 2.0, https://wwww.3.org/TR/xmldsig-core2/
6. Simple Object Access Protocol. https://en.wikipedia.org/wiki/SOAP
7. McIntosh, M., Austel, P.: XML signature element wrapping attacks and countermeasures. In: Proceedings of the 2005 Workshop on Secure Web Services, pp. 20–27 (2005)
8. Somorovsky, J., et al.: All your clouds are belong to us: security analysis o cloud management interfaces. In: Proceedings of the 3rd ACM Workshop on Cloud Computing Security, pp. 3–14 (2011)
9. Gajek, S., et al.: Analysis of signature wrapping attacks and countermeasures. In: Proceedings of IEEE International Conference on Web Services, pp. 575–582 (2009)
10. Jensen, M., et al.: On the effectiveness of XML schema validation for countering XML signature wrapping attacks. In: Proceedings of 1st International Workshop on Securing Services on the Cloud, pp. 7–13 (2011)
11. Kouchaksaraei, H.R., Chefranov, A.G.: Countering wrapping attack on XML signature in SOAP message for cloud computing. (IJCSIS) Int. J. Comput. Sci. Inf. Secur. 16, 1310–0441 (2013)
12. Gupta, A.N., Thilagam, P.S.: Detection of XML signature wrapping attack using node counting. In: Proceedings of the 3rd International Symposium on Big Data and Cloud Computing Challenge (ISBCC–2016), pp. 57–63 (2016)

Developing a Microservices Software System with Spring Could – A Case Study of Meeting Scheduler

Wen-Tin Lee[✉] and Tseng-Chung Lu

National Kaohsiung Normal University, Kaohsiung, Taiwan
wtlee@nknu.edu.tw, richard.lu889@gmail.com

Abstract. The traditional monolithic architecture is to implement the function modules of the application and execute them on the same service port. The entire application system must be repackaged and deployed if there is a program change. With the increase of requirements, the coupling relations between modules become more and more complicated which results in inflexible function expansion and difficult maintenance. Microservices architecture divides a single application into several independent executed services with its business logic and communicates with each other using the REST API or message queue. Therefore, the scalability and maintainability of the system are improved by features such as independent update and independent deployment.

This study implements a Meeting Scheduler monolithic software system using traditional object-oriented software development process. To transform the software into a microservices architecture, the Spring Cloud framework is used to implement the required microservices in the Meeting Scheduler. After completing the transformation from monolithic to microservices architecture, this study explores the development experience of microservices software and establishes a microservices architecture transformation process for software engineers to analyze, design, implement and deploy a software based on microservices.

Keywords: Microservices architecture · Monolithic architecture · Spring Cloud · Object-oriented software development · Service-Oriented Architecture

1 Introduction

Microservices is a service-oriented architectural style that structures an application as a collection of services that can be published, combined and used on the internet. The service communicates with each other through a well-defined contract and interface. Through the integration of components of heterogeneous systems by using microservices, the rapid functions expansion and code reuse can be achieved. Microservices and services-oriented architecture (SOA) are decentralized system architectures. Applications are composed of many distributed network components and communicated interactively by the REST application interface (API). SOA advocates linking multiple business services to implement new applications that are suitable for integrating complex large-scale applications. Microservices are used to build independent small

© Springer Nature Switzerland AG 2020
K.-M. Chao et al. (Eds.): ICEBE 2019, LNDECT 41, pp. 128–140, 2020.
https://doi.org/10.1007/978-3-030-34986-8_9

applications, limited to specific business functions, emphasizing fine-grained and loosely coupled service features, so the software system is split into multiple services that can be independently developed, designed, executed, and distributed to enhance application resiliency and performance.

This study is to explore the development process of monolithic software systems. In the process of transforming the application functionalities into microservices, it summarizes the software engineering practices to be followed to construct a microservices system and explains the benefits provided after conversion to microservices.

The rest of the paper is organized as follows: Sect. 2 discusses the related work, Sect. 3 introduces the research environment and development tools, Sect. 4 introduces the system architecture and development practices of the microservices system. The fifth section discusses the case study. Finally, the sixth section summarizes this study.

2 Related Work

This section explores the relevant literature on Service-Oriented Architecture, microservices architecture, Spring Boot and Spring Cloud.

2.1 Service-Oriented Architecture (SOA)

Service-Oriented Architecture (SOA) [1, 6] is a decentralized and component architecture that consists of different web services. SOA uses the service interfaces and service contracts such as services discovery and service specification as the basis for service interaction. The business implementation of the system uses a specific workflow language such as WS-BPEL [2] to define the complex behaviors of the service. In this way, the business process of service automation is established, and service discovery, service composition, service extension, and service substitution are provided. The purpose of this business process is accomplished by invoking and communicating web services.

Erl et al. [3, 4] proposed the service design principles which are (1) Loose coupling of services: reduce or decrease the dependencies between services. (2) Service contract: The service follows the service specifications defined in the description file. (3) Service abstraction: The service provider encapsulates its business logic, provides a service interface port externally, and hides the details of the internal business logic of the service. (4) Service reusability: A service can be used for multiple applications and business processes after it is established. (5) Service autonomy: Each service independently executes its business logic and has control over the encapsulated logic. (6) Service statelessness: The service provider does not need to store the user's context and status. (7) Service discoverability: When a service is published to the server, it is allowed to be automatically discovered and accessed by other consumers. (8) Service composability: A group of services can coordinate work and combine into another group of new services.

According to the definition of W3C [5], web service should be software to support the interaction of different machines between networks. A web service is usually composed of a number of application interfaces that are hosted by a remote server to handle service requests submitted by customers.

SOA is based on web service standards. It uses XML and web services to complete service communication. In addition, service interfaces are developed to implement SOA technology. The following describes the main communication protocols of SOA.

SOAP (Simple Object Access Protocol): A protocol for accessing and exchanging messages in a decentralized environment. Generally, messages are transmitted via HTTP, and the content of the message is described in XML format. By invoking a SOAP service call to a remote web service, other application services of different languages, platforms, and operating systems can be accessed and executed.

WSDL (Web Services Description Language): An XML-based description language used to describe the specification of service content and interaction. For example, interface operations for services, communication protocols, message formats, and data types.

UDDI (Universal Description, Discovery, and Integration): An XML-based registration protocol for publishing WSDL of services and allowing third parties to discover these services. UDDI is a directory service that allows web service providers to publish services on the internet and users to registers and discovers services through UDDI.

HTTP (or HTTPS): A communication protocol that is used to obtain an HTML web service for data exchange between a client and a server by sending a request/replying message.

2.2   Microservices Architecture

Fowler [7] defined the microservices architecture as a way to develop a single application into a small set of services, each running in its business process, with a lightweight communication mechanism (Usually HTTP/HTTPs). It can be deployed independently through an automated deployment mechanism. Microservices architecture has the following characteristics.

- Service componentization: The main purpose of componentization of the microservices architecture is to split the application into multiple services, each of which can be used and replaced independently. A request or response to a web service is realized through a service communication protocol.
- Organizing services according to business capabilities: The splitting of each service under the microservices architecture will be based on business functions and business processes.
- Decentralization: Microservices advocates splitting the components in the monolithic framework into different services and decentralizing the design of the services. Moreover, the microservices architecture tends to decentralize data and database management so that each service only manages its database, which is different from the traditional database centralized application.
- Infrastructure Automation: The characteristics of infrastructure automation are list below:

Service registration and discovery center: Under the service registration and discovery architecture, there are three types of microservices roles which are registration center, service provider (server), and service consumer (client). Each microservice needs to register with the registration center so that the service consumer can make service calls to the service provider through the registration center.

Service configuration center: The microservices configuration center is responsible for synchronizing with the configuration of each service node, configuring pushing, managing versions. The microservices connect the service nodes according to the configuration of the configuration center and realize the service discovery for the client and the server.

API gateway, service routing, service fault tolerance: API gateway is responsible for request invocation, identity authentication, load balancing, and routing rules. Besides, the gateway can handle fault tolerance such as downgrade, flow limit, and isolation.

Automated deployment and testing: Microservices split the system into separate small services that run independently, making the workload in deployment and testing larger, so automatically deploying and testing systems are needed to achieve continuous delivery and maintenance.

Service monitoring and tracking: Provides tracking and monitoring of service nodes and collects data and messages. Once a failure occurs, emergency processing is performed based on the collected data to improve the maintenance efficiency.

2.3 Spring Boot and Spring Cloud

Spring Boot [9, 10] is a framework which uses automated configuration functions to simplify the configuration of Spring applications, enable automatic management of dependent projects and provide application monitoring.

- Automatic configuration: Spring Boot uses the automatic configuration dependency function to import the project requirements, and set configuration items or custom attribute values in the configuration file (application.properties). Spring Boot uses the annotation declaration for automatic configuration to simplify the work of complicated configuration.
- Dependency management: Spring Boot provides a variety of different dependency management templates, each template has a corresponding function module and the application development tool. The required dependency template (such as spring-boot-starter-*) is added to the project pom file. The starter's dependency template will automatically configure the project dependencies, which can complete the rapid construction of the project dependent items.
- Actuator: Provides Spring Boot monitoring and management functions such as whether the application is in a normal state to meet more resource monitoring needs.

Spring Cloud [11, 12] is Spring's infrastructure for building microservices and distributed systems. The core components required for the development of the microservices system are listed below.

- Service Discovery Registration (Eureka): Eureka is used for registration and discovery of services. It consists of two components: Eureka server and Eureka client. The Eureka server is used for service registration and the Eureka client is an application. By connecting to the Eureka Server by Spring Cloud's microservices components (such as Eureka Client, Config, and Zuul), Eureka Server can be used to monitor whether the microservices in the system are working properly.

- Configuration Center (Config): The Spring Cloud Config project is a configuration management solution for distributed systems which is used to manage the environment configuration of microservices. It consists of two parts: the server and the client. The server is the microservices system configuration center. The client is a microservices application that manages the configuration of the distributed system by specifying a configuration center. In this study, Spring Cloud uses Github to store configuration files and Config to manage the configuration of the system.
- Service Invocation and Load Balancing (Feign): The purpose of Spring Cloud Feign is to make it easier for clients to make REST calls. The parameters, format, address and other information of the request can be defined by creating a server interface and configuring it with annotations. Feign will proxy the HTTP request, complete the service request and related processing.
- Message Queue (RabbitMQ): RabbitMQ is an open-source component that is used to decouple communication between services under the microservices architecture. Based on the message publishing and subscription pattern, the consumer subscribes to the topic queue, the producer establishes the message and then publishes the message to the queue, and finally, the producer sends the message to the listening consumer in an asynchronous manner.
- API-Gateway (Zuul): Spring Cloud Netflix's Zuul component provides service routing and load balancing functions in the microservices architecture. The Zuul service gateway calls the corresponding service according to the requested URL.

3   Microservice Architecture Transformation

The Meeting Scheduler case study is transformed from the traditional single system architecture to the microservices system architecture in this study. There are four steps to complete the transformation which are service modeling, service architecture design, service implementation and service deployment.

3.1   Service Modeling

Service modeling is a prerequisite for the microservices architecture [15]. This study referred to Domain Driven Design [13] to identify business domains and service functions, and define the boundaries of microservices and the interaction between services. Thus, in service modeling, the boundaries of services from the actual business logic are defined, and the principle of service abstraction and contract are used to develop service specifications. A service specification is a contract between a service provider and a consumer. The microservices are implemented according to the service specifications and service interface endpoints are provided that consumers can execute and invoke.

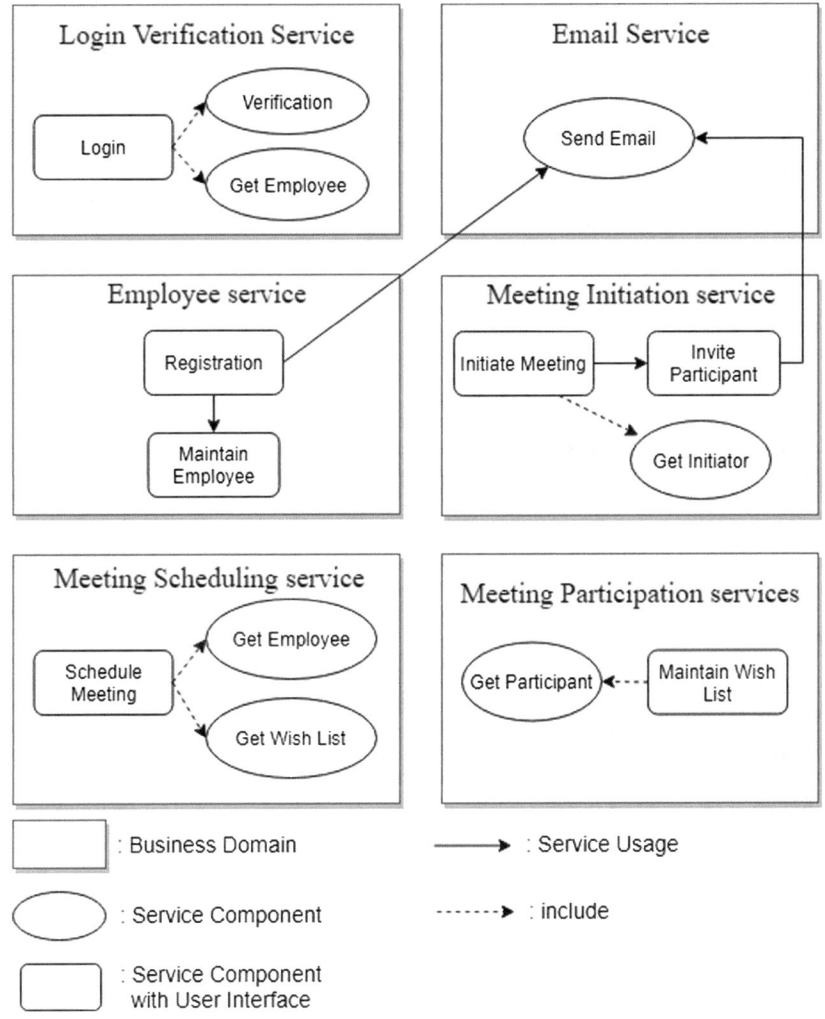

Fig. 1. Service boundary diagram

As shown in Fig. 1, according to the requirement analysis results of the Meeting Scheduler, the functional modules of each subsystem can be identified as the domain boundaries of the service. (1) The login verification service verifies employee using account and password, and obtains the employee's identity data after the verification is completed. (2) After logging into the system, the employee information can be maintained in the employee service, and the manager can register the new employee information. After the registration is completed, the system starts the email service and sends a notification to the registered employee. (3) In the meeting initiation service, the initiator initiates the meeting event, invites the meeting participants, and initiates the email service to notify the participants. (4) In the meeting participating service, the participants who are invited to the meeting fill in the information and resources required of the meeting. (5) In the meeting scheduling service, the system will perform scheduling according to the meeting date and resources in the wish list and the scheduling results will be displayed. (6) The email service will send a notification email when the employee registration is completed or the meeting participants are invited.

The following services are split from the context of the service boundary.

- Login Verification service: This service is to verify the employee's authority.
- Employee service: register and modify employee data.
- Meeting Initiation service: This service is to initiate meetings and set up meeting participants.
- Meeting Participation services: meeting participants fill the required information and resources to participate in the meeting.
- Meeting Scheduling service: The system displays the results according to the meeting date and resources.
- Email Service: send an email using external Gmail system.

Each service should design its interface with methods and data after the service modeling phase is completed. Table 1 shows the interface design with input and output data of each service.

Table 1. Service interface design

Service name	Interface name	Input	Output
Login verification	Login verification	Account and password	Identity verification (binding token)
Employee	Employee registration	Employee account, password, name, phone number, address, role, department, gender, email	Employee basic information
	Update employee	Employee account, password, name, phone number, address, role, department, gender, email	Employee basic information
	Delete employee	Employee ID	Employee basic information
	Query employee	Employee ID	Employee basic information
Meeting initiation	Initiate a meeting	Moderator, meeting room, meeting date range, discussion items, participants	Meeting information
	Update meeting	Moderator, meeting room, meeting date range, discussion items, participants	Meeting information
	Delete meeting	Meeting ID	Meeting information
	Query meeting	Meeting ID	Meeting information

(continued)

Table 1. (*continued*)

Service name	Interface name	Input	Output
Meeting participation	Add registration	Date and time range of the meeting	Meeting participation information
	Update registration	Date and time range of the meeting	Meeting participation information
	Delete registration	Registration ID	Meeting participation information
	Query registration	Registration ID	Meeting participation information
Meeting scheduling	Query schedule	Employee ID or meeting ID	Scheduling results
Email	Email	Email information	Email

3.2   Service Architecture Design

The overall microservices architecture of Spring Cloud is shown in Fig. 2. The client logs into the back-end system through the API Gateway, and the Login Verification service is performed. After the verification is successful, the identity token is generated and stored in the Redis database. Besides, the API Gateway provides a load balancing, traffic limiting mechanism, and the circuit breaker is set to isolate the fault to ensure system reliability.

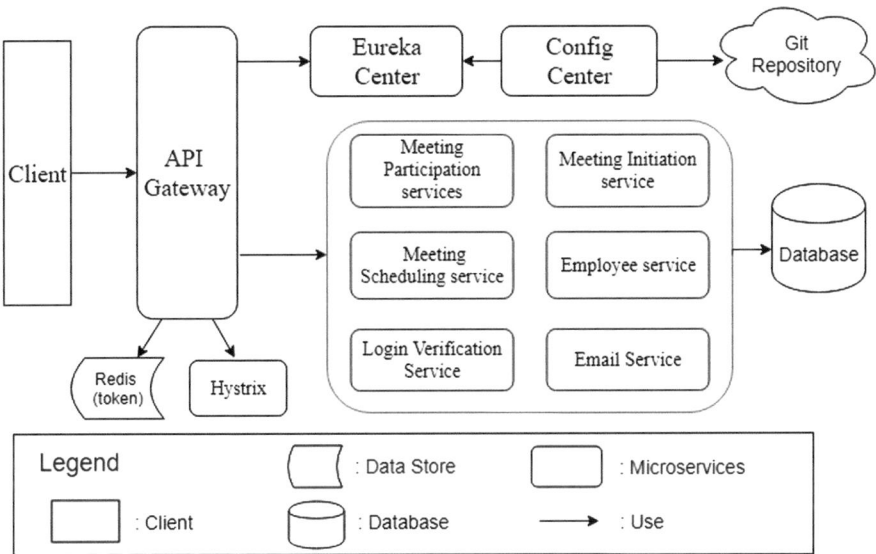

Fig. 2. Microservices architecture of meeting scheduler

The client uses the Feign component to call each microservices such as login verification service, employee service, meeting initiation service, meeting participation service, meeting scheduling service, and email service. The microservices are registered and discovered by the Eureka Registration Center. Each microservice synchronizes the configuration file to Git Repository for dynamic configuration and management.

3.3   Service Implementation

In the implementation of microservices, we use Spring Cloud as the architecture framework of microservices infrastructure. Spring Cloud uses Eureka to implement service registration and discovery, Hystrix to implement service fault tolerance, Zuul to implement API gateway and Config to implement the configuration center. Those infrastructure components used in this study are described below.

Service Registration Center

Figure 3 shows the registry related environment parameters in the Eureka Server configuration file (application.yml), such as server.port (port), application.name (server name), hostname (registered name), service-url (Eureka server address), etc. Figure 4 adds the @EnableEurekaServer annotation and start an Eureka server to enable registration and discovery of the service.

```
server:
    port: 8761
spring:
    application:
        name: eurekaServer

eureka:
    instance:
        hostname: localhost
    client:
        register-with-eureka: false
        fetch-registry: false
        service-url:
            defaultZone: http://${eureka.instance.hostname}:${server.port}/eureka/
```

Fig. 3. Configuration of Eureka Registration Center

```
@SpringBootApplication
@EnableEurekaServer
public class MeetingServerApplication {

    public static void main(String[] args) {
        SpringApplication.run(MeetingServerApplication.class, args);
    }
}
```

Fig. 4. Service initiation of Eureka Registration Center

Service Configuration Center

The configuration server is set according to the relevant environmental parameters of the server, such as server.port (connection port), application.name (server name), service-url (Eureka server address), and the storage address of specified Git Repository configuration file.

The Config service is a microservice, so you need to register to the Eureka server using the @EnableDiscoveryClient annotation to become the Eureka client, and adding the @EnableConfigServer annotationto start the Config Server configuration server (shown in Fig. 5).

```java
@SpringBootApplication
@EnableDiscoveryClient
@EnableConfigServer
public class ConfigServerApplication {

    public static void main(String[] args) {
        SpringApplication.run(ConfigServerApplication.class, args);
    }
}
```

Fig. 5. Service initiation of config center

Service API Gateway

Figure 6 uses @EnableDiscoveryClient annotation to become Eureka client, add @EnableZuulProxy and other annotationsto start service routing and service filter function. API Gateway is closely related to service invocation, service routing, and load balancing. @EnableDiscoveryClient annotation is added to become Eureka client.

```java
@SpringBootApplication
@EnableDiscoveryClient
@EnableZuulProxy
public class ZuulGatewayApplication {

    public static void main(String[] args) {
        SpringApplication.run(ZuulGatewayApplication.class, args);
    }
}
```

Fig. 6. API gateway service initiation

Custom Services

In this study, there are six microservices which use the @EnableDiscoveryClient annotation to become the Eureka client. Figure 7 uses @EnableFeignClients annotation to make the service call.

```
@SpringBootApplication(exclude = DataSourceAutoConfiguration.class)
@EnableDiscoveryClient
@EnableFeignClients
public class MeetingClientApplication {

    public static void main(String[] args) {
        SpringApplication.run(MeetingClientApplication.class, args);
    }
}
```

Fig. 7. Custom services initiation

3.4   Service Deployment

The dockerfile is a collection of docker command lines, mainly used to create a customized container image [8, 14]. When the program content changes, the container image can be re-created through the dockerfile. The following describes the deployment process of the microservices dockerfile in this study.

1. When creating a container image, use openjdk: 12.0.1-jdk as the base image, which contains the execution environment of the java runtime.
2. In the docker building process, copy the compiled jar file to the docker container using the copy command.
3. Use the expose command to specify the port that the container is open to.
4. Finally, the entrypoit command is used to set commands such as "java", "-jar" to be executed while starting the container.

4   Discussion

This study focuses on the transformation from the traditional monolithic application to the microservices software using the microservices architecture transformation process shown in Fig. 8. The traditional object-oriented analysis and design finds the functional modules of each subsystem. Then, the microservices architecture transformation process is performed for service modeling, service architecture design, service implementation and service deployment to complete the transformation from monolithic architecture to microservices architecture.

After the implementation of the microservices architecture transformation process, it can be found that this process has the following advantages. (1) The process is an extension of the object-oriented analysis and design method. Developers apply the traditional analysis design method first to find out the functional modules of the components, then follow the service-oriented design principles, clarify the boundaries and specifications of the services, and design the business logic of each microservice. It is easy to understand and follow so that developers can effectively shorten the learning curve. (2) In the monolithic system, the service modeling and service realization methods of the microservice can help the developer to correctly define each microservice and complete the transformation to microservices architecture. (3) The granularity of microservices is determined according to business needs. The services

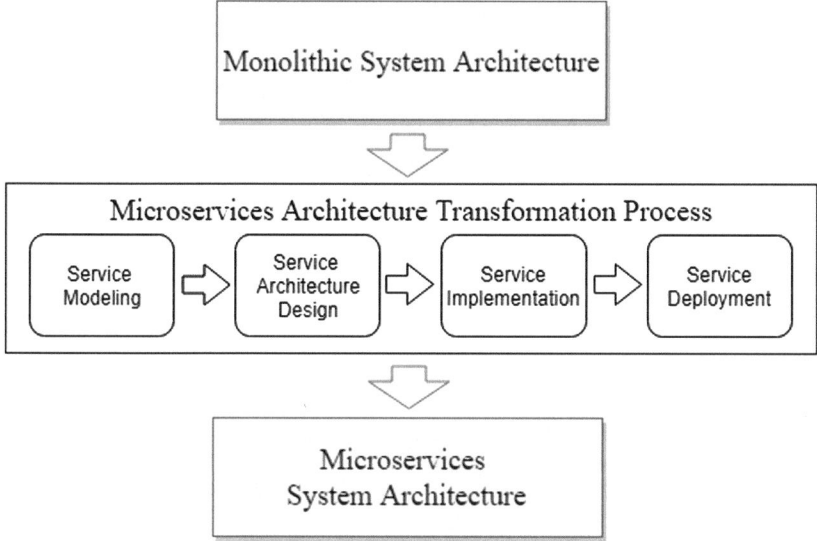

Fig. 8. Microservices architecture transformation process

are loosely coupled due to the progressive service splitting process, thus overcoming the lack of flexibility of the original single monolithic system, and the benefits of service reuse and the composition can be achieved. (4) For system maintenance, the maintainer only needs to modify and deploy the program for individual services. system anomalies caused by individual service changes can be reduced to improve the maintenance efficiency of the system.

5   Conclusion

Using microservices to divide a monolithic application into several microservices that are independent of each other can improve the scalability and maintainability of the system. Spring Cloud is a complete microservices architecture framework that provides a reliable service environment and ensures smooth communication between services. The software engineer defines each microservice according to different goals and requirements, and deploy the service into Spring Cloud to build the software system using microservices architecture.

In this study, we introduced a microservices architecture transformation process which is achieved by using service modeling, service architecture design, service implementation, and service deployment. The process can help developers to correctly describe each microservice, identify service tasks and relations, apply the service design principles to design and implement the microservices and finally, deploy the microservices in the docker container to efficiently complete the development of the microservices software system.

Acknowledgment. This study is conducted under the grant MOST 107-2221-E-017-002-MY2 which is subsidized by the Ministry of Science and Technology.

References

1. Erl, T.: Service-oriented Architecture: Concepts, Technology, and Design. Prentice Hall, August 2005
2. Jordan, D., Evdemon, J., Alves, A., Arkin, A., Askary, S., Barreto, C., Bloch, B., Curbera, F., Ford, M., Goland, Y.: Web services business process execution language version 2.0. OASIS Standard, vol. 11, no. 120, p. 5 (2007)
3. Erl, T.: SOA Principles of Service Design. Prentice Hall, July 2008
4. Erl, T.: SOA Design Patterns. Prentice Hall (2009). https://www.ibm.com/developerworks/cn/webservices/ws-soa-design/
5. Booth, D., Haas, H., McCabe, F., Newcomer, E., Champion, M., Ferris, C., Orchard, D.: Web Services Architecture, W3C11, February 2004
6. Erl, T.: Service-Oriented Architecture: A Field Guide to and Integrating XML and Web Services, Prentice Hall, 26 April 2004
7. Fowler, M., Lewis, J.: MicroServices, 25 March 2016. https://www.martinfowler.com/articles/microservices.html
8. Docker Inc.: Docker Documentation. https://docs.docker.com/. Accessed 2019
9. Pivotal Software: Spring Framework Documentation Version: 5.1.9.RELEASE. https://docs.spring.io/spring/docs/current/spring-framework-reference/. Accessed 2019
10. Webb, P., et al.: Spring Boot Reference Guide 2.1.7.RELEASE. https://docs.spring.io/spring-boot/docs/current/reference/html/. Accessed 2019
11. Pivotal Software:. Spring Cloud Reference Version: Greenwich.RELEASE. https://www.docs4dev.com/docs/en/spring-cloud/Greenwich.RELEASE/reference. Accessed 2019
12. Pivotal Software: Spring Cloud. https://spring.io/projects/spring-cloud. Accessed 2019
13. Evans, E.: Domain-Driven Design: Tackling Complexity in the Heart of Software. Addison-Wesley Professional, August 2003
14. Kane, S.P., Matthias, K.: Docker: Up & Running: Shipping Reliable Containers in Production., O'Reilly Media, October 2018
15. Newmain, S.: Building Microservices: Designing Fine-Grained Systems. Oreilly & Associates Inc., February 2015

Data Analytics for e-Business

Towards Identification and Validation
of Web Themes

Yan-Ting Chen, Bao-An Nguyen, Kuan-Yu Ho,
and Hsi-Min Chen[✉]

Department of Information Engineering and Computer Science,
Feng Chia University, Taichung, Taiwan
{m0705965,m0721053}@o365.fcu.edu.tw,
baoanth@gmail.com, hsiminc@fcu.edu.tw

Abstract. Nowadays, a typical website consists of multiple web pages to present interactive content and provide a variety of functions. To speed up web development process, websites are usually built by a team of developers. Although multi-page websites are developed constantly by different developers, web pages of a website should follow theme specifications in order to keep look and feel consistent. In this research, we propose an automated system to identify web themes from existing web pages as well as to validate whether newly added web pages adhere to specified themes. By means of this proposed system, look-and-feel consistency of websites can be maintained. Besides when someone commits a web page that does not follow a specified theme, the system will detect the violation of web theme automatically. Thus, the effort of manual validations can be dramatically reduced.

Keywords: Web theme template · Web programming · Software testing · Web similarity · Software quality

1 Introduction

Nowadays, a typical website consists of multiple web pages to present interactive content and provide a variety of functions. To speed up web development process, websites are usually built by a team of developers. Although multi-page websites are developed constantly by different developers, web pages of a website should follow design specifications in order to keep look and feel consistent. Web theme template is a scheme that can facilitate web developers to develop web pages with a consistent web theme by following a defined template. Figure 1 shows a basic theme template for several web sites. These pages of a site usually are identical in the block of the header and the block of the footer. Thus, based on a defined theme template, a group of web pages can present the same layout but different content.

For several websites developed from scratch, web theme templates, i.e., design specifications, are not taken seriously at early development stage because of small scale of websites, websites developed by a few developers, or short time to market. However, with the growth of websites, web themes become an indispensable concern that does not only provide professional look-and-feel presentations to users but also guide

K.-M. Chao et al. (Eds.): ICEBE 2019, LNDECT 41, pp. 143–160, 2020.
https://doi.org/10.1007/978-3-030-34986-8_10

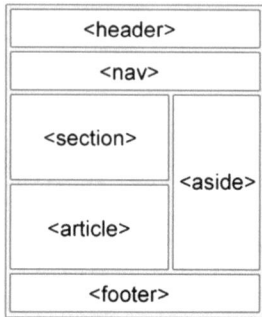

Fig. 1. Basic template of web theme [1].

developers to develop web pages with consistent style. In such context, how to identify a theme template from a set of web pages already developed becomes an important issue. On the other hand, when developers finish a web page, how do they make sure that this web page fully adhere to the specified web theme template? Without the support of tools, developers need to spend a lot of time in checking the compliance between web pages and templates. This causes another issue that the validation for newly developed web pages must be automated during the development of web sites.

To address the above issues, in the research, we proposed an approach to identifying theme templates from existing web pages and to validating newly added web pages in order to maintain the style consistent of websites. Through our approach, web developers can easily define templates from selected web pages without designing them from scratch and can get immediate feedbacks after the validation scheme is applied. We also developed a system based on the proposed approach to automate the identification and validation of web theme templates. By means of the system, the cost spent in developing web sites with consistent look-and-feel themes can be dramatically reduced.

The rest of the paper is organized as follows. Section 2 describes the related works. The proposed approach is presented in Sect. 3. Experimental results are conducted in Sect. 4 and finally in Sect. 5 we conclude our work.

2   Related Works

Web testing have been increasingly considered by researchers in the last decade due to the high impact of web applications in the Internet-based society. According to the systematic literature review about web testing in 2014 [2], in many testing schemes, the input models of testing cases can be classified in: navigation models, control or data flow models, DOM (Document Object Model) models, and others. In these approaches, DOM models were commonly used for GUI (Graphic User Interface) testing. In terms of testing approaches using DOM models, the vital task is to compare structures of input DOM tree to determine similarities between them. Tree Edit Distance measurement was used in [3] to measure the structural similarity of HTML pages. Yin *et al.* [4]

used Text Edit Distance between tags sequences extracted from the DOM trees as the main similarity measurement for template extraction using clustering web page. Pushpa *et al.* [5] adopted Minimum Description Length to manage unknown number of web templates. Buttler [6] proposed DOM's path with shingles to determine similarities between web pages according to the ratio of the intersection and union of the shingles path between two DOM. The methods proposed in [3–6] compare the similarities of web pages, but they are not suitable for template comparison because their goals are used to extract web page information, not to look for common themes in a website.

Webdiff [7] used the combination of structural analysis on DOM tree and visual analysis on web page layout to identify display errors in cross-browser testing. With Ajax web applications, the DOM tree of a single web page may change after any user action or event. Meshah *et al.* [8] proposed an automatic testing method using DOM tree invariants to detecting Ajax-specific faults. Matching method proposed in [8] tests variant of only one page after Ajax actions are performed. Different from the approaches [7, 8] applied in the same web page, the template identification takes place on multiple web pages. Thus, these approaches were not suitable to solve the mentioned issues.

Another feasible approach to matching trees, especially DOM trees, is finding common subtree or longest common template by using Suffix Trees [9]. Using Suffix Tree to find out longest common template can maintain the structural information of trees, however, for web testing this approach will be disturbed by the nodes that are not parts of templates

In addition to the DOM-tree-based analysis methods, computer vision techniques were used to identify errors in GUI of mobile applications by leveraging the design diagram [10]. Another approach for web layout matching [11] is converting two-dimension graph grammar of web page into one-dimension of string induction.

In this paper, we propose a novel approach for identify web theme template from DOM trees from given HTML documents. Our approach not only considers the content of nodes, but also maintains the structure of trees as well as position of nodes in trees.

3   Proposed Approach

In this section, we present our algorithms to identify and validate theme templates of a website. We adopt DOM trees as input models in our approach. The proposed algorithms of theme templates identify common nodes of DOM trees, starting from a given HTML document that serves as a base tree. This research targets two kinds of web theme templates called rigid template and soft template. Rigid template takes both order and position of nodes at the same level of a tree into account to identify template nodes. With respect to soft template, it only considers the order of common nodes at the same level of a tree. Our proposed approach is presented as follows. Firstly, we introduce notations with the corresponding descriptions used in the algorithms. Then, definitions and examples regarding two kinds of templates are also explained. Thirdly, we illustrate the algorithms of template identification. Finally, the algorithms for template validation is presented in the third subsection.

3.1   Theme Template Definitions

To illustrate the proposed algorithms, we first define notations with the corresponding descriptions as depicted in Table 1. Given a DOM tree T_i, which is a HTML document and serves as a base web page, and a set of DOM tree T, which are other pages selected to identify a theme template, the web templates are defined as follows:

Table 1. Notations used in the proposed algorithms

Notation	Description
T_j	A DOM Tree of the $HTML_j$
T	A set of n DOM trees $T = \{T_1, T_2, \ldots, T_n\}$
N_j	The set of all nodes $\{n_{j1}, n_{j2}, n_{jx}\}$ are belong to T_j
$n_{j,x}$	A node belonging to N_j
$C_{j,x}$	The set of all child $\{c_{j,x,1}, c_{j,x,2}, c_{j,x,y}\}$ are belong to $n_{j,x}$
$c_{j,x,y}$	A child belonging to $C_{j,x}$
$D_{j,x,y}$	The set of all descendant $\{d_{j,x,y,1}, d_{j,x,y,2}, \ldots, d_{j,x,y,m}\}$ are belong to $c_{j,x,y}$
$d_{j,x,y,z}$	A descendant belonging to $D_{j,x,y}$
$AL_{j,x,y,z}$	Name and level of $d_{j,x,y,z}$
$V_{y,q}$	Length of Longest Common Subsequence (LCS) of two trees node $c_{i,x,y}$ and $c_{j,x,q}$ *Example:* the length of LCS $\{div1,li2,a3\} \cap \{div1,li2\}$ is 2

a. Rigid template

Rigid template is defined based on the assumption that UI elements of a template should appear at fixed positions among all web pages of a website. In the other world, rigid template is a structure of common nodes that having the same order and fixed positions at the sample level in all DOM trees.

Rigid template nodes are picked at the same tree level from both two directions by a manner of approaching from left and from right. Formally, given a two DOM tree T_i and T_j in which T_i is the base tree, rigid template can be identified by following procedures:

From left: Two trees are traversed at each level from left with decreasing value of index y, which is the position of a node at that level. If $c_{i,x,y}$ of tree T_i and $c_{j,x,y}$ of T_j have the same index value, they are template nodes. As shown in Fig. 2, the first node from the left of both two trees is a div, so having the same index value of x and y. As a result, $c_{i,x,y}$ and $c_{j,x,y}$ should be the same that is selected in a template.

From right: Two trees are traversed concurrently from right. The position of a node is identified by inverse index which calculated by the subtraction of length the tree and y index value. If $c_{i,x,y}$ and $c_{j,x,z}$ are the same element and the same inverse index, they are template nodes. Take Fig. 2 as an example where the rightmost node at the second level with inverse index 0 of both two trees have the same value p. Hence, the p element is selected in template nodes.

In general, template nodes positioned in the bottom of a web page are linked, so the interruption in searching from the right indicates that next nodes do not belong to the template. As shown in Fig. 2, the second node from the right of two trees T_i and T_j is p and div, respectively. Because of this difference, the next nodes will not be counted as template nodes and the searching step is terminated. An overall example of rigid template is illustrated in Fig. 2.

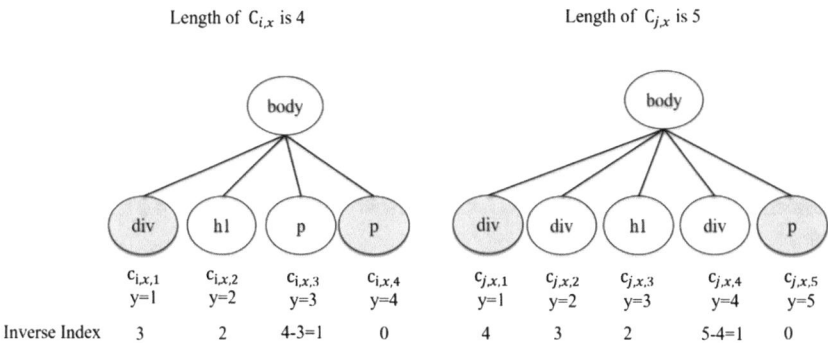

Fig. 2. Definition of the rigid template

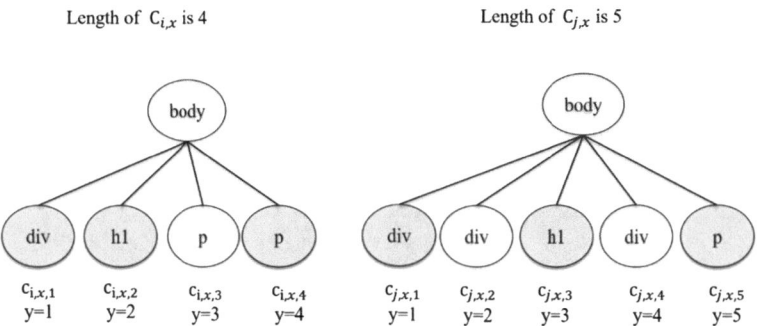

Fig. 3. Definition of the soft template

b. Soft template

Rigid template is suitable to maintain the structural information of the template, since the orders and indexes of a template node are fixed. In order to support a more flexible identification scheme, we propose soft template in which the orders of template nodes must be maintained but the corresponding index might vary among DOM trees.

As shown in Fig. 3, the template nodes, i.e. div, h1 and p, in the two DOM trees T_i and T_j have the same order but their indexes can be different. For instance, the two h1 nodes appearing in $c_{i,x,2}$ and $c_{j,x,3}$ can be consider as template nodes in soft template. We leverage LCS (Longest Common Subsequence) technique, to identify the soft template among DOM trees. All theme template identification algorithms will be discussed in the next section.

3.2   Template Identification Algorithms

The problem of template is defined as follows: Given a set $T = \{T_1, T_2, \ldots, T_n\}$ of DOM trees and a base tree T_i, we must identify the theme templates of T based on the matching procedures between T_i and each $T_j \in T$. The output template should be both rigid template and soft template.

We use Selenium [12] tool to solve the rendering problem of dynamic web pages and compare DOM Tree to identify which UI elements are templates. Figure 4 shows the flow chart of the identification process. Template trees is outputted at the end of the workflow.

The template identification flow can be explained as following: Let's start with the algorithm MultipleHTMLTemplateFind with the base tree T_i and a set T of DOM trees.

Step 1: Select an input T_j to compare. This step calls the LevelCompare algorithm.
Step 2: Start at the first level of DOM trees, LevelCompare identifies the template nodes of based on the comparison between T_i and T_j according two kinds of template: RigidLevelChildernCompare is used for rigid template; SoftLevel ChildernCompare is used for soft template.
Step 3: LevelCompare is operated with next levels of DOM tree until all nodes are considered.

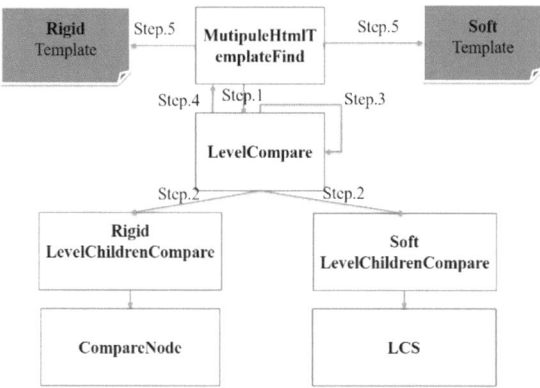

Fig. 4. Flow chart of template identification

Step 4: Return temporary results to the algorithm MultipleHTMLTemplateFind and select next tree T_j to continue.
Step 5: When all trees in T are compared, post-processing to final decision on template nodes and output the template files.

The overall workflow of our identification scheme is presented in the algorithm `MultipleHTMLTemplateFind` in Table 2 The rigid template is easily obtained

Table 2. Algorithm of `MutipleHtmlTemplateFind`

Algorithm Name : MutipuleHtmlTemplateFind
Input: A set of DOM trees T = { T_1 , T_2, ..., T_n}
Output：Web Template
Start
loss←{}
i← base web page number
// compare all webpages to base webpages to find the template
for each $T_j \in T$ **do**
if choose rigid method **then**
T_i←LevelCompare(T_i, T_j)
else if choose soft method **then**:
sameNodes←LevelCompare(T_i, T_j)
loss ←loss∪{T_i – sameNodes } *//add non-template node into loss*
end for
//get the template by deleting all non-template node
webTemplate ← {T_i – loss}
End

Table 3. Algorithm of `LevelCompare`

Algorithm Name : LevelCompare
Input: $n_{i,x}$, $n_{j,x}$
Output：HTML Contains Template
Start
$tempC_{i,x}$←the $C_{i,x}$ of $n_{i,x}$
$tempC_{j,x}$←the $C_{j,x}$ of $n_{j,x}$
//get the same child nodes of $C_{i,x}$ and $C_{j,x}$
if choose rigid method **then**
$C_{i,x}$← RigidLevelChildrenCompare($tempC_{i,x}$,$tempC_{j,x}$)
else if choose soft method **then**
$C_{i,x}$←SoftLevelChildrenCompare ($tempC_{i,x}$,$tempC_{j,x}$)
end if
//recursively get the child nodes of the next levels
for $\forall c_{i,x,y} \in C_{i,x}$ **do**
if $C_{i,x}$ is not empty **then**
$tempn_{i,x}$←take corresponding $n_{i,x}$ of $c_{i,x,y}$
$tempn_{j,x}$←take corresponding $n_{j,x}$ of $c_{j,x,y}$
LevelCompare($tempn_{i,x}$, $tempn_{j,x}$)
end if
end for
End

after a series of operations `LevelCompare` executed. In contrast, with soft template, we need a post-processing step to eliminate non-template nodes after each execution of `LevelCompare`. After all, the web template file is produced as the result of web template identification algorithm.

The `LevelCompare` method is executed for finding template nodes at specific level of the DOM tree. Like popular approaches for tree traversing, we use recursion method to reach all levels of the DOM trees (Table 3).

a. Rigid template detection

Table 4 shows the algorithm for rigid template node identification. Because the rigid template requires corresponding nodes in two trees must reside in the same position, we should consider the index or reverse index of nodes in template matching. Since the template nodes are usually located at the top and the bottom of web pages, we divide the page into two parts for comparison, from left and from right of the DOM node, as discussed in template definitions. The comparison of two leaf nodes is done by the algorithm `CompareNode` shown in Table 5.

First, the searching from right will be terminated when it encounters a difference, since template node at the bottom of web pages are usually continuous. As shown in Fig. 5, the only the right node p is identified as the template node when we match two trees from right.

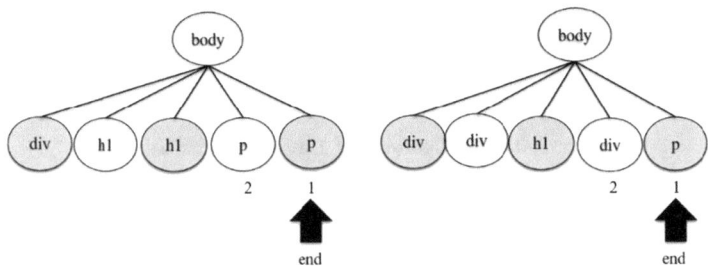

Fig. 5. Rigid method for the right template

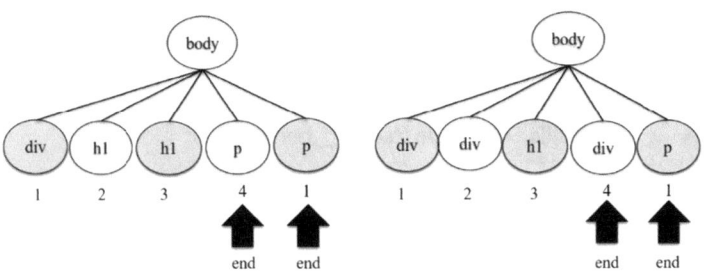

Fig. 6. Rigid method for the left template

Then, to identify the template nodes from left, the searching process starts from the left-most node and stop at the terminated point where the searching step from right have just stopped. As shown in Fig. 6, the two nodes `div` (index 1) and `h1` (index 3) are identified as template nodes after searching from left.

Table 4. Algorithm of `RigidLevelChildrenCompare`

Algorithm Name : RigidLevelChildrenCompare
Input: $C_{i,x}$, $C_{j,x}$
Output : new is the same child node list for $C_{i,x}$ and $C_{j,x}$

Start
$y \leftarrow$ the length of $C_{i,x}$
$z \leftarrow$ the length of $C_{j,x}$
 	
//---------traverse from right----------
while ($y > 0$ and $z > 0$) **do**
　　if CompareNode($c_{i,x,y}$, $c_{j,x,z}$) **then**
　　　　new add $c_{i,x,y}$;
　　　　if $y<z$ **then** *// decide the part form right*
　　　　　end$\leftarrow$y
　　　　else
　　　　　end$\leftarrow$z
　　　　end if
　　else then *// from right template encounters a differences*
　　　　break loop
　　end if
　　y-=1
　　z-=1
end while
 	
//---------traverse from left---------
y$\leftarrow$0
z$\leftarrow$ The original index of $c_{i,x,y}$ in $C_{i,x}$ before comparing T_i and T_1
while ($y <$ end and $z <$ end) **do**
　　z$\leftarrow$ The original index of $c_{i,x,y}$ in $C_{i,x}$ before comparing T_i and T_1
　　if CompareNode($c_{i,x,y}$, $c_{j,x,z}$) **then**
　　　　new add $c_{i,x,y}$
　　end if
　　y$\leftarrow$y+1
end while
return new
End

Table 5. Algorithm of CompareNode

Algorithm Name : CompareNode
Input: $n_{i,x}$, $n_{j,x}$
Output : Boolean. For $n_{i,x}$, $n_{j,x}$ equal or not
Start
if $n_{i,x}$ and $n_{j,x}$ is equal:
return True
else
return False
end if
End

b. Soft template identification

Soft template can be identified using matching methods to find LCSs. Here we traverse the tree with Node-Left-Right order (called PreOrder in short) and convert the trees into HTML tag sequences and do matching on these sequences.

We leverage LCS technique to find out the longest common subsequence between two trees and store length of LCSs in the array V. Given two set of nodes $C_{i,x}$, $C_{j,x}$, V_{yq} stores length of LCS between the yth node of $C_{i,x}$ and q^{th} node of $C_{j,x}$. As shown in Fig. 7, the comparison between 1st div node $c_{i,x,1}$ of the left tree and the 1st div node $C_{j,x,1}$ of the right tree resulted $V_{11}= 1$. In contrast, the comparison between 2nd div node $C_{i,x,2}$ of the left tree and the 1st div node $C_{j,x,1}$ of the right tree resulted $V_{21}= 5$. We can conclude that length the LCS between the two tree is 5 as marked in the Fig. 7.

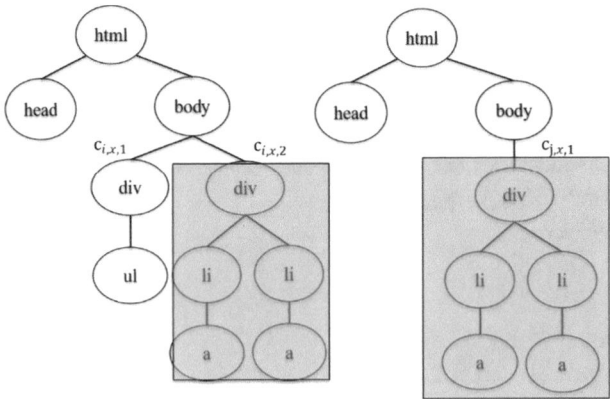

Fig. 7. Example of soft template

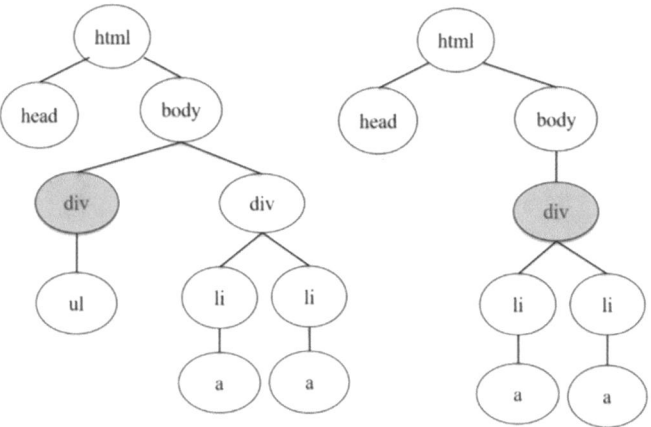

Fig. 8. Error matching case without referring child nodes

In practical, we found some error cases which must be prevented in matching phase as follows.

Error case 1: If the soft method compares two nodes without referring their child node, we could get template nodes which only match on the top level whose child nodes are completely different. As shown in the Fig. 8, the first div of the left tree and the div of the right tree matched, but their child nodes do not. To address this, propose method SoftLevelChildrenCompare compares both corresponding nodes and their descents as shown in Table 6.

Error case 2: Two different tree can be mismatch when two different tree structures produce the same tag sequences. As shown in Fig. 9, both two trees generated tag sequences are [div, div, div] and lead to a mismatch. To avoid this kind of error, we add level numbers to the tags when traversing tree nodes. For instance, by adding level number the left tree becomes [div1, div2, div3] and the right tree becomes [div1, div2, div2], so that the mismatch on div2 and div3 is prevented.

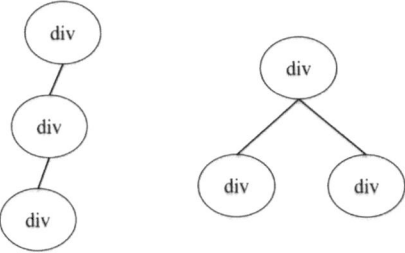

Fig. 9. Error matching case when different tree structures generates the same tag sequences

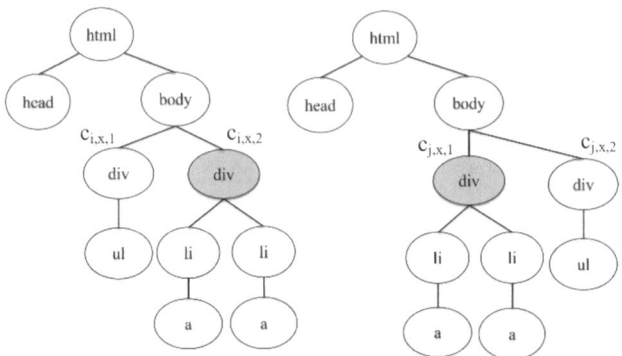

Fig. 10. Error case of incorrect order matching in soft template

Table 6. Algorithm of `SoftLevelChildrenCompare`

Algorithm Name : SoftLevelChildrenCompare
Input: $C_{i,x}$, $C_{j,x}$
Output : newTemp is the same child node list for $C_{i,x}$ and $C_{j,x}$

Start
    for each $c_{i,x,y} \in C_{i,x}$ **do** //get the all LCS for each node
      $Y \leftarrow$ PreOrder($D_{i,x,y}$) // get $AL_{i,x,y,z}$ of the each $d_{i,x,y,z}$
      while (q < the length of $C_{j,x}$) **do**
        $Q \leftarrow$ PreOrder($D_{j,x,q}$) //get $AL_{j,x,q,z}$ of the each $d_{j,x,q,z}$
        $V_{y,q} \leftarrow$ length of LCS of Y and Q
      end while
    end for
    while max($V_{y,q}$) > 0 **do** // *match the template node*
      indexY,indexQ$\leftarrow$Argmax($V_{y,q}$)
      newTemp.add($c_{i,x,indexY}$)
      newJ.add($c_{j,x,indexQ}$)
      sort newTemp by indexY of $c_{i,x,indexY}$
      sort newJ by indexQ of $c_{j,x,indexQ}$
 	
      //post-processing to prevent the wrong order of template node
        if index of $c_{i,x,indexY}$ in newTemp is different the index of $c_{i,x,indexQ}$ in newJ **then**
        remove $c_{i,x,indexY}$ from newTemp
        remove $c_{j,x,indexQ}$ from newJ
        $V_{indexY,indexQ} \leftarrow$ -1
        else then // *let the matched node wouldn't be matched again*
          set $V_{indexY,m}$ = -1 with all $0 \leq m < y$
          set $V_{n,indexQ}$ = -1 with all $0 \leq n < q$
        end if
    end while
    return newTemp
 End

Error case 3: Based on the soft template definition, the order of the template nodes cannot be different in two trees. As shown in Fig. 10, only the matching between $C_{i,x,2}$ and $C_{j,x,1}$ with $V_{2,1} = 5$ can be template node, $C_{i,x,1}$ and $C_{j,x,2}$ with $V_{1,2} = 2$ not matched as template nodes because of the wrong position error. We add the post-processing step into the algorithm to identify and remove this kind of matching error from the result. The `SoftLevelChildrenCompare` algorithm for identifying soft template is shown in Table 6.

3.3   Template Validation

After identifying theme template by the above algorithms for both rigid template and soft template, we introduce the template validation steps into the workflow as illustrated in Fig. 11. The validation steps output missing nodes of tested page after comparing it with the base page.

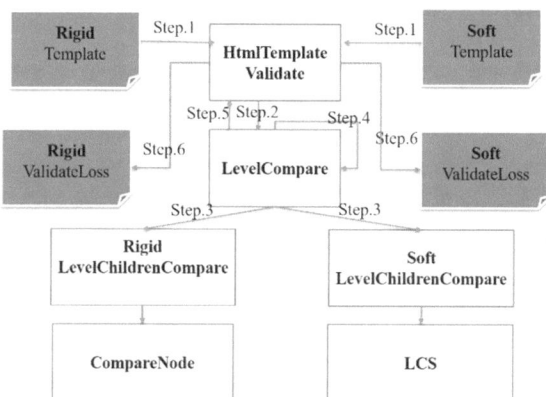

Fig. 11. Flow chart of validation

The algorithm `HtmlTemplateValidate` use the template to validate the newly added page and output the nodes which are required in the template but do not appear in T_j, as shown in Table 7.

Table 7. Algorithm of `HtmlTemplateValidate`

Algorithm Name : **HtmlTemplateValidate**
Input: webTemplate, a tested page T_j
OutputvalidateLoss contains the missing nodes

Start
    validateLoss ={}
    if rigid method selected **then** *//get the template node*
        sameNodes ←LevelCompare(webTemplate, T_j)
    elif soft method selected **then**:
        sameNodes ←LevelCompare(webTemplate, T_j)
//get the missing template node
validateLoss ← webTemplate – sameNodes
End

4 Experimental Results

In this experiment, we select a well-known website in communities of software developers which is StackOverflow [13] to evaluate our web template identification and validation methods. We compare our results with those of manual matching method and difflib tool [14]. With soft template identification, we also compare matching method based on LCS and Suffix Tree [9] to investigate the appropriate method in structural matching.

4.1 Experiment: StackOverflow Template

Figures 12 and 13 are two example web pages from StackOverflow in which the first is a typical question-answer page and the second is a question-answer page with a member registration form. The statistics of template identification result is shown in Table 8.

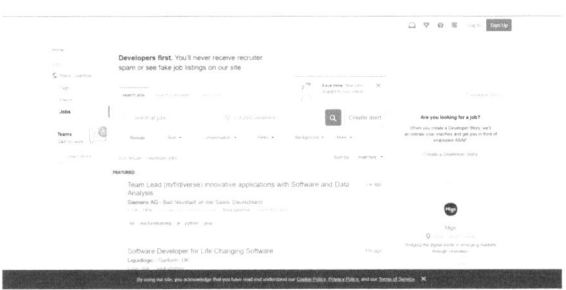

Fig. 12. Input web page 1 from StackOverflow

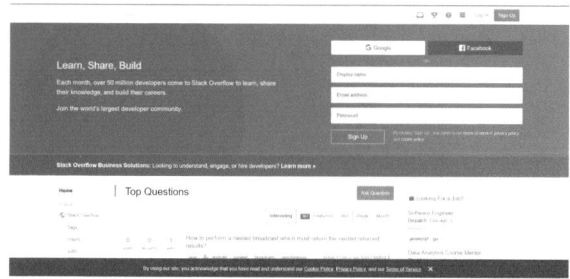

Fig. 13. Input web page 2 from StackOverflow

Table 8. Experimental result with StackOverflow website

Method	Number of nodes identified in the template
Manual identification	427
Rigid template	420
Soft template with LCS	**431**
Soft template with Suffix Tree	427
Difflib tool (before formatting)	58
Difflib tool (after formatting)	346

We manually found 427 nodes with the visual result shown in Fig. 14. Rigid method found 420 nodes. Since the second page in Fig. 13 contains a blue registration block inserted between the template nodes which causes the node position error, the indexed of below nodes are shifted and do not match any longer. It causes the nodes of the bottom black block to be missing. The result of rigid template is shown in Fig. 15. The soft method with LCS found 431 nodes with some position errors (error case 3) as marked in red rectangles in Fig. 16.

In order to evaluate robustness of matching method using LCS, we tried to replace LCS by Suffix Tree in soft method and run the experiments again. The resulted template is different from soft method with LCS. Since the Suffix Tree based matching method tries to find longest substring, the tag sequences must be continuous. If there is an extra node in the web page, the template will be interrupted by this node. For example, with two sequence [body div div] and [body div br div], we only find out the common sequence [body div], not [body div div]. It causes a central context node does not match in the template, as shown in Fig. 17.

Fig. 14. Manual template of StackOverflow

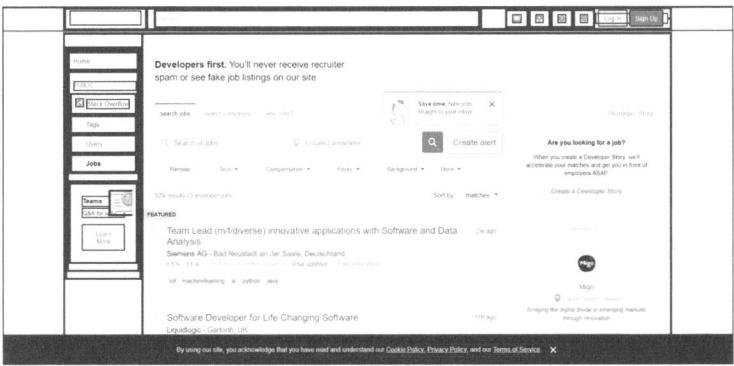

Fig. 15. Rigid template of Stack Overflow

Fig. 16. Soft template with LCS of Stack Overflow

Fig. 17. Soft template with Suffix Tree of Stack Overflow

Since the rigid method matches nodes by their indexes, whereas the soft method matches most similar structures, this reads to the difference between rigid method and soft method in Figs. 15 and 16. The soft method still successfully detects template node in the central region, whereas these regions are not matched in the rigid template.

The experimental results show that our approach is a potential method for web template identification and validation. Our method outperforms the difflib tool in all test cases. The rigid method has comparable performance with the manual method. With the flexible mechanism in structural matching, soft method with LCS based matching gave the most acceptable results, quite better than itself using Suffix Tree based matching. With these fruitful results, we believe that it is feasible to emerge our approach into a web GUI testing system.

5   Conclusion

In this paper, we present our method for web template identification and validation based on analyzing and matching DOM trees. The proposed rigid template is suitable for fixed layout website where the positions of GUI elements are often stable. The soft template can be well applied in flexible layouts where GUI elements can be occasionally added in some regions. By using our system, the efficiency and quality of web development can be improved. Web developers no longer need to manually check the web template to ensure website consistency. In future, we will integrate Continuous Integration tool, such as Jenkin [15] into our system to enable automated validation newly added or uploaded web pages during continuous integration.

References

1. W3Schools, HTML5 Semantic Elements. https://www.w3schools.com/html/html5_semantic_elements.asp
2. Doğan, S., Betin-Can, A., Garousi, V.: Web application testing: a systematic literature review. J. Syst. Softw. **91**, 174–201 (2014)

3. Gowda, T., Mattmann, C.A.: Clustering web pages based on structure and style similarity. In: Proceedings of IEEE 17th International Conference on Information Reuse and Integration (IRI), Pittsburgh, PA, USA, pp. 175–180, July 2016

4. Yin, G.-S., Guo, G.-D., Sun, J.-J.: A template-based method for theme information extraction from web pages. In: Proceedings of 2010 International Conference on Computer Application and System Modeling (JCCASM) (2010)

5. Pushpa, S., Kanagalatchumy, D.: A study on template extraction. In: Proceedings of International Conference on Information Communication and Embedded Systems (ICICES) (2013)

6. Buttler, D.: A short survey of document structure similarity algorithms. In: Proceedings of 5th International Conference on Internet Computing, Las Vegas, NV, USA, Jun 2004

7. Choudhary, S.R., Versee, H., Orso, A.: WEBDIFF: automated identification of cross-browser issues in web applications. In: Proceedings of 26th IEEE International Conference on Software Maintenance (2010)

8. Mesbah, A., Van Deursen, A., Roest, D.: Invariant-based automatic testing of modern web applications. IEEE Trans. Softw. Eng. **38**, 35–53 (2012)

9. Grossi, R.: On finding common subtrees. Theor. Comput. Sci. **108**, 345–356 (1993)

10. Moran, K., Li, B., Bernal-Cárdenas, C., Jelf, D., Poshyvanyk, D.: Automated reporting of GUI design violations for mobile apps. In: Proceedings of ACM/IEEE 40th International Conference on Software Engineering (2018)

11. Roudaki, A., Kong, J., Zhang, K.: Specification and discovery of web patterns: a graph grammar approach. Inf. Sci. **328**, 528–545 (2016)

12. Selenium. https://www.seleniumhq.org/. Accessed 15 Jun 2019

13. StackOverflow. https://stackoverflow.com. Accessed 12 Jun 2019

14. Difflib. https://docs.python.org/3/library/difflib.html. Accessed 15 Jun 2019

15. Jenkins. https://jenkins.io/. Accessed 20 Jun 2019

A Customer-Oriented Assortment Selection in the Big Data Environment

Morteza Saberi[1(✉)], Zahra Saberi[2], Mehdi Rajabi Aasadabadi[2,3], Omar Khadeer Hussain[2], and Elizabeth Chang[2]

[1] School of Information, Systems and Modelling, UTS, Sydney, Australia
morteza.saberi@uts.edu.au
[2] School of Business, UNSW Canberra, Canberra, Australia
[3] Australian National University, Canberra, ACT 0200, Australia

Abstract. Customers prefer the availability of a range of products when they shop online. This enables them to identify their needs and select products that best match their desires. This is addressed through assortment planning. Some customers have strong awareness of what they want to purchase and from which provider. When considering customer taste as an abstract concept, such customers' decisions may be influenced by the existence of the variety of products and the current variant market may affect their initial desire. Previous studies dealing with assortment planning have commonly addressed it from the retailer's point of view. This paper will provide customers with a ranking method to find what they want. We propose that this provision benefits both the retailer and the customer. This study provides a customer-oriented assortment ranking approach. The ranking model facilitates browsing and exploring the current big market in order to help customers find their desired item considering their own taste. In this study, a scalable and customised multi-criteria decision making (MCDM) method is structured and utilised to help customers in the process of finding their most suitable assortment while shopping online. The proposed MCDM method is tailored to fit the big data environment.

Keywords: Assortment selection · Online shopping · Big data · MCDM · Customer-oriented

1 Introduction

Online retailing is the process by which retailers sell their products to customers through an online webpage. According to a Forrester Research group's report, online retailing accounts for about 11% of total retailing in the United States [1]. At least three reasons are observable for this popularity: convenience, better prices, and greater product variety. One key activity typically planned carefully by online retailers is assortment planning. The importance of assortment planning is the attention it draws from customers by presenting a variety of products [2, 3]. There is no doubt that to be competitive, retailers have to put considerable effort into assortment planning.

The assortment planning process is the determination of the number of categories (the width) which will be carried, the number of different product lines in each category

© Springer Nature Switzerland AG 2020
K.-M. Chao et al. (Eds.): ICEBE 2019, LNDECT 41, pp. 161–172, 2020.
https://doi.org/10.1007/978-3-030-34986-8_11

(the breadth), and the number of variants to be kept for each product line (the depth) including different sizes and colours (stock keeping units or SKUs) [4, 5].

While there are many studies dealing with assortment planning, they commonly take the retailers' perspective into consideration for assortment selection and typically underestimate the importance of assortments identification. This can be a challenging process [2]. While it is important to present a variety of products to customers, customers confronted with endless options may feel overwhelmed and become confused and tired [3]. Therefore, potential customers may decide to defer the purchasing decision. Retailers may consider this choice overload conceptually in their planning but may not use any methodological approach to tackle that in practice. Assortment planning is crucial when considering customers cross-shopping (purchasing from multiple outlets) or discovery shopping (browsing without specific intent). While online retailers help their customers by providing them with search engine filters, the number of options may still be overwhelming for some customers.

In this work, we propose a multi-criteria decision making (MCDM) based approach, which semi-automatically ranks the product's assortment based on the customer's preference. We assume that the customer's way to evaluate the assortment is analytic rather than holistic or that the customer is open to change. People with an analytic thinking style see the product as a collection of components and features [4]. This assumption makes the choice of using a MCDM based approach reasonable through the consideration of weightings of importance for different components and features.

As with physical retailers transitioning from presenting products on the basis of the shop/business owner's preference, online stores have also focussed on the purchasers' preferences. This requires applying greater weighting to their customers' perspectives. As mentioned earlier, we acknowledge the existing literature on current online recommendation systems but their efforts target modifying and filtering the options from the search engine owner's perspective.

This paper aims to design a recommendation system taking into account the customer perspective to enrich the literature. The main aim of this proposed recommendation systems is to predict customer opinion on various items. This is performed using the utility function. The proposed recommendation system does not rank all the available items but instead recommends the most suitable items.

This paper is organised as follows. Section 2 presents a brief review of the most relevant work undertaken in this field. Section 3 discusses the methodology. After presenting the methodology, its application is examined. The results are then discussed, and direction for future studies are explained.

2   Literature Review

A retailer's assortment is the set of products it carries at a particular point in time [6]. The aim is to find an assortment that maximises the total income [7]. In physical stores, sellers are concerned with constraints such as limited shelf space [8].

Unlike this type of assortment planning, online assortment planning does not have to deal with such above-mentioned limitations [9]. Online stores are concerned with the priority in which the products are presented to customers [10]. While a variety of products need to be presented to the customer, overwhelming customers with too much variety may become confusing and/or encourage them to search for what they want elsewhere [11] or reduce purchase experience satisfaction. Therefore, we propose an optimised level of product variance should be attained. Amongst the growing literature regarding online product assortment, Li, Lu and Talebian [12] argued that online retailing has created a range of new challenges. They focussed on selecting the appropriate distribution channel for the retailer with regard to the given assortment, logistics, and consumer characteristics. Wang and Sahin [13] analysed the impact of customer enquiries about their potential choices before selecting a product and recommend the retailer with a methodology capable of finding the optimal product's final price. They asserted that the assortment problems with search cost are generally NP-hard, and proposed a k-quasi-attractiveness-ordered assortment and showed that it can be arbitrarily near optimal for the market share maximisation problem. Saure and Zeevi [14] focused on assortment planning problems where a customer purchasing decision is based on their utility and satisfaction, and the retailer must select subsets of products to present. In this case, the retailer observes the purchasing history of the customer and learns about customer decision making. Then, they develop a set of dynamic policies to balance the trade-off between exploration and exploitation and show limitation in possible improvement in a consumer's performance in a precise mathematical sense. Goyal, Levi and Segev [15] considered dynamic substitution with stochastic demands and provided near-optimal solutions to assortment planning for retailers to use. They showed that assortments with a relatively small number of product types could obtain almost all of the potential revenue. Pan and Honhon [16] focused on the case of a retailer managing a category of vertically differentiated products where the retailer has to pay for both the fixed cost for each assorted product and the variable cost per product sold. They showed that the optimal set of products depends on the distribution of customer valuations. They developed an efficient algorithm to identify an optimal assortment. Reviewing the literature, it becomes evident that researchers have analysed assortment problems and challenges from the retailer perspective.

Some retailers such as Amazon have worked on such presentation and use customisation, which is usually based on a customer's search history [12]. However, this is not sufficient for two reasons. Firstly, we believe that too much weight is still applied to what they (the retailers) want to sell rather than what suits customers. An alternative case, is where the online retailer could identify exactly what the customer wants but instead offers a 'near' perfect list in the hope of a secondary purchase upon a return visit by the customer. This case may involve the application of game theory and is excluded from the scope of this research [3].

This paper aims to address the current shortcoming of MCDM based techniques in handling a large number of alternatives, which is expected in the big data environment. The state-of-the-art methods in MCDM techniques did not address this important issue [17–22]. The next section of this paper will investigate how the Best Worst Method (BWM), which is one of most utilised techniques in decision making, can be extended for applying in the big data environment.

3 Methodology

In this section, we define some terms and list our assumptions, and then the structured method is presented.

a. Definitions and assumptions

The following main entities are associated with MCDM ranking models: decision maker, alternatives, and attributes (of alternatives).

- Decision maker: the decision maker is the individual, institution or group of individuals, who have the responsibility of making the decision. In this study, it is assumed that the decision maker is a prospective customer who is going to choose a suitable assortment of a given product.
- Alternatives: n this study, alternatives are an available assortment of a given product from which a decision maker may select.
- Attributes of alternatives: each alternative is characterised by a set of attributes. In this study, features such as brand, colour, size and the like are considered.

In MCDM based ranking models, two comparison matrices are constructed: (a) comparing features, and (b) comparing feature and alternatives. This process is either manual via the decision maker, or hybrid via mathematical techniques (Fig. 1).

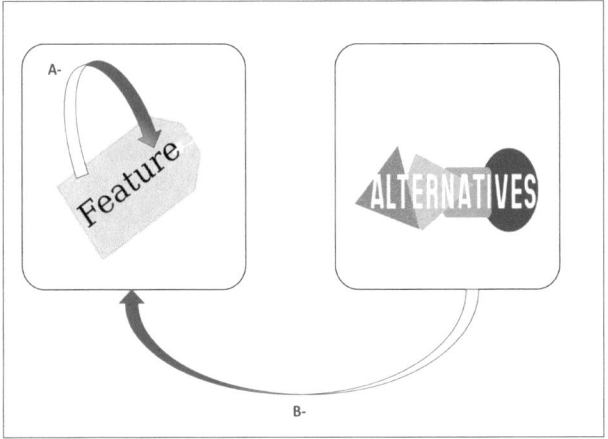

Fig. 1. Feature and alternatives

Assume that there are n alternatives and m features. The total number of comparisons for both cases A and B are at least: $\binom{m}{2} + \binom{n}{2}$. It can be discerned from Table 1 that by increasing the number of assortment alternatives, the number of preference comparisons exponentially increases. This shows that MCDM based ranking models cannot be applied to this environment without proper modification. In fact, it is unreasonable to expect decision makers to provide their preference on every alternative and answer many questions. It is assumed that a decision maker's preferences for each feature are independent of the values of other features. Despite the limitations of this assumption, applying MCDM techniques over a big data environment is a relatively unexplored research area.

Table 1. Fast growth of comparison matrix's size by increasing the size of alternatives

	10	20	50	100
$\binom{n}{2}$	45	190	1225	4950

b. Consumer Utility Functions

If the decision maker's preference can be extracted for each feature to determine the utility function, the large possible size of available alternatives would be able to be addressed. To this end, we consider the following cardinal utility functions which have been listed by Rezaei [23]: (1) Increasing, (2) Decreasing, (3) V-shape, (4) Inverted V-shape, (5) Increase-level, (6) Level-decrease, (7) Level-increase, (8) Decrease-level, and (9) Increasing-level-decreasing. Using cardinal utility function rather than the ordinal one allows the consideration of many alternatives. It is not possible to compare the utility of a pair of alternatives, which is the core concept in ordinal utility functions. Table 2 shows the eleven cardinal utility functions along with their parametric formula.

Table 2. Cardinal utility functions

	Name	Functional form	To be specified
1	Increasing	$u_{ij} = \begin{cases} \frac{x_{ij}-a_j^l}{a_j^u-a_j^l}, & a_j^l \leq x_{ij} \leq a_j^u \\ 0, & otherwise \end{cases}$	a_j^l, a_j^u
2	Decreasing	$u_{ij} = \begin{cases} \frac{a_j^u-x_{ij}}{a_j^u-a_j^l}, & a_j^l \leq x_{ij} \leq a_j^u \\ 0, & otherwise \end{cases}$	a_j^l, a_j^u
3	V-shape	$u_{ij} = \begin{cases} \frac{a_j^m-x_{ij}}{a_j^m-a_j^l}, & a_j^l \leq x_{ij} \leq a_j^m \\ \frac{a_j^m-x_{ij}}{a_j^m-a_j^l}, & a_j^m \leq x_{ij} \leq a_j^u \\ 0, & otherwise \end{cases}$	a_j^l, a_j^m, a_j^u
4	Inverted V-shape	$u_{ij} = \begin{cases} \frac{x_{ij}-a_j^l}{a_j^m-a_j^l}, & a_j^l \leq x_{ij} \leq a_j^m \\ \frac{a_j^u-x_{ij}}{a_j^u-a_j^m}, & a_j^m \leq x_{ij} \leq a_j^u \\ 0, & otherwise \end{cases}$	a_j^l, a_j^m, a_j^u
5	Increase-level	$u_{ij} = \begin{cases} \frac{x_{ij}-a_j^l}{a_j^m-a_j^l}, & a_j^l \leq x_{ij} \leq a_j^m \\ 1, & a_j^m \leq x_{ij} \leq a_j^u \\ 0, & otherwise \end{cases}$	a_j^l, a_j^m, a_j^u

(continued)

Table 2. (*continued*)

	Name	Functional form	To be specified
6	Level-decrease	$u_{ij} = \begin{cases} 1, & a_j^l \leq x_{ij} \leq a_j^m \\ \frac{a_j^u - x_{ij}}{a_j^u - a_j^m}, & a_j^m \leq x_{ij} \leq a_j^u \\ 0, & otherwise \end{cases}$	a_j^l, a_j^m, a_j^u
7	Level-increase	$u_{ij} = \begin{cases} 0, & a_j^l \leq x_{ij} \leq a_j^m \\ \frac{x_{ij} - a_j^m}{a_j^m - a_j^m}, & a_j^m \leq x_{ij} \leq a_j^u \\ 0, & otherwise \end{cases}$	a_j^l, a_j^m, a_j^u
8	Decrease-level	$u_{ij} = \begin{cases} \frac{a_j^m - x_{ij}}{a_j^m - a_j^l}, & a_j^l \leq x_{ij} \leq a_j^m \\ 0, & a_j^m \leq x_{ij} \leq a_j^u \\ 0, & otherwise \end{cases}$	a_j^l, a_j^m, a_j^u
9	Increasing-level-decreasing	$u_{ij} = \begin{cases} \frac{x_{ij} - a_j^l}{a_j^m - a_j^l}, & a_j^l \leq x_{ij} \leq a_j^{m1} \\ 1, & a_j^{m1} \leq x_{ij} \leq a_j^{m2} \\ \frac{a_j^u - x_{ij}}{a_j^u - a_j^m}, & a_j^{m2} \leq x_{ij} \leq a_j^u \\ 0, & otherwise \end{cases}$	$u_0, a_j^l, a_j^{m1}, a_j^{m2}, a_j^u$

Now it faces two sorts of uncertainty being the type of utility function and the parameters of the utility function for each decision maker. As utility functions should reflect a customer's preference for different attributes of a product, communication with customers or users plays a key role in getting the right utility function. However, customers may not prefer to spend time answering many questions. To avoid such issues, an interactive approach needs to be designed for locating the right utility and specifying its associated parameters.

We design an interactive approach to simultaneously address both of these types of uncertainty. This interactive approach is designed in order to specify a customer's utility function and its associated parameters by asking two simple questions obtaining that specific customer's opinion. For example, consider 'distance to work' as a criterion for the home renting MCDM problem. It is evident that for most people looking for a place to rent, less distance for travelling to work is an attractive attribute. Therefore, 'increasing' utility function can be removed from the list. However, as some customers may prefer some distance from their workplace for a variety of reasons, this can be achieved through choosing the 'Increasing-level-decreasing value' function. Thus, to get the right utility function, communication plays a key role.

c. Heuristic Algorithm for Users' Preference Elicitation

In this section, we propose two interactive questions in order to determine a user's preference elicitation. For the ease of explanation, these two questions are written for a 'rental decision-making case'. The first question has been written for a number of bedrooms in Fig. 2.

Question 1: What are the most satisfiable values for the number of bedrooms?
- Below … and Above …
- Below …
- Between … and …
- Above …

Fig. 2. First sample question

The decision maker selects one option and fills the associated blanks. After this, the second one depends on the user's response to the first question. For instance, if the user chooses the second option, the second question is written as follows:

Question 2: The number of bedrooms should not be above …?

Fig. 3. Second sample question

The first question finds the preferred range of a given feature of a product or service for a user and the second question finds the range that the user is uncomfortable with it. These customised questions enable efficient and effective communication with the user. We have designed the two-interactive-questions based algorithm by having four paths, which cover all the four cardinal utility functions. The figure depicts these four paths along with the associated cardinal utility function (Fig. 3).

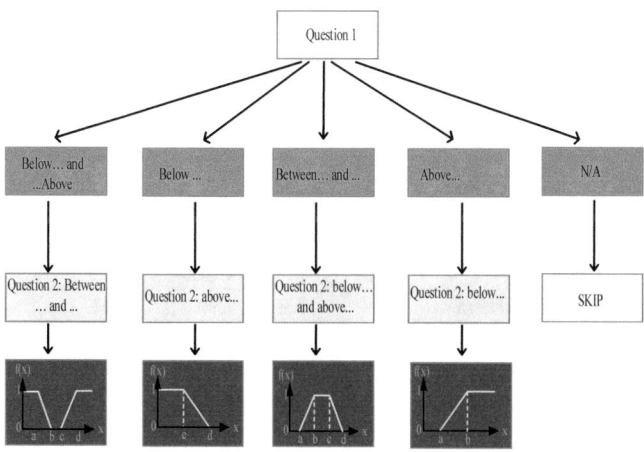

Fig. 4. Two-interactive-question based algorithm for user's preference elicitation

d. A Scalable Assortment Ranking Method

   This method is relying on three modules. The first module elicits the user preference, utility function, with respect to each feature using two-interactive-questions based
algorithm. The second module finds the weights of each feature using BWM to specify
array W. The last module completes the ranking process by specifying the matrix A by
using the specified utility function. This occurs by populating the utility function using
the available data set, which assigns the score (Fig. 4).

Algorithm.	
Input: User response to TNQ algorithm: *User*, User feedback on best, worst features, and associated scores: *PC*. **Output:** Overall score of alternatives: *Score*	
1:	$C \leftarrow assings\ a\ set\ of\ features$
2:	$Utility \leftarrow$ TNQ(User) (Module 1)
3:	$W \leftarrow$ BWM (PC) (Module 2)
4:	$A \leftarrow$ Populate(Utility) (Module 3)
5:	$Score_i \leftarrow \sum_{j=1}^{m} w_i * a_{ij}$ (final ranks)
6:	Return *Score*

4   Real Case Study

The rental decision-making process involves the customer's assessment before she
starts the negotiation with the owner. We have utilised the Allhomes.com.au website
for collecting the real data set. To find the process complexity, the following example
has been highlighted in Table 3 and Fig. 5.

Table 3. Assortment size example for 'rental decision'

Suburb name	#Options	Suburb name	#Options
Belconnen	236	North Canberra	335
Greater Queanbeyan	104	Central Coast	1066
Gungahlin	235	North Coast	1627
South Canberra	259	South Coast	418

   The following features are used for this decision making:

- Rent: F1
- Number of Bedrooms: F2
- Number of Bathrooms: F3
- Number of Parking: F4
- Distance to Work: F5

'Distance to Work' is selected as the best criterion while the 'Number of Bathrooms' is regarded as the worst criterion between the capabilities criteria. The preferences of a given decision maker for the best criterion over all the criteria and vice versa are listed in Tables 4 and 5, respectively.

Table 4. Best-to-Others vector

Criteria	Rent	#Bedroom	#Bathrooms	#Parking	Distance
Best criterion-Distance	2	4	8	3	1

Table 5. Others-to-Worst vector

Criteria	Worst criterion - #Bathrooms
Price	6
#Bedrooms	3
#Bathrooms	1
#Parking	2
Distance	8

An initial question has been customised for each feature.

Price:

Question 1: What is your most satisfactory range for the rent cost?

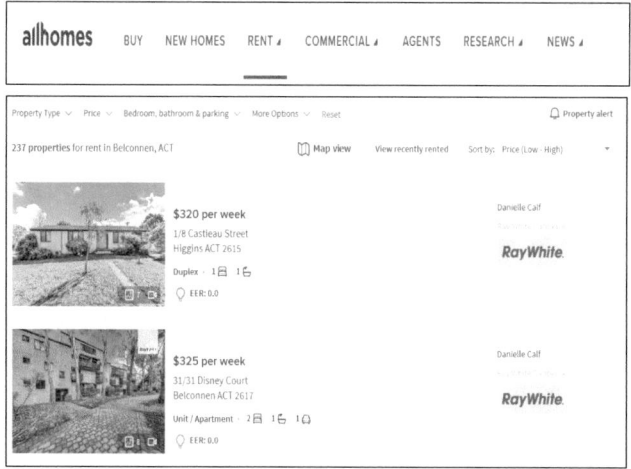

Fig. 5. Allhomes website

Number of Bedrooms

Question 1: What is the most satisfactory range for a number of bedrooms?

Number of Bathrooms

Question 1: What is the most satisfactory range for a number of bathrooms?

Number of Parking

Question 1: What is the most satisfactory range for a number of parking bays?

Distance to Work

Question 1: What is your most satisfactory range for the distance of the rental place to your workplace?

In the same way, Question 2 can be customised for the case of rental decision making. Table 6 and Fig. 6 show the preference of a given user for every five attributes by answering the two questions.

Table 6. A given user's preference for the five attributes

Criteria	Rent	#Bedroom	#Bathrooms	#Parking	Distance
Question 1	Between **300** and **350**	Between **1** and **1**	Between **1** and **2**	Between **1** and **1**	Between **0.2** and **0.3**
Question 2	Below **250** and above **400**	Below **0** and above **3**	Below **0** and above **2**	Below **0** and above **3**	Below **0.1** and above **0.5**

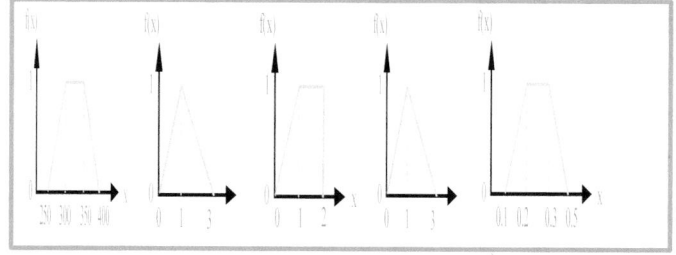

Fig. 6. Five attributes utility for a given user

By using a web scraping technique and Python language, the associated data for rental options in Canberra, Australia, based on allhomes.com.au was extracted [24]. The proposed method was able to rank all 1550 alternatives presented at the time.

5 Conclusion

While the focus of organisations in their everyday decision making has shifted from hum judgement towards being data-driven or AI-driven, there are areas in which the involvement of the human is essential. Due to this focus, the MCDM based approach in the era of big data is still in its infancy. To the best of our knowledge, this is the first work that aims to address the issue of scalability of MCDM techniques for use in the big data environment. To this end, BWM has been used which is one of the more recently proposed methods in the area of human-based decision making. In this paper, we showed how the new method could handle numerous alternatives, which ultimately makes the method scalable. The model has been tested in a 'renting decision making' using a real data set.

References

1. Saberi, Z., et al.: Stackelberg game-theoretic approach in joint pricing and assortment optimizing for small-scale online retailers: seller-buyer supply chain case. In: 2018 IEEE 32nd International Conference on Advanced Information Networking and Applications (AINA). IEEE (2018)
2. Saberi, Z., et al.: Online retailer assortment planning and managing under customer and supplier uncertainty effects using internal and external data. In: 2017 IEEE 14th International Conference on e-Business Engineering (ICEBE). IEEE (2017)
3. Saberi, Z., et al.: Stackelberg model based game theory approach for assortment and selling price planning for small scale online retailers. Future Gener. Comput. Syst. **100**, 1088–1102 (2019)
4. Hart, C., Rafiq, M.: The dimensions of assortment: a proposed hierarchy of assortment decision making. Int. Rev. Retail Distrib. Consum. Res. **16**(3), 333–351 (2006)
5. Flamand, T., et al.: Integrated assortment planning and store-wide shelf space allocation: an optimization-based approach. Omega **81**, 134–149 (2018)
6. Kök, A.G., Fisher, M.L., Vaidyanathan, R.: Assortment planning: review of literature and industry practice. In: Retail Supply Chain Management, pp. 175–236. Springer (2015)
7. Kunnumkal, S., Martínez-de-Albéniz, V.: Tractable approximations for assortment planning with product costs. Oper. Res. **67**, 436–452 (2019)
8. Flores, A., Berbeglia, G., Van Hentenryck, P.: Assortment optimization under the sequential multinomial logit model. Eur. J. Oper. Res. **273**(3), 1052–1064 (2019)
9. Kautish, P., Sharma, R.: Managing online product assortment and order fulfillment for superior e-tailing service experience: an empirical investigation. Asia Pac. J. Market. Logistics **4**, 1161–1192 (2019)
10. Melacini, M., et al.: E-fulfilment and distribution in omni-channel retailing: a systematic literature review. Int. J. Phys. Distrib. Logistics Manag. **48**(4), 391–414 (2018)
11. Argouslidis, P., et al.: Consumers' reactions to variety reduction in grocery stores: a freedom of choice perspective. Eur. J. Mark. **52**(9/10), 1931–1955 (2018)
12. Li, Z., Lu, Q., Talebian, M.: Online versus bricks-and-mortar retailing: a comparison of price, assortment and delivery time. Int. J. Prod. Res. **53**(13), 3823–3835 (2015)
13. Wang, R., Sahin, O.: The impact of consumer search cost on assortment planning and pricing. Manag. Sci. **64**(8), 3649–3666 (2017)
14. Sauré, D., Zeevi, A.: Optimal dynamic assortment planning with demand learning. Manuf. Serv. Oper. Manag. **15**(3), 387–404 (2013)
15. Goyal, V., Levi, R., Segev, D.: Near-optimal algorithms for the assortment planning problem under dynamic substitution and stochastic demand. Oper. Res. **64**(1), 219–235 (2016)
16. Pan, X.A., Honhon, D.: Assortment planning for vertically differentiated products. Prod. Oper. Manag. **21**(2), 253–275 (2012)
17. Azadeh, A., et al.: Z-AHP: A Z-number extension of fuzzy analytical hierarchy process. In: 2013 7th IEEE International Conference on Digital Ecosystems and Technologies (DEST). IEEE (2013)

18. Shokri, H., et al.: An integrated AHP-VIKOR methodology for Facility Layout design. Ind. Eng. Manag. Syst. **12**(4), 389–405 (2013)
19. Aboutorab, H., et al.: ZBWM: the Z-number extension of Best Worst Method and its application for supplier development. Expert Syst. Appl. **107**, 115–125 (2018)
20. Nawaz, F., et al.: An MCDM method for cloud service selection using a Markov chain and the best-worst method. Knowl.-Based Syst. **159**, 120–131 (2018)
21. Zhang, Y., et al.: Ranking scientific articles based on bibliometric networks with a weighting scheme. J. Informetr. **13**(2), 616–634 (2019)
22. Asadabadi, M.R., Chang, E., Saberi, M.: Are MCDM methods useful? A critical review of analytic hierarchy process (AHP) and analytic network process (ANP). Cogent Eng. **6**, 1623153 (2019). (just-accepted)
23. Rezaei, J.: Piecewise linear value functions for multi-criteria decision-making. Expert Syst. Appl. **98**, 43–56 (2018)
24. Mitchell, R.: Web Scraping with Python: Collecting More Data from the Modern Web. O'Reilly Media Inc, Sebastopol (2018)

Behaviorial-Based Network Flow Analyses for Anomaly Detection in Sequential Data Using Temporal Convolutional Networks

Wen-Hui Lin[1], Ping Wang[1(✉)], Bao-Hua Wu[1], Ming-Sheng Jhou[1], Kuo-Ming Chao[2], and Chi-Chun Lo[3]

[1] Department of Information Management,
Kun Shan University, Tainan, Taiwan
{linwh, pingwang}@mail.ksu.edu.tw, weq498aa@gmail.com,
s7920378@gmail.com
[2] Engineering and Computing, School of MIS, Coventry University,
Coventry, UK
k.chao@coventry.ac.uk
[3] Institute of Information Management, National Chiao Tung University,
Hsinchu, Taiwan
cclo@faculty.nctu.edu.tw

Abstract. In many applications the capability of Temporal Convolutional Networks (TCNs) on sequence modelling tasks has been confirmed to outperform classic approaches of recurrent neural networks (RNNs). Due to the lack of adequate network traffic flow analyses, anomaly-based approaches in intrusion detection systems are suffering from accurate deployment, analysis and evaluation. Accordingly, this study focused on network intrusion detection for DDoS threats using TCNs with network flow analyzer, CICFlowMeter-v4.0 to classify the network threats using behavior feature analyses. The experimental results reveal that that the prediction accuracy of intrusion detection goes up to 95.77% for model training with $N = 50,000$ for sizing (N) of samples using the IDS dataset CIC-IDS-2017.

Keywords: Intrusion detection · Temporal convolutional networks · Recurrent neural networks · DDoS · CICFlowMeter

1 Introduction

Recently, many proactive attacks continued to occur with the proliferation of automated attack tools. It is expected that future cyber warfare will become *Fool Automated Attacks and Defenses*. In fact, the individual defense system for security protection has been unable to respond promptly to proactive attacks from high-risk security concerns. The security protection for organizations has gradually moved towards a highly integrated platform using mechanical learning (MA) approaches with and cognitive computing capability to assist defenders reduce the complexity of security management and manpower cost. Specifically, it is achieved by enhancement of threat identification, interpretation and prediction accuracy from filtering and learning behavioral features

© Springer Nature Switzerland AG 2020
K.-M. Chao et al. (Eds.): ICEBE 2019, LNDECT 41, pp. 173–183, 2020.
https://doi.org/10.1007/978-3-030-34986-8_12

from collected information flows, to improve the efficiency of real-time security monitoring and awareness of network threats through predictive analysis and visualization.

Additionally, information security engineers tried to apply the deep learning networks (DLNs) with the MA approaches to secure the network management in an automated way by analyzing the statistical network behavior based threat detection results. Typically, the DLNs are categorized into four major categories, i.e., Deep Neural Network (DNNs), Convolutional Neural Networks (CNNs), Recurrent Neural Networks (RNNs) and Deep Belief Networks (DBNs). Basically, some of the deep learning models are specific to solve the specific problems but not well suitable for everything else, i.e., a single deep learning model often fails to deal with real complex problems effectively. To improve the classification accuracy of threat detection and reduce its false positive rate for NIDS, the present study proposes a behavior-based model for network anomaly detection by training a Temporal Convolutional Network (TCN) [1, 2] learning classifier. And then the feature vectors are transformed into the feature matrices to form the images as the inputs of the TCN to accurately categorize cyber threats according to collected behavioural features derived from analyses of network traffic, enabling the defence system to quickly respond to high-risk threats.

In model learning, network flow analyzer CICFlowMeter-v4.0 was used to filter the behavioral feature vectors of high frequency fields for information flows, and the feature vectors are formed into time sequential data as input for training TCNs, and obtain optimal weights of the hidden layers for class mapping. In implementing automated network anomaly detection, the NIDS can effectively classify threat categories by using image recognition capability of trained TCNs for prompt reacting to threats from different untrusted sources. To evaluate the effectiveness of TCN classifier, the cloud computing platform was used for simulating the distinct attack scenarios associated with the Top 10 threat in 2018 published in the Open Web Application Security Project (OWASP) on vulnerabilities of distinct operating systems as testing inputs. Especially, a benchmark IDS dataset, namely CIC-IDS-2017 collected from global threats are used for pre-training to learn the behavioral features of network flows in advance.

Most existing DLN-based approaches for threat detection involve RNNs to perform the feature extraction and learning of malicious activities from time series data. However, it might raise the vanishing gradient problem [3] in training artificial neural networks with gradient-based learning methods and backpropagation that results in problem of learning long-term dependencies (LTD). Compared to RNN, the TCN is good at modeling for processing long sequential data, especially in complex behavioral-based network flow analyses, where the RNNs might have a vanishing gradient problem.

In the mode, the behavioral feature vector of the high frequency is extracted by TCN from sequence data with parallel computing, and then the feature vectors are formed into multiple feature vectors and transferred to the Full Convolutional Network (FCN) with residual connections to perform class mapping in network anomaly detection scenarios.

The rest of this article is organized as follows. Section 2 reviews previous studies on TCNs. Section 3 presents the proposed approach for network intrusion detection model with TCN architecture in an online information security management system. The experimental results are presented in Sect. 4. Finally, Sect. 5 concludes the work.

2 Relate Work

Basic RNN architectures are notoriously difficult to train and more elaborate architectures for recurrent networks are commonly used instead, such as the LSTM (Long Short Term Memory) [4] the Gated recurrent unit (GRU) [5] and TCN [6] (Fig. 1). Compared to different types of recurrent units (LSTM vs. GRU), the TCN uses a combination of temporal networks (augmented with residual layers) and dilated convolutions by several fully connected or sparsely connected layers followed by a spatial pooling layer to determine the class of an input data, as shown in Fig. 2.

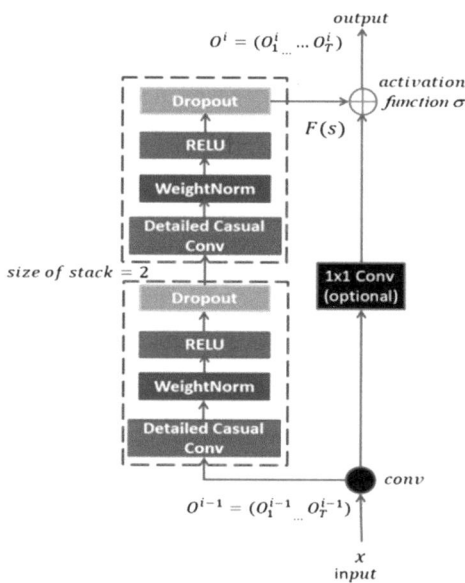

Fig. 1. Basic architecture of TCN [2]

As shown in Fig. 1, the TCN model is composed of one or more dilated casual convolutional layers with RELUs and an activation function. In [2], it reported that TCN used a simple convolutional architecture outperforms canonical recurrent networks such as the LSTM and the GRU across a diverse range of tasks and datasets, while demonstrating longer effective memory. To improve classification accuracy in real-time detection, the TCN was regularly selected. The distinguishing features of TCNs are listed as follows.

(i) Similar to those of RNN, the TCN architecture can input a time sequential data of any length and length of the output sequences is same as input.

(ii) The TCNs construct very long effective sizing of history data (i.e., the ability for the networks to look very far into the past to make a prediction) using a combination of very deep networks augmented with residual layers and dilated convolutions.

(iii) The convolutions in the TCN architecture are causal, meaning that there is no information leakage from future to past.

To accomplish the first point, the TCN uses a 1D fully-convolutional network (FCN) architecture, where each hidden layer is the same length as the input layer, and zero padding of length (kernel size − 1) is added to keep subsequent layers the same length as previous ones. To achieve the second point, the TCN uses causal convolutions, convolutions where an output at time t is convolved only with elements from time t and earlier in the previous layer [6].

$$TCN = 1D\ FCN + causal\ convolutions \qquad (1)$$

As shown in Fig. 1, the basic architecture can look back at history with size linear in the depth of the network and the filter size. Hence, capturing LTD becomes really challenging. One simple solution to overcome this problem is to use dilated convolutions as shown in Fig. 2.

In particular, TCNs can change flexible receptive field size for input sequential data and use relatively low memory requirement for training compared to other recurrent units. To improve the classification accuracy of threat detection and reduce its false positive rate for NIDS, this study developed an improved behaviour-based classifier learning model for network anomaly detection by training an TCNs with TensorFlow developed by Google to extract the enhanced behaviour features and identify the class of threats by using collected statistical data.

3   Network Intrusion Detection Model with Temporal Convolutional Networks

A behaviour-classification approach for network intrusion detection model with time sequential data as input is presented. Detailed workflow from suspicious network flows for behaviour classification using the revised Encoder-Decoder Temporal Convolutional Network (ED-TCN) model [7] is shown in Fig. 2; In Fig. 2 illustrates the revised ED-TCN model incorporating the encoding and the decoding design to extract different features of malicious network flows in the behaviour-classification process for NIDS. The former is encoder which is used for feature extraction and recording, and the latter is decoder which is used for feature classification.

In the training phase of the method, initially filter the time sequences data marked with time stamp from malicious network flows as process behavior. Then extract features from the trained TCN by using the flow analyzer with CICFlowMaker-v4.0 to generate feature images with normal and intrusive pattern for NIDS. In the validation phase, we evaluate the predicted accuracy of TCN model for overcoming over-training problem by using various n-folds of the cross-validation scheme. Firstly, obtain a

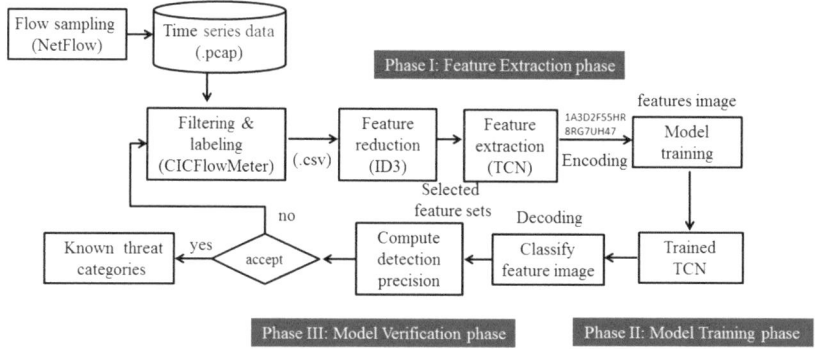

Fig. 2. Operation flow of network intrusion detection by using revised TCN model.

feature image by trained TCN with test data. Then predict the accuracy and false alarm rate using trained CNN model parameters. Figure 3 illustrates the proposed model incorporating three subphases in the behaviour-classification process for NIDS: (i) the feature extraction phase, (ii) the model training phase, and (iii) the model verification phase.

A. *Feature extraction phase.*

In data preprocessing phase, the training sample data were obtained from data source: (i) the archive dataset for behavioural features download from CIC-IDS-2017 [8] calculated for all benign and intrusive flows by using CICFlowMeter [9], 2017). Basically, CIC-IDS-2017 dataset contains total 80 features which cover all the 11 necessary criteria with common updated attacks such as DoS, DDoS, Brute Force, XSS, SQL Injection, Infiltration, Port scan and Botnet [8].

Step 1.1 Data analyses using CICFlowMaker

Firstly, a series of experiments were performed to filter 80 network traffic features from traffic flows by using the CICFlowMeter-v4.0, that automatically converts flow records (text) into .csv format for further processing. Basically, CICFlowMeter is a flow analyzer filtering crucial features from a pcap file, and label the specific packets where the flow label in this application includes SourceIP, SourcePort, DestinationIP, DestinationPort and Protocol (Fig. 3).

B. *Model learning phase.*

In model learning phase, the present study generate the feature images thru the use of TCN encoding-decoding process for model training from normal and intrusive flows and illustrated as shown in Fig. 3 and Table 1. Inspired by ED-TCN [7] and RNN Encoder-Decoder [11] designed by Google Brain, TCN encoding-decoding structure is designed as shown in Fig. 4.

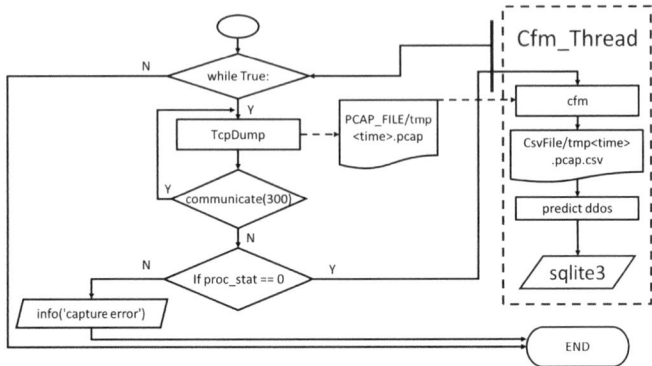

Fig. 3. Feature filtering process using CICFlowMeter and TcpDump

Step 1.2 Feature reduction

In the second step, to find the best feature set for detecting each attack from 80 extracted features, used Information Gain (IG)-based ID3 algorithm with scikit-learn library (Pedregosa et al. 2011) to calculate weight (relative importance) of each feature in the whole dataset. The ID3 algorithm begins with the original set S as the root node. In each iteration of the solution procedure, the algorithm will repeat through every unused attribute of the set S and calculates the information gain IG(S) of that attribute. It can effectively assist the defenders determine the required attribute which has the largest information gain value.

The attribute with the largest information gain is used to split the set S in the current iteration. Finally, the set S is then split by the selected attribute to produce data subsets. Let the information gain of attribute A be represented as $IG(S,A)$, which is the measure of the difference in entropy before and after the set S is split on an attribute A as [10]:

$$IG(S,A) = E(S) - \sum_i p(S_i)E(S_i) = E(S) - \sum_i \frac{|S_i|}{|S|} E(S_i), \qquad (2)$$

$$E(S) = -\sum_{i=1} p(S_i) \log_2 p(S_i), \qquad (3)$$

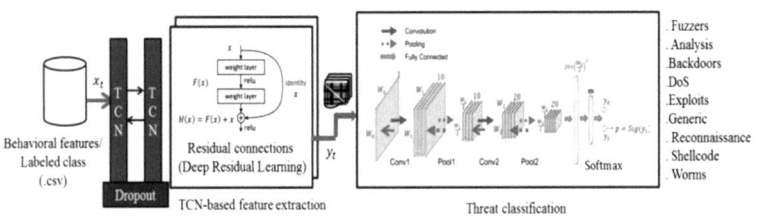

Fig. 4. Basic structure of ED-TCN for feature learning and class recognition

where entropy E(S) is a measure of the amount of uncertainty in the dataset S; S_i represents the subsets created from splitting set S by attribute A such that $S = U_i(s_i)$, $p(s_i)$ is the ratio of the number of elements in S_i ($|S_i|$) to the number of elements in set S ($|S|$), and $E(S_i)$ is the entropy of subset S_i. Equation (1) shows the uncertainty in S, which was reduced after splitting set S on attribute A.

As shown in Fig. 5, the proposed model also incorporates the residual connection structure to fine-tune the model parameters for features learning using error derivatives of back-propagation with the learning rate. That is, the classification is used to determine the learning error of multiple layer neural nets and then adjust the weights of neural nets to minimize it in the learning process of TCNs.

C. Model validation phase.

We adopt a cross-validation scheme to evaluate the predicted accuracy of TCN model for overcoming over-training problem by using various n-folds of the cross-validation scheme; for example, $k = 8$ means that 70% of the dataset collected was used in the training experiment, and the remaining 10% of the dataset was used for alternative testing repeated 8 times. In model validation phase, the system provides the benefit of quick-responding for threat classification thru the use of weights of neural nets using the trained TCNs in model learning phase.

4   Experimental Results

In this section, the performance of the proposed TCN-based intrusion detection model is demonstrated by means of a cyber security example for intrusion detection system which is shown in Fig. 5.

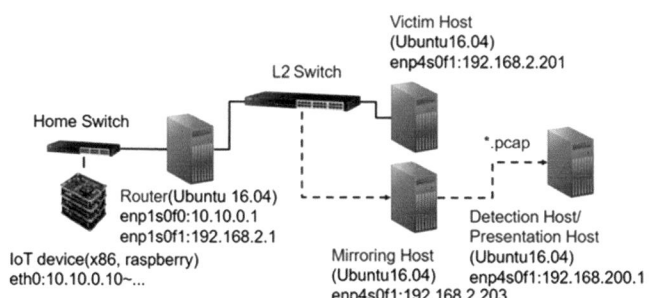

Fig. 5. Experiment environment of intrusion detection

As shown in Fig. 5, the experimental network environment using DDoS attacks is used to verify the performance of the developed system for real-time DDoS detection. In the environment, there are three major hosts including the attacker, the victim and the network detection associated with network switches. Basically, the attacker uses the Raspberry Pi 4 model to perform DDoS attacks on victim hosts using CIC-IDS-2017 dataset (Table 1). The attack network flows were received by mirroring host through the L2 Switch in.pcap format, and then transmits the pcap files to the.csv using CICFlowMaker for network intrusion detection.

In the following, conduct an experiment for classifying major intrusion threats by selecting DOS attack in training process, the statistical results of threat types for CIC-IDS-2017 for the following training and test are illustrated in Table 1. In Table 1, only six major types of threats were screened to be classified in the experiment. Basically, the CIC-IDS-2017 data set is typically classified into two categories: (i) DoS and (ii) non-DoS. The total number of 1,000,000 of dataset for TCN training and testing which are randomly selected as shown in Table 1.

Table 1. A statistical analysis for threat types in CIC-IDS-2017 dataset for model training and testing

Attacks	Records		
	No of records in CIC-IDS-2017	Training data	Testing data
Non-DoS	868,592	435,571	433,021
DoS Hulk	80,275	39,762	40,513
DoS slowloris	1,409	439	970
DoS Slowhttptest	1,573	611	962
DDoS	45,114	22,426	22,688
DoS GoldenEye	3,037	1,191	1,846

Step 1. Feature extraction phase

To examine the model efficiency, the first example incorporates a revised TCN model with CIC-IDS-2017 dataset. In the experiment, feature extraction involved a pre-process including two steps: (i) flow filtering and feature labeling, and (ii) feature reduction by IG and convert features to an image matrix.

Step 1.1. Data analyses using CICFlowMaker

This step is to preprocess using CICFlowMaker-v4.0 to filter and transfer the information flows by DDoS attacks for the experiment data, including (i) the symbol conversion of the network packets and (ii) normalization for numeric data, and (iii) convert features in time sequential data to feature vectors with feature labelling. The information capturing period started at 04:06 on Thursday, December 27, 2018 and continuously ran for an exact duration of 3 days, ending at 12:00 on Sunday December 30, 2018. DDoS attacks similar to those of CIC-IDS-2017 were subsequently executed and recording during this period. Partial records in the experiment were shown in Table 2.

Table 2. Data analyses results using CICFlowMaker

Records set for DDoS attacks	Records by Tcpdump (pcap)	Filtering records by CICFlowMeter (.csv)
303,844	2018-12-27 04:06:14.pcap (16,579,258)	2018-12-27 04:06:14.pcap_Flow.csv (3,030,068)
309,789	2018-12-27 04:10:14.pcap (14,502,914)	2018-12-27 04:10:14.pcap_Flow.csv (2,607,136)
291,604	2018-12-27 04:14:14.pcap (14,409,525)	2018-12-27 04:14:14.pcap_Flow.csv (2,600,716)

Step 1.2. Feature reduction

The features of the CIC-IDS-2017 dataset were ranked according to the score assigned by IG measure. To improve classification speed, a set of reduced feature was regularly selected using information gain (IG) scheme where the number of feature selected (16 at IG ≥ 0.0002) from top 80 ranked features are listed in Table 3.

Table 3. Ranked features using information gain

Feature title	Weighting
B.Packet Len Std	0.2028
B.Packet Len Min	0.0479
Flow Duration	0.0443
FIAT Min	0.0378
Flow IAT Min	0.0317
B.IAT Mean	0.03
F.IAT Mean	0.0265
Fwd IAT Min	0.0257
Active Min	0.0228
Flow IAT Std	0.0227
Active Min	0.0219
Flow IAT Mean	0.0214
Avg Packet Size	0.0162
Subflow F.Bytes	0.0007
Total Len F.Packets	0.0004
F.Packet Len Mean	0.0002

Step 2. Model learning phase

In the following, the TCN was trained to detect network intrusion using the trained weights of networks. Table 4 showed that the prediction accuracy for training is 94.25% with the loss value of 0.1559, while the verification loss value was 0.1121, and the verification accuracy was 95.77%.

Table 4. Training accuracy associated with loss value

Loss	Training accuracy (%)
0.1559	94.25%
0.1121	95.77%

Step 3. Model validation phase

The total number of 500,000 of other dataset (no overlap with training dataset) for TCN testing is randomly selected. This sampled validation set is excluded from the training data. The experimental results reveal that the prediction accuracy (%) by using the cross-validation method (k = 8) for testing is 92.38% with the loss value of 0.2924 with normal and intrusive flows.

An experiment for sensitivity analysis was conducted by only used DoS attacks, i.e., omitted the non-DoS attacks. Unfortunately, the precision rate of model will be decreased to 73.37% or less, if the numbers of samples for non-DoS type were initially screened out for CIC-IDS-2017 dataset that produces the learning bias compared to both samples of both the DoS and the non-DoS attacks.

5   Conclusion

This paper presents an intrusion detection model based on a TCN-based classifier for enhancing the precision of model with sequential data as input. Importantly, the proposed approach revises the ED-TCN model with the encoding-decoding architecture to extract and preserve behavioural features with traffic flow analyser that fine-tunes the model parameters and quick response to intrusion detection using Tensor flow. The proposed method improves the accuracy of intrusion detection for threat classification by using enhanced behaviour features from trained TCNs. Overall, the proposed approach can reach the precision with approaches proposed in [9] (approximately 96%) for behavioural-based detection approach.

Acknowledgement. This work was supported jointly by the Ministry of Science and Technology of Taiwan under Grant Nos. MOST 108-3114-E- 492-001 and MOST 108-2410–H -168-003.

References

1. Firat, O., Oztekin, I.: Learning deep temporal representations for brain decoding. In: First International Workshop, (MLMMI 2015), France (2015)
2. Bai, S., Kolter, J.Z., Koltun, V.: An Empirical Evaluation of Generic Convolutional and Recurrent Networks for Sequence Modeling (2018). arXiv:1803.01271
3. Gupta, D.: Fundamentals of Deep Learning – Introduction to Recurrent Neural Networks. Analytics Vidhya (2107)
4. Hochreiter, S.: Long-short term memory. Neural Comput. **9**(8), 1735–1780 (1997)
5. Cho, K., Merrienboer, B., Gulcehre, C., Bahdanau, D., Bougares, F., Schwenk, H., Bengio, Y.: Learning Phrase Representations using RNN Encoder-Decoder for Statistical Machine Translation (2014). arXiv:1406.1078
6. Yu, F., Koltun, V.: Multi-Scale Context Aggregation by Dilated Convolutions (2015). arXiv: 1511.07122
7. Lea, C., Flynn, M.D., Vidal, R., Reiter, A., Hager, G.D.: Temporal convolutional networks for action segmentation and detection. In: Proceedings - 30th IEEE Conference on Computer Vision and Pattern Recognition, CVPR 2017, pp. 1003–1012 (2017)
8. University of New Brunswick, CIC-IDS2017 on AWS (2017). https://www.unb.ca/cic/datasets/ids-2017.html
9. Sharafaldin, I., Lashkari, A., Ghorbani, A.: Toward generating a new intrusion detection dataset and intrusion traffic characterization. In: Proceedings of the 4th International Conference on Information Systems Security and Privacy (ICISSP 2018), pp. 108–116 (2018)
10. Kotsiantis, S.B.: Supervised machine learning: a review of classification techniques. Informatica **31**, 249–268 (2007)
11. Cho, K., Merrienboer, B., Gulcehre, C., Bahdanau, D., Bougares, F., Schwenk, H., Bengio, Y.: Learning Phrase Representations using RNN Encoder-Decoder for Statistical Machine Translation (2014). arXiv:1406.1078I

Software Engineering for e-Business

Software Requirements Prioritization with the Goal-Oriented Requirement Language

Abderrahmane Leshob[1,2]([⊠]), Pierre Hadaya[1], and Laurent Renard[1]

[1] LATECE Laboratory, University of Quebec at Montreal, Montreal, Canada
[2] UQAM School of Management (ESG UQAM), Montreal, Canada
leshob.abderrahmane@uqam.ca

Abstract. Requirements prioritization is an important activity of the software development process. It is performed to rank information systems (IS) requirements in the order in which they will be implemented to maximize the value delivered to customers. A number of methods and techniques have been proposed to help organizations prioritize IS requirements. However, they all suffer from a number of limitations and their implementation are mostly informal. This work aims to design a novel method that automates the requirements prioritization process. The proposed method uses the Goal-oriented Requirement Language (GRL) that permits the: (i) modeling of requirements and linking them to the business objectives/goals, and (ii) evaluation of the impact of requirement choices on business objectives/goals. The method attaches contribution and preference values to GRL edges (e.g., contribution links) and GRL nodes (e.g., goals and solutions) to compute the score of IS requirements. The attached values can be based on any set of business and IT factors (e.g., value for the customer, cost of implementation, risk). This paper also presents the principles underlying the proposed IS requirements prioritization method and discusses its possible implementation in practice.

1 Introduction

Requirements prioritization is an important activity that is performed in early phases of the software development process. It consists of managing the relative importance of different requirements to maximize the value delivered to customers. Requirements prioritization enables organizations to focus on the most important requirements by ranking them in their order of implementation releases [1]. Effective and accurate requirements prioritization ensures that the most critical requirements are addressed immediately in case the organization faces budget or time constraints. For Berander and Andrews [2], requirements prioritization plays a pivotal role in Information System (IS) planning. For Achimigu *et al.* [1], requirements prioritization is a major step taken in making crucial decisions to increase the economic value of IS. For Mulla and Girase [3], requirement prioritization helps organizations optimize Information Technology

© Springer Nature Switzerland AG 2020
K.-M. Chao et al. (Eds.): ICEBE 2019, LNDECT 41, pp. 187–198, 2020.
https://doi.org/10.1007/978-3-030-34986-8_13

(IT) investments by allowing them to invest money, time and resources on the most important features.

Although prioritization plays a crucial role in the requirement engineering process, existing prioritization techniques suffer from a number of limitations. Amongst these, we note their complexity, lack of scalability and of coordination among stakeholders and requirements dependency issues [1].

This paper proposes a method for IS requirements prioritization. The proposed method uses the Goal-oriented Requirement Language (GRL) [4] that permits the: (i) modeling of requirements and linking them to the business objectives/goals, and (ii) evaluation of the impact of requirement choices on business objectives/goals. This paper extends the work presented in [5] by enabling organizations to prioritize IS requirements using any set of business and IT factors (e.g., value for the customer, value for the organization, cost of implementation, risk, and difficulty of implementation). We argue that by using GRL, a simple and graphical notation, and by supporting any set of factors, we are able to design a method that is generic (can be applied to different types of IS projects regardless of the business domain), usable, flexible, and adaptable according to the specific needs of the organizations.

To attain our research objective, we followed the first steps of the design science methodology as proposed by Peffers *et al.* [6] to design the conceptual ingredients of our artifact. Future work will complete the last steps of the methodology: demonstration and evaluation. This study establishes the guidelines that will enable us to advance towards the long-term goal of our research program: automating the requirements prioritization process.

The remainder of the paper is organized as follows. Section 2 describes the proposed requirements prioritization method. Section 3 surveys related work. Section 4 draws our conclusions and summarizes our future work in the area.

2   The Artifact: A Method to Prioritize IS Requirements

This work focuses on the design and development of a new method that automates the prioritization of IS requirements. Our goal is to build an easy-to-use method that is flexible (e.g., easy to add new prioritization factors), adaptable (i.e., can be adapted according to the specific needs of the organizations), and can be applied to different types of IS projects, regardless of the organization's business domain.

To design the prioritization method, we needed a language that (i) is easy to use by stakeholders who are not software requirements specialists (e.g., business analysts, product owners), (ii) links the requirements to business goals/objectives, and (iii) allows users to easily apply prioritization factors (e.g., value, cost, risk) to evaluate the impact of requirements choices on business goals/objectives. Among the various goal-oriented languages available, including KAOS (Keep All Objects Satisfied) [7], i^* [8] and the NFR Framework [9], we found that the Goal-oriented Requirement Language (GRL) that is part of the User Requirements Notation (URN) standard [4], is the most suitable language to design our method.

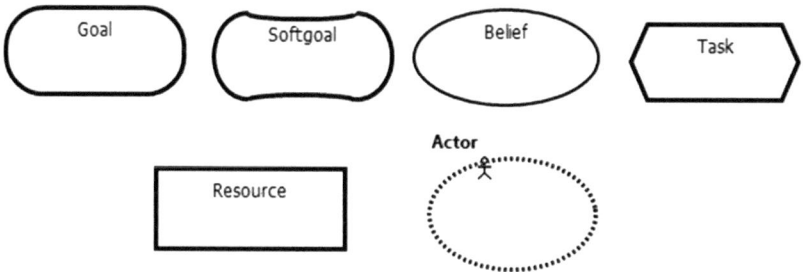

Fig. 1. GRL intentional elements (adapted from [10]).

2.1 Goal-Oriented Requirements Language

GRL is a goal-oriented modeling language. It is part of the User Requirements Notation (URN) standard [4]. GRL makes it possible to explicitly model the objectives, requirements, alternatives and their relationships. GRL also provides support for evaluations to analyze appropriate trade-offs between divergent stakeholders goals, allowing stakeholders to observe and understand, for example, why some requirements should be prioritized.

The basic elements of the GRL language are shown in Fig. 1. Section (a) of the figure presents the intentional elements of GRL, such as goals, softgoals, tasks, and resources. A goal (or hard goal) is quantifiable while a soft goal refers to qualitative aspects that cannot be measured directly (e.g., customer satisfaction) [10]. Unlike in the concept of (hard) goal, there are no clear-cut criteria for whether the softgoal is achieved. Softgoals are usually related to *Non-Functional Requirements* (NFR), while goals are related to *Functional Requirements* (FR). Tasks are solutions to goals or softgoals [10]. Beliefs are used to represent design rationale. An actor represents a stakeholder of the system or another system [10].

GRL links and GRL contributions types are shown in Fig. 2. The section (a) of the figure presents the GRL links, such as the decomposition, contribution, correlation or dependency links [10]. These links are used to connect GRL elements, such as goals, softgoals, tasks, and resources. An intentional element can be decomposed into sub-elements using decomposition links. These links support the logical operators And, Or, and Xor [10]. The correlation links show side effects between the intentional elements. GRL uses the *Means-End* links to describe how goals are achieved. Each task provided is an alternative means for achieving the goal. Graphically, a Means-End link connects an end node with the means node achieving it. In GRL, only goals are originally applicable to means-ends link. The relationships between actors are illustrated using dependency links.

The section (b) of the figure presents the contribution types used to model the qualitative or quantitative impacts an element has on another. These impacts are propagated through the contribution links presented in section (c) [10]. These contribution links can be labeled using only icons, texts, or numbers.

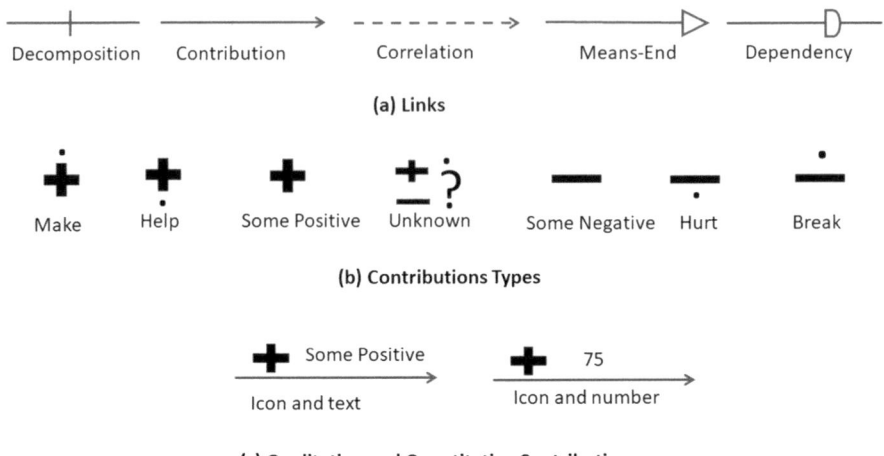

Fig. 2. GRL links and contributions types (adapted from [10]).

2.2 The Prioritization Process

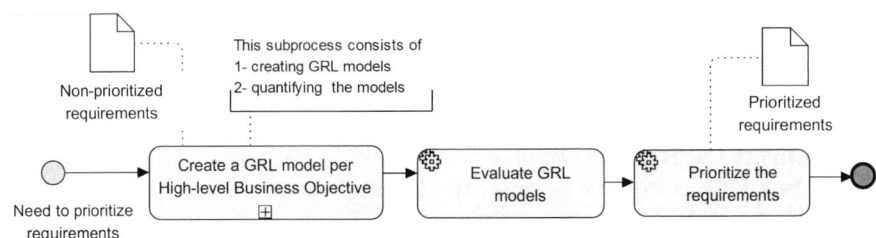

Fig. 3. Requirements prioritization process.

We propose a tree-step process to prioritize IS requirements (see Fig. 3). The first step builds a GRL model (goal graph) based on higher business objectives. During this step, the user creates a GRL goal model per high-level business objective and then attach quantitative values to GRL edges and nodes. This model shows the high-level business objectives/goals, functional and non-functional requirements, and the alternatives for achieving these high-level objectives. The second step computes the score of each functional and non-functional requirement by evaluating GRL models using the quantitative values attached to the models during the previous step. The third and last step ranks the requirements according to the score obtained in the previous step. As shown in the BPMN model of Fig. 3, the first step requires the intervention of the user (e.g., business analyst or product owner). The last two steps can be fully automated.

The following subsections detail the process in the context of an e-commerce application. Note that the prioritization process may be performed by a group of stakeholders (e.g., business analysts).

Create a GRL Model by Higher Business Objective: The goal of the first step is to create a GRL model per High-level Business Objective (HBO) that the system must meet. An HBO (e.g., Efficient payment, System security, High performance) is represented by a GRL softgoal. An HBO may have multiple subgoals. For example, in a B2C e-commerce application, *Efficient payment* softgoal may have three subgoals: *Secure Payment, Reduce payment fees*, and *Reduce payment delay* (see Fig. 4).

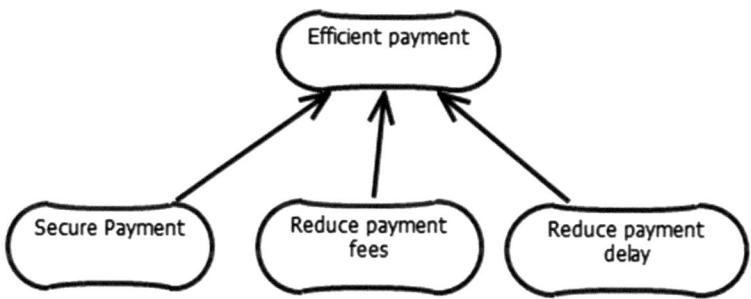

Fig. 4. Example of the efficient payment HBO in a B2C e-commerce application.

The resulting GRL models link the requirements, business objectives and solutions (Tasks) that achieve the goals (means-end) or satisfy softgoals (contribution). These models also allow stakeholders to elaborate the requirements, detect possible conflicts and reach a common understanding of the requirements and the prioritization criteria being used. Figure 5 shows an example of a GRL model for the *Efficient payment* HBO according to the use case diagram of Fig. 6.

After creating the GRL model, the user must quantify it by assigning initial values to the intentional elements (goals, softgoals and solutions) and contribution links. To do this, the user adopts the following strategy:

- By default, assign an importance value of 1 (neutral value) to the leaf GRL solutions (Tasks). An importance attribute is used when evaluating strategies for the goal model. The importance is shown between parentheses in intentional elements (see [10]). This does not impose any constraint on the importance values types or ranges. The user can use non-integer values from any range he/she wants. An importance value of 1 does not represent a limit of our method even if we can think about other values that can be assigned to the solutions, such as the implementation costs and the technical risks associated with implementing them. In doing so (i.e., assigning an importance value

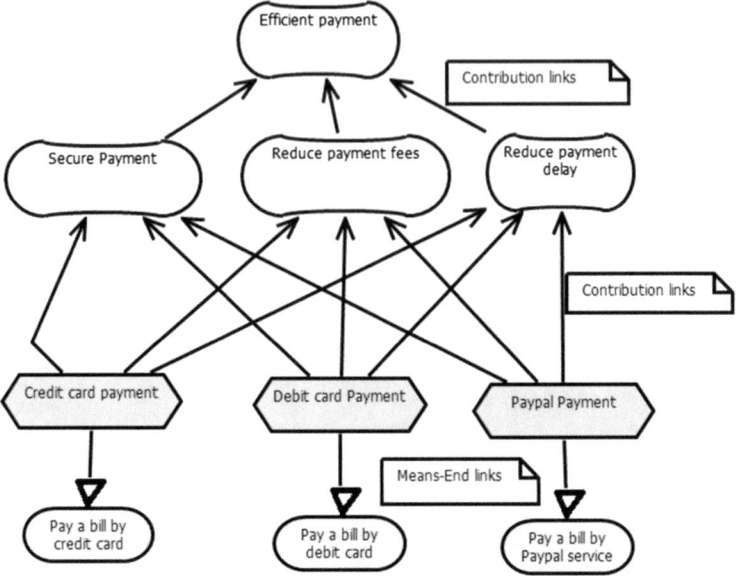

Fig. 5. GRL model for the *Efficient payment* HBO.

of 1 to the leaf GRL solutions), the method can apply any prioritization formula (e.g., Wiegers prioritization formula [11]) when assigning quantitative
values to contribution links in the next step. Furthermore, the value of 1
allows our method to reuse the quantitative evaluation algorithm proposed
in [10]. More precisely, we adapted slightly the button-up propagation and
the calculation of actors satisfaction to compute requirements score.

– Assign importance values to goals and softgoals of the GRL models. The user
can use any prioritization factor when assigning values to the importance
attribute. The importance values can be based on a single value representing
the importance of the goals and softgoals to the customer. The importance
values can also be based on two values. For example, the first value may
represent the importance of the goal to the customer while the second represents the importance of the goal to the organization. In this case, the user
can use different formulas depending on the importance of each factor (i.e.,
importance of the goal to the customer and the importance of the goal to the
organization). For example, let's say the value representing the importance of
the goal to the customer is VALUE1 and the value representing the importance
of the goal to the organization is VALUE2. The importance value of the goal
(IMPORTANCEVALUE(GOAL)) may be computed as follows:

$$value1 + value2$$

$$\text{or } \frac{(value1 + value2)}{2}$$

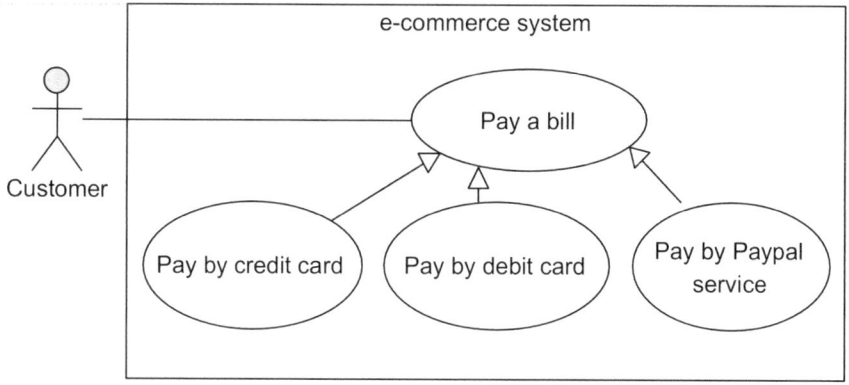

Fig. 6. Use case diagram for an online payment.

if the importance of the goal to the customer and the organization are weighted equally,

$$\text{or } \frac{(f1 \times value1 + f2 \times value2)}{f1 + f2}$$

where f1 and f2 are the weights of the importance of the goal to the customer and organization respectively.

- Assign values to the contribution[1] links between solutions and softgoals according to the used prioritization factors. By default, a single value attached to the contribution link is sufficient. The user may choose to use more factors such as the cost and the technical risk associated with implementing the solution. In this case, the value associated to the contribution link ($contribValue(contribution)$) is computed as follows:

$$contribValue(contribution) = \frac{value}{cost + risk}$$

The user may also decide to use different weights if the prioritization factors (e.g., value, cost, and risk) are not weighted equally. In that case,

$$contribValue(contribution) = \frac{f1 \times value}{(f2 \times cost) + (f3 \times risk)}$$

- Assign values to the contribution links between softgoals. The value of the contribution indicates the desired impact of one softgoal on a parent softgoal.

Figure 7 shows an example of quantified GRL model for the *Efficient payment* HBO.

---

[1] Correlations are treated the same way as contributions.

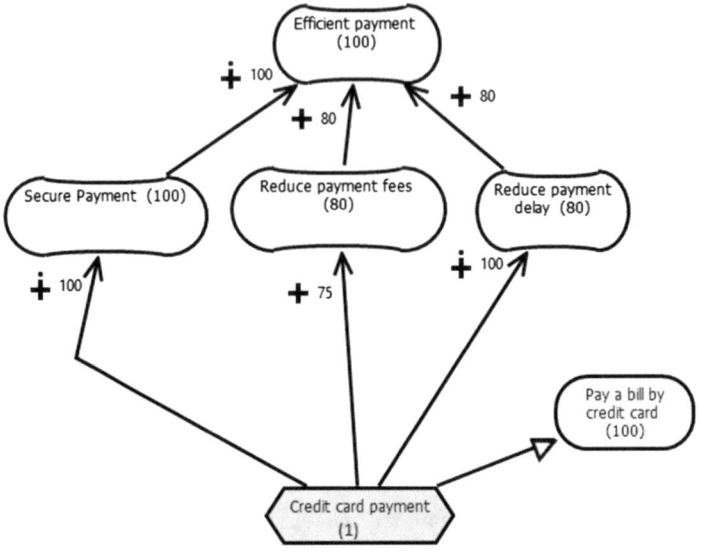

Fig. 7. The quantified GRL model for the *Efficient payment* HBO.

Evaluate GRL Models: To evaluate GRL models, we use an adaptation of the quantitative evaluation algorithm described in [10]. We compute the score of each requirement based on:

1. The quantitative values of the contribution links between the solutions and softgoals,
2. The quantitative values of the contribution links between the softgoals,
3. The quantitative importance value of the intentional elements.

The algorithm starts by propagating values using a bottom-up approach to obtain *evaluation values* for intentional elements (see [10]). By default, we assign an evaluation value of 1 to leaf solutions. The evaluation values of the solutions will be propagated to goals through Means-end links. Figure 8 shows the GRL model after propagating the values. For example, the evaluation value of the softgoal *Reduce payment fees* is computed by multiplying the evaluation value of the solution *Credit card payment* (i.e., 1) by the value of the contribution link that connects them together (i.e., 75).

The quantitative evaluation value of a softgoal ($SgoalEvalValue(softgoal)$) reached by N contribution links is the sum of the products of the evaluation value of each source element ($EvalValueSourceElement_i$) by its contribution value to the element ($ContribValueElement_i$). More precisely, the evaluation value of a softgoal ($softgoal$) is computed as follows.

$$SgoalEvalValue(softgoal) = \sum_{i=1}^{N} EvalValueSourceElement_i \times ContribValueElement_i$$

For example, the quantitative evaluation value of the softgoal *Efficient payment* is the sum of the products of the evaluation value of each source element (i.e., softgoals Secure payment, Reduce payment fees, and Reduce payment delays) by its contribution value to the softgoal *Efficient payment* element. Therefore, the evaluation value of the *Efficient payment* softgoal is $((100 \times 100) + (75 \times 80) + (100 \times 80))$. The same strategy is applied to GRL decomposition links.

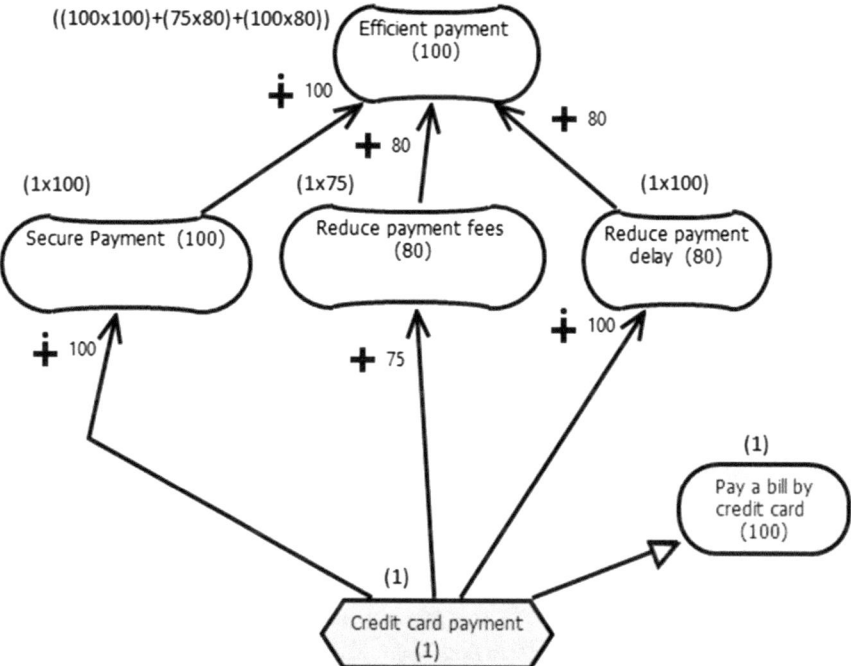

Fig. 8. The GRL model for the *Efficient payment* HBO after propagating values.

Let's call the propagated value to intentional elements *evaluation value*. Then, we compute the score of each requirement using the higher score of its solution(s). Let's say that there is N intentional elements from a solution element(*solution*) to an HBO. Each intentional element ($Element_i$), has its evaluation value ($EvalValue_i$) computed by propagation and importance value ($ImportanceValue_i$) set by the user. Therefore, the score of a solution SCORE(SOLUTION) is computed as follows.

$$Score(solution) = \frac{\sum_{i=1}^{N} EvalValue_i \times ImportanceValue_i}{N}$$

Prioritize the Requirements: The goal of the last step is to prioritize the requirements using the scores obtained during the GRL evaluation performed in the previous step. Thus, to prioritize the requirements, the user must simply rank them from the highest to lowest scores.

3   Related Work

A number of approaches have been proposed in the past to prioritize software requirements. In [12], Karlsson and Ryan proposed a cost-value approach based on the value of each requirement and its relative cost of implementation. Their method is based on the Analytic Hierarchy Process (AHP), a well-established analytical technique. In [11], Wiegers proposed an eight-step process to prioritize requirements using four factors: (1) the relative benefit that each feature provides to the customer or the business, (2) the relative penalty the customer or business would suffer if the requirement is not included, (3) the relative cost of implementing each requirement, and (4) the risk associated with each requirement. In turn, Azar *et al.* proposed in [13] a process called Value-Oriented Prioritization (VOP) that ranks requirements according to their contributions to five business factors (Sales, Marketing, Competitive, Strategic, and Customer retention) relative to technical and business risks.

In agile software projects, the prioritization process is performed by actively involving the client (client-centric) [14–16]. Essentially, the requirements prioritization process is based mostly on *business value* (i.e., value-driven). For [14,15], the main requirement prioritization criterion in agile software development is the business value of the requirement/story in the product and the requirement prioritization process is *continuous*. According to [14,17], the continuous prioritization of requirements during agile projects plays a central role in achieving value creation while the *Must have, Should have, Could have* and *Won't have* (MoSCoW) prioritization technique is particularly useful in such contexts [18,19]. MoSCoW provides a way to reach a common understanding on the relative importance of each requirement/story in the product. It classifies requirements/stories into four broad categories that designate the overall level of priority of requirements. *Must have* are critical to the current product, *Should have* are important but not necessary for the product, *Could have* are desirable but not necessary, and finally, *Won't have* are not appropriate for the product.

4   Conclusion and Future Work

In this paper, we propose a generic method to prioritize software requirements using the GRL language. GRL is a graphical and simple language that permits the: (i) modeling of requirements and linking them to the business objectives/goals, and (ii) evaluation of the impact of requirement choices on business objectives/goals. The proposed method is *flexible* because it can easily be extended to include any new *factor* or *technique* to prioritize requirements.

It is also *adaptable* as it can be adapted to fit the specific needs of any organizations. For example, an organization may decide to use a home-made evaluation algorithm to compute requirements scores.

This study establishes the guidelines that will enable us to advance towards the long-term goal of our research program: automating the requirements prioritization process. Much work remains to be done, both in regards to the core *functionality* of the method dealing with factors evaluation/computation (e.g., assisting the user in the process of computing the cost of a solution) and the technical dependencies that exist between requirements, and in regards to the method *usability* before we can make it into a tool that business analysts will readily use. In this respect, the next challenges we face are to: (i) build an end-to-end *configurable* tool that supports the method (e.g., so that we can configure the tool to use specific techniques, such as the AHP technique, a cost estimation technique or other prioritization factors, such as the penalty the customer or business would suffer if the feature is not included), (ii) elaborate a precise and simple *cost-value prioritization method* where the cost to implement the requirements uses a software-based measurement method/model, (iii) take into consideration different types of dependencies between requirements (e.g., technical dependencies), and (iv) conduct experiments to validate the usability and the soundness of the proposed method.

Acknowledgments. This research was supported by the Natural Sciences and Engineering Research Council of Canada (NSERC).

References

1. Achimugu, P., Selamat, A., Ibrahim, R., Mahrin, M.N.: A systematic literature review of software requirements prioritization research. Inf. Softw. Technol. **56**(6), 568–585 (2014)
2. Berander, P., Andrews, A.: Requirements prioritization. In: Engineering and Managing Software Requirements (2005)
3. Mulla, N., Girase, S.: Comparison of various elicitation techniques and requirement prioritisation techniques. Int. J. Eng. Res. Technol. IJERT **1**(3), 1–8 (2012)
4. ITU-T, User Requirements Notation (URN)-Language definition (2012)
5. Barbouch, M.C., Leshob, A., Hadaya, P., Renard, L.: A goal-oriented method for requirements prioritization. In: The Ninth International Conference on Business Intelligence and Technology BUSTECH, pp. 15–20, May 2019
6. Peffers, K., Tuunanen, T., Rothenberger, M.A., Chatterjee, S.: A design science research methodology for information systems research. J. Manage. Inf. Syst. **24**(3), 45–77 (2007)
7. Van Lamsweerde, A.: Goal-oriented requirements engineering: a roundtrip from research to practice. In: 12th IEEE International Requirements Engineering Conference (RE04), pp. 4–7 (2004)
8. Yu, E.: Towards modelling and reasoning support for early-phase requirements engineering (2002)
9. Chung, L., Nixon, B.A., Yu, E., Mylopoulos, J.: Non-Functional Requirements in Software Engineering (2011)

10. Amyot, D., Ghanavati, S., Horkoff, J., Mussbacher, G., Peyton, L., Yu, E.: Evaluating goal models within the goal-oriented requirement language. Int. J. Intell. Syst. **25**(8), 841–877 (2010)
11. Wiegers, K.: First things first: prioritizing requirements. Softw. Dev. (1999)
12. Karlsson, J., Ryan, K.: A cost-value approach for prioritizing requirements. IEEE Softw. (1997)
13. Azar, J., Smith, R.K., Cordes, D.: Value-oriented requirements prioritization in a small development organization. IEEE Softw. (2007)
14. Daneva, M., Van Der Veen, E., Amrit, C., Ghaisas, S., Sikkel, K., Kumar, R., Ajmeri, N., Ramteerthkar, U., Wieringa, R.: Agile requirements prioritization in large-scale outsourced system projects: an empiricalstudy, J. Syst. Softw. (2013)
15. Racheva. Z., Daneva, M., Herrmann, A.: A conceptual model of client-driven agile requirements prioritization: results of a case study. In: ESEM 2010 - Proceedings of the 2010 ACM-IEEE International Symposium on Empirical Software Engineering and Measurement (2010)
16. Racheva, Z., Daneva, M., Herrmann, A., Wieringa, R.J.: A conceptual model and process for client-driven agile requirements prioritization. In: 2010 Fourth International Conference on Research Challenges in Information Science (RCIS) (2010)
17. Ramesh,B., Cao, L., Baskerville, R.: Agile requirements engineering practices and challenges: an empirical study. Inf. Syst. J. (2010)
18. Craddock, A., Roberts, B., Richards, K., Godwin, J., Tudor, D.: The DSDM agile project framework for scrum. In: Dynamic Systems Development Method (DSDM) Consortium (2012)
19. IIBA, BABOK. A guide to the business analysis body of knowledge® (2015)

Theoretical Aspects of Music Definition Language and Music Manipulation Language

Hanchao Li[1][✉] ⓘ, Xiang Fei[1][✉] ⓘ, Ming Yang[1][✉] ⓘ,
Kuo-ming Chao[1][✉] ⓘ, and Chaobo He[2][✉] ⓘ

[1] The Faculty of EEC, Coventry University, Coventry, UK
lih30@uni.coventry.ac.uk,
{aa5861,ab2032,csx240}@coventry.ac.uk
[2] School of Information Science and Technology,
ZhongKai University of Agriculture and Engineering, Guangzhou, China
hechaobo@foxmail.com

Abstract. As there is an increase in the amount of on-line music and digital albums, the way of express the music has become more and more important nowadays, so it can be used in several Music Information Retrieval tasks. For example, the music search engine and the plagiarism detection tool. From the existing symbolic approaches, there are a lot of music notation and relevant theories so the musician can understand and follow easily. Thus, we need to build a theoretical system for our proposed coding scheme, naming Music Definition Language and Music Manipulation Language. Therefore, this paper is focused on some of the important theories that derived from MDL and MML. After the theoretical analysis and practical discussion, we have showed the relationship between the proposed coding scheme and the existing audio or symbolic formats. These proved the feasibility of using the new coding scheme for those MIR tasks.

Keywords: Audio · Music Definition Language · Music Manipulation Language · Theoretical aspects · Symbolic

1 Introduction

Music Information Retrieval (MIR) systems discover useful information from music pieces. Hence, there are various tasks for an MIR system can do. For example, pattern recognition, genre detection, mood evaluation, instrumentation, music recommendation, plagiarism detection tool based on similarity, and music search engines. These MIR systems can be either tag-based or content-based. Tag-based systems are normally based on artist, album or the music title, whereas typical content-based systems can categorized into either audio or symbolic approach [1, 2].

For the audio-based MIR systems, the input is normally the audio file. Thus, the audio fingerprint method is often used, from the audio fingerprint catalogue. This approach can be used for tasks involving genre, lyrics, mood and timber. As each music track corresponds to a unique fingerprint, it is ideal for exact matching. Hence, it would be difficult to find the original music from its variations or with background

© Springer Nature Switzerland AG 2020
K.-M. Chao et al. (Eds.): ICEBE 2019, LNDECT 41, pp. 199–207, 2020.
https://doi.org/10.1007/978-3-030-34986-8_14

noise. Finally, the longer the audio length, the larger the fingerprint, the longer the recognition time for an on-line audio MIR systems [3–5].

For the symbolic-based MIR systems, the key step is how to code music using various symbols. MIDI files, for example, can easily be stored and manipulated compared to the audio files, as they take less space by just storing the musical events and parameters. As a result of this, an on-line symbolic MIR systems have short processing time and high retrieval accuracy. Most importantly, we are able to recognize similar melody patterns and carry out the exact or approximate music content-based search for different musical features. However, it is hard to deal with lyrics and only suitable for music-sheet-convertible music tracks [6–9].

Moreover, there exist a mapping between the audio format and the symbolic format, as we can generate audio sounds by playback the symbolic file, as well as symbolic feature extraction from the analogue using signal processing [10–14].

Therefore, we designed an alternate coding scheme, naming Music Definition Language (MDL) and Music Manipulation Language (MML), so the advantages from the audio and symbolic approaches can be retained, and those shortcomings from both approaches can be addressed theoretically. In this paper, we are focused on the discussion to the theories behind the MDL and MML.

2   Definition of MDL and MML

In this section, we define the MDL and MML with examples.

2.1   MDL and MML

Music Definition Language, MDL, is a collection of vector sequences which represents the basic stream flow of the musical melody. There are two basic form of MDL.

The first form is the symbolic-based MDL, S-MDL. It has designed to have the similar concepts as the standard MIDI format and the music notations. Thus, S-MDL should include these musical information:

- The position of the musical note in a modulo 12 system N_{mod12}.
- The set (index) number of the position of the musical note, S_I, such that the middle C (261.626 Hz) lies in the set '0'.
- The relative dynamic value, A_d, also known as the relative amplitude.
- The bar number, B. In terms of the bar number from the music sheet, this will distinguish the musical notes that appears in different bars.
- The beat number, B_T. This refers to the specific beat time, the musical note is played inside the bar number.
- The duration of the note, D. This is a relative time value which relate to the time signature.

As a conclusion, the first two features together defines the note we should played, followed by the relative dynamic value. The next two identifies the beginning time of the note. The final feature represents how long the note should be hold. Therefore, we have the following mathematical definition for the symbolic based MDL.

Definition 1. S-MDL: *Symbolic-based Music Definition Language (for a music melody) is a stream of vector sequence which describe the (symbolic) flow of the music. Each 6 × 1 vector has stored the main six-tuple features of single musical note in the melody sequence, which relates to the music notations. Hence, S-MDL can be expressed as:*

$$
S\text{-}MDL_{Melody} := (S\text{-}MDL_{Note})_i := \left(\begin{pmatrix} N_{mod12} \\ S_I \\ A_d \\ B \\ B_T \\ D \end{pmatrix} \right)_i
\tag{1}
$$

where $N_{mod12} \in \{0, 1, \ldots, 11\}$, $S_I \in \mathbb{Z}$, $A_d, B_T, D \in \mathbb{Q}^+$, $B \in \mathbb{N}$ *and* $i \in \mathbb{N}^+$.

Theoretically, a rest is in the form of $\begin{pmatrix} N_{mod12} \\ S_I \end{pmatrix} = \begin{pmatrix} 0 \\ -\infty \end{pmatrix}$.

The second form is audio-based MDL, A-MDL, which is an extension of S-MDL that allows the MDL to describe the flow of the melody when facing much complicated accoustic sounds. Thus, A-MDL should include these information:

- The fundamental frequency of the sound, f_p. Each musical note has its own fundamental frequency (pitch), thus allowing it to extend from the discrete frequency set.
- The amplitude of the sound, A. Unlike S-MDL, is regardless of the dynamic notations. However, this can vary due to the volume, hence we treat $A = k * A_d$, for some constant k.
- The beginning time of the sound, t. By default, it will last until the next A-MDL. Thus, we do not need to code the duration nor the ending time.

Therefore, we have the following mathematical definition for the audio based MDL.

Definition 2. A-MDL: *Audio-based Music Definition Language (for a music melody) is a stream of vector sequence which describes the flow of the music in acoustic waveform. Each 3 × 1 vector has stored the main three-tuple features from the sinusoidal wave. Thus, A-MDL can be expressed as follows:*

$$
A\text{-}MDL_{Melody} := (A\text{-}MDL_{Note})_i := \left(\begin{pmatrix} f_p \\ A \\ t \end{pmatrix} \right)_i
\tag{2}
$$

where $f_p, A, t \in \{0\} \cup \mathbb{R}^+$ *and* $i \in \mathbb{N}^+$.

On the other hand, MML is known as Music Manipulation Language. It is used to describe how different artists perform the MDL in more technical details, either in the collection of topological vector form or as a mapping function. The topological vector form is defined as Topological Music Manipulation Language, T-MML. Every MDL should follow at least one T-MML with similar format. Therefore, there are two forms of T-MML, one for A-MDL and one for S-MDL.

Definition 3. T-MML: *Topological Music Manipulation Language is a sequence of vectors to describe how, topologically, the notes were played under the defined MDL interval (A-MDL or S-MDL). Thus, T-MML can be expressed as:*

$$
\left\{
\begin{array}{l}
T\text{-}MML_{A\text{-}MDL} := \left(\begin{pmatrix} \Delta f_p \\ \Delta A \\ \Delta t \end{pmatrix} \right)_j \\[2em]
T\text{-}MML_{S\text{-}MDL} := \left(\begin{pmatrix} \Delta N_{mod12} \\ \Delta S_I \\ \Delta A_d \\ \Delta B \\ \Delta B_T \\ \Delta D \end{pmatrix} \right)_j
\end{array}
\right. , \forall j \geq 1
\tag{3}
$$

where 'Δ' indicates the continuous change between the start and the end of the MDL intervals.

Finally, when MML is acting as a function, it is defined as O-MML.

Definition 4. O-MML: *Operational Music Manipulation Language is a 'meaningful' function, either in matrix, vector or the combination form, such that it can be applied to certain consecutive MDL vectors.*

2.2   Examples of MDL and MML

Using track [1] as an example, Fig. 1 shows the music sheet and its corresponding S-MDL and A-MDL.

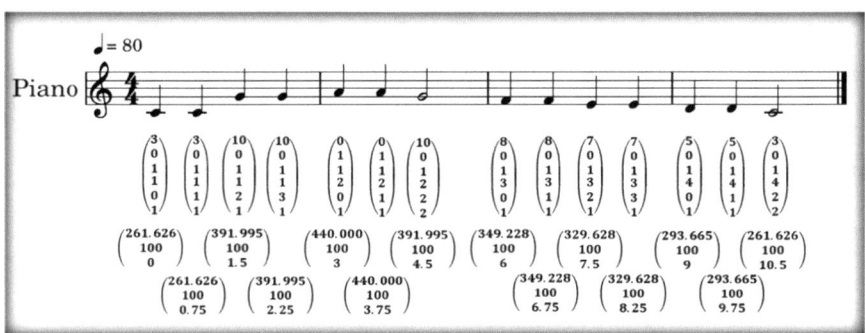

Fig. 1. Music sheet and S-MDL/A-MDL for the music track [1].

Using track [2] as an example, shows an example of A-MDL and the corresponding T-MML, $T\text{-}MML_{A\text{-}MDL}$. For clarity, the vector without any signs is the A-MDL and the vector where the three entries all having the + or – sign is the T-MML.

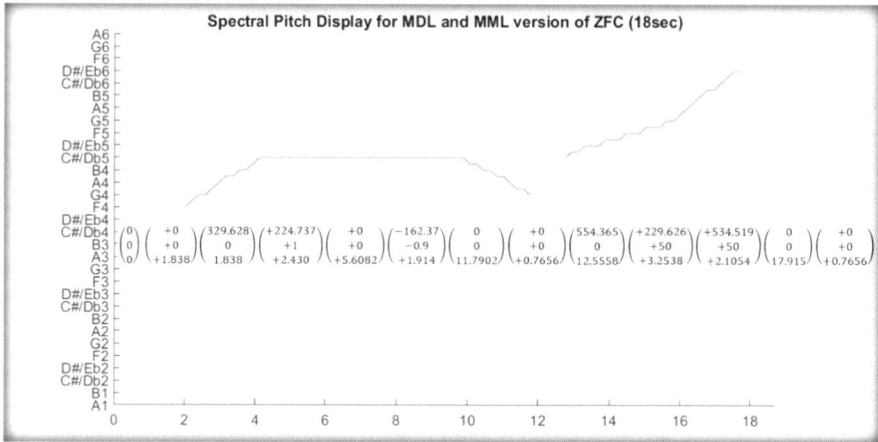

Fig. 2. A-MDL and T-MML for the siren-sound from the intro of [2] and the corresponding spectral pitch display.

3   Theoretical Aspect of MDL and MML

In this section, we discuss some typical features for the MDL and MML.

3.1   Mathematical Lemma and the Theoretical Proof

From the relationship between the symbolic coding and the audio, we can derive the following lemma:

Lemma 1. The A-MDL and S-MDL Conversion for Monophonic Melody: *Given the the standard amplitude, tempo measurement and the time signature, every S-MDL for a monophonic melody can be converted into A-MDL. The opposite direction holds iff* $\log_2\left(\frac{f_p}{440}\right) \in \mathbb{Z}$ *&* $t \in \mathbb{Q}$.

Proof: Firstly, Eqs. (4) and (5) shows the conversion between the N_{mod12} and S_I from S-MDL and the f_p from A-MDL. Since $S_I \in \mathbb{Z}$ by definition, then the condition we need is $\log_2\left(\frac{f_p}{440}\right) \in \mathbb{Z}$.

Secondly, $A = k * A_d$ for some non-zero constant k, can convert the relative amplitude into the real amplitude with respect to the standard amplitude, and vice versa.

Finally, Eqs. (6) and (7) shows the conversion between the B, B_T and D from S-MDL and the t from A-MDL, given the duration of a standard note from the tempo measurement (metroneme marks) and the time signature. More specifically, x beat per minute means $60/x$ s for one relative duration length, $\frac{1}{l}$, and define $D = 1$ for one standard duration length, $\frac{1}{\beta}$. The time signature is given in the form of $\frac{\alpha}{\beta}$. Note both β and l were in powers of 2. Thus, for perfect beated music, the condition we need is $t \in \mathbb{Q} \subseteq \mathbb{R}$.

$$f_p = 440 * 2^{\left(\frac{12(S_I - 1) + N_{mod12}}{12}\right)} \tag{4}$$

$$\begin{cases} N_{mod12} = \left(12 \log_2\left(\frac{f_p}{440}\right)\right) mod\, 12 \\ S_I = \frac{12 \log_2\left(\frac{f_p}{440}\right) - N_{mod12}}{12} + 1 \end{cases} \tag{5}$$

$$\begin{cases} t = \left(\frac{60}{x}\right) * \left(\frac{l}{\beta}\right) * ((B - 1) * \alpha + B_T) \\ or \\ t_i = \left(\frac{60}{x}\right) * \left(\frac{l}{\beta}\right) * \sum_{j=0}^{i-1} D_j, \forall i \in \mathbb{N}^+ \, (mono.) \end{cases} \tag{6}$$

$$\begin{cases} B = \frac{t * \left(\frac{x}{60}\right) * \left(\frac{\beta}{l}\right) - B_T}{\alpha} + 1 \\ B_T = \left(t * \left(\frac{x}{60}\right) * \left(\frac{\beta}{l}\right)\right) mod\, \alpha \\ D_i = (t_{i+1} - t_i) * \left(\frac{x}{60}\right) * \left(\frac{\beta}{l}\right), \forall i \in \mathbb{N}^+ \, (mono.) \end{cases} \tag{7}$$

where $x, l, \alpha, \beta \in \mathbb{N}^+$ and $D_0 = 0$.   ∎

This lemma can be extended to polyphonic music. Thus, we have:

Lemma 2. The A-MDL and S-MDL Conversion for Polyphonic Music: *Given the the standard amplitude, tempo measurement and the time signature, every S-MDL can be converted into A-MDL. The opposite direction holds iff* $\log_2\left(\frac{f_p}{440}\right) \in \mathbb{Z} \,\&\, t \in \mathbb{Q}$.

Proof: Result follows once we break down the polyphonic music into multi-channelled monophonic melodies with a lot of 'dummy' rests.   ∎

From these two lemmas, we have the following theorem:

Theorem 1. A-S Theorem: *S-MDL is a subset of A-MDL in terms of music descriptivity.*

The proof is from those conversion conditions, as $\log_2\left(\frac{f_p}{440}\right) \in \mathbb{Z} \subseteq \mathbb{R}$ and $t \in \mathbb{Q} \subseteq \mathbb{R}$.   ∎

We also can derive the following lemma, as O-MML is defined to be a function that applied to a number of MDLs:

Lemma 3. The Extended A-MDL and S-MDL Conversion Lemma: *Given the the standard amplitude, tempo measurement and the time signature, there exists a non-linear O-MML such that every S-MDL melody can be converted into A-MDL melody. The opposite direction holds iff* $\log_2\left(\frac{f_p}{440}\right) \in \mathbb{Z} \,\&\, t \in \mathbb{Q}$.

This means depends on the type of musical variation, we can convert the original music track into a correct MDL format, such that it will be benefit for the similarity calculation, partially or fully.

Moreover, we have some of the typical lemma with O-MML:

Lemma 4. The C Key Transformation Lemma: *There exists an O-MML which allows the music in any key to change into the C key.*

Lemma 5. The Set Transformation Lemma: *There exists an O-MML which allows the music to change tone while staying in the same key.*

Lemma 6. The Volume Tuning Lemma: *There exists an O-MML which allows the music to change the dynamic/volume.*

Lemma 7. The Tempo Tuning Lemma: *There exists an O-MML which allows the music to change the tempo.*

Lemmas 4, 5, and 6 relate to the S-MDL whereas Lemmas 6 and 7 relates to the A-MDL.

Combining Lemmas 3, 4, 5, 6 and 7, the similarity calculation of an on-line MIR system will down to comparing different O-MML functions and the core MDL files.

3.2   Practical Discussion

Back to the A-S Theorem, A-MDL is based on the sinusoidal wave, which means it is close to the audio. On the other hand, S-MDL is based on the existing symbolic music notations. Therefore, this theorem proved one statement mentioned in the introduction: "it (symbolic-based) is hard to deal with lyrics and only suitable for music-sheet-convertible music tracks". However, there is a need to review the relationship between the A-MDL, S-MDL and the audio, MIDI respectively.

Using music track [1] as an example, we built a proof-of-concept prototype using MatLab to illustrate the relationship between S-MDL and the MIDI file. The algorithm has outlined in Algorithm 1.

Algorithm 1: MIDI MDL Conversion
Input: MIDI file.
Output: Similarity scores
1. Read MIDI file, manually create the A-MDL /S-MDL file and parameters initialization (e.g. the threshold frequency, ϑ).
2. Fetch the acoustic value.
3. Extract relevant features using signal processing
4. Convert into A-MDL file.
5. Convert into S-MDL file using Equation (5) & (7). Apply rounding when necessary.
6. Extract the melody line, $m_1\,m_2$, and the rhythm line $r_1\,r_2$ for the manually and automatically generated (A-/S-)MDL file, respectively.
7. $Sim_C = \frac{|A|}{l} * e^{-MSD} = \frac{|A|}{l} * e^{-(|A|+|B|-|C|)^{-1}*\sum_{t=1}^{l}(m_2(t)-m_1(t))^2}$ where $$\begin{cases} A = \{i\,|\,|m_1(i) - m_2(i)| \le \vartheta,\ i \in \{1, \dots, l\}\} \\ B = \{i\,|\,|m_1(i) - m_2(i)| > \vartheta, i \in \{1, \dots, l\}\} \\ C = \{i\,|\,|m_1(i) - m_2(i)| = 0, i \in \{1, \dots, l\}\} \end{cases}$$
8. $Sim_R = \frac{|R|}{l}$ where $R = \{i\,|\,|r_1(i) - r_2(i)| = 0\}$
9. Output Sim_C & Sim_R.

The practical result shows that the melody accuracy for A-MDL is 85.93% when set $\vartheta = 1.0\,\mathrm{Hz}$, 100% for S-MDL and for both, the rhythmic accuracy is 100%. We are assuming the manually created MDL file perfectly matches the MIDI file, as we manually insert the relevant feature values into the correct position of the MDL file.

On the other hand, Fig. 2 and track [2a] illustrate that the A-MDL & T-MML is able to code the sound wave considers continuous frequency set, which is hard for MIDI to code that easily. This indicates the feasibility of applying the coding scheme to the modern electronic music. As we only using 5 A-MDLs and 8 T-MMLs to code in that example, despite the size is larger than MIDI, but is still a lot less than audio files, whereas the flow of the melody is maintained. This shows the feasibility of using the MDL and MML format at the MIR systems. For example, we can speed up the various content-based music search engine with high accuracy and user satisfaction.

4 Conclusion

As a summary, we have defined the MDL and MML in this paper, proved some mathematical theories and demonstrated that the performance of the proposed coding scheme lies in between the audio and symbolic methods. These results will be benefit the on-line MIR systems in the future, as in [15], we use MDL and MML to identify original music from its variation, based on SOM and various similarity measurements.

Acknowledgment. This paper involves the melodies from the following music track(s):

1. Mozart, W. A. (1806) Twinkle, Twinkle, Little Star (a.k.a. Twelve Variations On "Ah Vous Dirai-Je, Maman", K.265).

2. BASTARZ (Block B) (Lee, M., Kim, Y. and Pyo, J.) (2015) 품행제로 *(Zero For Conduct).*

2a. D.J.S 137 (covers Bastarz of Block B) (2015) *BASTARZ (Block b) - Zero For Conduct* 품행제로 *(Piano Tutorial) [Sheets + MIDI].*

This work was partially supported by the following projects:

The Science and Technology Support Program of Guangdong Province of China under Grant 2017A040405057, and in part by the Science and Technology Support Program of Guangzhou City of China under Grant 201807010043 and Grant 201803020033.

References

1. Casey, M., Veltkamp, R., Goto, M., Leman, M., Rhodes, C., Slaney, M.: Content-based music information retrieval: current directions and future challenges. Proc. IEEE **96**(4), 668–696 (2008)
2. Lamere, P.: Social tagging and music information retrieval. J. New Music Res. **37**(2), 101–114 (2008)
3. Typke, R., Wiering, F., Veltkamp, R.: A survey of music information retrieval systems. In: Proceedings of 6th International Conference on Music Information Retrieval. pp. 153–160 (2005)

4. Duong, N., Duong, H.: A review of audio features and statistical models exploited for voice pattern design. In: Seventh International Conferences on Pervasive Patterns and Applications (PATTERNS 2015, Nice, France) (2015)

5. Fan, Y., Feng, S.: A music identification system based on audio fingerprint. In: 2016 4th International Conference on Applied Computing and Information Technology/3rd International Conference on Computational Science/Intelligence and Applied Informatics/1st International Conference on Big Data, Cloud Computing, Data Science & Engineering (ACIT-CSII-BCD), pp. 363–367 (2016)

6. Lubiw, A., Tanur, L.: Pattern matching in polyphonic music as a weighted geometric translation problem. In: Proceedings 5th International Conference on Music Information Retrieval, pp. 289–296 (2004)

7. McKay, C., Fujinaga, I.: Jsymbolic: A feature extractor for MIDI files. In: Proceedings of International Computer Music Conference, pp. 302–305 (2006)

8. Huang, T., Xia, G., Ma, Y., Dannenberg, R., Faloutsos, C.: MidiFind: fast and effective similarity searching in large MIDI databases. In: Proceedings of the 10th International Symposium on Computer Music Multidisciplinary Research, pp. 209–224 (2013)

9. Saxena, A., Jasola, S., Barik, K., Joshi, A., Dubey, A., Jauhari, M.: Unit-9: the computer system: hardware for educational computing. In: Block-3: Introduction to Computers in Education, pp. 5–18 (2018)

10. Dixon, S.: On the computer recognition of solo piano music. In: Proceedings of Australasian computer music conference, pp. 31–37 (2000)

11. Chen, C., Jang, J., Liu, W., Weng, C.: An efficient method for polyphonic audio-to-score alignment using onset detection and constant Q transform. In: 2016 IEEE International Conference on Acoustics, Speech And Signal Processing (ICASSP), pp. 2802–2806 (2016)

12. Cogliati, A., Temperley, D., Duan, Z.: Transcribing human piano performances into music notation. In: Proceedings of 17th International Conference on Music Information Retrieval, pp. 758–764 (2016)

13. Fournier-S'niehotta, R., Rigaux, P., Travers, N.: Is there a data model in music notation? In: International Conference on Technologies for Music Notation and Representation (TENOR 2016), pp. 85–91 (2016)

14. Yang, N., Usman, M., He, X., Jan, M., Zhang, L.: Time-frequency filter bank: a simple approach for audio and music separation. IEEE Access **5**, 27114–27125 (2017)

15. Li, H., Fei, X., Chao, K., Yang, M., He, C.: Towards a hybrid deep-learning method for music classification and similarity measurement. In: 2016 IEEE 13th International Conference on e-Business Engineering (ICEBE), pp. 9–16 (2016)

Evaluation of Varying Visual Intensity and Position of a Recommendation in a Recommending Interface Towards Reducing Habituation and Improving Sales

Piotr Sulikowski[✉]

Faculty of Computer Science and Information Technology, West Pomeranian University of Technology, ul. Zolnierska 49, 71-210 Szczecin, Poland
psulikowski@wi.zut.edu.pl

Abstract. The abundance of advertising in e-commerce results in limited user attention to marketing-related content on websites. As far as recommender systems are concerned, presenting recommendation items in a particular manner becomes equally relevant as the underlying product selection algorithms. To enhance content presentation effectiveness, marketers experiment with layout and visual intensity of website elements. The presented research investigates those aspects for a recommending interface. It uses a quantitative research methodology involving gaze tracking for implicit monitoring of human-website interaction in an experiment instrumented for a simple-structure recommending interface. The experimental results are discussed from the perspective of the attention attracted by recommended items in various areas of the website and with varying intensity, while the main goal is to provide advice on the most viable solutions.

Keywords: E-commerce · Recommendation system · Recommending interface · Eye tracking

1 Introduction

Rapid development of electronic commerce justifies the need to investigate online sales enhancing solutions, in particular recommender systems. Their aim is to substitute salesmen present in a traditional shopping environment. Whereas in a physical shopping environment, a salesperson may recommend products thanks to direct communication, in an online environment it is the recommending interface that promotes items interesting to the customer. While convenience of online shopping is an attractive benefit for customers, it is often perceived as lacking personal touch or too complicated, especially when a customer is to choose products from a wide selection. Therefore, user experience including personalization and recommending interfaces has a vital role in e-commerce website design. Recommendation systems play an increasingly important role in purchase decisions and they often prove successful in improving sales [1].

© Springer Nature Switzerland AG 2020
K.-M. Chao et al. (Eds.): ICEBE 2019, LNDECT 41, pp. 208–218, 2020.
https://doi.org/10.1007/978-3-030-34986-8_15

In such a system a user model is often created – a description of a user, created or selected by a system to ease interactions between the system and the user [2]. A digital representation of a user model is a user profile, which reflects their preferences, online behavior and transactions [3]. Online systems collect a wide spectrum of user data [4–7] necessary to build those profiles and recommend products optimal in terms of fit and resulting sales. A lot of effort has been made to discover user preferences from that stream of data [8, 9]. Early solutions were based on collaborative filtering algorithms [10] and were extended towards explanation interfaces [11] with the use of context [12] and other techniques [13], including social media data input [14].

The effectiveness of a recommender system, however, depends on factors that go beyond the recommendation algorithm itself [15]. In comparison with vast research in the field of such algorithms, there is less research in the area of the next steps of the recommending process, including the ways products should optimally be presented to the customer. Those steps are based on human-computer interaction with recommender interfaces, which can be analyzed with DOM-events-based solutions [4] or eye tracking. The number of recommendations, the descriptions and associated images, as well as layouts etc. can be analyzed in order to optimize the interface [16]. In the era of customers inundated with information, in particular marketing content, the habituation effect often comes into place resulting in the banner blindness phenomenon. Thus, even recommendations which are best from the algorithm perspective may have little effect unless they are wisely presented in the right part of a website, at the right moment of the decision process, and with the right level of content intensity [17–20].

This paper focuses on a human-website interaction study conducted with the help of eye tracking. The general aim of the study was to evaluate visual intensity levels and positions of recommendations in a simple-structure recommending interface, while the attention drawn and the resulting product purchase interest were being examined. The results are presented and discussed in the paper and they may inspire to perform similar experiments on other e-commerce sites in order to improve layout and content presentation of recommending elements. The manner of presentation of data delivered by recommending interfaces cannot be overstated.

The remainder of the article is structured as follows: Assumptions for the study and its conceptual framework are presented in Sect. 2. The structure of the experiment and empirical results are provided in Sect. 3 and conclusions are presented in Sect. 4.

This work was supported by the National Science Centre of Poland (Decision No. 2017/27/B/HS4/01216). The eye tracking sessions were performed to order of the author by a graduate student, Dobrowolski, Ł., at the Faculty of Computer Science and Information Technology of the West Pomeranian University of Technology in Poland and the initial results were included in an unpublished Master's thesis with the permission granted by the author [21].

2   Conceptual Framework

The main objective of the presented study is to assess the attractiveness of different positions of a recommendation within a recommending interface of an e-commerce website as well as varying visual intensity of a recommendation with regard to attracting customer interest, from the perspective of marketing goals and user experience. Although users are usually most interested in the main editorial content of a website, marketing goals could be accomplished by such positioning and modifying intensity of the most profitable recommendations so that they would attract highest user attention. That in turn could lead to boosting sales.

There are a few ways of determining user interest, including asking the user explicitly, or observing the interaction implicitly. Unfortunately, explicit questioning may disrupt natural behavior and adds and additional burden on the user [22, 23]. Furthermore, in case of fast Internet browsing there could be a lot of unconscious attraction caused by some parts of a webpage. Therefore, inobtrusive implicit measures are better suited for the purpose of the study, allowing the monitored subjects to focus normally on tasks performed, not causing extraneous cognitive load and not requiring special motivation to continuously provide explicit ratings [24–27].

The methodology of the research assumes the use of eye tracking for implicit user behavior observation and analysis. Gaze tracking is the most popular technique of observing human-computer interaction and a powerful method generating implicit feedback. Within the scope of this study gaze-based data are used and interpreted in a simple e-commerce scenario. Eye movements are used to discover which parts of a shopping website are mostly looked at, and so which of them attract user attention the most or are the most relevant to the user. Raw data is processed and analyzed with the help of popular eye tracking software and data analysis algorithms.

Although eye movements are often unconscious and unordered in nature, they are generally strictly connected with cognitive processes [28]. Therefore, basing on gaze data from eye tracker it is possible to infer about user attention and interest. Gaze is an excellent data source for providing information about how much attention the user pays to what locations and contents on the screen [29]. In this study, total fixation duration is the main gaze-based measure used. This parameter is used as an indicator of attractiveness by a lot of other research [30, 31]. It is calculated as the sum of durations of fixations aggregated on a section of a website, in particular one with recommendation content RC and the main section with editorial content EC.

Apart from experimenting with positions of a recommending interface on a website as well as testing positions of a particular recommendation item within that interface, the aspect of content visual intensity is also taken into consideration in the study. Changing visual intensity is a known marketing technique to counteract habituation and attract attention [32]. Three basic levels of intensity were used.

3   Experimental Results

3.1   Experiment Structure and Procedure

Task. All participants were given the same task, which was to shop online in order fully furnish a studio apartment with six types of furniture. A subject was asked to go from one product category to another and choose one item from each category, according to their individual preference.

Website. The experiment was composed of a recommending interface within a dedicated e-commerce website developed using Drupal CMS. Each subject was provided with the same website. The pages were written in Polish and consisted of title, menu, product images and little descriptory text. The website functionally covered such functions as product list, buying cart and recommendations.

The editorial content *EC* was placed in the central part of the screen, under the main menu and included product lists which were approximately 3 screen page long and contained 10 products for each product category. Each product had three unique features: name, image of the product and price. Products in a category were quite similar visually and similarly priced. There were six product categories PC_j: wardrobes, chests of drawers, beds, bedside cabinets, tables and chairs. In addition, "Add to Cart" button was placed under the description of a piece of furniture allowing to store customer selections in a database. From the moment a product is selected, its short description is available in cart preview as well as the main cart page. Naturally it is possible to remove a product from the cart in order to allow a user make necessary changes to the final selection.

Recommending Interface. There were two alternative recommendation layouts, i.e. horizontal and vertical recommending interface mode, which means that the recommendation content *RC* section was anchored in one of two dedicated parts of the screen below the main menu, either on the left side of the page next to the general product list (vertical mode), or at the top of the page, above the general product list (horizontal mode). Only one recommendation layout is available at a time, i.e. when the horizontal mode is on, the vertical one is deactivated on a page and vice versa. Figures 1 and 2 show two variants of the location of recommendation content *RC*.

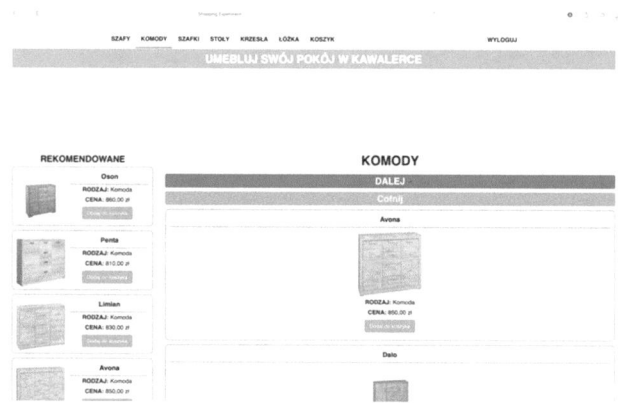

Fig. 1. Website with vertical recommendation layout (*RC* on the left).

Fig. 2. Website with horizontal recommendation layout (*RC* at the top).

The *RC* section always consisted of 4 recommendation items RC_1 to RC_4 randomly selected from all products in a given category. The section in each variant did not change its location on the screen throughout browsing products in a product category, regardless of user scroll of the *EC* section. Only the general product lists were actually made scrollable, in order to ensure reliable subject exposure to recommendations.

It was ensured that product features, *i.e.* name, image and price did not stand out from other products in a category. It was intended that the distinction of a particular RC_i location would be accomplished only by means of visual intensity VI. Three levels of intensity were used: standard (no highlight) – VI1, flickering (slowly disappearing and reappearing every 1–2 s) – VI2 and background in red – VI3. For each product category there was maximum one RC_i at VI2 or VI3. Example of visual intensity of the last kind (VI3) is presented in Fig. 3.

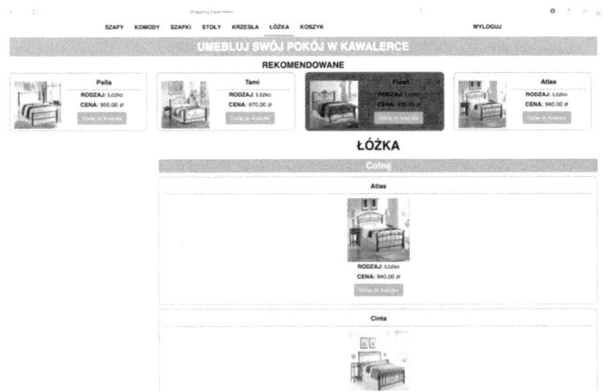

Fig. 3. Example of visual intensity of the third recommendation item – red background (RC3, VI3).

Procedure. An experimental run proceeded as follows. First, the examined person sat at the test stand in such a way that their eyes were in the optimal range of the eye tracking camera. It was explained what the device for tracking eyeball movements is, and then the eye tracker was calibrated using *Gazepoint Control* software and a 9-point calibration method. For better accuracy, the calibration was always performed twice, the first time to familiarize the subject with the process. There was a double monitor setup and the operator's screen was invisible to the participant. Thanks to correct calibration the device allowed to determine the coordinates of the place the user was looking at.

Next, the participant was informed what their task was, but they were not told about the goal of the study. After this introduction, the subject had to furnish the apartment. After choosing one item from a category, the subject clicked 'Next' and was automatically moved to the next category. Category after category, the visual intensity of recommendation items changed each time. Furthermore, for the first three categories, the layout of *RC* was vertical, and after moving to the fourth category it changed to horizontal and remained this way for the following categories. On the whole, there were at least 6 subsequent pages with different recommendation variants presented to each participant.

Every session was recorded and controlled live using *Gazepoint Analysis* software, and it was constantly double-checked on the operator's monitor that the eyes of the subject were in optimal position in relation to the camera etc. After completing the task by the participant, basic data such as age were collected and a question was asked whether a subject felt they were influenced by recommendations. Finally, all data was saved and stored by the eye tracking system for further analysis. A typical experimental run lasted around 12 min.

Participants. The experimental group of users consisted of 52 people and all of them produced valid eye tracking data. Most of them were undergraduate or graduate students invited in person or attracted by advertisements for the study, and they were native Polish speakers. They ranged in age from 14 to 54 years (mean = 25.2, $\sigma = 8.0$).

3.2   Results and Analysis

Please note that Analysis of gaze tracking data shows that completing the task took 2.3 min on average. In total 312 products were selected for purchase by 52 study participants. Fixation time on the recommending interface was averagely 16.3 s per person, which is 12% of average task completion time. The average amount of time devoted by subjects to observing *RC* was 8.2 s and 8.1 s for vertical and horizontal layout, respectively. Therefore, in terms of fixation times the two presented layout variants of recommending interface performed equally.

Table 1 shows more detailed distribution of fixation times for all recommendation item locations. It was found that the first three RC_i locations were most favourable, irrespective of the layout. The least eye-catching locations were fourth in the list, next to the bottom bar of a website for the vertical layout or next to the right edge of the screen for the horizontal layout. The most popular of all was RC_3 location in the horizontal layout (average fixation time 3.9 s). This might have been influenced by the fact that this recommendation item was positioned directly above the general product list. The second popular location was RC_2 and the third one was RC_1, both in the vertical layout. The evident attractiveness of RC_2 in this layout was influenced by that in one product category this location was presented as flickering (VI2). The popularity of RC_1 location although always presented with standard visual intensity (VI1) may be resulting from that many people tend to perceive the first location in a list as the best one. It should be noted that for vertical layout this first position still performed better than RC_3 which for one product category was presented with dazzling intensity VI3. Item RC_3 in vertical layout performed on a par with item RC_2 in horizontal layout, the latter being supported by flickering effect (VI2) for one product category.

Table 1. Average fixation time (s) for each recommendation location

Recommendation location	Time (s)	
	Vertical RC	Horizontal RC
RC_1	2.4	1.3
RC_2	3.1	2.1
RC_3	2.1	3.9
RC_4	0.6	0.8
Total	8.2	8.1

An aggregated gaze heatmap for all participants in the study is presented in Fig. 4. It depicts the views of the users in a given area. The areas that received the most attention get a warmer color, while those less "attractive" get a "colder" color. The map shows that in relation to the whole time spent on completing the task, the recommending interface received a lot of attention, but less than the general product list. The differences in attractiveness of recommendation items on different positions to the disadvantage of RC_4 for both layout variants can also be noticed.

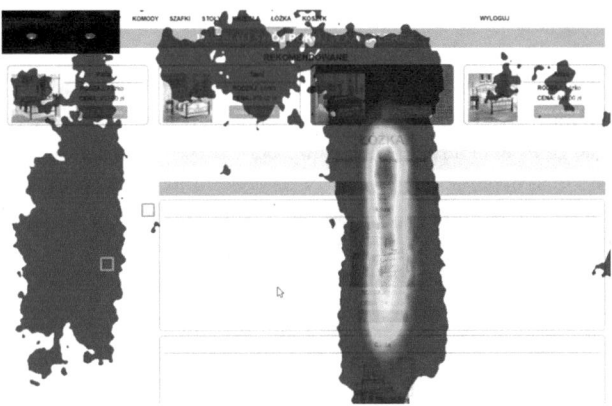

Fig. 4. Heatmap aggregated for all participants in the study.

From sales perspective, 12% of products in all carts were added there directly from recommendation items. It is exactly the same proportion as the one of recommending interface fixation time to task completion time. The number is coincidental but it justifies the importance of focusing attention on recommended items.

Vertical RC layout was responsible for 62% of product choices from recommendations, while the rest were thanks to horizontal RC layout. Vertical mode turned out to be almost twice as effective as the other one, despite in terms of average fixation times there was practically no difference between them.

The inferior effects of horizontal RC may be connected with banner blindness, since banner ads historically have usually been placed in the same parts of a website. In case of vertical layout: for RC with all RC_i at standard intensity level (VI1), the recommendation-driven purchases (RDP's) were quite evenly distributed between recommended products (each item attracted 17–33% of all RDP in the product category); for RC with RC_2 at flickering intensity level (VI2), the item attracted 44% RDP's in the product category; for RC with RC_3 in red background, the VI3 item surprisingly attracted only 13% of RDP's in the product category. All in all, it was RC_2 that was the best performing, so the second recommendation in the vertical list resulted in most sales (48% of RDP's for vertical RC, and 30% of all RDP's).

In case of horizontal layout: for RC with all RC_i at standard intensity level (VI1), 100% of identified RDP's resulted from RC_1 (RC_2, RC_3 and RC_4 generated no RDP's); for RC with RC_2 at flickering intensity level (VI2), the item attracted 50% of RDP's in the product category; for RC with RC_3 in red background (VI3), the item attracted only 33% of RDP's in the product category. On the whole, it was RC_1 that was the most popular here, so the first recommendation in the horizontal list resulted in most sales (43% of RDP's for horizontal RC, and 16% of all RDP's).

It is interesting that in a survey directly following each gaze tracking session, 33% of the participants responded that they felt their selections were influenced by recommending interface of the website, and 6% felt strongly about it. The rest claimed the opposite, including 52% who strongly felt they did not care about recommendations on the website. The last group indeed seemed to show resistance to recommendations – some of those participants when shown the RC sections after the test were surprised they might have neglected them at all, treating them comparably to advertisements, which confirms the need to seek ways to overcome habituation effect resulting from advertising clutter.

4   Conclusions

Marketing experts and e-commerce engineers need solutions to attract web user attention and encourage them to commit purchase, such as optimally set up recommending interfaces. The presented study showed the influence of recommending interface layout, position of the recommending item and different levels of its visual intensity on user behavior in a simply structured e-commerce website. Thanks to gaze tracking methodology as well as tracking each participant's purchase decisions the attractiveness of selected recommendation areas was analyzed. Research limitations such as a relatively small sample size, limited representativeness and the constraints connected with the simple website format should be noted, which justifies a larger-scale experiment in the future.

There are several important conclusions. In the experiment averagely 12% of task completion time was devoted to looking at the recommending interfaces, and the same percentage of goods were bought with direct use of recommendations. While comparing the vertical and horizonal recommending interface layouts, from the perspective of fixation time they had equal performance, however from the perspective of purchase commitments the vertical layout performed almost twice as good as the horizontal one. The inferior sales result of the horizontal layout might be directly connected with banner blindness since banners tend to occupy similar space in the upper parts of the screen. In the superiorly performing vertical layout the most eye catching in terms of fixation time was the position in the list, where a slow flickering effect was used to increase visual intensity. The first position in the list was the second best attractive, despite the lack of any visual highlight, which may result from the conception that the first one is always the best, similarly to reactions to search results. The level of attractiveness of dazzling red background was relatively low probably due to excessively high content intrusiveness which turned out to be counterproductive. It was also found that the first three locations in a recommending interface were most attractive irrespective of the layout; the least popular locations being the last ones, bordering with the bottom or right edge of the website, for vertical and horizontal layout, respectively.

The study justifies that vertical rather than horizontal layout is worth considering while designing a recommending interface and that it is necessary to seek balanced rather than radical visual intensity solutions to counteract habituation effects without adversely affecting buyers.

References

1. Castagnos, S., Jones, N., Pu, P.: Recommenders' influence on buyers' decision process. In: 3rd ACM Conference on Recommender Systems (RecSys 2009), pp. 361–364 (2009)
2. Allen, R.B.: User models: theory, method, and practice. Int. J. Man-Mach. Stud. 32(5), 511–543 (1990)
3. Kelly, D.: Implicit feedback: using behavior to infer relevance. In: Spink, A., Cole, C. (eds.) New Directions in Cognitive Information Retrieval. The Information Retrieval Series, vol. 19, Sect. IV, pp. 169–186 (2005)
4. Sulikowski, P., Zdziebko, T., Turzyński, D., Kańtoch, E.: Human-website interaction monitoring in recommender systems. Procedia Comput. Sci. 126, 1587–1596 (2018)
5. Sulikowski, P., Zdziebko, T., Turzyński, D.: Modeling online user product interest for recommender systems and ergonomics studies. Concurrency Comput.: Pract. Experience 31 (22), e4301, s. 1–9 (2019). http://doi.org/10.1002/cpe.4301
6. Wątróbski, J., Jankowski, J., Karczmarczyk, A., Ziemba, P.: Integration of eye-tracking based studies into e-commerce websites evaluation process with eQual and TOPSIS methods. In: Wrycza, S., Maślankowski, J. (eds.) Proceedings of Information Systems: Research, Development, Applications, Education. 10th SIGSAND/PLAIS EuroSymposium 2017, Gdańsk, Poland, 22 September 2017, Lecture Notes in Business Information Processing, vol. 300, pp. 56–80. Springer, Cham (2017)

7. Jankowski, J., Ziemba, P., Wątróbski, J., Kazienko, P.: Towards the tradeoff between online marketing resources exploitation and the user experience with the use of eye tracking. In: Nguyen, N.T., Tawiński, B., Fujita, H., Hong, T.P. (eds.) Intelligent Information and Database Systems. 8th Asian Conference, ACIIDS 2016, Da Nang, Vietnam, 14–16 Mar 2016, Proceedings, Part I. Lecture Notes in Artificial Intelligence, vol. 9621, pp. 330-343. Springer, Berlin (2016)

8. Adomavicius, G., Tuzhilin, A.: Toward the next generation of recommender systems: a survey of the state-of-the-art and possible extensions. IEEE Trans. Knowl. Data Eng. **17**(6), 734–749 (2005)

9. Peska, L., Vojtas, P.: Estimating importance of implicit factors in e-commerce recommender systems. In: Proceedings of the 2nd International Conference on Web Intelligence, Mining and Semantics, Article No. 62, New York (2012)

10. Yang, X., Guo, Y., Liu, Y., Steck, H.: A survey of collaborative filtering based social recommender systems. Comput. Commun. **41**, 1–10 (2014)

11. Tintarev, N., Masthoff, J.: A survey of explanations in recommender systems, In: 2007 IEEE 23rd International Conference on Data Engineering Workshop, pp. 801–810. IEEE (2007)

12. Verbert, K., Manouselis, N., Ochoa, X., Wolpers, M., Drachsler, H., Bosnic, I., Duval, E.: Context-aware recommender systems for learning: a survey and future challenges. IEEE Trans. Learn. Technol. **5**, 318–335 (2012)

13. Lu, J., Wu, D., Mao, M., Wang, W., Zhang, G.: Recommender system application developments: a survey. Decis. Support Syst. **74**, 12–32 (2015)

14. He, J., Chu, W.W.: A social network-based recommender system (SNRS). In: Data Mining for Social Network Data, pp. 47–74 (2010)

15. Swearingen, K., Sinha, R.: Beyond algorithms: an HCI perspective on recommender systems. In: Herlocker, J.L. (eds.) Recommender Systems, papers from the 2001 ACM SIGIR Workshop, New Orleans (2001)

16. Bortko, K., Bartków, P., Jankowski, J., Kuras, D., Sulikowski, P.: Multi-criteria evaluation of recommending interfaces towards habituation reduction and limited negative impact on user experience. Procedia Comput. Sci. **159**, 2240–2248 (2019)

17. Portnoy, F., Marchionini, G.: Modeling the effect of habituation on banner blindness as a function of repetition and search type: Gap analysis for future work. In: CHI 2010 Extended Abstracts on Human Factors in Computing Systems, pp. 4297–4302. ACM (2010)

18. Ha, L., McCann, K.: An integrated model of advertising clutter in offline and online media. Int. J. Advertising **27**, 569–592 (2008)

19. Jankowski, J., Hamari, J., Watróbski, J.: A gradual approach for maximising user conversion without compromising experience with high visual intensity website elements. Internet Res. **29**, 194–217 (2019)

20. Jankowski, J.: Modeling the structure of recommending interfaces with adjustable influence on users. In: Intelligent Information and Database Systems, Lecture Notes in Computer Science, vol. 7803, pp. 429–438 (2013)

21. Dobrowolski, Ł., Sulikowski, P.: Prezentacja tresci w systemach rekomendacyjnych, Unpublished master's thesis. West Pomeranian University of Technology, Szczecin, Poland (2019)

22. Nichols, D.M.: Implicit ratings and filtering. In: Proceedings of the 5th DELOS Workshop on Filtering and Collaborative Filtering, Hungary, pp. 31–36 (1997)

23. Middleton, S.E., Shadbolt, N.R., De Roure, D.C.: Capturing interest through inference and visualization: ontological user profiling in recommender systems. In: Proceedings of the Second Annual Conference on Knowledge Capture (2003)

24. Zdziebko, T., Sulikowski, P.: Monitoring human website interactions for online stores. Adv. Intell. Syst. Comput. **354**, 375–384 (2015)

25. Kim, K., Carroll, J.M., Rosson, M.: An empirical study of web personalization assistants: supporting end-users in web information systems. In: Proceedings of the IEEE 2002 Symposia on Human Centric Computing Languages and Environments, Arlington, USA (2002)
26. Kelly, D., Teevan, J.: Implicit feedback for inferring user preference: a bibliography. SIGIR Forum **37**(2), 18–28 (2003)
27. Avery, C., Zeckhauser, R.: Recommender systems for evaluating computer messages. Commun. ACM **40**(3), 40–88 (1997)
28. Liversedge, S.P., Findlay, J.M.: Saccadic eye movements and cognition. Trends Cogn. Sci. **4**(1), 6–14 (2000)
29. Buscher, G., Dengel, A., Biedert, R., Van Elst, L.: Attentive documents: eye tracking as implicit feedback for information retrieval and beyond. ACM Trans. Interact. Intell. Syst. **1**(2), 9 (2012)
30. Xu, S., Jiang, H., Lau, F.C.: User-oriented document summarization through vision-based eye-tracking. In: Proceedings of the 13th International Conference on Intelligent User Interfaces (IUI 2009), pp. 7–16. ACM, New York (2009)
31. Ohno, T.: Eyeprint: support of document browsing with eye gaze trace. In: Proceedings of the 6th International Conference on Multimodal Interfaces (ICMI 2004), pp. 16–23. ACM, New York (2004)
32. Lee, J., Ahn, J.-H., Park, B.: The effect of repetition in internet banner ads and the moderating role of animation. Comput. Hum. Behav. **46**, 202–209 (2015)

Social Network Analytics for e-Business Management

Mechanics and Quality of Agent-Informational Clustering in Social Networks

Daria A. Yakovleva$^{(\boxtimes)}$ and Olga A. Tsukanova$^{(\boxtimes)}$

National Research University Higher School of Economics, Moscow, Russia
dayakovlevaa@gmail.com, otsukanova@hse.ru

Abstract. The present paper is devoted to the study of the mechanics of agent-informational clustering in a social network on the example of user segmentation tasks taking into account an influence criterion. The main features of data generated by social networks (social big data) and metrics that characterize influential network nodes are considered. A review of community-building algorithms based on the theory of social networks, as well as clustering methods based on machine learning, is carried out. Metrics for assessing the quality of segmentation are presented. The results of the application of methods (selected on the basis of the performed analysis) to a test dataset are shown. The limitations of the applicability of considered approaches and possible problems during the implementation of algorithms in the field of social network analysis are described. Evaluation of the effectiveness is performed.

Keywords: Social networks · Social big bata · Clustering · Segmentation

1 Introduction

People all over the world generate enormous amounts of data on regular basis – 2,5 quintillion bytes per day [1]. In 2018 the population of Internet users achieved 4 billion humans, more than 3 billion people now use social media each month and this number is still growing by 13% year-on-year [2].

This growth of internet usage, production and consumption of online information create opportunities for enterprise representatives to manage their relationships with potential and existing customers in a more effective way on the base of customers' insights. The open data of users' opinions about a company and its products can help to develop a strategy for future growth and work for PR purposes, sales planning or marketing needs. That means estimation of present reaction to offers of the company (some comments about organization itself and its services or goods – quality, amount, requirements for some features, etc.), as well as forecasts within scenarios like "if …, then …". Insights can also be used for creating targeted offers for some particular audience.

Thus, the variety of different tools is arising because of the constant need for increase in the efficiency of contact with the user while reducing its cost – to push the user to make a purchase, while creating as little expenses as possible. Targeting with the help of mentioned instruments solves this issue to some extent, further developments can be achieved by applying a new approach – to contact not with the entire

© Springer Nature Switzerland AG 2020
K.-M. Chao et al. (Eds.): ICEBE 2019, LNDECT 41, pp. 221–235, 2020.
https://doi.org/10.1007/978-3-030-34986-8_16

audience at once, but to select certain groups, or users segments inside and contact with their most prominent representatives who are able to spread the message further on their own. In this case, the cost for user engagement dramatically decreases due to fewer amount of contacts, while the number of purchases and their average check grows. It states the relevance of the research.

Literature review shows that there are a lot of different academic methods of forming separate segments inside a big audience for different purposes, including marketing. The most popular, properly studied and rather simple approaches are machine learning clustering methods – k-means, mean-shift, DBSCAN, expectation-minimization, etc. But as the present research is focused on social networks, it is vital to take into account their features as well. Social network theory with its community detection algorithms processes this kind of data with the respect to peculiar properties of graphs and relations between them, as it is believed to appear in real networks. That is why it is important to identify which methods work more accurate with social networks data and produce stricter results.

The object of the research is agent-informational clusters, the subject – segmentation techniques in large social networks. The main goal of the research is to find out different users' segmentation methods proper for extracting knowledge from social networks for future information spread. Social networks theory algorithms and machine learning techniques are being compared.

To achieve the main goal, the following tasks are to be completed:

- To examine main terms;
- To consider the methods of community detection in social networks theory and machine learning clustering techniques and try the most appropriate on the test example;
- To compare obtained results using models of information spread to answer the question which model distributes messages within network in faster way.

2   General Terms

2.1   "Social Networks" and "Big Data"

The term "agent-informational" is used in the present research because it is important to consider both 2 sides of a social network: its resource is represented by a set of agents (actors) from the one hand and a set of informational resources that is either staying behind or being generated by the network agents from the other hand.

A social network, in general, has two meanings:

(1) a network of individuals (such as friends, acquaintances, and coworkers) connected by interpersonal relationships [3];
(2) a website or computer program that allows people to communicate and share information on the internet using a computer or mobile phone [4].

Both definitions consider essential components of social networks: individuals/people who can also be viewed as agents (actors) and relationships/communications/

sharing of information resources between the agents. These "sum of professional, friendship, and family ties is the fabric of the society and determines the spread of knowledge, behavior and recourses" [5] via agent-informational relations – how each agent connected to any other and what information he can deliver.

Social big data is a definition which was determined from social networks research and big data applications. Big data can be defined as the sphere of data analysis which uses more sophisticated methods for information collection, processing, storage, analysis and visualization for too complex data for standard data processing procedures [6]. In addition, big data can be a definition for data itself with several vital issues – volume (the size of datasets is enormous), variety (this data is from a variety of sources and is represented via different types), velocity (time spend of data generation, collection and analysis is small); moreover, some additional features are validity (high quality) and value (the value which this data bring to its owners) [7]. As a result, social big data is the combination of networks and big data – all the data that is collected from social networks and corresponds with all vital big data specifications.

2.2 "Communities" (in Social Network Theory) and "Clustering" (in Machine Learning Theory)

Segment on the whole is considered as one of the parts into which something is naturally separated or divided. Segments inside social networks are also called as communities – the groups of people in the condition of sharing or having certain attitudes and interests in common [8]. A network concept of community also considers community as a set of relations between actors [9]. These relations can be friendship, colleagues, business partners, as well as general passion for one common object of interest, goal, some preference for a lifestyle, etc. In general, an individual relates to several communities that can also overlap. Sometimes within a community "pre-existing interpersonal relationship between members of a community" can be [10], but new members do not necessary know all other members.

Social network science defines network communities as "groups of vertices such that inside vertices the group is connected with many more edges than between groups". A typical community has the following most vital features:

- mutuality of ties – each member of a community must have connection (edge) with any other;
- compactness – any member can reach another member within short number of steps (in addition, they are not necessarily adjacent);
- density of edges – as everybody within a community has link to one another, the frequency of ties is very high;
- separation – this frequency of ties within a community is very high compared to the frequency of ties to non-members [11].

Besides, communities can be overlapping – an actor can be a member of several communities at once – and non-overlapping. Community detection itself is the process of deciding to which community to assign a particular actor. Within the present research for community detection non-overlapping communities will be considered.

Segments in machine learning sphere are called clusters – the same group of objects or individuals which are more similar comparing to the rest of objects or individuals in a sample. The clustering itself is the process of breaking down a large dataset into a smaller group of some components with resembling characteristics. To form a cluster, its components properties are needed to be represented into numerical way.

3 Theoretical Basis. Overview of Existing Algorithms for Segmentation

3.1 Considered Algorithms of Communities Detection in Social Network Theory

There are three types of community detection algorithms: graph partitioning, hierarchical clustering, partition clustering [12].

Graph partitioning is about vertices separation into some groups with fixed size in such way, that the number of edges lying between the groups (cut size) is minimal. This approach is a base for a variety of more sophisticated computing methods; its result relies on the specification of the number of partition clusters and their size.

Hierarchical clustering goal is to find vertices with high similarity and group them together. Depending on the exact algorithm, hierarchical clustering methods can be divided into agglomerative and divisive algorithms. Agglomerative algorithms use bottom-up approach – at the initial stage each node has its own community and while moving through further stages these communities are merged together if vertices have similarities; divisive algorithms have top-down approach – all nodes are in one community in the beginning and further they become spitted into different communities. For estimation of communities which are to be merged or divided can be used the following metrics for dissimilarity measurement – Euclidean distance, squared Euclidian distance, Manhattan distance, Maximum distance and Mahalanobis distance. Among the examples of hierarchical algorithms are Edge betweenness (or Girvan–Newman algorithm), Fast greedy algorithm, the Leading eigenvector algorithm, the Walk trap algorithm, Spin glass algorithm, Label propagation algorithm. The advantage of hierarchical clustering is the absence of necessity to set some valuable preliminary parameters (for instance, size of communities as in graph partitioning), but at the same time there is result dependence on the selected dissimilarity measure and some primary arrangement.

Partition clustering is another class of methods used for building communities. It uses some settled number of clusters (for example, k) and defines each node as a point and distance between pairs of vertices as some metric space. This distance is the main measure of similarity between points within network; the main achievement is to divide all the space into k communities in the way, so that the cost function is maximized or minimized on the base of distance between vertices and defined positions within the space. The most often used functions are minimum k-clustering; k-clustering sum, k-center and k-median. Partition clustering methods have the same disadvantages and restrictions as graph partition methods – they cannot recommend the number of communities within a network, a researcher have to specify it at the beginning. Moreover, embedding to a metric space can be not proper for some graph due to their specific structure.

3.2   Considered Algorithms of Machine Learning Clustering

Clustering is an unsupervised type of methods. It builds outcomes without having "right answers" to compare with. They can be divided into three groups: density-based, partition-based and more sophisticated non-parametric methods [13].

Density-based methods are focus on identification areas with a big amount of point (high concentration) and areas where they are divided with some emptiness. One of the most popular density-based clustering algorithms is DBSCAN (density-based spatial clustering of applications with noise). It uses distance measurements (the most often is Euclidean distance) and minimum number of points to group nodes that are close to each other. This approach is the one of the fastest machine learning clustering algorithms. It can also be transformed into HDBSCAN, or Hierarchical DBSCAN which has the same start, but then HDBSCAN form groups slightly different. It discards all points that cannot be emerged with others and forms big clusters that are equal or more that minimal cluster size. It makes DBSCAN closer to real-world data and increase the speed of its application.

All partition-based algorithms use the same principle – they take some given measures, try to minimize them by relocating points between different clusters. One the most famous algorithms due to its simplicity and quality of results is k-means clustering.

Instead of k-means clustering, Expectation–Maximization (EM) Clustering using Gaussian Mixture Models (GMM) can also be used. It avoids some naivety of k-means in choosing the center of cluster; it uses the mean and standard deviation while assuming that points are Gaussian distributed.

Finally, there are Mean-Shift clustering algorithm which belongs to non-parametric sliding-window-based algorithms. Its main focus is to find some dense areas with big amount of points and locate the central nodes for each group. Central point C and radius r for a kernel are to be selected; further, kernel is continuously replaced to the areas with higher density until there is no direction at which the kernel can accommodate more points and all modes are inside the window. There is no need to select the number of clusters, but at the same time the selection of the center C and radius r is not so simple.

4   Metrics Used for Segmentation Quality Estimation

One of the proper ways to estimate the quality of obtained communities is to assess the speed and capacity of information spread – what are the maximum number of actors who will receive the information and in what time.

There are several network epidemic models [14] that mathematically describe not only infectious diseases, but also information units' propagation for other different areas, for example, messages or ideas.

A community has some number of potential contacts; each node can have one of three states:

- susceptible – a node that has not received any information yet; the probability for i vertex to have susceptible state at some time t is $s_i(t)$;

- infected – a node that has received information and spreads it further; the probability is $x_i(t)$;
- recovered – a node that received information, but no longer spreads it further; the probability is $r_i(t)$.

There are also two vital models parameters: the infection rate β (the probability to get infected on a contact within some time δt) and recovery rate γ (the probability to recover (to stop spreading information) on a contact at time δ t). It is necessary for nodes to be reachable, so that there is always a possible way to communicate from one vertex to another even via other different nodes.

The first model is the simplest one – SI model, or getting from susceptible to infected state $(S \rightarrow I)$. It states that once an actor receives the information, it starts to spread it further to neighbors without an ability to recover and stop being an information distributor. There is β, the infection rate (the probability to get infected in a unit time), where $x_i(t + \delta t) = x_i(t) + \beta s_i(t) \sum_j A_{ij} x_j(t) \delta t$. Every node within the model has one of the states {S, I}. In the beginning, some nodes become infected and after this each node with state I has β probability to spread the information to its nearest neighbors.

In addition, there is SIS model, which allows nodes to recover. The algorithm starts with the same number of vertices that are infected and the probability for them to spread information further is the same – β, but each node stays infected $1/\gamma$ time steps, where γ is the recovery rate. So, after some time steps node recovers $(I \rightarrow S)$, or a person no longer pays attention to/forgets about the gained information just like he never received it.

Important addition here is that there is some predefined threshold R_0, so that if $\frac{\beta}{\gamma} < R_0$, and the information dies over time; if $\frac{\beta}{\gamma} > R_0$, the information reaches everybody or almost everybody in a community.

Finally, there is SIR model, where nodes have three states {S, I, R}. The model is very similar to SIS model, but has only one vital difference – after $\frac{1}{\gamma}$ time steps node does not become susceptible again, but starts to be recovered, i.e. a person knows the information, can use it for his own need, but does not spread it. The model dynamics is following: the information reaches everybody or almost everybody in a community, but after some time most of the infected nodes become recovered and stop spreading the information.

5   Potential Software Tools

Generally, there are two main instruments used for data analytics – Python and R programming languages with their built-in libraries. Both of these languages are open-source and, as a result, allow development of different approaches, including simple data processing and sophisticated machine learning models. They also have some packages with visualization tools which can be proper for some basic data representation. These languages and internal libraries can work with social big data due to the variety of possible methods, simplicity of preprocessing methods and calculation abilities in terms of power and speed.

In addition, there is some social networks-specific software being used for descriptive analytics and visualization of large amounts of social network data. It can be Gephi, NodeXL, Graphviz, Social Networks Visualizer, Cytoscape, etc.; these instruments provide such functionality as good scaling of large graphs, full setup of visualization, studying social and mathematical parameters of the graph.

6   Potential Limitations of the Research

All real networks which are utilized by regular users have strict users' information employment rules due to General Data Protect Regulation (GDPR) in Europe and its analogs in different countries. This law regulates collection, storage and protection of information about users which assembled via their activities within a website or app [15]. This data is personal data – "any information that relates to an identified or identifiable living individual" [16]. The processing of personal data acquisition cannot be started without a user's approval, he needs to accept an informational message warning about that process and to approve its further handling. In addition, social networks as information keepers have to inform users about the fact of sharing personal data with third parties and purposes for doing that.

These limitations bring difficulties into real networks scientific analysis: social networks create special algorithms which are aimed at resistance for collecting users' data via crawlers' robots or any other techniques for automatized data collection; event if the data can be gathered bypass protection, all the data must be properly anonymized in order to stop being personal information.

For the present research there was selected a prepared dataset collected by the researchers from Arizona State University [17] – an anonymized data about a set of actors from a network, primarily created for video sharing. Besides the legit character of the data, it has another advantage – its size. It has fewer records to simplify calculations, but at the same time it retains all important features of the whole network. As a result, it makes easier to apply processing and analytics algorithms (e.g., calculation of betweenness centrality requires huge computing power and takes a lot of time even for several millions of edges) and at the same time allows to draw conclusions about inner users relations, as all obtained outcomes that are fair to a sample, fair to the general data, too.

7   Tests

The main instruments used within the present research are Python programming language, an environment for its usage – its distribution package Anaconda, and Jupyter Notebook, an application for Python projects creation.

In addition, for drawing a network, its main metrics distribution and performing community detection algorithm Gephi can be used – an open-source platform for visualizing and analyzing large networks graphs. It has user-friendly interface and a variety of features that allow to perform different kinds of analytics.

7.1 Basic Analysis of a Dataset

The open dataset selected for the research provides an ability to build a graph on the base of "networkx" library. The obtained graph has 15 088 nodes and 76 747 edges, the average degree is 10,1733, in other words, an average vertex has ten closest neighbors. Network density is 0,0006743057112157054 (the percentage of real connections between actors divided by the number of all the connections that could exist between all possible vertices in case of full graph). This metric shows how tightly all the nodes are connected – often the values between 0 and 1 depict that the network is not so dense. The network is illustrated on the Fig. 1 – nodes are black dots, while edges are grey lines; the size of a dot corresponds to its degree – the higher degree, the more the size of the dot.

Fig. 1. The illustration of the network user within the research

7.2 Testing a Community Detection Algorithm

After the performed analysis of above-mentioned algorithms, the greedy one is selected for the present research. It does not require any parameters to be set in advance, which is very proper for the chosen volumetric dataset. In addition, its possible disadvantage of merging very small communities into one can be transferred into advantage – the data includes a lot of different vertices with complex inner relations, and they are likely to create a lot of different communities which are to be analyzed and tested with information spread models; their unification into «the rest» type is the way to make the research simpler and more convenient for further comparison with machine learning techniques.

Selected 4 biggest communities cover 70% of all the nodes of the network on the whole: 1st (violet color) – 21%, 2nd (red color) – 20%, 3rd (green color) – 16% and 4th (pink color) – 14%. This partition can be illustrated in structured way on the Fig. 2. All the results gathered within work with Gephi software were added to the dataset for further use in clustering machine learning techniques and information spread models realization.

Fig. 2. Structured illustration of gathered communities

7.3 Testing Clustering Algorithms

From the variety of clustering methods described previously only two were selected – k-means and DBSCAN. The choice is determined by the performance of these algorithms within big volumes of data – the speed of application, its simplicity and necessity for some preset variables.

To start with, the simplest algorithm is k-means; it works fast, but at the same time it requires determination of clusters' number and some centroids. DBSCAN is also considered as a rather simple one; it does not require any preset parameter, but it requires more time and calculation capacity. Its more sophisticated version – HDBSCAN – is more complex and consuming from the point of view of calculation recourses. Mean Shift requires additional parameters established in advance; this process is much more complex than decision of the clusters' number, so some non-trivial analysis before launching the clustering algorithm itself is needed to be performed. Moreover, it takes appreciable amount of computing power, which can increase exponentially with the size of dataset. EM using GMM does not need to have some preset argument, as it uses mean and standard deviation; but at the same time, it uses additional resources for recalculation mean, standard deviation and the probability of affiliation to any cluster at each iteration. This brings flexibility for complex array of data, but makes the calculations energy consuming.

As a result, the most proper algorithms for the present research are k-means and DBSCAN [18, 19] (Table 1):

Table 1. Comparison of machine learning clustering algorithm within real data implementation.

Algorithm	Complexity	Time costs	Preset parameters
K-means	Low	Low	+ (simple to determine)
DBSCAN	Low	Low	–
HDBSCAN	Moderate	Moderate	–
Mean Shift	High	High	+ (complex determination)
EM using GMM	Moderate	High	–

7.4   Comparison of Results

After realization of the chosen methods, the results were compared to evaluate the quality of segmentation.

For further comparison of the results the SIR model is used, as it more fully represents dynamics of social networks – there are not only the moment when actors start spreading the information, but also the situations when they remain to be aware of some message/idea/product and at the same time stop sharing information about it to the nearest neighbors.

For each method for building segments (fast greedy from social network theory, k-means and DBSCAN from machine learning) there were performed three types of model – when infection rate β is higher than recovery rate γ; when infection rate β equals recovery rate γ; when infection rate β is lower than recovery rate γ. Number of iterations at the same time stay the same – 1000.

These are the outcomes (Figs. 3, 4 and 5):

- Case 1 – infection rate β is higher than recovery rate γ;

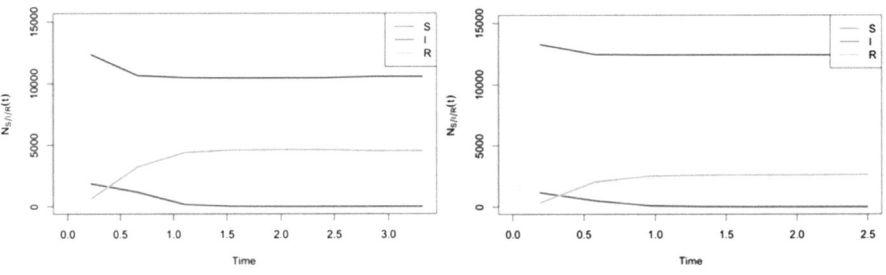

Fig. 3. Graph of SIR model for fast greedy, $\beta > \gamma$

Fig. 4. Graph of SIR model for k-means, $\beta > \gamma$

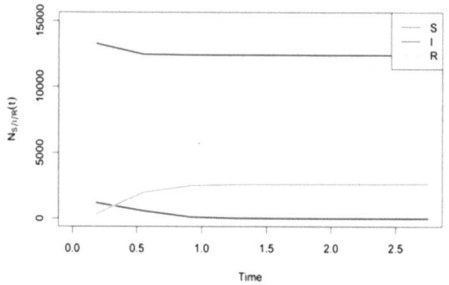

Fig. 5. Graph of SIR model for DBSCAN, $\beta > \gamma$

From the graphs the following conclusions can be drawn: the longest process of information disseminating is in fast greedy – it ends up to very low levels at 1,2 time, while k-means stop at 1 and DBSCAN at 0,9; in addition, the fast greedy also stops with the lowest number of susceptible, as a consequence, the information reaches more actors.

- Case 2 – infection rate β equals recovery rate γ (Figs. 6, 7 and 8);

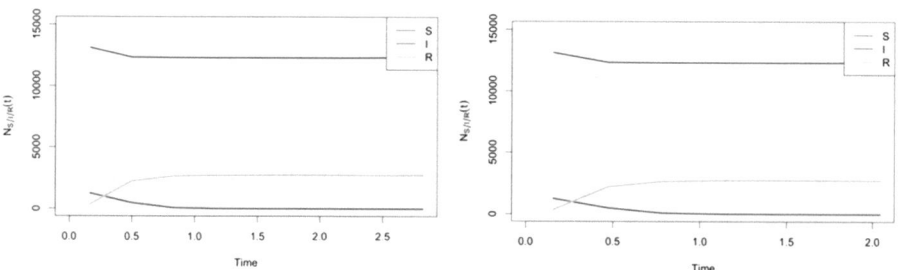

Fig. 6. Graph of SIR model for fast greedy, $\beta = \gamma$

Fig. 7. Graph of SIR model for k-means, $\beta = \gamma$

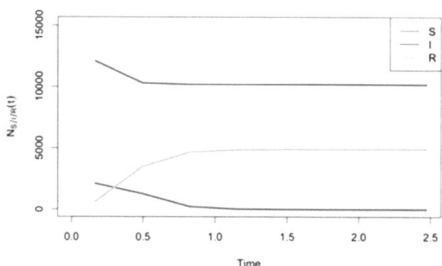

Fig. 8. Graph of SIR model for DBSCAN, $\beta = \gamma$

Within this case the longest spread of information is performed in DBSCAN communities (ends up at 0,8, while fast greedy and k-means – at 0,7) and, as a result, it provides much less number of actors who has not been reached by the message (around 10000 for DBSCAN and around 12500 for k-means and fast greedy).

- Case 3 – infection rate β is lower than recovery rate γ.

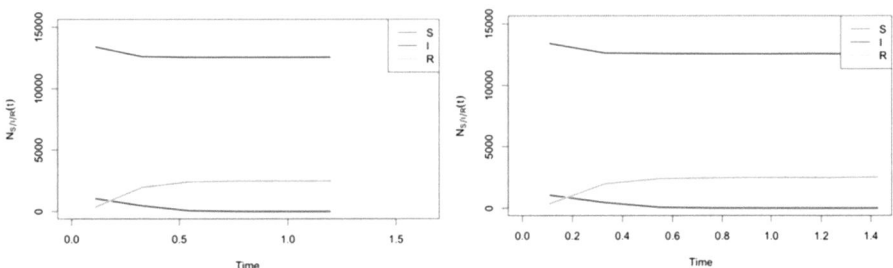

Fig. 9. Graph of SIR model for fast greedy, $\beta < \gamma$

Fig. 10. Graph of SIR model for k-means, $\beta < \gamma$

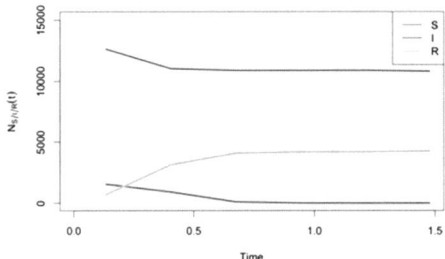

Fig. 11. Graph of SIR model for k-means, $\beta < \gamma$

For this case (Figs. 9, 10 and 11) the best result is also shown by DBSCAN – it ends up spreading at the latest point (0,7, while k-means at 0,5 and fast greedy at 0,6) and provides less actors without knowledge or message (around 11000, while 12500 for fast greedy and 13000 for k-means).

To sum up, in two cases out of three DBSCAN showed best result, while k-means had the worst effectiveness in all situations. It gives the conclusion that machine learning can be a good instrument of social network users' segmentation, but the clustering method has to be properly chosen.

8   Analysis of Potential Applicability of the Results

As the effectiveness of obtained results is estimated on the base of the speed of information, final outcomes of the research are applicable to the tasks where the idea or message are to be transmitted in a very fast way via digital resources.

First of all, the technique for using influential agents for further spread of some innovation or idea can be beneficial in cases where traditional word of mouth does not work and the decision about adopting new product can be realized only via personal communications with closest individuals (an actor starts to inherit some product/service/message only if he received the knowledge about it from a significant in terms of interpersonal communications neighbor). For example, online services which can be spread as a part of communication or some information sharing can use this technique for marketing needs. If some individual use some of these services and shares information via it (send emails, share files via cloud, etc.), he provides other actors not only with general information about a service, but also with knowledge of some best practices of its usages and cases of implementation in everyday life. If the neighbor, who is shared with this message, appreciates his relationship with the spreader in a high enough way, he is going to adopt the usage of this service with a high probability.

Another suitable case – marketing of products which are considered to be niche. This kind of products cannot be advertised in an effective way using traditional marketing methods, but their fans and regular consumers can be a good source for marketing by themselves. A brand can spread the knowledge about itself and its products via its persistent customers. These users with high probability have friends or acquaintance who share with them some particular hobby or interest and may adopt utilization of some niche relevant product. It solves problem of unpopular sales units for e-commerce and allows to make the variety of sold items larger – people start to buy non-mass products in more frequent way, and it creates an opportunity for realization their substitutes.

The previous cases also create possibility for implementation another one – building recommendations like "if an actor bought X, he also with high probability will be interested in Y". Via analysis of interpersonal relations and their buying activities, companies can employ customers each other's authorities and influence for new products adopting. People are interested in their friends' purchases and tend to be affected by their buyer activities. As a result, once a company finds out that an actor started to repeat friends' purchases, it can make further appropriate recommendations on the base of previous neighbors' performance.

9   Conclusion

To sum up, there was a study of community building methods in the spheres of social network analysis and machine learning – some methods use sophisticated preset parameter, some not, so there is the need to apply these methods with tools that include the required features. The most proper algorithms were chosen out of the set of reviewed ones. This knowledge was implemented on the test social network data and the results were estimated on the base of information spread theory which answers the main question – the model that provides the fastest way to spread information to the largest amount of actors is DBSCAN in two cases out of three.

Obtained results can be interesting for IT services marketing or advertisement of sophisticated products and building recommendation for future purchases. At the same time, there are vital obstacles for application of considered techniques: data collection and processing issues, partnership fees and data privacy issues.

Further research may include the following aspects:

- considering more machine learning clustering algorithms after adding more calculation capacity or dataset reduction;
- usage of other social networks' data; in this case some specifics and important structural features of new social big data sources can be investigated and, as a consequence, the final results can have significant variations;
- building new metrics for actors' partitioning; there can be another way of individuals interpersonal relations estimation, on the basis of which the segmentation can be built.

References

1. How Much Data Do We Create Every Day? The Mind-Blowing Stats Everyone Should Read. The Forbes. https://www.forbes.com/sites/bernardmarr/2018/05/21/how-much-data-do-we-create-every-day-the-mind-blowing-stats-everyone-should-read/#6c90f85760ba. Accessed Feb 2019
2. Digital in 2018: world's Internet users pass 4 billion mark. We are social. https://wearesocial.com/blog/2018/01/global-digital-report-2018. Accessed Feb 2019
3. Definition of social network. Merriam-Webster dictionary. https://www.merriam-webster.com/dictionary/social%20network. Accessed Feb 2019
4. Definition of social network. Cambridge dictionary. https://dictionary.cambridge.org/ru/%D1%81%D0%BB%D0%BE%D0%B2%D0%B0%D1%80%D1%8C/%D0%B0%D0%BD%D0%B3%D0%BB%D0%B8%D0%B9%D1%81%D0%BA%D0%B8%D0%B9/social-network. Accessed Feb 2019
5. The Network Science Book. Albert-Laszlo Barabashi. http://networksciencebook.com. Accessed Feb 2019
6. Oussous, A., Benjelloun, F.Z.: Big data technologies: a survey. J. King Saud Univ. Comput. Inf. Sci. **30**, 431–448 (2018)
7. What is Big Data? – A definition with five Vs. The * umBlog. https://blog.unbelievable-machine.com/en/what-is-big-data-definition-five-vs. Accessed Feb 2019
8. Definition of segment. Oxford dictionary. https://www.dictionary.com/browse/segment. Accessed Mar 2019
9. Definition of community. Oxford dictionary. https://en.oxforddictionaries.com/definition/community. Accessed Mar 2019
10. Williams, K.: Social networks and social capital. J. Commun. Inform. (2017)
11. Community vs Social Network. Khoros Community. https://lithosphere.lithium.com/t5/Science-of-Social-Blog/Community-vs-Social-Network/ba-p/5283. Accessed Mar 2019
12. Newman, M.E.J., Girvan, M.: Finding and evaluating community structure in networks (2004)

13. Usama, M., Qadir, J., Raza, A.: Techniques, applications and research challenges (2017)
14. Keeling, M.J., Eames, K.T.D.: Networks and epidemics models. J. R. Soc. Interface (2005)
15. The EU General Data Protection Regulation (GDPR) is the most important change in data privacy regulation in 20 years. EU GDPR.ORG. https://eugdpr.org. Accessed Mar 2019
16. What is personal data? European Commission. https://ec.europa.eu/info/law/law-topic/data-protection/reform/what-personal-data_en. Accessed Mar 2019
17. Dataset YouTube 2. Arizona State University. http://socialcomputing.asu.edu/datasets/YouTube2. Accessed Apr 2019
18. Benchmark Performance and Scaling of Python Clustering Algorithms. HDBSCAN documentation. https://hdbscan.readthedocs.io/en/latest/performance_and_scalability.html. Accessed Mar 2019
19. Comparing Python Clustering Algorithms. HDBSCAN documentation. https://hdbscan.readthedocs.io/en/latest/comparing_clustering_algorithms.html. Accessed Mar 2019

Knowledge Graph Construction for Intelligent Analysis of Social Networking User Opinion

Tingyu Xie[1], Yichun Yang[2], Qi Li[3], Xiaoying Liu[4],
and Hongwei Wang[1(✉)]

[1] ZJU-UIUC Institute, Zhejiang University, Haining, China
emmaxty@163.com, hongweiwang@intl.zju.edu.cn
[2] School of Journalism and Communication, Xiamen University, Xiamen, China
616538770@qq.com
[3] College of Science, North University of China, Taiyuan, China
liqi_lqq@163.com
[4] College of Biomedical Engineering and Instrument Science,
Zhejiang University, Haining, China
1014634062@qq.com

Abstract. Microblogging is a popular social networking tool on which people tend to express their views and opinions. As such, the massive data on microblogging platforms mean abundant research value to social science researchers. To help them better analyze these data, a framework for understanding diverse user opinions and identifying complex relationships in the form of knowledge graphs is proposed in this paper. The two main tasks in the framework are sentiment analysis and knowledge graph construction. In the first task, the Skip-gram model is employed to obtain the word embedding matrix and the Bi-LSTM model is adopted to perform stance classification. It is found in this paper that Bi-LSTM showed better performance in classifying different sentiments, compared with Naive Bayes and SnowNLP. In the second task, relations between different users are extracted from their micro-blogs through recognizing specific strings, and on this basis user attitudes are integrated into the knowledge extracted. A knowledge graph of user opinions is constructed with the Neo4J tool. With the knowledge extracted by this framework, social science researchers can more easily observe rules of perspective communication and perform further analysis of the data.

Keywords: Sentiment analysis · Natural language processing · Knowledge graph · User opinion

1 Introduction

As Web2.0 develops rapidly, end-users become an important part in the creation and sharing of content on the Internet. Among various kinds of social media platforms, microblogging is one of most popular platforms among users, such as Twitter, Weibo and so on. Different from traditional blogs, micro-blogs have content with a typically smaller size. Micro-blogs allow users to exchange small elements of content such as short sentences, individual images, or video links [1], which may be the major reason

© Springer Nature Switzerland AG 2020
K.-M. Chao et al. (Eds.): ICEBE 2019, LNDECT 41, pp. 236–247, 2020.
https://doi.org/10.1007/978-3-030-34986-8_17

for their popularity [2]. Thus, microblogging becomes a convenient platform for users to express opinions and views on varied topics. With the explosive growth of user numbers, the content users post on microblogging platforms has become a massive data set. Mining and analysis of social media data using advanced computational methods has become a popular research field. With the huge amount of data, researchers can analyze potential social media models, including elections, emotional analysis, information dissemination, and topic trends [3]. Below are three specific applications of social media data mining.

(1) e-commerce enterprises can gain users preferences by analyzing their micro-blogs. Then these enterprises adjust their marketing strategies to achieve more accurate delivery of advertisements.

(2) Hotels and restaurants can make improvements on their services based on consumer evaluations and suggestions posted on social media.

(3) Government departments can observe views of citizens on specific social events or proposed policies. By grasping the trend of public sentiment, government departments can take measures immediately to avoid social turmoil.

The big data in social media tend to reflect the dissemination mode and diffraction speed of hot topics, behavior patterns of people, the evolution of social network, and so on. Therefore, these data have great research value for researchers from social sciences. However, most social science researchers do not have the technology and experience to perform data mining – it is especially the case for the massive text data on social media.

To address this issue, this paper proposes a framework to perform automatic analysis of user opinions on specific topics and on this basis construct knowledge graphs to show important relations and trends. Microblogging is taken as an example to demonstrate and evaluate the framework and methods proposed in this paper. Specifically, we first adopt the semantic analysis model to get users opinion, then extract the knowledge of opinion dissemination, and finally construct knowledge graphs from the information extracted. With this knowledge graph, data become easily to interact with and retrieve, and thus researchers can visualize the relationship and dissemination of user opinions conveniently.

In 2002, Pang et al. first put forward emotional analysis in [4]. In this field, stance detection is a common research topic, which aims to automatically determine from the text whether the author is in favor of, against, or neutral towards a given target [5]. Most traditional stance detection methods are based on feature engineering and traditional machine learning. Liu et al. [6] utilized TF-IDF and sentiment dictionary to represent sentences as sparse vectors, and used ensemble classifier to detect stance in a supervised and a semi-supervised way. However, building a sentiment dictionary costs a great quantity of manual labor. Sun et al. [7] designed four types of features including Lexical, Morphology, Semantics, and Syntax to detect stance and analyzed the effectiveness of different features for stance detection. The system proposed by Sun et al. get a good performance in stance classifier with Chinese micro-blogs. However, the feature-based methods depend heavily on sophisticated features of large amount of domain knowledge, and require repeated tuning for parameters in ensemble classifier. With the rapid development of deep learning, stance detection based on deep neural network achieved satisfactory results. Thus, the method we applied are based on deep learning.

Google officially proposed Knowledge Graph in 2012. The original intention is to optimize Google search engine to enhance user experience. Knowledge Graph is a semantic network that reveals the relationship between entities, which organizes the knowledge in an very advanced way. It enables the computer to understand the way humans communicate in language, thus feedback the desired output to users efficiently [8]. A knowledge graph of social networking user data can offer social science researchers a more convenient and efficient way to interact with and retrieve social media knowledge. In this paper, we construct a knowledge graph of user opinions in social platform, providing researchers with a powerful tool to analyze opinion dissemination and human behavior in social media. A general representation of knowledge Graph is a set of triples. Its basic form is entity-relation-entity or entity-relation-attribute. Knowledge extraction is the precondition of knowledge graph construction, including entity extraction, attribute extraction and relation extraction [9]. In this paper, the data we analyze consist of structured part and unstructured part. Entities and attributes are in the structured data, and relations are in the unstructured data. Hence, we focus on relation extraction. After that, we adopt Neo4J to construct the Knowledge Graph.

The rest of this paper is organized as follows. Section 2 introduces the framework for constructing knowledge graph of user opinions in microblogging. The methods for semantic analyzing and knowledge graph construction are detailed in Sect. 3. Section 4 describes experiment settings and results. Finally, a conclusion is drawn and future work is discussed in Sect. 5.

2 Framework

As shown in Fig. 1, the framework of user opinion knowledge graph construction consists of three tasks: Data collecting, natural language processing and knowledge graph construction.

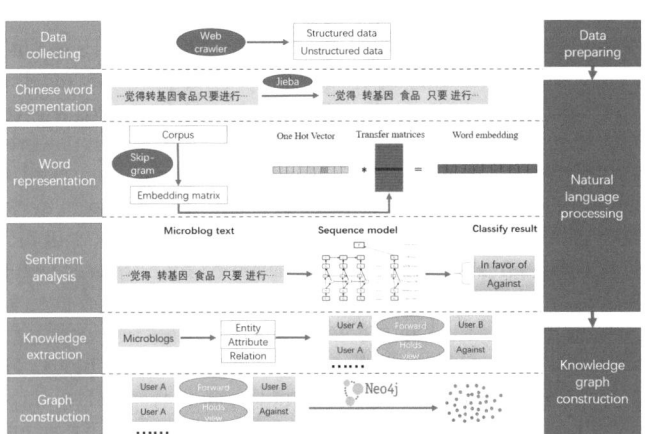

Fig. 1. Framework of constructing knowledge graph for user opinions.

The data we analyzed were micro-blogs on the Internet. These micro-blogs were analyzed through a series of natural language processing methods to obtain attitude of each user. Next, the necessary knowledge for graph construction were extracted from data, and finally a knowledge graph was constructed through a power tool.

A. Data Collecting

There are massive data on social media platforms. We adopted crawling techniques to obtain thousands of micro-blogs posted by users. After that we labeled part of data to be our training set and test set.

B. Natural Language Processing

There are three subtasks, Chinese word segmentation, word representation and sentiment analysis.

(1) *Chinese word segmentation*: Since there is no blank space in Chinese sentence to separate words, we need to achieve Chinese word segmentation according to the semantic of sentences. After that, machine can recognize words in sentences and learn how to represent them.

(2) *Word representation*: Machine can not represent words in language as human beings do, but they can represent words as vectors. The simplest way to represent words in vectors is to represent them as one hot vectors. But this way wastes much memory and have low efficiency in computation. Thus, we adopted word embeddings to represent words in our task.

(3) *Sentiment Analysis*: This task is to make them recognize different sentiment in sentences. Many sequence models were proposed to do semantic analysis, among which Bidirectional LSTM (Bi-LSTM) is a classic model and performed well on sentiment analysis tasks. Thus, we adopted Bi-LSTM as our method do classify between different sentiments.

C. Knowledge Graph Construction

The subtasks are knowledge extraction and knowledge graph construction.

(1) *Knowledge extraction*: We user triples of knowledge to construct graph. To obtain a graph with high expressiveness, we defined entities, attributes and relations according to the significant information in the data. Then we extracted the defined elements from micro-blog texts.

(2) *Graph construction*: With triples of knowledge prepared, we put them into powerful tool Neo4J to construct knowledge graph.

3   Methodology

A. Chinese Word Segmentation

Chinese word segmentation is one of the foremost tasks in the field of natural language processing, which aims to split a given Chinese string into linguistic units [10]. With the widespread use of Chinese, many researchers have made significant progress in

Chinese word segmentation [11]. In this paper, the Chinese word segmentation tool called 'jieba' is applied to identify suitable breaks in sentences.

B. *Word Representation*

Words need to be represented as n-dimensional numeric vectors that is sufficient to encode all semantics of our language. Each dimension of word vectors would encode some meaning that we transfer using language. In this way, the similarity and difference between words are also captured by word vectors. In this paper, Skip-gram model with negative sampling is applied to obtain the word vectors, which are arranged into an embedding matrix with shape (number of words, n). The embedding matrix is used as the parameters of the embedding layer in the network, which is detailed in the next subsection.

C. *Sentiment Analysis*

To achieve stance detection, the sequence model consists of four layers are adopted, which are an embedding layer, a Bi-LSTM layer, a LSTM layer, and a dense layer. The architecture of this network is shown in Fig. 2. First, a micro-blog is represented as a sequence of one hot vectors, and transferred into the embedding layer. Second, word vectors are generated by multiplying the one hot vectors with the embedding matrix in the embedding layer. Third, contextual representation of each word is obtained through Bi-LSTM layer. Fourth, a vector representing the sentence encoding is output by LSTM layer. Finally, a probability is output by the dense layer.

(1) *Preprocessing*

First, punctuation and emoticons are removed in the sentences. Second, Chinses sentences are separated into Chinese words by open-sourced Chinese segmentation tool 'jieba' as mentioned in subsection *A*. Next, words are represented as one hot vectors. These one hot vectors are the input of the embedding layer.

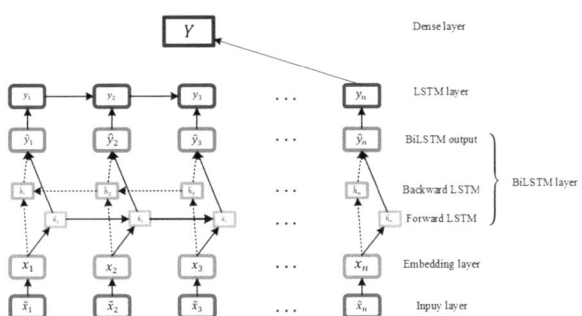

Fig. 2. Architecture of the sequence model adopted.

(2) *Embedding layer*

The function of this layer is to convert one hot word vectors into embedding word vectors. Word embedding matrix mentioned in subsection *B* is the parameters of the embedding layer. As shown in Fig. 3, by multiplying the input one hot vectors with the embedding matrix, embedding word vectors are generated and transferred into the Bi-LSTM layer.

(3) *Bi-LSTM and LSTM layer*

Different from traditional neural network, Recurrent Neural Networks (RNN) are able to remember the information carried by all of the previous words. However, RNN has the problem of gradient explosion and vanishing. An improved network proposed by Hochreiter and Schmidhuber in 1997 named Long Short-Term Memory (LSTM) provides an easier way to preserve the information over many timesteps. Based on LSTM, an advanced model called Bidirectional LSTM (Bi-LSTM) is proposed by Augenstein et al. [12], which not only carry semantic information forward, but also backwards. This model achieved state-of-the-art results in stance detection for microblogs on twitter. Thus, Bi-LSTM is applied to catch the contextual information of words in this paper.

Bi-LSTM is evolved from LSTM. In LSTM, to capture long-term dependencies, complex units for activation are designed with four fundamental operational stages. The architecture of LSTM unit is showed in Fig. 4, and the calculations in one such unit is listed below.

Input gate:

$$i_t = \sigma\left(W^{(i)}x_t + U^{(i)}h_{t-1}\right) \tag{1}$$

Forget gate:

$$f_t = \sigma\left(W^{(f)}x_t + U^{(f)}h_{t-1}\right) \tag{2}$$

Output/Exposure gate:

$$o_t = \sigma\left(W^{(o)}x_t + U^{(o)}h_{t-1}\right) \tag{3}$$

New memory cell:

$$c_t = \tanh\left(W^{(c)}x_t + U^{(c)}h_{t-1}\right) \tag{4}$$

Final memory cell:

$$c_t = f_t \circ c_{t-1} + i_t \circ c_t \tag{5}$$

Output:

$$h_t = o_t \circ \tanh(c_t) \tag{6}$$

Where x_t represents the input at time-step t. i_t, f_t, o_t indicate the output of input gate, forget gate, output gate respectively, $\tilde{c}_t, c_t$ indicate the new memory and final memory respectively, h_t represents the hidden state at each time-step, t. W, U are parameters to be learned through backpropagation procedure.

The architecture of Bi-LSTM model is shown in Fig. 2. There are two hidden layers in one Bi-LSTM layer, one for the left-to-right propagation and another for the right-to-left propagation. The only difference between these two hidden layers is in the direction of recursing through corpus. The formula to calculate the outputs of hidden layers are listed below.

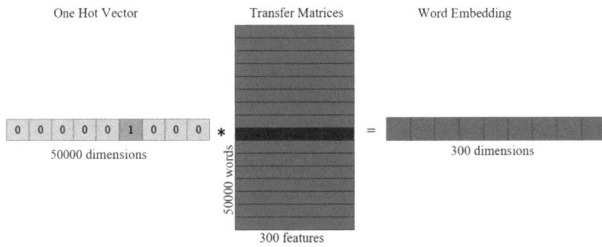

Fig. 3. The process of transfering a one hot vector into an embedding vector.

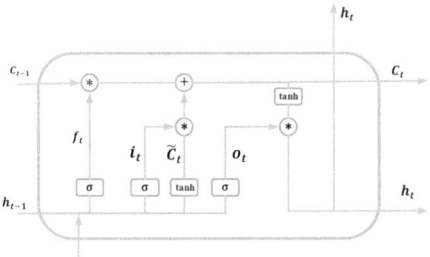

Fig. 4. The architecture of the LSTM unit.

$$\vec{h}_t = f\left(\vec{W}x_t + \vec{V}\vec{h}_{t-1} + \vec{b}\right) \tag{7}$$

$$\overleftarrow{h}_t = f\left(\overleftarrow{W}x_t + \overleftarrow{V}\overleftarrow{h}_{t-1} + \overleftarrow{b}\right) \tag{8}$$

$$h_t = \vec{h}_t; \overleftarrow{h}_t \tag{9}$$

Where $\vec{h}_t$ and $\overleftarrow{h}_t$ are the outputs in time step t of the forward layer and the backward layer respectively. Similarly, $\vec{h}_{t-1}$ and $\overleftarrow{h}_{t-1}$ are the outputs in the last time step of the forward layer and the backward layer respectively. x_t is the same as in LSTM units. $\vec{W}$, $\vec{V}$ and $\vec{b}$ are the parameters learned in the forward layer, $\overleftarrow{W}$, $\overleftarrow{V}$ and $\overleftarrow{b}$ are the parameters learned in the backward layer. The output h_t of time step t in Bi-LSTM layer is generated through concatenating $\vec{h}_t$ and $\overleftarrow{h}_t$, catching the context of a certain word in the sentence.

The contextual information of each word is generated through Bi-LSTM and is then transferred to the LSTM layer. Then the semantics of the whole sentence is extracted as a vector by LSTM layer and passed to the dense layer.

(4) *Dense layer with sigmoid activation*

A fully connected layer is employed as the final layer of our model, whose input is the sentence encoding vectors. Sigmoid function is applied as the activation function, which outputs a numerical value between 0 and 1. This value represents the probability of the user being supportive to the target. The numeric result with roundness equal to 1 is defined to have a positive attitude towards the certain topic, and vice versa. Sigmoid function is as follows.

$$s(x) = \frac{1}{1 + e^{-x}} \tag{10}$$

where x indicates the input of sigmoid function, and the result $s(x)$ is a numerical value between 0 and 1.

D. *Knowledge Extraction*

Each user is defined as an entity. Number of fans and attitudes towards the target topic are defined as two attributes of a user. The attitudes of users if either in favor of or against towards the target topic. If user A reposts a micro-blog of user B, then user A and user B have the relation of "forwards". Table 1 summarizes the triples in our knowledge graph.

Entities, attributes and relations between users and attributes are already in structured form in the data we crawled off Weibo. Thus, this paper only focuses on extracting relations between users. When user A reposts a micro-blog from user B, the reposted micro-blog will be specially formatted as "Content created by user A // @username of user B: content created by user B". We let machine go through each micro-blog and recognized those containing string "//@username:". Thus, the original poster of this micro-blog is found by 'username' in the string, and the relations between users are extracted.

E. *Graph Construction*

Neo4J is adopted in this paper to construct knowledge graph, which is a mainstream database for analyzing knowledge graph at present. It stores data in graph network rather than in tables. Neo4J can also be regarded as a high-performance graphics engine with all the features of a mature database. In Neo4J, entities and attributes are represented as nodes, while the relationships between entities and attributes are represented as lines between nodes. At the same time, this graphical structure with semantic relationship can be queried according to nodes and edges, instead of the need for multiple tables in SQL database, which greatly improves query efficiency.

4   Experiment

A. *Data Set*

Weibo is the most popular microblogging platform in China. Thus, we took Weibo as an example to analyze user data. Whether to accept genetically modified food is a typical hot topic among Weibo users. Hence, we chose this topic to do our experiment. We collected data through crawling on the Weibo web pages. And we labeled 3806 samples of micro-blogs to be our training and test data. The attitude of those micro-blogs is either in favor of, or against genetically modified food, which account for 30% and 70% respectively. Micro-blogs showing supportive attitude were labeled as category "1", and the opposite were labeled as category "0". Our task in stance detection is to make machine learn to classify between category 1 and 2. Table 2 displays examples for each category.

B. *Cost Function*

As we mentioned in last subsection, the distribution of different categories is imbalanced. To address this problem, we adopted weighted cross entropy to be our loss function. In this improved cross entropy function, we took the proportion of negative samples to be the weight of the loss term of positive samples, and vice versa. The formula of the loss function is as follows.

Table 1. Triples in the knowledge graph

Triple	Example
User A – Forwards – User B	黄若宾 – Forwards - 清风茶语
User A – Number of fans - Number	清风茶语 – Number of fans - 68
User A – Holds view - Attitude	蓝黑 – Holds view – Against

Table 2. Micro-blogs of opposite attitude

Micro-blogs	attitude
我觉得转基因食品只要进行明示和标注，可以在市场上出售，供消费者个人选择。	In favor of
想了想，还是什么季节吃什么季节生长的菜最好，应该是符合养生之道的吧？大棚、进口新秀少之，转基因避之。	Against

$$\text{loss} = - \sum_{i=1}^{n} \alpha y \log y + (1 - \alpha)(1 - y)\log(1 - y) \tag{11}$$

where α is the proportion of negative samples.

C. *Training Details*

Since the lengths of different micro-blogs varies, we took the mean plus one standard deviation of lengths to be the uniform length. Micro-blogs shorter than the uniform length were padding by zeros at the start of sentences, and those longer than the uniform length were truncated at the end of sentences. In our experiment, the uniform length is 1031.

The quality of word vectors depends deeply on the scale of training corpus. In this paper, more than 880 thousand micro-blogs on Weibo were collected to be the corpus for training word vectors. After the training, the most frequently used 50 thousand words were chosen to form the vocabulary of the sentiment analysis task.

The database with 3806 samples was split into a training set and a test set with respect to the proportion of 70% and 30% respectively. The hyper parameters of our experiments are listed in Table 3.

D. *Experiment Results*

(1) *Sentiment Analysis*

Naive Bayes model and SnowNLP python library were adopted as baselines. F1 score was adopted as the metric to evaluate these three methods. Table 4 lists the experiment results.

According to this table, all the methods gained better results on negative samples. Although Naive Bayes scored over 0.9 on both training and test sets of negative samples, it performed much worse on positive samples. The performances of SnowNLP on positive and negative samples were not very different, but both were very poor. Bi-LSTM did slightly better on negative samples, and scored very well on both sides. What's more, it scored higher in all terms than any other methods. According to the results of comparative experiments, the method adopted in this paper is superior.

(2) *Knowledge Graph Construction*

The knowledge graph constructed in this paper is partly shown in Fig. 5. The dissemination of user opinions can be visually expressed though this graph. With this knowledge graph, researchers from social science can analyze social media data conveniently, regardless whether they know about data mining technology.

Table 3. Values of hyper parameters

Hyper parameter	Value
Dimension of word vectors	300
Number of units in Bi-LSTM layer	64
Number of units in LSTM layer	16
Batch size	225
Number of epochs	5

Table 4. Experiment results on sentiment analysis

Method	Data set	$F_{against}$	F_{favor}	$F_{overall}$
Naive Bayes	Training	0.95	0.67	0.81
	Test	0.92	0.46	0.69
SnowNLP	Training	0.26	0.22	0.24
	Test	0.26	0.22	0.24
Bi-LSTM	Training	0.99	0.94	0.97
	Test	0.97	0.84	0.91

Fig. 5. A part of the knowledge graph constructed based on user opinions from Weibo.

5 Conclusion

Micro-blogs are one of the most important social networking platforms today. There is a huge volume of user data on these platforms, which contain abundant information for researchers from social sciences. In order to help social science researchers better analyze social media user data, a framework for automatically understanding social networking user opinions and constructing knowledge graphs is proposed. First, sequence models are used to analyze user attitudes. Second, relations between users are extracted from micro-blog text. Third, the attitudes of users are integrated into the knowledge extracted. Finally, the powerful tool Neo4J is employed to construct user opinion knowledge graphs. With this graph, social science researchers can observe the dissemination of user opinions intuitively, and interact with data efficiently.

The performance of the algorithms developed will be further improved in our future work. There are many other kinds of social platform besides micro-blogs. The data on these platforms have different features from micro-blogs, and also contain valuable information for social science researchers. Therefore, it is necessary to study the methods of applying knowledge graph on user data of these social platforms in the future.

Acknowledgment. This work was supported by the Zhejiang University/University of Illinois at Urbana-Champaign Institute, and was led by Principal Supervisor Prof. Hongwei Wang. The authors would also like to thank Dr. Gongzhuang Peng from University of Science & Technology Beijing for his guidance and help on this work.

REFERENCES

1. Kaplan, A.M., Haenlein, M.: The early bird catches the news: nine things you should know about micro-blogging. Bus. Horiz. **54**(2), 105–113 (2011)
2. You, Q., Luo, J.: Towards social imagematics: sentiment analysis in social multimedia. In: Proceedings of the Thirteenth International Workshop on Multimedia Data Mining, p. 3. ACM (2013)
3. Aichner, T., Jacob, F.H.: Measuring the degree of corporate social media use. Int. J. Market Res. **57**(2), 257–275 (2015)
4. Bo, P., Lee, L.: Seeing stars: exploiting class relationships for sentiment categorization with respect to rating scales. In: Proceedings of the 43rd Annual Meeting on Association for Computational Linguistics, pp. 115–124. Association for Computational Linguistics (2005)
5. Matt, T., Pang, B., Lee, L.: Get out the vote: determining support or opposition from congressional floor-debate transcripts. In: Proceedings of the 2006 Conference on Empirical Methods in Natural Language Processing, pp. 327–335. Association for Computational Linguistics (2006)
6. Liu, L., Feng, S., Wang, D., Zhang, Y.: An empirical study on Chinese microblog stance detection using supervised and semi-supervised machine learning methods. In: Natural Language Understanding and Intelligent Applications. Springer, Cham, pp. 753–765 (2016)
7. Sun, Q., Wang, Z., Zhu, Q., Zhou, G.: Exploring various linguistic features for stance detection. In: Natural Language Understanding and Intelligent Applications. Springer, Cham, pp. 840–847 (2016)
8. Cao, Q., Zhao, Y.: Technology realization process of knowledge map and related applications. Intell. Theory Pract. **38**(12), 127–132 (2015). (In Chinese)
9. Xu, Z., Sheng, Y., He, L.: Overview of Knowledge mapping technology. J. Univ. Electron. Sci. Technol. China **45**(4), 589–606 (2016). (In Chinese)
10. Luo, G., Zhang, Z.: Principles and Technical Realization of Natural Language Processing. Electronic Industry Press, Berlin (2016)
11. Chen, X., Qiu, X., Zhu, C., Liu, P., Hua, X.: Long short-term memory neural networks for Chinese word segmentation. In: Proceedings of the 2015 Conference on Empirical Methods in Natural Language Processing, pp. 1197–1206 (2015). (In Chinese)
12. Augenstein, I., Rocktäschel, T., Vlachos, A., Bontcheva, K.: Stance detection with bidirectional conditional encoding. arXiv preprint arXiv:1606.05464 (2016)

The Study of Predicting Social Topic Trends

Sung-Shun Weng[(✉)] and Huai-Wen Hsu

National Taipei University of Technology, Taipei 10608, Taiwan
wengss@ntut.edu.tw

Abstract. The rapid growth of the social media leads people participate in the popular topics that have been discussed in our daily lives by the social networks. Large amounts of word-of-mouth and news event have flood the social media. Recognizing the trends of the main topics that people care about from the huge and various social messages, grasping the business opportunities and adopting appropriate strategies have become an important lesson for business, governmental and non-governmental organizations. Previous research on social topic detection has focused on sentiment analysis for content. This study integrates the hidden markov model and latent dirichlet allocation topic model to forecast trends of the social topics based on time series data of user reviews. Experimental results on real dataset showed that the approach proposed by this study are able to recognize the latent social topics, keywords and forecast the trends of social topics effectively on the social media.

Keywords: Social media · Topic detection · Time series · Trend prediction

1 Introduction

With the rapid development of the internet, multiple social networking sites came into being, such as Facebook, Twitter, Mobile01, Reddit, PTT, etc. These social networking sites become more popular because of the increasing use of the social network. Moreover, these platforms are often filled with a lot of word of mouth information and news articles. People have begun to understand or participate in many hot issues and focus on them through these social networking sites. E.g. Enterprise brand information, social events, government, social organization of policy advocacy and user generated content.

Users of social networking sites master the power of the community's message passing. Many enterprises and government organizations have begun to invest in mining social networking sites. They leverage social network mining techniques to develop potential customers and strengthen enterprise service quality in order to expand the market. Chen et al. [1] pointed out that there are two reasons for mining social data: (1) The information generated by the social networking sites usually contain the public interest, attention and comments. These messages can be turned into a reference index for the improvement of products and services. (2) Social data used to connote some invaluable insights into the market trend which often closely related to corporate organizations.

© Springer Nature Switzerland AG 2020
K.-M. Chao et al. (Eds.): ICEBE 2019, LNDECT 41, pp. 248–262, 2020.
https://doi.org/10.1007/978-3-030-34986-8_18

Gartner (2012) pointed out that an enterprise might lose their customers about 15% if they cannot respond to their customers. However, the article information on the social networking site has the characteristics of explosive growth and instant. How to identify the subject and preference of public or specific groups from huge and complicated social text messages and understand the main topic trend to grasp the market business opportunities and take the corresponding marketing strategy has become one of the important topics in the enterprises and government organizations.

In the past, topic detection and tracking techniques are widely applied to the public opinion monitoring system or document classification, but mostly focus on news articles. In recent years, with the rise of Facebook, Twitter and other social networking sites, many studies have used the technology for topic detection and tracking in social media [2]. The main approaches can be broadly classified into three types. The first approach is based on the TF-IDF technique, which extracts keywords and identifies new words which occur at different time. It usually establishes a graph for keywords to observe the relationship between keywords. The second approach is based on clustering technique, which through the different features to propose more newly clustering algorithms. The last approach is based on the topic modeling techniques, which through the probability distribution and sampling to estimate the extent of the topic distribution or topic conversion in the document and combines the lexicon and dictionary learning to enhance the description ability of the topic model.

Based on the above-mentioned approaches, many studies have integrated sentiment analysis or solving the short text problem. Previous researches rarely track the popular topics of social networking sites and forecast their future trends. Therefore, in order to help marketing to carry out the relevant topic detection and trend insight then support their marketing strategy. This study focus on the characteristics of the PTT social networking site. We combine the topic and time series model to detect the popular issues on the PTT and forecast their future trends to know whether these issues will be popular or declined.

To sum up the research background and motivation, the purpose of this study has three main objectives:

1. Establish a social topic detection mechanism for the PTT.
2. Extract time series pattern and model different social topics for the PTT.
3. Use open source corpus to extend the lexicon and enhance the description ability of the topic model.

2 Related Works

2.1 Social Media Marketing

According to the characteristics of different websites, Kaplan and Haenlein [3] defined that social media is a network application system platform based on Web2.0 concept and technology, which enables internet users to communicate with each other through the platform and establish the content and information. Social media can provide people through various forms of media, share each other's life experience and the network platform of opinions and views of things, including text, photos, video and music, etc. [4]. It is a big characteristic of the social media that social networking site users can through the many-to-many information transmission and the community interaction to rich and generate a lot of content. According to Kotler and Keller [5] definition, marketing is a series of and organized processes that provide value to customers through creation and communication, and to maintain a better return to the stakeholders in relation to the customer. Marketing research is nothing more than to make data collection and analysis of the market and through the analysis of the results to make marketing decisions. In order to meet the trend of social media services, Wollan et al. [6] considered that social media marketing is to find potential customer groups through the social data, listening the voices of customers in social media and understanding them to seek products and services that are suitable for development, sales for these products and services.

2.2   Social Word-of-Mouth and Social Mining

In recent years, data mining technology has become one of the analytical methods of social media marketing. Hemann et al. [7] pointed out that social media marketing analysis can be divided into four categories, social listening, search analytics, content analytics, engagement analytics. Liu [8] believes that Word-of-Mouth Social is a sharing of experiences and opinions of a certain thing, which represents an indicator of corporate brand and product service reputation, has an important value to the enterprise organization. Many enterprises would collect the customer's word of mouth and reviews in social media, and leverage the social mining technique to understand the acceptance of their own market. Therefore, there are many researchers study on different aspects of social mining. E.g. As an important reference of the viral marketing, Cha et al. [9] research on the user`s influence of public opinion on Twitter and leverage three features In-Degree, Retweets and Mentions to identify the high influence users. Li et al. [10] used the Markov Random Fields to detect the specific issue promoters in Twitter. Singh et al. [11] use the SVM to do Sentiment Analysis for the movie reviews.

2.3   Topic Detection

Topic Detection can be applied to the document data of different domain, by using different adjusted model to do advanced application, such as image recognition, gene-sequencing, social network analysis, recommender system, etc. Topic Detection can be divided into three types of methods:

1. Keyword-based Topic Detection

   For the research on Keyword-based Topic Detection, the main approach is to use TF-IDF to obtain the keywords from the articles and combine other approaches to conduct the topic detection. E.g. Chen et al. [12] used TF-IDF to extract the keywords from the news articles and combine the Aging Theory to build up a life cycle model for topic detection. Sayyadi et al. [13] used TF-IDF to extract the keywords, and generate a keyword graph to cluster Twitter articles by using different distance similarity methods, and finally identify some latent topics. Cataldi et al. [14] transformed the Twitter's tweets into text vectors and defined that a new word has high frequency in the same given time interval, but it was not high frequency in the past, and based on the co-occurrence words to use these features to generate a keyword graph to represents an emerging topic. The Keyword-based Topic Detection mainly builds up the keyword relation graph to observes the variations of the topic in details. However, to build a graph often requires a very large computational space.

2. Clustering-based Topic Detection

The Clustering-based Topic Detection approach is to combine the traditional document clustering methods and other methods to conduct topic detection. E.g. You et al. [15] regards each series response message on the Yahoo BBS as a related topic. With the articles and the associate of author`s influence to filter the articles, using the Two Pass K-means to cluster the articles into topic, finally use the Back-Propagation Neural Network to classify different popularity levels of the topic. Xie et al. [16] use the hierarchical clustering to cluster the Weibo's articles into different topics, which is based on the distance similarity of the articles' vector. Becker et al. [17] obtain some representative features from social media articles' cluster by using the Cluster Ensemble approach, and then put forward a suitable clustering approach for these features. Chen et al. [1] identify relevant articles and user of the government agencies and non-profit organizations on the social media by using the co-training Learner, and leverage the Single-Pass Incremental Clustering to cluster the articles into different topics and presented to the relevant organizations to track. The Clustering-based Topic Detection can greatly reduce the time of the human labeling and labor costs. To choose an appropriate number of clusters, these approaches still need to through the large scale experiments and experience to adjust, and spend a huge amount of computing time and iteration when facing large data sets.

3. Topic Model-based Topic Detection

Topic Model is a type of probabilistic generative model, it can identify the latent topic structure from a large amount of structured and unstructured documents, and use an abstract way to describe the topics of these documents [18]. Topic model is commonly used in the topic detection and identification research. In related studies, the Probabilistic Latent Sematic Analysis (PLSA) proposed by Hofmann [20] and the Latent Dirichlet Allocation (LDA) proposed by Blei et al. [21] are the most widely used approaches; such as the Topic Over Time (TOT) proposed by Wang et al. [21], which not only consider the distribution of words and articles but also including the articles' timestamp into the model to observe the topic's variation in each times. Wang et al. [22] proposed the TM-LDA, which consider the time factor and identify the topic on the Twitter's articles, then explore and study the association between topics. Topic Model is an unsupervised learning approach, it doesn't need the labeled data when facing a large amount of data. However, these type of probabilistic model's design can't suit for every domain, they often need some adjustment and improvement for the specific domain.

2.4   Topic Popularity and Trend Forecasting

Topic popularity and trend forecasting mainly focus on a specific event or issue, and consider one or more point of view to forecast the trend. There are two types of approaches: The first is Statistical Models-based Trend Forecasting approaches, they focus on the specific features of an event or issue to be a forecasting index. e.g. Szabo and Huberman [23] research on two social media platforms, YouTube and Digg; they found that the popularity of the current online videos or news articles have a high positive correlation with the popularity of history. They forecast the popularity of some online videos and news articles by using the Logarithmically Transformed Linear Regression Model. Achrekar et al. [24] collected some H1 N flu epidemic from Twitter, they combine the Auto-Regressive and Moving Average (ARMA) and Vector Auto-Regression (VAR) to forecast the trend of the flu epidemic. Finally, they found that the correlation between the flu epidemic and the records of the Centers for Disease Control and Prevention (CDC) can reach positive correlation about 0.98; the approach can forecast the trend of flu epidemic early with Twitter's data and help related agents to make property decision and pick up right strategy. Kim et al. [25] use the click of news articles and combine the saturation theory and life cycle theory of the economics to predict the news' popularity and reach the accuracy about 86%.

The second is Machine Learning-based Trend Forecasting, due to most of the traditional statistical models are depend on lots of variable's hypothesis, many researches collect a large amount of historical data and build up a model for forecast the popularity and trend of topics by using machine learning. E.g. Ritterman et al. [26] predicted the foot-and-mouth disease by collecting a large amount of Twitter's data by using the N-gram model to extract some features and build up a Support Vector Regression (SVR) model. The comparison of the experimental result and the index predicted by the prediction market show that adding these features can obtain more accurate prediction results. Bandari et al. [27] classify the Twitter's hot news and reach the accuracy about 84% by using the times of Named Entities, Link's tweets, etc. Zaman et al. [28] build up a social network that include the relation between the Tweet and Retweet; they combine the Bayesian-based Approach to predict the retweet trend of a tweet. Figueiredo et al. [29] leveraged the K-Spectral Clustering (KSC) to find out he different trend's time series pattern, and choose some different pattern as a feature to forecast the trend. Fernandes et al. [30] constructed an Intelligent Decision Support System (IDSS), which can predict the popularity of the Mashable's news through the keywords, contents, cited articles, etc.

3   Research Methodology

3.1   Research Architecture

This study proposes a trend forecasting for social topics mechanism based on topic and time series model. This mechanism can be divided into five parts: data collection, data pre-processing, topic finding and detection, topic popularity and trend forecasting. In this study, we developed an automatic crawler to collect the articles and comments from PTT. PTT is a kind of social networking site like Reddit, which is a popular social networking site in Taiwan. We leveraged the Chinese Knowledge Information Processing System (CKIP) to segment the contents of the articles and filtered the stop-words in these articles. After the word segmentation and stop-words filtering, we obtained the articles' vector and used the LDA model to generate the topics, topics' related articles, and keywords. We forecasted the trend of the topics' articles by extracting the time series data of the users' comments from each article; compared the time series data pattern between the current reviews and historical comments. The overview of our mechanism is shown in Fig. 1.

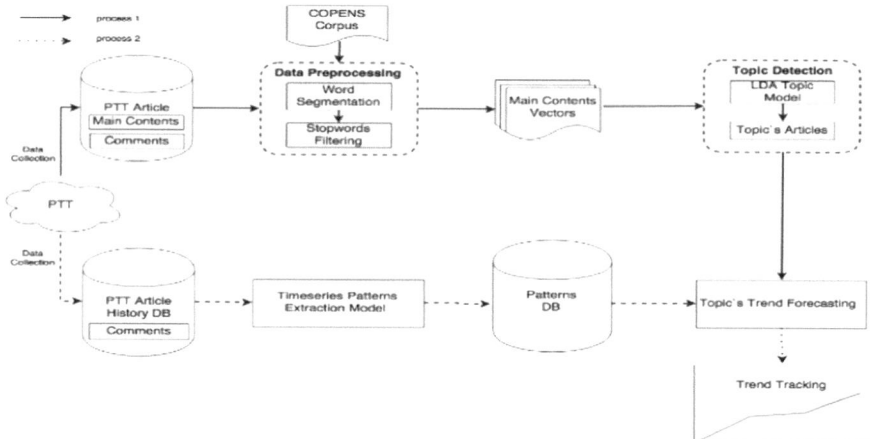

Fig. 1. Research mechanism.

3.2   Data Collection and Pre-processing

In this study, we develop a distributed crawler to collect the articles and user comments from the PTT Gossiping thread. In order to generate the input format for the LDA model, we leveraged the CKIP to segment the contents of the articles and then obtain the vectors of the articles. In practical, stop-words, adverbs, conjunctions are meaningless for the topic model and it could be a kind of noise. However, CKIP only provide a simple stop-words list, it doesn't enough to this study. Therefore, after the word segmentation, we used the COPENS[1] corpus to extend the lexicon and stop-words list to filter more stop-words and meaningless words to enhance the description ability of the topic model.

3.3   Topic Finding and Detection

There are a number of different types of discussion thread in the PTT, and a large number of users publish daily life related events in articles or discussion series every

---

[1] http://lopen.linguistics.ntu.edu.tw/.

day. However, in the face of this instant and rapid growth of the community platform, the topics generate continuously, the cost of time and labor is very high when using the artificial way to identify popular topics from these large numbers of articles and discussions; therefore, in order to adapt to the phenomenon, we apply the LDA model proposed by Blei et al. [19] to identify the latent social topics on the PTT. LDA topic model regarding the words based on the bag of words (BOW) model; therefore, LDA ignores the order of the words into consideration of sampling model and greatly reduce the complexity of the model, so it is often used in solving the problem of large-scale document topic detection.

As shown in Fig. 2, we regard all the PTT's articles as a collection of documents $D = \{d_1, d_2, \ldots d_m, \ldots, d_M\}$; the parameter W represents a collection of each word in the corpus set, which is processed by the pre-processing; the parameter N is the total number of words in the corpus; the parameter K is the number of topic identification; the parameter α is the prior Dirichlet distribution of the topic assigned to each document d_i; the parameter β is the prior Dirichlet distribution of the assignment of words to each topic; the parameter θ is a $M * K$ matrix, then the topic distribution of the mth document can be expressed as $\overrightarrow{\theta_m}$; φ the parameter θ is a $K * W$ matrix, then the word distribution of the kth topic can be expressed as $\overrightarrow{\varphi_k}$; the parameter Z determines the corresponding topic k of the nth word in document d_m.

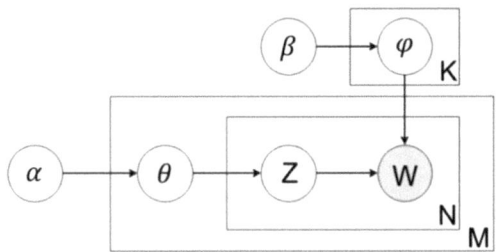

Fig. 2. LDA Model (Source: Blei et al. [21]).

LDA is a probability distribution based unsupervised model, it does not require prior manually labeling for the articles; LDA only needs to identify the scope of the PTT's articles, and select the appropriate number of topics and data pre-processing, then it can identify latent topics from a large number of discussed articles. By using the cosine similarity calculation as Eq. (1) shows, we can group the articles into clusters by similarity.

$$doc\_sim(D_x, D_y) = \frac{\vec{V}(D_x) \cdot \vec{V}(D_y)}{\left|\vec{V}(D_x)\right|\left|\vec{V}(D_y)\right|} \tag{1}$$

The parameter D_x and D_y represent two different ariticles, and $\vec{V}(D_x) \cdot \vec{V}(D_y)$ is the dot product of the article D_x and D_y; then doc_sim represent the angle between the vectors of two articles. Thus, the larger the value of the doc_sim, the smaller the angle is, and the higher cosine similarity of these two articles, the higher the similarity of the two articles.

3.4   Topic Popularity and Trend Forecasting

In the major social media, PTT not only a high degree of discussion and usage, but also an intensive and instant platform. We can classify a topic or issue into five states: (1) Slow Growth: the topic discussion trend growth rate is slow, the time interval is longer; (2) Fast Growth: the topic discussion trend growth rate is fast, the time interval is longer; (3) Stable: the topic discussion trend is maintained at a level that does not rise nor decreases; (4) Fast Recession: the topic discussion trend in a short period of time showed a rapid decline; (5) Slow Recession: the topic discussion trend of decline is slow and the time interval is longer. We leverage the output of the topic finding and detection module to forecast the trend of the topics routinely; and observe the growth situation of the popular degree of the article under different topic. The topic of PTT and the trend of the article is a non-stationary time series. The trend of the topic and the topic discussed in the article is only influenced by the current topic and the popularity of the article. In other words, the state of time point t is only related to the time point T −1; every state of the transition is random. Therefore, this stochastic process is considered as a Markov Process, and then we forecast the topic popularity and trend based on the hidden markov model (HMM). The parameters of hidden markov model structure can be represented as follows:

- $S = \{s_1, s_2, \ldots, s_N\}$ represents a collection of hidden states s;
- $O = \{o_1, o_2, \ldots, o_M\}$ represent collection of observation o;
- A represents a $N * N$ Transition Probability Matrix (TPM);
- B represents a $M * N$ Emission Probability Matrix (EPM);
- π represents a N dimension Initial Probability Matrix;
- H_t represents the hidden state at time point t, and $H_t \in S$;
- E_t represents the observation at time point t, and $E_t \in O$;
- A, B, π must obey the following conditions:
- $\sum_{j=1}^{N} a_{ij} = 1$ and $1 \leq i, j \leq N$, where a_{ij} represents the element of the matrix A, indicates the probability of the hidden state s_i transfer to the hidden state s_j;
- $\sum_{k=1}^{M} b_i(o_k) = 1$ and $1 \leq i \leq N, 1 \leq k \leq M$; where $b_i(o_k)$ represents the element of the matrix B, indicates that in the case of the hidden state s_i, the probability of observation value of o_k;
- $\sum_{i=1}^{N} \pi_i = 1$ and $1 \leq i \leq N, 0 \leq \pi_i \leq 1$; where π_i represents the element of the matrix π, the probability of each hidden state s_i at the initial time point t = 1;

According to the popularity of the PTT articles, there are five hidden states, and transfer between the five hidden states. That is, N = 5. s_1 represents the slow growth state, s_2 represents the fast growth state, s_3 represents the stable state, s_4 represents the fast recession state, s_5 represents the slow recession state. These five kinds of hidden state in this study can't be directly observed, we can only through the observations of the output sequence of these hidden state to estimate them. In the hidden markov model, the initial hidden state probability distribution can be the uniform distribution, so the initial hidden state probability distribution is set to:

$$\pi = (s_1 = 0.2, s_2 = 0.2, s_3 = 0.2, s_4 = 0.2, s_5 = 0.2).$$

In this study, we use topic finding and detection module to get the topic and related articles, and establish the hidden state transition matrix. We can generate a Transition Probability Matrix A for the hidden states. As shown in Fig. 3, where $1 \leq i, j \leq 5$, a_{ij} represents the probability that the hidden state s_i transfer to the hidden state s_j from time point t to time point $t + 1$, and $0 \leq a_{ij} \leq 1$.

$$A = \begin{bmatrix} a_{11} & a_{12} & a_{13} & a_{14} & a_{15} \\ a_{21} & a_{22} & a_{23} & a_{24} & a_{25} \\ a_{31} & a_{32} & a_{33} & a_{34} & a_{35} \\ a_{41} & a_{42} & a_{43} & a_{44} & a_{45} \\ a_{51} & a_{52} & a_{53} & a_{54} & a_{55} \end{bmatrix}$$

Fig. 3. Hidden state transition diagram.

In this study, we defined the observation as the growth situation of the comments score, and defined the observation sequence as the rate of the article's comments score. As shown in Eq. (2), we collect the growth rate of the article's comments score from time point t to time point $t - 1$.

$$\frac{score_t - score_{t-1}}{score_t} \times 100\% \tag{2}$$

Then we assumed the observation at s_i are obey the Gaussian distribution as shown in Eq. (3). Where the μ is the average value of the observation, and $\sum$ is the co-variance matrix.

$$P(E_t | H_t = s_i) = \frac{1}{det(2\pi \sum)} exp \left[-\frac{1}{2} (E_t - \mu)^T \sum^{-1} (E_t - \mu) \right] \tag{3}$$

To forecast the trends, we need to train a model for each article and maximum $P(E|\lambda)$ when we get a series data of observation $E = \{E_1, E_2, \ldots, E_t\}$. We refer Rabiner's research [31] and train the forecasting model by applying the forward-backward algorithm and Baum-Welch algorithm to find the most probable state transition probability and the probability of the observed output, constantly update and adjust the parameters of the model; then find the best model and parameter of E, which is the most consistent to the sequence data. We then train the model for each series data of historical comments by Baum-Welch algorithm, and record the likelihood of the series data at each time point. We assumed that some trends might be similar to the trends in the past, and the future trend should follow about the same past trend pattern. Therefore, we can find the most similar pattern for the current series data, and then follow the same behavior to forecast the trend of the current series data.

4 Experiments and Results

4.1 Design of Experiments

To measure the effectiveness of the topic finding and detection module and topic popularity forecasting model; we split each time series data by the time interval from 1 to 10 min, then use the topic popularity forecasting model to forecast the comment score of each article. Thus, we design some experiments as follows:

1. To observe whether the topic model is affected by the difference of the number of topics, and the description ability of the social topics.
2. To observe whether the forecasting model is affected by the difference of the length of the time interval, then compare what the length of the time interval is the most suitable for the forecasting model.
3. To observe whether the forecasting model is affected by the difference of the percentage of the training data, and the more training data we collect, the more accurately we can forecast.
4. To observe the effectiveness of the forecasting model on real dataset and each topic.

4.2 Data Sets and Pre-processing

Due to the reason that PTT has been developed based on the bulletin board system (BBS), collect the data from the BBS version is more complicated than web version, so we develop a web crawler to collect the gossiping thread data from 2016/01 to 2016/03. To make sure the time series data are long enough, we filter out the time series data of the comments score, which is longer than 100. Finally, there are 2273 articles in total. The details as shown in Table 1.

Table 1. Information about the dataset.

Number of articles	2273
Start time	2019-01-01 00:00:00
End time	2019-03-31 23:59:59

4.3 Evaluations

To measure the effectiveness of the topic popularity forecasting model, we use the mean absolute percent error (MAPE) to be our evaluation index, as shown in Eq. (4); the lower the MAPE value we get, the smaller the error between the predicted value and the actual value is, which means that they are in highly consistency. Where the act_i

represents the actual value, pre$_i$ represents the predicted value, and N represents the number of the comments score data.

$$MAPE = \frac{1}{N} \sum_{i=1}^{N} \left| \frac{act_i - pre_i}{act_i} \right| \times 100\% \qquad (4)$$

Among many evaluation indexes, Lewis [32] considered the MAPE index the most effective one. Therefore, many previous works have evaluated their study by this Lewis's definition.

4.4   Experimental Results and Discussion

To determine the number of the topics in the dataset, we refer Griffiths and Steyvers [33] study; choose the appropriate number of the topics by computing the perplexity value. We can find the most appropriate number of the topics when the perplexity began to converge; the setting can be more descriptive. As shown in Fig. 4, we found the appropriate number of the topics is about 20 and the perplexity value was 3370. According to the experiment of perplexity, we got 20 topics and compared the trend topics manually and the results.

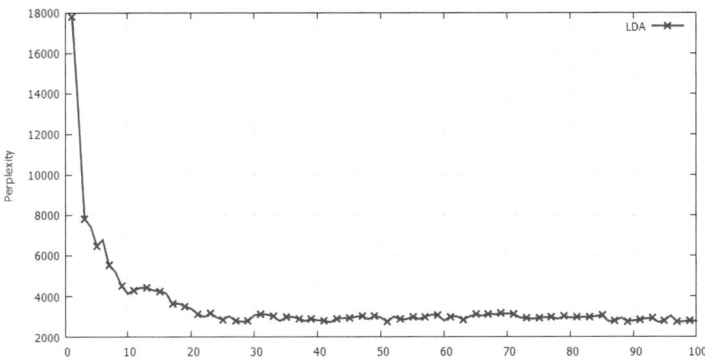

Fig. 4. Experiment of perplexity

In order to know what length of the time interval can be more effective and suitable for the PTT's behavior. We forecasted the series data of the comments score with different time interval; computed the MAPE at each time point. As shown in Fig. 5, the shorter time interval length we use, the more effective model can adapt to the change of the trend and 1 min is the best setting.

In this study, we assumed that we might forecast the popularity when we get more training data. Therefore, we built up the forecasting model with 10% training data to 90% and based on the 1 min time interval. As shown in Fig. 6, when the percentage of the training data up to 70%, the MAPE and forecasting can be more accurate.

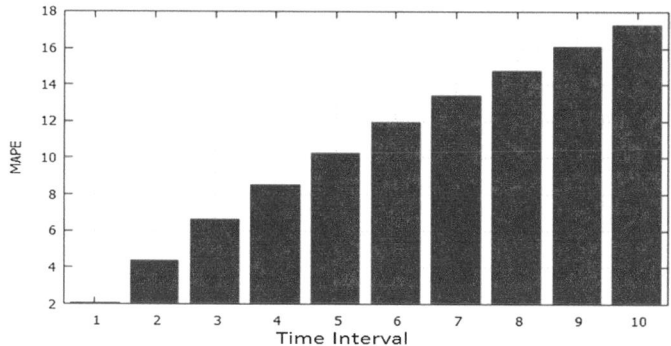

Fig. 5. Experiment of perplexity.

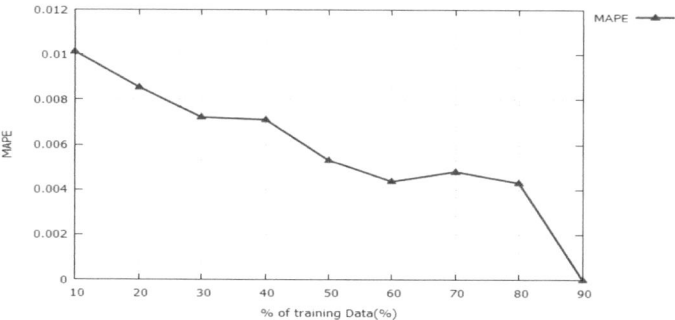

Fig. 6. Experiment of different percentage of training data.

After the topic finding and detection, we got 20 topics and many clusters of their related articles. We computed the average MAPE for every topic related clusters. As shown in Fig. 7, most of the topics' MAPE are lower than 0.1, which is belong high accuracy level. Although there is one topic's average MAPE is higher than 0.1, but it is still on accuracy level.

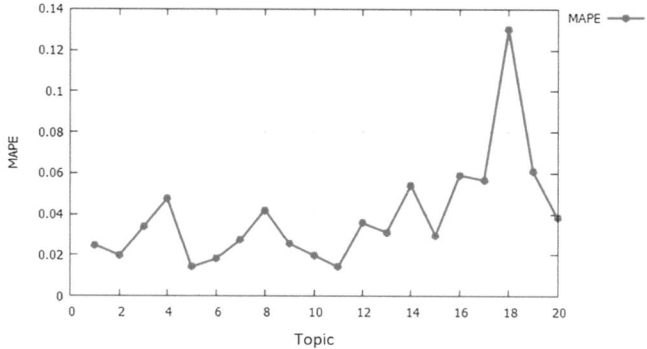

Fig. 7. Average MAPE of each topic.

5 Conclusions

In this study, we detected the latent social topics from the PTT social network site and found topic related articles by the distance similarity approach. We defined five states for the trend of the articles to build up forecasting model. We integrated the hidden markov model and latent dirichlet allocation topic model to forecast trends of the social topics based on time series data of user reviews. Experimental results on real dataset, this study can recognize the latent social topics, keywords and forecast the trend of each topic effectively on the PTT. We conclude this study and result as follows:

1. Applying the LDA topic model combined with COPENS corpus filtering more meaningless stop-words, find out the potential of PTT social topics and related key words from PTT, and let the model`s keyword description clearer.
2. Applying hidden Markov model, predict the trend of the social topics based on time series, can reach a pretty good MAPE accuracy about 3.9%. No need to use any other features of the past to predict the trend and to achieve trend tracking and forecasting results.
3. Propose a mechanism of social topic trends forecasting that combining hidden Markov time series model and LDA topic model. Not only from a large number of community discussions to identify important social issues, but also forecasting and monitoring the trend.

References

1. Chen, Y., Amiri, H., Li, Z., Chua, T.-S.: Emerging topic detection for organizations from microblogs. In: Proceedings of the 36th International ACM SIGIR Conference on Research and Development in Information Retrieval, pp. 43–52. ACM, New York (2013)
2. Ritter, A., Etzioni, O., Clark, S.: Open domain event extraction from Twitter. In: Proceedings of the 18th ACM SIGKDD International Conference on Knowledge Discovery and Data Mining, pp. 1104–1112. ACM, New York (2012)
3. Kaplan, A.M., Haenlein, M.: Users of the world, unite! the challenges and opportunities of Social Media. Bus. Horiz. 53(1), 59–68 (2010)
4. Turban, E., King, D.R., Lang, J.: Introduction to Electronic Commerce. Prentice Hall, Upper Saddle River (2009)
5. Kotler, P., Keller, K.L.: Marketing Management. Pearson Prentice Hall, Upper Saddle River (2009)
6. Smith, N., Wollan, R., Zhou, C.: The Social Media Management Handbook: Everything You Need to Know to Get Social Media Working in Your Business. Wiley, Hoboken (2011)
7. Hemann, C., Burbary, K.: Digital Marketing Analytics: Making Sense of Consumer Data in a Digital World. Que, Indianapolis (2013)
8. Liu, B.: Web Data Mining: Exploring Hyperlinks, Contents, and Usage Data. Springer, Berlin (2011)
9. Cha, M., Haddadi, H., Benevenuto, F., Gummadi, P.K.: Measuring user influence in Twitter: the million follower fallacy. In: ICWSM, vol. 10, no. 10-17, p. 30 (2010)

10. Li, H., Mukherjee, A., Liu, B., Kornfield, R., Emery, S.: Detecting campaign promoters on Twitter using markov random fields. In: 2014 IEEE International Conference on Data Mining, pp. 290–299 (2014)
11. Singh, V.K., Piryani, R., Uddin, A., Waila, P.: Sentiment analysis of movie reviews and blog posts. In: 2013 IEEE 3rd International Advance Computing Conference (IACC), pp. 893–898 (2013)
12. Chen, C.C., Chen, Y.-T., Sun, Y., Chen, M.C.: Life cycle modeling of news events using aging theory. In: Lavrač, N., Gamberger, D., Blockeel, H., Todorovski, L. (eds.) Machine Learning: ECML 2003, pp. 47–59. Springer, Berlin Heidelberg (2003)
13. Sayyadi, H., Hurst, M., Maykov, A.: Event detection and tracking in social streams. In: ICWSM, May 2009
14. Cataldi, M., Di Caro, L., Schifanella, C.: Emerging topic detection on Twitter based on temporal and social terms evaluation. In: Proceedings of the Tenth International Workshop on Multimedia Data Mining, pp. 4:1–4:10. ACM, New York (2010)
15. You, L., Du, Y., Ge, J., Huang, X., Wu, L.: BBS based hot topic retrieval using back-propagation neural network. In: Su, K.-Y., Tsujii, J., Lee, J.-H., Kwong, O.Y. (eds.) Natural Language Processing – IJCNLP 2004, pp. 139–148. Springer, Berlin Heidelberg (2004)
16. Xie, J., Liu, G., Ning, W.: A topic detection method for Chinese microblog. In: 2012 International Symposium on Information Science and Engineering (ISISE), pp. 100–103 (2012)
17. Becker, H., Naaman, M., Gravano, L.: Learning similarity metrics for event identification in social media. In: Proceedings of the Third ACM International Conference on Web Search and Data Mining, pp. 291–300. ACM, New York (2010)
18. Blei, D.M.: Probabilistic topic models. Commun. ACM **55**(4), 77–84 (2012)
19. Hofmann, T.: Unsupervised learning by probabilistic latent semantic analysis. Mach. Learn. **42**(1–2), 177–196 (2001)
20. Blei, D.M., Ng, A.Y., Jordan, M.I.: Latent Dirichlet Allocation. J. Mach. Learn. Res. **3**, 993–1022 (2003)
21. Wang, X., McCallum, A.: Topics over Time: a non-Markov continuous-time model of topical trends. In: Proceedings of the 12th ACM SIGKDD International Conference on Knowledge Discovery and Data Mining, pp. 424–433. ACM, New York (2006)
22. Wang, Y., Agichtein, E., Benzi, M.: TM-LDA: efficient online modeling of latent topic transitions in social media. In: Proceedings of the 18th ACM SIGKDD International Conference on Knowledge Discovery and Data Mining, pp. 123–131. ACM, New York (2012)
23. Szabo, G., Huberman, B.A.: Predicting the popularity of online content. Commun. ACM **53**(8), 80–88 (2010)
24. Achrekar, H., Gandhe, A., Lazarus, R., Yu, S.-H., Liu, B.: Predicting flu trends using twitter data. In: 2011 IEEE Conference on Computer Communications Workshops (INFOCOM WKSHPS), pp. 702–707 (2011)
25. Kim, S.D., Kim, S.H., Cho, H.G.: Predicting the virtual temperature of web-blog articles as a measurement tool for online popularity. In: 2011 IEEE 11th International Conference on Computer and Information Technology (CIT), pp. 449–454 (2011)
26. Ritterman, J., Osborne, M., Klein, E.: Using prediction markets and Twitter to predict a swine flu pandemic. In: 1st International Workshop on Mining Social Media, vol. 9, pp. 9–17, November 2009
27. Bandari, R., Asur, S., Huberman, B.A.: The pulse of news in social media: forecasting popularity (2012)
28. Zaman, T., Fox, E.B., Bradlow, E.T.: A Bayesian approach for predicting the popularity of tweets. Ann. Appl. Stat. **8**(3), 1583–1611 (2014)

29. Figueiredo, F.: On the prediction of popularity of trends and hits for user generated videos. In: Proceedings of the Sixth ACM International Conference on Web Search and Data Mining, pp. 741–746. ACM New York (2013)
30. Fernandes, K., Vinagre, P., Cortez, P.: A proactive intelligent decision support system for predicting the popularity of online news. In: Pereira, F., Machado, P., Costa, E., Cardoso, A. (eds.) Progress in Artificial Intelligence, pp. 535–546. Springer, Berlin (2015)
31. Rabiner, L.R.: A tutorial on hidden Markov models and selected applications in speech recognition. Proc. IEEE **77**(2), 257–286 (1989)
32. Lewis, C.D.: Industrial and Business Forecasting Methods: A Practical Guide to Exponential Smoothing and Curve Fitting. Butterworth-Heinemann, Oxford (1982)
33. Griffiths, T.L., Steyvers, M.: Finding scientific topics. Proc. Nat. Acad. Sci. **101**(Suppl. 1), 5228–5235 (2004)

IoT and Blockchain Driven Techniques for e-Business Management

Intelligent Car Parking System Based on Blockchain Processing Reengineering

Quanyi Hu[1(✉)], Simon Fong[1], Peng Qin[1], Jingzhi Guo[1], Yong Zhang[2], Dan Xu[3], Yuanlan Chen[3], and Jerome Yen[3]

[1] Faculty of Science Technology, University of Macau, Macau, China
{yb77476,ccfong,yb77428,jzguo}@um.edu.mo
[2] Faculty of Law, University of Macau, Macau, China
sb75115@um.edu.mo
[3] Faculty of Business Administration, University of Macau, Macau, China
{sb645017,sb84545,jeromeyen}@um.edu.mo

Abstract. Due to guide the industries or enterprises leveraging new technologies and witnessing reengineering, we propose a scheme based on Business Process Reengineering and Blockchain Technology named Blockchain Process Reengineering (BCPR). Different from the technical push mechanisms that generally improve the development of blockchain, BCPR is a new solution from a business perspective that can address the pain points of industry. We aim to provide a detail theoretical method of BCPR and draw support from it to reengineer the parking industry. In contrast with the existing methods, it is proved that BCPR is a feasible solution to solve the problem of pain points in industry. It is also conducive to the development of the industry and blockchain technology with the realization of resource integration while taking into account the issue of trust.

Keywords: Blockchain Process Reengineering · Parking industry

1 Introduction

Nowadays, the blockchain technology has been developed rapidly [26], but many industries still follow the traditional model to survive while confronting the challenges brought by blockchain and confusing how to integrate with them to strive for transformation. So far, the research on blockchain technology mainly focuses on technical architecture, the performance of platforms' algorithm and information security protection [2,21]. It is obvious that such research is based on the technical level, besides, the scale-up innovative applications are scarce that cannot be combined with industries to generate value [19]. Therefore, it is necessary to analyze the urgent needs of transformation from a business perspective, which enable blockchain technology to better serve various industries. Thus, we propose a new theory aims to promote the development of blockchain technology with "commercial push": Blockchain Process Reengineering (BCPR).

The core attributes of the blockchain, which are *transaction* (value) [3,5], *deposit* [4,7], *trust* [6,14], *intelligent* [7,8], and *traceability* [12,22], can match

© Springer Nature Switzerland AG 2020
K.-M. Chao et al. (Eds.): ICEBE 2019, LNDECT 41, pp. 265–273, 2020.
https://doi.org/10.1007/978-3-030-34986-8_19

the indicators of industries development. For example, trust attribute and traceability attribute can match and solve the guarantee and foundation indicators of industries promotion. Therefore, BCPR is a new solution for the development of the industry. This paper will discuss the feasibility and provide the theoretical methods of BCPR and take the parking industry as a case to illustrate the process of BCPR.

[16, 18] provide the indoor localization system that can be effectively applied in the parking industry, however, they can only be used in a single indoor environment with limited information exchange, inefficient resource utilization and imperfect trust mechanism just as others indoor localization system. Blockchain technology can be acted as a real-time and secure business-to-business (B2B) data synchronization existing in the parking industry. Therefore, we will carry out effective process reengineering and give complete solutions to the parking industry based on the theoretical guidance of BCPR.

In the next of this paper is included as follows. Part 2 will explore the relevant literature, introduce the significance and explain the advantages of BCPR theory. Part 3 focuses on the concept and theory of BCPR. The Part uses the BCPR as a theoretical guide to reconstruct the parking industry. Section 5 will summarize the paper. Sections 6 and 7 are acknowledgment and references respectively.

2 Analysis of Relevant Research and the Significance of BCPR

The platform performance, technical solutions and information security have always been the focal point of researchers since the development of blockchain technology. [27] provides a distributed database with the ability to operate in a decentralized system without relying on trusted intermediaries. [28] stores a record of transactions in distributed nodes that are linked in a peer-to-peer network. [29] utilizes immutability and traceability of blockchain to help avoid online maliciousness, which includes identity theft, online fraud, and hacking [29]. The transaction throughput of the latest platform can reach 5000 pens/second [9, 10, 21], indicating that the performance of blockchain system has been significantly improved [1]. The diversified development of development tools [13, 20], the breakthrough of cross-chain interconnection and multi-parties operations [10] indicate that the solution of technology is gradually improving [2, 3]. The emergence of two-way authentication shows that information security protection has been enhanced [4, 11, 12]. The above method is technical push, however, on the one hand, blockchain technology unable to play its due value in practical applications due to the high technical threshold and development cost, on the other hand, the technical pull has been overemphasized which will hinder the development of the blockchain with the dispersed characteristics. It is not conducive to the development of blockchain technology [24]. Therefore, different from the above literature, our theoretical method found a new way to prevent the technical dominant shift of blockchain research.

Fig. 1. The process of BCPR

BCPR is also different from traditional BPR. The success rate of BPR is less than 30% [17,23] as it cannot be kept up to date in practice. As an innovation of BPR, BCPR inherits the essence of BPR, separately discusses the significance of blockchain technology in industry or enterprise reengineering and narrows the scope of BPR theory. BCPR will assess the digitization extent of the process management, judge the conditions of using blockchain technology, calculate the expected goals and the resources (time) investment for reengineering the process rationally.

3   The Theory of BCPR

The five significant attributes of blockchain are *transaction, deposit, trust, intelligent,* and *traceability,* which can be applied to industries' pain points and solve the challenges of the industry. Industry development needs to reengineer certain processes. Also, it needs a guide to apply blockchain technology properly and make it appear in the process of reengineering.

We divide the processes existing in enterprises in different industries into internal and external parts, including six closed-loop structures separately, which are Purchase, Financing, Manufacture, Sell, Logistics, After-sales, and Finance, HR, R&D, Internal control, Management, Administration as is shown in Fig. 1. We judge the key points of blockchain based on the indicators and data of each enterprise. It needs to be explained that not all processes are suitable for BCPR (Airbnb is suitable for full-scale BCPR in shared economy), so in practice, it is necessary to combine with case analysis.

The BCPR needs to go through five stages. The first stage is the analysis and diagnosis. The main tasks are: 1. Understand the current situation and requirement of the industry. 2. Select the appropriate process to re-engineer according to the inner and outer circle. The second stage is the startup of the project. The main tasks including draw a business flow chart and select the significant nodes. The third stage is the process design. The main tasks are: 1. Design BCPR solutions based on the second stage. 2. Follow the characteristics and operation mechanism of blockchain technology, combine the existing scientific and technological methods to carry out process reengineering. The fourth stage is reengineering implementation. The main tasks including reorganize the structures and implement pilot reengineering. The fifth stage is monitoring and evaluation.

4   Case Analysis of BCPR in the Parking Industry

BCPR must match the characteristics of the industry itself with the blockchain technology, so we use the BCPR theory in the parking industry as a case study.

4.1   Analytical Diagnosis of the Parking Industry

Analysis of the parking industry's present situation is the primary segment of BCPR. First, we list the drawbacks and pain points of the traditional parking industry. Second, we review the existing transformation methods of traditional parking industry to reflect the advantage of BCPR in the parking industry. The traditional parking industry of China confronts the following problems:

- The market scale is large and the parking lot is scattered.
- The tidal phenomenon of car parking is sharp.
- The degree of digitization is low.
- Research and development of auxiliary equipment and enhanced parking technology.
- Establishing the staggered time mechanism.
- Digitalize only parts of the process.

4.2   Project Start-Up Plan for the Parking Industry

The parking industry has effectively solved some of the problems and has completed the corresponding digitization stage after the above process is improved. However, these types of digital improvements are too fragmented which can only be applied to the enterprise-level frameworks and hardly be used in the industry [16]. Therefore, some researchers have proposed a solution based on big data or Cloud technology which will establish a city-level smart parking management platform that connects car owners, parking lots, car service providers, city managers and other participants of parking industry gradually [15, 25]. However, trust issues and efficiency (real-time) problems emerge. In summary, we list the pain points that still exist after the industry digitalization as follows and give the reengineering plan:

- Digital system independence and the difficulty in data integration
- Data security issues and the efficiency.

4.3   BCPR Process Design for Parking Industry

Based on the analysis of the above, we propose a solution, the reengineering focus on the following parts:

- Resolving the resource integration after digitalizing the industry
- Accelerate the background processing of information
- Strengthen the security of user's personal data

The reengineering methods are mainly separated into two aspects: hardware and software construction. The hardware construction is basically mature at present and there's no need to be modified (e.g., through the construction of three-dimensional parking equipment to expand the urban parking space), while the software construction is based on the existing digital solution to establish a unified traffic management platform "Transport & Parking Alliance (TPA) Chain". The prototype of the TPA Chain is shown in Fig. 2.

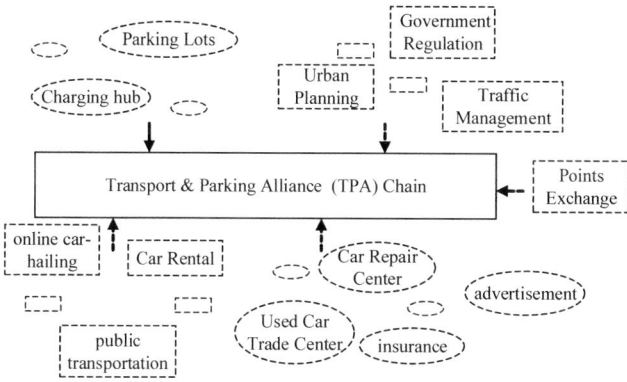

Fig. 2. TPA chain prototype

TPA Chain is designed and used in the public domain of transportation based on blockchain technology. It has an independent "trunk chain" to fully protect personal privacy while providing an open platform and source tools, as shown in Fig. 3.

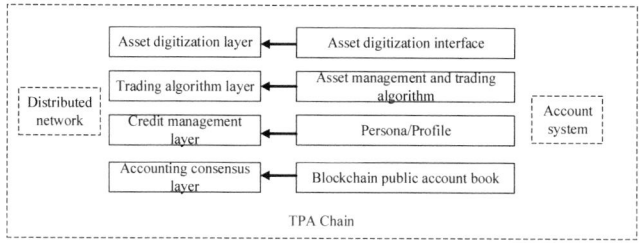

Fig. 3. Design of the transportation ecosystem

4.4   Implementation of Reengineering

The parking industry reengineering process based on BCPR theory and the solution of TPA Chain will meet the following features based on user experience.

(1) Inquiry and reservation - query and reserve all parking lots in the ecosystem
(2) Parking space sharing - owning the function of publishing the parking space individually to share the unused parking spaces
(3) Digital assets - users can obtain parking warrants, asset trading and exchange
(4) Credit parking - own credit parking system, realize credit level and enjoy the rights

4.5   Evaluation of the Reengineering Based on BCPR Theory

We judge the implementation of the process reengineering of the parking industry from three dimensions according to the theoretical guidance of BCPR.

(1) Strategic choices of the parking industry.

We conducted a preliminary analysis of 18 business cases related to the parking industry, including 12 parking management (Parkme, JustPark, MonkeyParking, LUXE, Feibotong, PP Parking, ParkBees, Anjubao, 51Park, AWCloud, Easy-Parking, WaletParking), 4 smart payment (WeChat pay, Baidu wallet plus Baidu-map, Alipay plus Amap, CareWallet), and 2 service of government (Shenzhen Road Traffic Management Center and Shanghai Parking Management System). Intelligent payment companies have already deployed blockchain technology entirely. 71% (10 companies) of the parking management and government service companies are in "transformation" status and the rest are on the "wait-and-see" status. The analysis shows that blockchain technology has great opportunities in the industry and the reengineering based on BCPR theory is extremely urgent.

(2) Business benefits.

The purpose of the BCPR is to reduce costs and improve efficiency and it has the two characteristics: (1) one-time cost; (2) continuous income. The traditional BPR method often requires continuous capital investment, but the BCPR can achieve continuous benefits even though the initial cost is larger than BPR. Besides, the blockchain technology eliminates the influence of third-party intermediaries within enterprises in industry and deletes too many cumbersome procedures or operations which will improve the efficiency. The analysis shows that the BCPR will bring great value to business benefits.

(3) The complexity of BCPR applied to the parking industry.

For the parking industry, the key point is to ensure that blockchain technology can solve the most important business problem and to support new growth opportunities. Also, whether the industry ready to adopt blockchain technology and understand the technology needs for reengineering. Business benefits will increase significantly as the complexity of BCPR increases, but it will be saturated after reaching a certain level. Therefore, BCPR will inevitably have a large cost overhead in the early stage. We expect that the BCPR should be concentrated in technical reengineering. It is estimated that the team needs about 80–100 people in the technical area which are responsible for the design and development of the TPA Chain with emphasis on the establishment of the transportation ecology. The entire industry involves in many parts including government departments and insurance companies. The previous period can be carried out in batches and passed the test step by step.

5   Conclusion

This paper proposes a blockchain process reengineering method from the business point of view named BCPR which adding blockchain technology to the process reengineering theory of industry or enterprises. It makes up for the lack of relevant theories and improves the development speed of blockchains in practical application areas. Although blockchain technology can promote the transformation of the industry, it needs to formulate a clear strategic plan. The complexity and value of the blockchain will be different from the different strategies of each industry. It is expected to have a greater impact if the industry combined the BCPR with other technologies (5G, Internet of Things, big data, artificial intelligence, etc.).

References

1. Sompolinsky, Y., Zohar, A.: Secure high-rate transaction processing in bitcoin. In: International Conference on Financial Cryptography and Data Security, San Juan, Porto Rico, pp. 507–527. Springer, February 2015
2. Androulaki, E., Cachin, C., De Caro, A., Kind, A., Osborne, M.: Cryptography and protocols in hyperledger fabric. In: Real-World Cryptography Conference (2017)
3. Chraibi, H., Houdebine, J.C., Sibler, A.: Pycatshoo: toward a new platform dedicated to dynamic reliability assessments of hybrid systems. In: International Conference on Probabilistic Safety Assessment and Management (PSAM), Seoul, South Korea, October 2016
4. Androulaki, E., et al.: Hyperledger fabric: a distributed operating system for permissioned blockchains. In: 13th EuroSys Conference, Porto, Portugal, p. 30. ACM, April 2018
5. Tschorsch, F., Scheuermann, B.: Bitcoin and beyond: a technical survey on decentralized digital currencies. IEEE Commun. Surv. Tutorials $18(3)$, 2084–2123 (2016)

6. Li, W., Sforzin, A., Fedorov, S., Karame, G.O.: Towards scalable and private industrial blockchains. In: Proceedings of the ACM Workshop on Blockchain, Cryptocurrencies and Contracts, pp. 9–14. ACM (2017)
7. Abdellatif, T., Brousmiche, K.-L.: Formal verification of smart contracts based on users and blockchain behaviors models. In: IFIP NTMS International Workshop on Blockchains and Smart Contracts (BSC), Paris, France, February 2018
8. Nehai, Z., Piriou, P.-Y., Daumas, F.: Model-checking of smart contracts. In: IEEE International Conference on Blockchain, Halifax, Canada, August 2018
9. Garay, J., Kiayias, A., Leonardos, N.: The bitcoin backbone protocol: analysis and applications. In: Advances in Cryptology - EUROCRYPT 2015, pp. 281–310. Springer, Heidelberg (2015)
10. Kiayias, A., Panagiotakos, G.: On trees, chains and fast transactions in the Blockchain. Cryptology ePrint Archive, p. 545 (2016)
11. Cimino, M.G.C.A., Marcelloni, F.: Enabling traceability in the wine supply chain. In: Methodologies and Technologies for Networked Enterprises, pp. 397–412. Springer, Berlin (2012)
12. Christidis, K., Devetsikiotis, M.: Blockchains and smart contracts for the IoTs. In: IEEE Access, Special section on the plethora of Research in IoT, pp. 2292-2303 (2016)
13. Kraft, D.: Difficulty control for blockchain-based consensus systems. Peer-to-Peer Netw. Appl. **9**(2), 19 (2016)
14. Chraibi, H.: Dynamic reliability modeling and assessment with PyCATSHOO: application to a test case. In: International Conference on Probabilistic Safety Assessment and Management (PSAM), Tokyo, Japan, April 2013
15. Carlos, S., Garrido, N.: A low-cost smart parking solution for smart cities based on open software and hardware. In: Conference on Intelligent Transport Systems (2018). At Guimarães
16. Hu, Q., Zhu, J., Chen, B., Zou, Z., Zhai, Q.: Deployment of localization system in complex environment using machine learning methods. IEEE RFID-TA (2016)
17. Benelli, G., Pozzebon, A.: An automated payment system for car parks based on near field communication technology. In: International Conference for Internet Technology and Secured Transactions (2010)
18. Ferilli, G.: An analysis of the city centre car parking market: the supply side point of view. A thesis submitted in partial fulfilment of the requirements of Edinburgh Napier University, for the award of Doctor of Philosophy, December 2008
19. Groh, B.H., Friedl, M., Linarth, A.G., Angelopoulou, E.: Advanced real-time indoor parking localization based on semi-static objects. In: 17th International Conference on Information Fusion (FUSION), Salamanca, pp. 1–7 (2014)
20. Palmer, A.: Customer experience management: a critical review of an emerging idea. J. Serv. Mark. **24**(3), 196–208 (2010)
21. Macdonald, M., Liu-Thorrold, L., Julien, R.: The blockchain: a comparison of platforms and their uses beyond bitcoin. Adv. Comput. Netw. Secur. (2017)
22. matiszz/SmartID. https://github.com/matiszz/SmartID
23. Habib, M.N.: Understanding critical success and failure factors of business process reengineering. Int. Rev. Manag. Bus. Res. **2**(1), 1–10 (2013)
24. Zheng, Z., Xie, S., Dai, H.N., Chen, X., Wang, H.: Blockchain challenges and opportunities. Int. J. Web Grid Serv. **14**, 4 (2018)
25. Qureshi, K.N., Abdullah, A.H.: A survey on intelligent transportation systems. Middle East J. Sci. Res. **15**(5), 629–642 (2013)
26. Nakamoto, S.: Bitcoin: a peer-to-peer electronic cash system. Consulted **1**, 2012 (2008)

27. Bano, S., Sonnino, A., Al-Bassam, M., Azouvi, S., McCorry, P., Meiklejohn, S., Danezis, G.: Consensus in the age of blockchains. arXiv preprint arXiv:1711.03936 (2017)
28. Viriyasitavat, W., Hoonsopon, D.: Blockchain characteristics and consensus in modern business processes. J. Ind. Inf. Integr. **13**, 32–39 (2018)
29. Cachin, C.: Architecture of the hyperledger blockchain fabric. In: Workshop on Distributed Cryptocurrencies and Consensus Ledgers, vol. 310, July 2016

A Smart Contract-Based Risk Warning Blockchain Symbiotic System for Cross-border Products

Bin Wu, Yinsheng Li$^{(\boxtimes)}$, and Xu Liang

Software School, Fudan University, Shanghai 200433, People's Republic of China
{bwu18,liys}@fudan.edu.cn

Abstract. In the current supervision mode of cross-border products, the government supervises insufficiently due to incomplete and untrustworthy risk data, non-autonomous and human intervened risk evaluation models. A smart contract-based risk warning blockchain symbiotic system is proposed to reform the issues of the current system. The system is a new third-party system that provides risk warning data services for the government. A permissioned blockchain ecosystem has been developed to provide open, equal, and credible services for the government and enterprises. A risk warning model is implemented by smart contracts to provide a non-intervention evaluation for cross-border products. The autonomy of the system is realized through smart contracts such as enterprise access audit, risk data acquisition, risk assessment and feedback. The system effectively improves the science and intelligence of supervision, cut down the customs clearance time and sampling proportion, and has been verified in the Administration of Inspection and Quarantine in Shanghai Airport.

Keywords: Blockchain · Smart contract · Risk warning · Symbiotic

1 Introduction

The new entry-exit inspection and quarantine policy issued by the General Administration of Quality Supervision and Inspection (AQSIQ) calls for a significant reduction in customs clearance time and sampling proportion. The increase of cross-border trade disputes calls for scientific and intelligent supervision. However, at present, there are many links and enterprises involved in products, many kinds of paper documents to be transferred, which lead to inefficient processing, difficult accountability, and challenging in reducing clearance time.

The mediation dependency on the current system leads to incomplete risk data. The risk warning services are lack of supporting information. The information fragmentation and inefficient collaboration between enterprises lead to untrustworthy risk data. On risk evaluation models, the current calculation of the risk levels of products and the credit level of enterprises is one-time and human-made. The results are not convincing enough.

© Springer Nature Switzerland AG 2020
K.-M. Chao et al. (Eds.): ICEBE 2019, LNDECT 41, pp. 274–289, 2020.
https://doi.org/10.1007/978-3-030-34986-8_20

The technology of blockchain and smart contracts have brought opportunities. Blockchain is a distributed ledger technology. Each block is "linked" to the next block using cryptographic signatures, shared and collaborated among anyone with sufficient authority [1], and collaboratively maintains the authenticity of ledgers through consensus algorithms. A smart contract is an executable code that runs on the blockchain to facilitate, execute and enforce the terms of an agreement [2]. Blockchain and smart contracts make it possible to develop a credible, autonomous and symbiotic system because of the characteristics of non-tampered data and trustworthy code execution.

The proposed symbiotic system solves the incompleteness of risk data caused by mediation dependency through the mode of permissioned chain and the method of permission control. The participants of the ecosystem include producers, consignees, the government and multiple intermediate service providers. Participants can control the uploaded data without trusting a single subject. The blockchain provides an equal environment that eliminates concerns about sharing sensitive and private business data. The unalterable record and validation of data, access audit of subject realized by blockchain and smart contract ensure the credibility of risk data. The main functions of risk warning such as risk data acquiring and processing, risk assessing and updating are executed through smart contracts to ensure equality and autonomy.

The remainder of this paper is organized as follows: Sect. 2 presents the literature work; Sect. 3 presents the scenario and problems of the system. Section 4 describes the design of the risk warning blockchain symbiotic system; Sect. 5 presents the system implementation. Section 6 presents the experiments. Section 7 concludes the paper.

2   Literature Work

Many customs around the world attempt to use blockchain in business scenarios. In June 2017, the Directorate-General for Taxation and Customs Union launched a proof-of-concept project to see whether an application based on the blockchain could add an extra layer of trust to the digitization of ATA Carnets [11].

In March 2018, the Inter-American Development Bank has been supporting the development of a blockchain solution to enable automated, secure and efficient information sharing on Authorized Economic Operators among the Customs administrations of Mexico, Peru and Costa Rica to ensure the efficient implementation of Mutual Recognition Arrangements/Agreements [12].

In September 2018, The Korea Customs Service (KCS) entered an agreement with Samsung SDS to implement the latter's blockchain technology for an export customs clearance system to enable efficiency and curb fraud in the industry. Forty-eight other organizations and companies, including relevant government authorities, shipping operators, logistics firms and insurance providers joined the agreement [3].

In risk assessment and warning, Xiao [14] presents a risk decision-making model and apply it to the historical customs declaration data set in customs targeting, based on the proposed dynamic K-means clustering algorithm.

Kim et al. [9] present risk-based categorization for each country-commodity combination based on the probability of quarantine pest interceptions and the uncertainty in that assessment.

A blockchain-based Production Credit Mechanism (PCM) for manufacturing services is put forward to regulate the cross-enterprise collaborations among socialized manufacturing resources [10].

3   Scenarios and Problems

The risk warning in the current situation makes the supervision of the government insufficiently under the following scenarios. The proposed system is to solve the issues and improve the efficiency and quality of supervision.

3.1   Scenarios

Data Fragmentation. According to statistical data and calculations by the United Nations Commission on Trade and Development and UN/CEFACT, every shipment involves an average of 27 trade participants, 40 kinds of documents and the copies of nearly 400 documents [13]. In Shanghai, in order to ensure the transfer of products from exporters to importers, as many as 20 different documents are required. Besides, most of the data is transferred by paper documents due to the separate system and sensitive business data.

Mediation Dependency. In the current working mode, multiple links and participants are involved in the supply chain of cross-border products. The participants not only include producers, consignees and the government. Multiple intermediate service providers such as logistics service providers, freight companies, tallying warehouses, agents, customs declaration enterprises are involved in the process. However, the government mainly gets limited data from the custom brokers. At present, the data uploaded to the system includes only basic declaration information, such as HS code, component, country of origin and the total value of the products.

Intervened Risk Evaluation. Researches and explorations on the management mode of products have been carried out very early in the inspection and quarantine system of China. The government inspects every batch of products in the category list at first [4]. "Classified Management" first appeared in 1984, which includes three parts: examination and confirmation of qualified product type, sample inspection of export batches, and supervision of factory quality system [16]. The "Shanghai Free Trade Port Construction Plan" points out that the management mode has changed from mainly managing products to enterprises. The current evaluation of enterprises considers indicators such as credit status, management level, and business ability [6]. The evaluation methods are not autonomous and intervened by the human.

3.2   Problems

Data flow is the central part of the business processes. In the current supervision mode, the enterprises do not want to open data to other systems due to the risk of leaking sensitive business data to others. The government can not get enough data to realize the early warning of risks. Besides, it takes too much time to check the authenticity of data because of the limited data source. The risk assessment of enterprises and products is done by the human, which faces the problem of human intervention. The inefficiency of the government has led to a long clearance time, which brings losses to enterprises too, especially those that produce products with short shelf life, such as fresh milk. The main issues of the current system are as follows:

Incomplete Risk Data. The products and consignees, which are the most important roles of the cross-border products, rely on the intermediate service providers such as custom brokers to deal with the government. Plenty of vital information related to the process is not collected and used, such as subject information, product information, traceability information, early warning information and consensus information.

Usually, a platform that includes all services is good at solving data fragmentation. Traditionally, a centralized platform is always controlled by one institution that has more power than general participants. However, it is not suitable in this case because enterprises involved in cross-border products often span at least two different countries. Neither institution is willing to give control of sensitive business data to other institutions.

Untrustworthy Risk Data. The data source of the government is quite single. The intermediate service providers can deceive upstream or downstream participants. It costs much time for the government to check the authenticity of data, and it is hard to find the responsible participant. The low-trusted data transmission and inefficient process collaboration have a negative effect on both enterprises and the government, for example, long customs clearance time.

Non-autonomous Risk Evaluation. The risk evaluation method is not accurate and convincing based on incomplete and untrustworthy risk data. Besides, the calculation of risk levels is done off-line, which is easy to be intervened. The non-autonomous risk evaluation has damaged the principles of open, fair and just in the supervision mode.

3.3   Features

The risk warning blockchain symbiotic system is proposed to solve these issues. The system is to make the government more effective and enterprises that participate in it more competitive in the market. The enterprises join the alliance and upload their data by blockchain clients in the information system or public platform. They can get simpler and faster procedures, also a shorter customs clearance time, especially when the risk degree of the products and enterprises evaluated by the model is low. The government needs to provide more convenient customs clearance environment as an incentive mechanism for the enterprises. The risk warning system can reduce the proportion of sampling inspection under the condition of ensuring quality. The features of the system are as follows:

Blockchain. As mentioned earlier, it is difficult to use a centralized system to involve all participants. We choose to use blockchain technology to build a system that is running by multiple participants and satisfies every participant. The system provides not only supervision but also services for the enterprises.

The data produced in the business process is uploaded to the blockchain by the clients. The authenticity is ensured by non-tampered data recorded by blockchain. Through the mode of permissioned chain and the method of permission control, enterprises can prevent their private and sensitive business data from leaking to third-party institutions or competitors.

Smart Contract. Smart contracts are complementary to the blockchain. If we store the data on-chain but deal with the data off-chain, the issues of security and equality appear again. The smart contract guarantees autonomy, equality and trustworthy of processing. We convert manual assessment to automatic assessment by combining the risk warning model with smart contracts. Also, we use smart contracts to audit access and validate the data uploaded by enterprises.

3.4   Indicator Requirements

The system should satisfy the indicator requirements in both function and system design to prove the effect in solving the issues. The system should realize the data communication between the blockchain system and the clients, including business clients of the government and enterprises, third-party creditworthiness cloud, which deal with the incomplete and untrustworthy data. The smart contracts of enterprise access audit and data cross validation should be realized to ensure data authenticity. Risk calculation should give out correct risk score and warning information. Besides, it should meet the requirements of reliability, stability, compatibility and maintainability as a system.

4   A Risk Warning Blockchain Symbiotic System

4.1   Application Architecture

As shown in Fig. 1, at the business level, the business process starts from the producers, through several intermediate service providers, such as warehousing enterprises, customs declaration enterprises, to the inspection and quarantine department, finally reaches the consignees, with the logistics service providers involved in the process. The batch is used as the tracing unit, and the essential

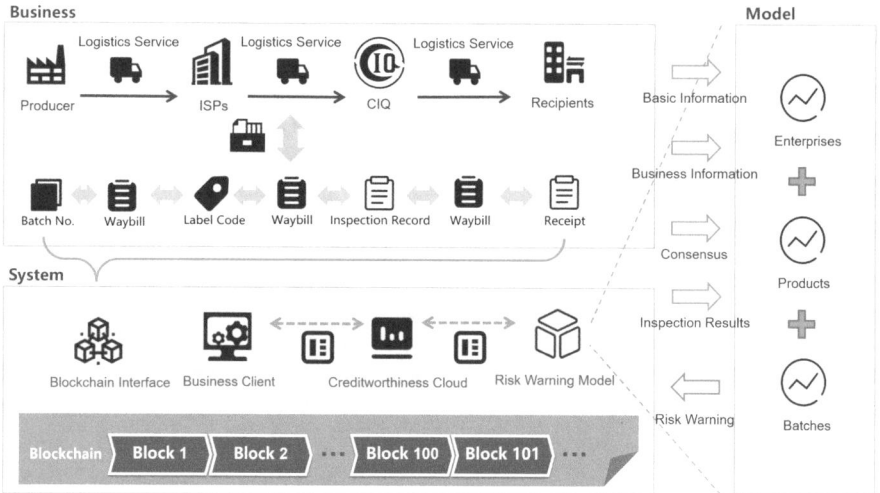

Fig. 1. Application architecture.

data and files such as the qualification certificate, inspection report, waybill are associated with the tracing unit based on business logic. The products are traced by batch through the close association of data carriers in each link. The business-level accesses the system through the interface of the blockchain platform, and the data related to each link of the product is stored on the blockchain. Business clients, creditworthiness cloud and risk warning model interact with the blockchain. The risk warning model obtains the required data from inside and outside the blockchain. It sends out risk warning to relevant inspection and quarantine departments when risks are detected.

4.2   Participants and Nodes

In order to deal with the mediation dependency problem, the participants should cover the whole supply chain of the products. According to the critical processes of the supply chain, we sort out the main participants of the system:

– Producer. The producer is the provider of products and the starting point of the product flow, which is usually in a different country from the consignee.
– Consignee. The consignee is the receiver of products and the endpoint of the product flow.
– Intermediate Service Provider. ISPs connect the producers and consignees, which are critical roles in product flow. ISPs include such as logistics service providers, freight forwarders, customs brokers.
– The Government Departments. There are several departments responsible for different types of products and laboratory responsible for quarantine. The duty of the government is ensuring the authenticity, correctness and legality of products.

In general, not each participant in the system runs a node. Some small enterprises do not want to or can not afford to run a node by themselves. They can upload and query data from nodes that are running by big enterprises and the government.

4.3 Mode Selection

According to the application architecture, participants of the blockchain system are limited to specific institutions. We choose the permissioned chain, which is limited to the participants of alliance members. The permissions of the blockchain are formulated according to the alliance rules. Different participants in the system are in equal positions. Nobody controls the system by oneself. Furthermore, we apply permission control based on their identity so that they can control their data.

4.4 Permission Control

There are two types of roles in the system, participants and network administrator. For participants, enterprises and the government have different permission. The enterprises can set create, read, update and delete actions on different data objects according to their needs. The risk results are controlled by the government to prevent from intervening by other participants.

A business network administrator is a participant who is responsible for configuring the business network after the business network is deployed, and is responsible for on-boarding other participants [7]. The specific permission control of the system is shown in Table 1.

Table 1. Permission control of system

Role		Permission
Network administrator		All operations on user resources and system resources
Participants	Enterprises	1. Link-first participants create batch resources;
		2. Participants supplement information on their links;
		3. Participants modify their resources;
		4. The current owner of a batch resource can be modified only if the batch owner is himself.
	The Government	1. Participants read batch resources;
		2. Participants add results of inspection and quarantine to batch resources;
		3. Participants call transactions that calculate risks

4.5   Risk Warning Model

In order to realize the early warning of products' risks, plenty of articles [5,14,15,17] have used historical inspection data to train and predict. However, this kind of method covers limited risks due to insufficient supporting information. By combing the risk warning model with the proposed ecosystem, we can obtain more useful information to evaluate the risk. The risk warning model comprehensively considers historical data and current situation, and comprehensively considers the risks of enterprises and the risks of the products. The data includes subject information, product information, traceability information, early warning information and consensus information. The main functions of the risk warning model including risk data acquisition and processing, risk assessment and updating are realized through smart contracts to give out an autonomous risk warning service without manual intervention.

5   System Implementation

5.1   System Infrastructure

Hyperledger Fabric is an open source framework implementation for private and permissioned business networks, where the member identities and roles are known to the other members. It leverages container technology and delivers enterprise-ready network security, scalability, and confidentiality [8]. We choose it for higher speed, lower cost and better privilege control over other platforms such as Ethereum.

The smart contract we are using is the Chaincode in Hyperledger Fabric. Chaincode is software that defines assets and related transactions; in other words, it contains the business logic of the system. Chaincode is invoked when an application needs to interact with the ledger. Chaincode can be written in Golang or Node.js [8].

Swagger is a normative and complete framework for generating, describing, invoking and visualizing RESTful-style web services. By integrating Swagger, some interfaces provided by Fabric can be invoked in real time to query data directly.

5.2   System Architecture

We use the system architecture shown in Fig. 2 to realize the proposed system. The symbiotic system based on blockchain is designed hierarchically, including the blockchain bottom platform, smart contract, business layer and application layer. The main functions of each layer are as follows:

Blockchain bottom platform: It provides blockchain functions such as maintenance of distributed shared books, maintenance of state databases, and full life cycle management of smart contracts to realize data integrity and business logic of smart contracts. Besides, member registration and cancellation functions are provided to control the rights of participants with different identities.

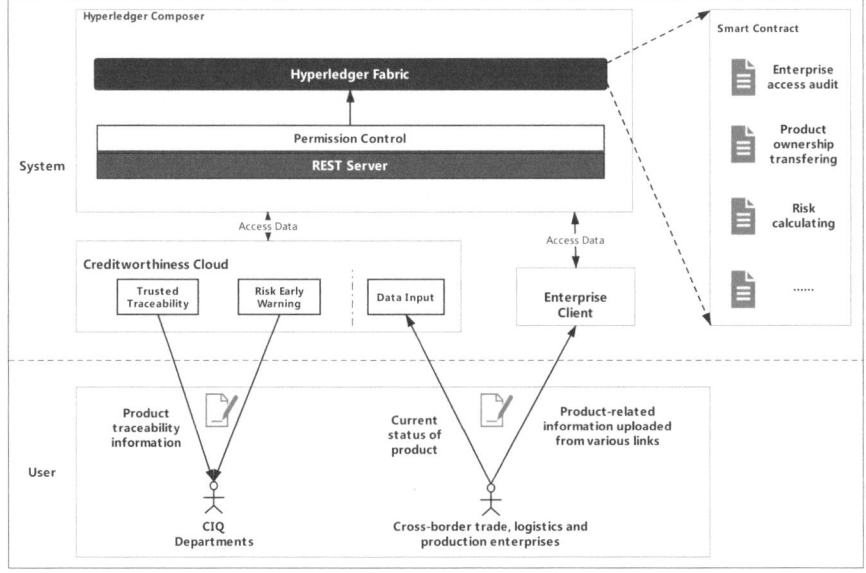

Fig. 2. System architecture.

Smart contract: Smart contract is realized through Chaincode, including Chaincode calling functions such as enterprise access audit, product ownership transfer, data cross validation and risk calculating.

Business layer: The business layer is the back-end service of the application program, providing an interface to the web application and processing the business request of the front-end. The service layer communicates with the blockchain network, and interact with other service systems, mainly including creditworthiness cloud system and target enterprise clients.

Application layer: Web application has the characteristics of modularization, automatic data binding, etc. It provides user interactive interface operations, including user and business operations. User operations include functions such as user registration, review, login, exit. Business operations include functions such as creating new product templates, creating new product batches, supplementing product traceability information, and viewing risk information.

5.3   Data Acquisition

All the data required by the risk warning model should be stored on blockchain since the smart contracts can only get data from the blockchain. There are some different sources and methods to get the original data.

The sources for collecting risk information of inspection and quarantine include internal sources and external sources [1]. The internal data includes relevant regulations such as composition for dairy products or design requirements for clothes, information obtained from business links such as enterprise declaration, auditing, and results of inspection and quarantine. The external data includes such as conformity assessment data given by inspection certification agencies, epidemic announcements given by government websites, and consensus information in news media.

The acquisition method includes acquiring data uploaded by users through clients, acquiring data of relevant institutions through business database API, acquiring image-based declaration data through intelligent image processing method, and collecting consensus and announcement data through the network crawler.

5.4   Data Model

Data Design. The data design of the framework mainly includes three aspects: participants, assets and transactions.

- Participants represent the organizations or personnel participating in the digital business network. In this system, participants are enterprises and government departments. The enterprises have different roles corresponding to different links. The government has different departments responsible for different types of products and work such as the department responsible for fresh goods and laboratory responsible for quarantine.
- Assets include any logistics or non-material assets. In this system, the main assets are information assets, including such as warning information, subject information, product information, traceability information and consensus information. The traceability information is designed to be template-based to suit for different types of products that have different traceability data items. The rule refers to the rules that are defined by the human for products, for example, the transportation temperature of the milk can not exceed $4\,^\circ$C.
- Transactions are submitted by participants to influence the behavior of assets in the blockchain, mainly including enterprise access audit, product ownership transfer, data cross validation and risk calculating.

The data design of the system is shown in Table 2.

Risk Result. For every indicator, the result is stored in a triple $\{Tag,\ Score,\ Msg\}$.

- The tag stands for the sign that the products must be inspected when it equals 1. In some cases, for example, when the product is declared for the first time, this batch must be inspected. We set this tag to prevent the weight of indicators influencing the results in some particular cases.
- The score stands for the risk degree. When the tag equals 0, the score defines the degree of risk of the batch. It is similar to most models.
- The msg stands for the message that shows to the inspectors. In most evaluation models, they just give out a binary classification. In order not to lose the information of the model, the msg defines the risk content description used to assist inspectors in their judgment.

Table 2. Data design of the system

Type	Object	Description
Participant	Enterprise	Different roles of enterprises
	Department	Different departments in the government
Asset	ProductInfo	The basic information of product
	Template	Define the traceability template of product
	Batch	Record the related, enterprises, status, information and risk results of the batch
	Rule	The human-defined rules for the product
	Access	The access information of enterprises
	Consensus	The collected consensus and announcements with corresponding score and product
Transaction	AuditAccess	Audit the enterprise that want to access the alliance
	OwnerTransfer	Transfer the owner of a batch to the next link of the enterprise
	CrossValidation	Cross validation of uploaded batch data
	RiskCalculate	Calculate the risk degree of a batch

The final triple of the risk warning model is $\{\sum Tag > 0?1 : 0, \sum Score, Msg\,[]\}$.

The final msg of the model contains the messages of indicators whose tag equals 1 or score is in the top K;

5.5 Smart Contracts

In Fabric, smart contracts are implemented by Chaincode. Chaincode stores the relevant data structure model of the code on the blockchain, which maintains a tamper-proof record of all state values. The status transition on the blockchain is the result of the participants calling the transaction of Chaincode. Each transaction in each block produces a set of modified key-value pairs for the assets, which will be submitted to the blockchain.

The smart contract interfaces implemented are defined in Table 3, including enterprise access audit, product ownership transfer, data cross validation and batch risk calculation.

- Enterprise access audit: We use smart contracts to audit the enterprises that want to join in the alliance automatically. The enterprises are required to upload the data items. The authenticity of the data is audited by comparing with official data. Rules are predefined to audit the uploaded data items to identify their qualification.

- Product ownership transfer: The owner of the batch needs to be transferred to the enterprise of the next link in the supply chain after finishing the current link. The smart contract is limited to be executed only if the owner of the batch is the caller.
- Data cross validation: The government departments can use predefined rules to validate the data between different links to prevent data fraud situation. For example, the quantity and weight data uploaded by the producer and the freight forwarder are supposed to be the same. Cross-validation between different enterprises makes it more difficult to falsify data.
- Batch risk calculation: The smart contracts get the risk data from blockchain by the corresponding id. The data is processed and calculated to give out the results, which is stored in a triple.

Table 3. Smart contract definition

Enterprise access audit	Content	Automatically verify the access status of enterprises applying for participating
	Input parameters	Enterprise Id
	Output parameters	Access or not
	Caller	The administrator
Product ownership transfer	Content	Transfer the owner of a batch to the next link of the enterprise
	Input parameters	Batch number. Enterprise ID. The link of the enterprise
	Output parameters	None
	Caller	The current owner of the batch
Data cross validation	Content	Cross validation of uploaded batch data
	Input parameters	Batch ID
	Output parameters	Incorrect data items
	Caller	The government departments
Batch risk calculation	Content	Calculate the risk degree of a batch
	Input parameters	Batch Number
	Output parameters	{Tag, Score, Msg[]}
	Caller	The government departments

6 Experiment

6.1 Test Indicators

In order to test the performance of the proposed system in dealing with incomplete and untrustworthy data, non-autonomous and human intervened risk evaluation models. The test is divided into three aspects, namely:

Functional Testing. The goal of functional testing is to verify that data is accepted, processed and retrieved correctly, and business rules are implemented appropriately. The main test method is that users interact with applications through GUI (graphical user interface), and analyze the output or acceptance of interaction, in order to verify the consistency of requirement functions and implementation functions. The functional indicators include are as follows:

- Upload/Store/Update data: Upload and update data objects, including participant classes and asset classes.
- Query the data objects on the blockchain, and the returned results conform to the corresponding retrieval rules.
- User privilege allocation: When users operate as different roles, it meets the corresponding privilege requirements.
- Smart contracts realize corresponding functions.
- The platform can obtain correct scoring and warning information through the interface.
- Support inspectors define corresponding decision rules.

System Testing. For the system testing, the hardware, software and operators are regarded as a whole. This kind of testing can find errors in system analysis and design. It mainly includes system reliability, stability, compatibility and maintainability. The system design indicators are as follows:

- Reliability: The software system can perform the required functions within the specified time and under the specified environmental conditions.
- Stability: The system is stable and error-free in long-term operation.
- Compatibility: The system achieves correct responses on different browsers.
- Maintainability: The ability of a system to correct, improve, or adapt the software to changes in the environment, requirements, and functional specifications.

Application Effect. Application effect refers to the effect in the actual business scenario of the government and enterprises.

6.2 Test Results

Functional Test. We test the functional indicators by user scenarios. The main steps are as follows:

  i Enterprise users apply for participant identities in the blockchain;
 ii Alliance members review the relevant information of the enterprise;
iii Administrators create a new product traceability information template;
 iv Administrators define corresponding decision rules of the product;
  v Producers of products create a new batch;
 vi Associated enterprises supplement product traceability information;
vii Inspectors choose the high-risk batches and inspect specific batches.

The process covers the functional indicators. The test completes the main functions from uploading data by enterprises to querying data and risk warning results by the government. The test results are in line with the expected results, indicating that the smart contract-based risk warning blockchain symbiotic system proposed in the paper is technically feasible.

System Test. According to the system design requirements, we have tested in four aspects:

- Reliability: For the user scenarios of both enterprises and the government, ten sets of complete experiments are conducted, and no program collapse occurred. It meets the requirement of reliability.
- Stability: The system has been running for months in stable and error-free. It meets the requirement of stability.
- Compatibility: We test compatibility in four popular web browsers, Google Chrome, Internet Explorer, Firefox and Safari. The data communication between the platform and the blockchain system gets no error. All the operations pass the test. It meets the requirement of compatibility.
- Maintainability: The data model and smart contracts of the blockchain system support updating. The models are expected to change and evolve over time. However, some care and discipline must be applied when making model changes to ensure that existing instances are still valid with respect to the new model [19]. It meets the requirement of maintainability.

Application Effect. We use the blockchain system to integrate services, digitize the core documents of all critical links in cross-border trade, record the important information of contract signing, remittance, bill of lading circulation and customs supervision in the trade process. It significantly enhances the completeness and authenticity of data. The risk warning model can give out more comprehensive and accurate risk results.

Through the evaluation of the smart contract-based risk warning model, the autonomous risk assessment of each batch is realized. According to the tag, score and messages given by the model, the inspectors can choose the high-risk batches accurately, and reduce the proportion of sampling inspection under the condition of ensuring quality.

According to the information reflected by the participating enterprises, it used to take at least three days to get the inspection and quarantine certificates. Now, through the risk warning blockchain symbiotic system, it can be acquired in one day, thus realizing the "one-day tour" of product clearance, which provides great help for participating enterprises to enhance their competitiveness.

7   Conclusions

The paper analyzes some shortcomings of current supervision mode of inspection and quarantine for cross-border products. For the government departments, the incomplete and untrustworthy risk data, non-autonomous and human intervened risk warning model leads to insufficient supervision for enterprises and products. For this reason, we proposed a smart contract-based risk warning blockchain symbiotic system for cross-border products.

A blockchain ecosystem involving producers, consignees, intermediate service providers and the government departments has been developed to provide credible services. The symbiotic system combines the risk warning model with smart contracts on the blockchain. It is a two-way innovation of risk warning and blockchain technology. We use smart contracts to realize the main functions of the risk warning system such as enterprise access audit, data cross validation, risk data acquisition and processing, risk assessment and updating to ensure the autonomy and fairness. The risk warning model comprehensively considers historical data and current situation, and comprehensively considers the risks of enterprises and products. The model returns a triple {Tag, Score, Msg[]} to assist in inspections. The system cuts down the clearance time and sampling proportion under the condition of ensuring quality, which pleasures both the government and enterprises.

On the experiments of the system, by simulating user scenarios, the user's operations and data communication with the blockchain system are tested, which validates the technical feasibility of the system. The application effect is verified in the Administration of Inspection and Quarantine in Shanghai Airport.

References

1. Adviser of the UK Government Chief Scientific: Distributed ledger technology: beyond block chain, January 2016
2. Alharby, M., van Moorsel, A.: Blockchain-based smart contracts: a systematic mapping study. arXiv preprint arXiv:1710.06372 (2017)
3. Barley, M.: Samsung SDS to create Korean customs blockchain, September 2018. https://www.ledgerinsights.com/samsung-sds-creates-korean-customs-blockchain/
4. Dai, Y.: A research on a planning on entry-exit inspection and quarantine conformity conditions screening and inspection & risk pre-warning. Doctor , Nanjing University of Science and Technology (2009)

5. Feng, W., Huang, J.: Early warning for civil aviation security checks based on deep learning. Data Anal. Knowl. Discovery **2**(10), 46–53 (2018)
6. General Administration of Quality Supervision, Inspection and Quarantine: Circular of the general administration of quality inspection and quarantine on publishing measures for credit management of entry-exit inspection and quarantine enterprises. http://www.gov.cn/gongbao/content/2013/content_2509242.htm
7. Hyperledger: Hyperledger Composer Docs
8. IBM: Blockchain basics: Hyperledger Fabric, July 2018
9. Kim, B., Hong, S.C., Egger, D., Katsar, C.S., Griffin, R.L.: Predictive modeling and categorizing likelihoods of quarantine pest introduction of imported propagative commodities from different countries. Risk Anal. **39**(6), 1382–1396 (2019)
10. Liu, J., Jiang, P., Leng, J.: A framework of credit assurance mechanism for manufacturing services under social manufacturing context. In: 2017 13th IEEE Conference on Automation Science and Engineering (CASE), pp. 36–40, August 2017
11. Saadaoui, Z.: Digitization of ATA Carnets: how the Blockchain could enhance trust
12. Santamaria, S.C.: CADENA, a blockchain enabled solution for the implementation of mutual recognition Arrangements/agreements
13. Shenzhen OneConnetc Smart Technology Co., Ltd and Ping An Blockchain Research Institute: Blockchain Whitepaper For Cross-border Trade, Feburary 2019
14. Xiao, H.: Ensemble learning models and algorithms for risk decision-making problems. Doctor, Chongqing University (2017)
15. Yu, Y.: Mining and analysising of chongqing customs' import and export data. Master, Chongqing University (2008)
16. Yuan, C., Xu, Y.: Research on implementing classified inspection supervision for export mechanical and electrical products. Modern Commod. Inspection Sci. Tech. (05), 7–9+17 (2001)
17. Zhou, X., Zhang, C.: Customs risk classification and forecasting model based on data mining. J. Customs Trade **38**(02), 22–31 (2017)

BYOD Security and Risk Challenges in Oman Organisations

Khoula Al Harthy[1(✉)] and Nazaraf Shah[2(✉)]

[1] Middle East College, Muscat, Oman
khoula@mec.edu.om
[2] Coventry University, Coventry, UK
aa0699@coventry.ac.uk

Abstract. Bring Your Own Device (BYOD) is an environment where the end users use their own devices to complete their organization's tasks. However, with the growth of a number of mobile devices, especially with rise of IoT based solutions, the BYOD environment has become even more challenging from security and privacy perspective. Hence, the extant information security management approaches and procedures need to be revised to be able to deal with the new risks presented by BYOD. This paper aims to study the current BYOD security frameworks and procedures being adopted by Omani organisations in order to identify the security gaps and effectiveness of the security measures being employed. Moreover, this, paper follows a primary data collection methods in order to understand the challenges from both users and professional perspectives. The both surveys of users and professionals aim to assess the current security frameworks and risk identification mechanisms used by the organization in Oman. This survey will also help to understand the level of BOYD user's awareness. This attempt will help to identify potential threats in BYOD environment.

Keywords: BYOD · Security · Risk management · Data mining

1 Introduction

BYOD is an environment where the end users use their own devices to complete their organization's work. BYOD has several advantages such as enhancing work performance, increasing productivity and cost reduction [1, 2]. Cost saving and enhanced performance are the two main advantages of BYOD which have been discussed in the following subsection of the paper. However, the security management in the BOYD has been hot research topic in recent years. Hence, this study has been conducted in Oman in order to understand the level of BYOD adoption and it associated security risks. We have selected a number of organisations from both public and private sectors for our study. The rationale behind selection of these organisations is to have an in-depth understanding of challenges posed by BYOD environment in different sectors. The paper is organised as follow: Sect. 1 presents literature review and research gaps in current research. Section 2 discuss the methodology. Section 3, provides results and proposed framework. Section 4 concludes the paper.

© Springer Nature Switzerland AG 2020
K.-M. Chao et al. (Eds.): ICEBE 2019, LNDECT 41, pp. 290–301, 2020.
https://doi.org/10.1007/978-3-030-34986-8_21

2 Related Work

The BOYD concept started appearing literature in the late 2010s. BYOD is basically an approach which allows the employees to use their own personal smartphones and tablets rather than the devices supplied by the workplace [2]. With an integration of personal mobile devices in a business environment resulted in a change in security landscape. Moreover, the critical condition of security had been raised within a BYOD environment as noted by [3, 4] the authors believe that BYOD presents two levels of security challenges. The two levels are personal levels such as privacy breach to user's data and business level breaches such as losing control of business information, reputation and money and this view is also endorsed other researchers [5, 6].

Moreover, Lounsbury indicated that the smartphones are running mobile applications and services from cloud has increased level of web attacks in the BYOD environment [7]. Hence, well-known attacks such as malwares, phishing attack, and signature attack and authorizations attack present mounting security challenge [8]. This demonestrates the importance of the security challenges faced by the BOYD network administrators [9, 10]. Therefore, the traditional security countermeasures in the networks such as firewall, IPS, IDS and router filtering, are used to secure both users and services under one domain either LAN or WAN [11, 12]. However, the concept of a centralize management has limited applicability to current technological environments such as cloud computing, virtualizations and BYOD. Due to connecting personal mobile devices such as laptops, smartphones and tablets through remote accessing the number of attacks has also increased within the network [13]. The effect of these attacks has reflected in web traffic security. Hence, the environment of cybersecurity has increased in its complexity due to rise in adoption of IoT and BYOD associated security vulnerabilities and risks [14, 15]. Downer and Bhattacharya Analysed the BYOD security challenges by dividing them into in four categories as shown in Fig. 1.

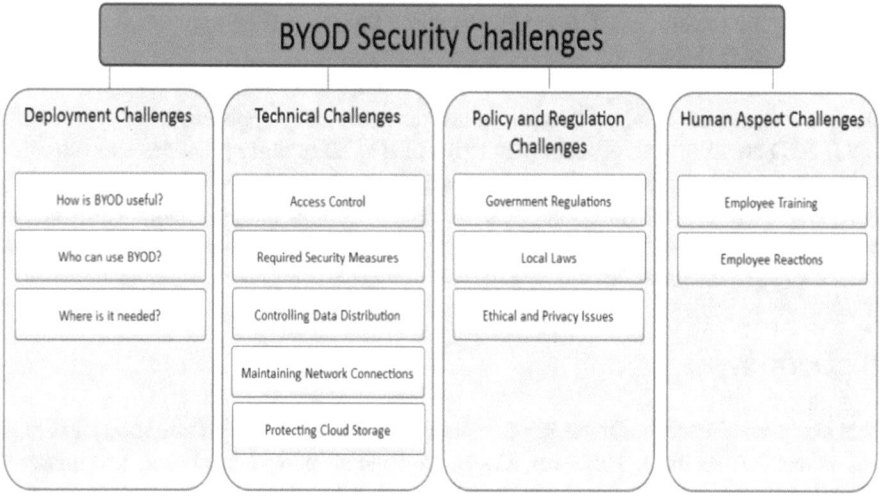

Fig. 1. BYOD security challenges [15]

The challenge is not only detecting the security threats from the smartpones hotspot attack but also having effective methods in place to prevent them.

3 Methodology

It is important that after the research gap in literature review these gaps have to be related in BOYD operational environments by through primary data collection methods. The data collection is be based on primary data which is obtained by conducting surveys.

These surveys' main aim was to find the current challenges in the most well-known security control methods and application of security framework in BYOD environments in Omani organisations. The survey was also an attempt to evaluate the technical challenges involved in the support mechanisms offered in risk management currently adopted in Omani government and private sectors. Additionally, the survey also aims at identifying the mobile users' awareness level of the security in BYOD environment in Omani organisations.

Two survey has been distributed to achieve the research objective regarding identifying users' awareness and finding the currently used risk management frameworks. The first survey focuses at the organizational level for evaluating the effectiveness of the frameworks and risk management approaches which are currently employed in the BYOD environment. This survey aims to find the current challenges in the most well-known security control methods and applied security framework in BYOD environment. Participants were asked about the challenges of technical support to manage the risk in current frameworks. This survey was designed to extract professional knowledge. Therefore, the survey includes both open and closed questions. The closed questions aim to get the direct answers without any possibilities of confusion. The open questions are for getting additional inputs that have potential to benefit this research and provide opportunity for the professionals to provide their knowledge and experience of security threats and risks in BOYD. This will help in identifying the expertise of the professional and their way they apply their experience to manage BOYD risks.

The second survey targeted the users to evaluate their awareness level. Moreover, the aim is to measure the mobile user's awareness level in applying security measures in a BYOD environment, especially in terms of BYOD security practices and activities. This survey is designed for the end users who have less knowledge of the security concepts. Therefore, the questions were written in a simple language with less technical jargon. All questions in survey forms are closed questions which will help to get needed precise information.

4 Analysis

This study conducted in Oman in order to understand the level of adoption of BYOD and associated security. However, the sectors which had been chosen in Oman are educational, governmental, private, banking and, oil and gas. The selected organisations had been chosen in order to have an in-depth understanding of challenges posed by BYOD environment in different sectors. The participated organisations have agreed to participate in the survey because they allow smartphones users to access organisational resources as a part of their work. The educational institute has agreed to participate as they allow the state of the art teaching methods using smart devices for the smart education environment. The government organisation are a bit restrictive in allowing BYOD with a specific role. The banking sector relies on M-banking applications and mobile transactions. Oil and gas sector allows the use of smart devices in the field work and allow tracking over mobile applications.

However, these organisations have shown interest to enhance their BYOD risk management by willing to participate in the survey. The Tables 1 and 2 shows each organisation participation percentage in surveys.

Table 1. Organisation participation in professional level (survey 1).

Questions	Responses	Respondents	Percentage
How to describe your organizations area?	Oil and gas	4	20%
	Educational	5	25%
	Government	9	45%
	Banking	2	10%

Table 2. Participation in user awareness (survey 2)

Questions	Responses	Respondents	Percentage
How to describe your organizations area?	Oil and gas	60	26.66
	Educational	50	22.22
	Government	58	25.77
	Banking	17	7.55
	Private	40	17.77

4.1 Survey 1 Results and Findings

The first survey targets the IT and security experts in order to find real BYOD challenges and risks in their daily work environment. This survey highlights the currently adopted frameworks and countermeasures within their organization. Overall 20 security and IT expert from 5 different organisations had participated in this survey. The participants was working as network administrator, security manager, CEO, IT and security consultant and IT Technician as shown in Table 3.

Table 3. Professional work affiliation.

Questions	Responses	Respondents	Percentage
What is your job role?	IT technician	6	30%
	Network admin	4	20%
	Security manager	1	5%
	CEO	1	5%
	Security/IT consultant	8	40%

Within the survey, it was necessary to know the current risk management steps that have been followed by information security experts. Therefore, the Fig. 2 contains the options that have given to the participants which include risk assessments, risk analysis, disaster recovery, risk treatment, risk identifications and reactions from IT team. The participants were allowed to indicate management steps. Hence, the Fig. 2 shows the percentage for each applied steps of overall participants answers.

Fig. 2. Risk management steps used in BYOD

Was considered important to know the results of the used risk management framework in selected organisations. Hence, two participants out of 20 tasted that they are not using any framework or risk management standards in their organisations as is shown in Table 4. However, 12 participant's stated that they use of ISO 27001 which is information security management framework and only 2 (10%) follow ISO 27002 for risk management framework and standards. Four participants state that they are following COBiT approach which is responsible for providing connection of business and IT environments. The level of the participants' awareness of the frameworks shown in Table 5. However the applicability of these frameworks in these organisations varies.

Table 4. Current information security frameworks/standards used in participants organizations

Questions	Responses	Respondents	Percentage
What are the current information security framework or standards used in your organizations	CobiT	4	20%
	ISO 27001	12	60%
	ISO 27002	2	10%
	ITIL	2	10%

Table 5. Level of compatibility of current organizations IT strategy with BYOD IS strategy

Questions	Responses	Respondents	Percentage
What is the level of effectiveness and control of these frameworks in a BYOD environment	10–30% compatibility	1	5%
	30–50% compatibility	7	35%
	50–80% compatibility	12	60%
	80–100% compatibility	0	0%

In Table 5 60% of the participants were agreed that the current frameworks and standards do not match the BYOD security requirements. Moreover, 35%% stated that there is a average compatibility between 30 to 50% match of effectiveness in risk management and finally 0% stated that these framework have capability deal with BYOD risks and threats. As indicated in the literature there is a high possibility that the current standards and principles do not match the appearing risk and threats [3, 16–18].

Moreover, the participant had been asked of how often they update their risk management system and framework and the response is as shown the Fig. 3. 30% never update their risk management system which reflect likelihood of the high risk.

How often you review/update your business risk management system

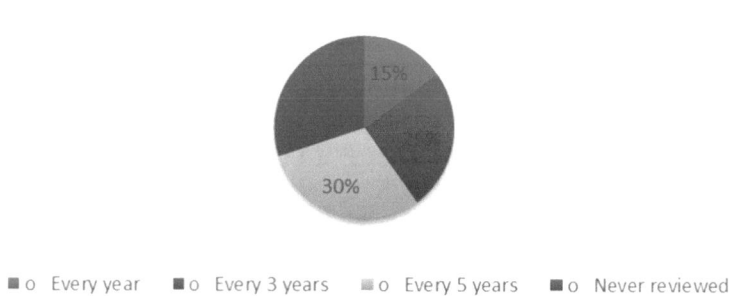

■ o Every year   ■ o Every 3 years   ▥ o Every 5 years   ■ o Never reviewed

Fig. 3. How often risk management process and frameworks are updated

4.2 Survey 2 Results and Findings

The critical part of building any security programs for an organisation is end user awareness of the nature of security systems. However, in BYOD environment the end users are the controller of the traffic and information. Therefore, it mandates a significant requirement to understand the smartphones user awareness about the current threats and risks. This survey has 225 participants from 5 different organisations in order to assess the user's security awareness. These are same organisations where the professionals participated in the first survey. The following table shows the level of understanding of security threats. The results shows that 44% user are not aware of the security measure in the organisation (Table 6).

Table 6. Level of understanding of security threats

Questions	Responses	Respondents	Percentage
Are you aware or use any security applications or features to protect your smartphones?	Yes	126	56%
	No	99	44%

Hence, the lack of knowledge and no action taken to secure the devices it is reasonable to see high risk impacts and increase the possibility of data loss is in Omani organisations. Therefore, 56% indicate that they have experienced data loss over their smartphones. However, data loss happen due to multiple reasons. The participants had been asked to select one or many reasons of their data loss from their smartphones. The answers was labelled as following: fake applications, unknown reason, improper storage, device damage, online services device loss and attacks. The participants in the survey highlights one of the BYOD attacks which is applications attack. They stated that the fake and untrusted applications which used from unknown source are considered threats to their personal and business data. However, this type of attack was also mentioned in the literature in Krombholz [21] Spoorthi and Sekaran [22] and Romer [23] (Table 7).

Table 7. Data loss in smartphones

Questions	Responses	Respondents	Percentage
What was the cause of the data loss?	Attack	15	8%
	Fake application	30	34%
	Online services malwares	28	16%
	Device damage	56	24%
	Improper storage	36	12%
	No data loss	71	6%

5   Discussion and Recommendations

This section reflects on the findings from participants response obtained through surveys. The finding of these survey's demonstrate the gaps in adoption of effective security and privacy measures in Omani organisations, which we address by proposing a new framework. The Fig. 4 provides an overview of the process of findings and drawing requirements for a new framework.

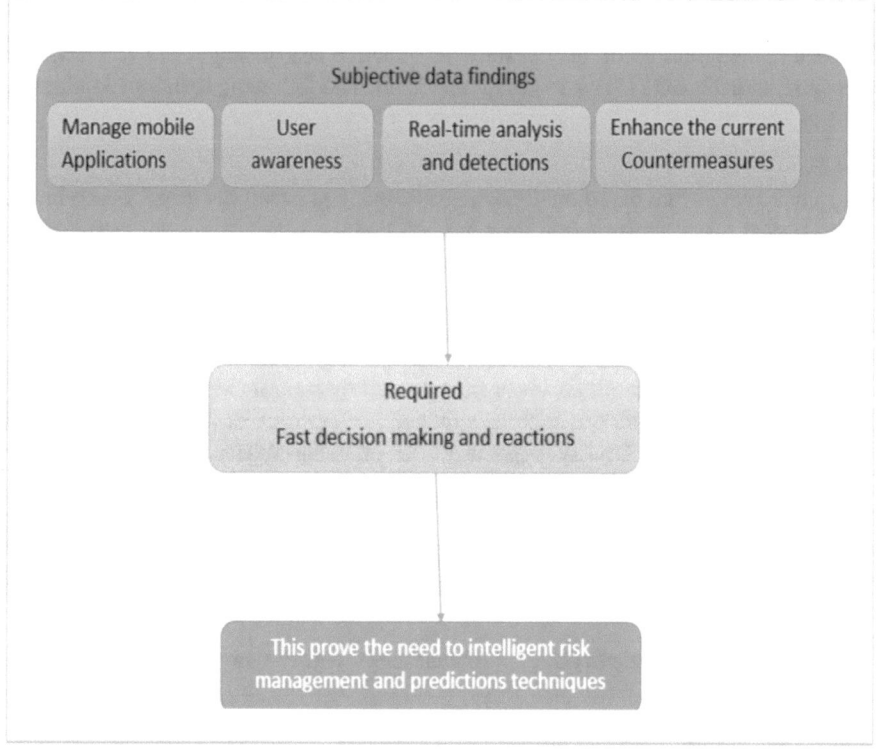

Fig. 4. Survey data findings

The findings highlight the gaps into following categories:

- Mobile applications management
- User security awareness
- Real time analysis and detections
- Enhancement in current countermeasures.

These gaps lead into major requirements of the framework
Accurate detection and timely detection of risks in BOYD

- Timely response to detected BYOD risks
- Fast decisions making when attacks accrues.

Therefore, it becomes imperative to propose a new risk management framework to with the BOYD risk both from server side and client side perspectives.

5.1   Recommendations

The recommendation focus on the need of effective risk management framework in place to reduce the BOYD risk to acceptable level. The following recommendations are based on the findings which has been obtained through surveys data insight.

User Security Awareness
Bring your own device based on the concept of using personal devices for carrying out organisational tasks, which mean user have to use the smartphones for business purposes. However, the users usually forget or ignore their responsibility of protecting their devices as it contains the organisational data. As indicated in the users' survey, the lack of user's awareness of BYOD security risks causes many risks such as data loss and data leakage. BOYD users' security awareness play a pivotal role in overall BOYD security. Waters et al. Identified seven risks caused by lack of users awareness by [19]. These risks are devices loss, malicious applications, connection through public Wi-Fi, internet browsing, no security apps installed on smartphones, keep changing the devices and lack of data encryption [5].

Watson and J, Zheng has proposed a prime division of risk solution which can be applied to user's smartphones. The proposed list of solutions focuses on user's lack of awareness, user behaviours or even risks reaction plan for the risks caused by mobile applications [20]. Hence the major recommendation for the users are:

- Avoid installing unnecessary applications with many access permission requirement.
- Avoid public Wi-Fi connectivity.
- Report any business data loss.
- Make sure of wiping the business data in case of device loss, stolen or sealing the device.

Recommendations for the Organisation
The organizations adopting BYOD, should keep reviewing the access policies and maintain the legal liabilities. They should maintain the security control by having effective BOYD security assessment in place. The security control means the responsibilities of each area of the business environment, which includes IT, Management, and users. In each of these areas, there are a set of control methods required such as policies, risk management, frameworks, agent-based devices control, training of MDM adoption. For the policy level, the organizations should take care of building the policy to serve the different layers that are interacting with a BYOD environment. BYOD policy should focus on three layers, which are operational, tactical and strategical policies layers. However, the major challenge over the literature was controlling human resources, the survey finding and analysis of literature agreed that the users need education and training.

Moreover, to enhance the BYOD management environment the business needs to update their legal terms and policy to protect the BYOD parties' access privileges, roles and responsibilities.

6 Proposed Framework

The proposed framework focuses on enhancing risk management by improving risk detection and predictions capabilities. It should also enable BYOD risk management system to help in generating notifications about possible preventive/mitigation actions which can be taken to deal with the threats (Fig. 5).

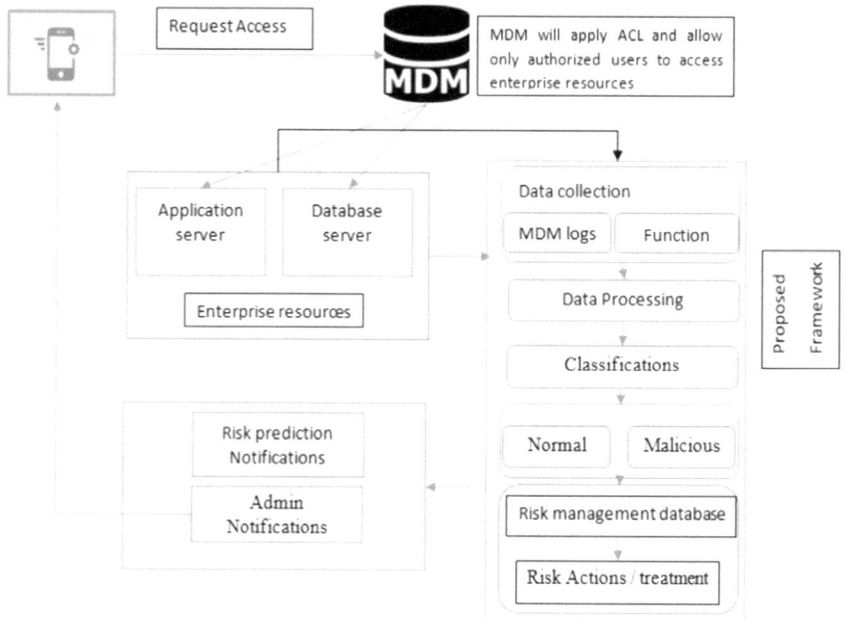

Fig. 5. Proposed risk management framework

The proposed framework is presented in Fig. 3. The flow of information and actions taken in various components of framework discussed in the following steps:

1. User installs MDM agent in advance before connecting to network resources.
2. User connects to enterprise resources through MDM
3. MDM admin applies ACL and allow only authorised users to access enterprise resources
4. User access to network database and applications.
5. Administrator extract the log records of MDM activities and user access through resources.
6. The framework use these log records to initiate its functioning
7. The framework classifies the log entries using machine learning approaches. The output is either normal or abnormal event.

8. Based on classification result the framework generates required actions for prevention/remedy.
9. The final outcome is sent to smartphone users as notifications and also used by the system to take remedial actions.

Key components of the proposed framework:
The users access resources through MDM server which create record of activities in MDM log file. Some servers also make the record of activities carried out by them these records are recorded in system functions log.

MDM Log: The MDM log records are user's activities carried out through the MDM server. These two components are primarily concerned with the log history of the users' actions. After extracting log records and cleaning and labelling data, classification approach is used to classify events into two categories known as normal or malicious. The outcome of classification is used by the decision component to decide the action plan. Various kind of alerts and notification are also generated as result of events classification.

7   Conclusion

Now a days BYOD is being widely used in business environments, and this environment bring high security with it. It is important to have effective security framework in order to reap the benefits of the BOYD. Security and privacy of data is a top priority for BYOD environment. Therefore, in this paper we present the BYOD adoption concerns in Oman and current challenges. We have presented the results of surveys and recommendation a actions can followed by professionals and users in context of Omani organisations.

References

1. Jones, J.: Beginner's Guide to BYOD (Bring Your Own Device) (2012). Accessed 9 Feb 2014
2. Brooks, T.: Classic enterprise IT: the castle approach. Netw. Secur. **2013**(6), 14–16 (2013)
3. Eslahi, M., Salleh, R., Anuar, N.B.: Bots and botnets: an overview of characteristics, detection and challenges. In: 2012 IEEE International Conference on Control System, Computing and Engineering. IEEE (2012)
4. Ghosh, A., Gajar, P.K., Rai, S.: Bring your own device (BYOD): security risks and mitigating strategies. J. Glob. Res. Comput. Sci. **4**(4), 62–70 (2013)
5. Ismail, K.A., Singh, M.M., Mustaffa, N., Keikhosrokiani, P., Zulkefli, Z.: Security strategies for hindering watering hole cyber crime attack. Procedia Comput. Sci. **124**, 656–663 (2017)
6. Assing, D., Calé, S.: Mobile Access Safety: Beyond BYOD. Wiley, Hoboken (2013)
7. Lounsbury, J.: Application security: from web to mobile. Different vectors and new attacks. Secur. Knowl. 2–30 (2013)
8. Howard, F.: Modern web attacks. Netw. Secur. **2008**(4), 13–15 (2008)
9. Atallah, E., Chaumette, S.: A smart card based distributed identity management infrastructure for mobile ad hoc networks. In: IFIP International Workshop on Information Security Theory and Practices, pp. 1–13. Springer, Heidelberg (2007)

10. Conti, M., Giordano, S.: Mobile ad hoc networking: milestones, challenges, and new research directions. IEEE Commun. Mag. **52**(1), 85–96 (2014)
11. Roberts, L.G., Wessler, B.D.: Computer network development to achieve resource sharing. In: Proceedings of the Spring Joint Computer Conference, 5–7 May 1970. ACM (1970)
12. Andrea, I., Chrysostomou, C., Hadjichristofi, G.: Internet of things: security vulnerabilities and challenges. In: 2015 IEEE Symposium on Computers and Communication (ISCC), pp. 180–187. IEEE, July 2015
13. Eslahi, M., Naseri, M.V., Hashim, H., Tahir, N., Saad, E.H.M.: BYOD: current state and security challenges. In: 2014 IEEE Symposium on Computer Applications and Industrial Electronics (ISCAIE). IEEE (2014)
14. Siboni, S., Shabtai, A., Elovici, Y.: An attack scenario and mitigation mechanism for enterprise BYOD environments. ACM SIGAPP Appl. Comput. Rev. **18**(2), 5–21 (2018)
15. Downer, K., Bhattacharya, M.: BYOD security: a new business challenge. In: 2015 IEEE International Conference on Smart City/SocialCom/SustainCom (SmartCity). IEEE (2015)
16. Sitnikova, E., Asgarkhani, M.: A strategic framework for managing internet security. In: 2014 11th International Conference on Fuzzy Systems and Knowledge Discovery (FSKD). IEEE (2014)
17. Boehmer, W.: Cost-benefit trade-off analysis of an ISMS based on ISO 27001. In: 2009 International Conference on Availability, Reliability and Security. IEEE (2009)
18. Huang, Z., Zavarsky, P., Ruhl, R.: An efficient framework for IT controls of bill 198 (Canada Sarbanes-Oxley) compliance by aligning COBIT 4.1, ITIL v3 and ISO/IEC 27002. In: 2009 International Conference on Computational Science and Engineering. IEEE (2009)
19. Waters, E.K., Sigh, J., Friedrich, U., Hilden, I., Sørensen, B.B.: Concizumab, an anti-tissue factor pathway inhibitor antibody, induces increased thrombin generation in plasma from haemophilia patients and healthy subjects measured by the thrombin generation assay. Haemophilia **23**(5), 769–776 (2017)
20. Watson, B., Zheng, J.: On the user awareness of mobile security recommendations. In: Proceedings of the SouthEast Conference, pp. 120–127. ACM, April 2017
21. Krombholz, K., Hobel, H., Huber, M., Weippl, E.: Social engineering attacks on the knowledge worker. In: Proceedings of the 6th International Conference on Security of Information and Networks. ACM (2013)
22. Spoorthi, V., Sekaran, K.C.: Mobile single sign-on solution for enterprise cloud applications. In: 2014 First International Conference on Networks & Soft Computing (ICNSC). IEEE (2014)
23. Romer, H.: Best practices for BYOD security. Comput. Fraud Secur. **2014**(1), 13–15 (2014)

IoT-Driven Business Model Innovation

A Case-Study on the Port of Barcelona in the Context of the Belt and Road Initiative

Ricardo Henríquez[1]($\boxtimes$), F. Xavier Martínez de Osés[1],
and Jesús E. Martínez Marín[2]

[1] Catalonia Polytechnic University, Barcelona, Spain
ricardo.daniel.henriquez@upc.edu,
fmartinez@cen.upc.edu
[2] Pompeu Fabra University, Barcelona, Spain
jmartinezma@tecnocampus.cat

Abstract. During the last decade, new technologies have pushed organizations to innovate their business models. In particular, Industry 4.0 and the Internet of things (IoT) have called for their own business models, in order for them to live up to their potential as sources of value and revenues. Seaports have not been alien to this trend. As key nodes in the global supply chain and logistic networks, some of them have evolved towards a new value-generating and community-focused model, categorized as the fifth generation (5G) port. This research paper explores what impact does IoT technology will likely have in the evolution of seaports business models, as well as the role that institutional initiatives play in such evolution. To do so, a case-study is presented about the Port of Barcelona, and the strategic policies that it has adopted in the context of the Belt and Road Initiative developed by the Chinese government. The case shows, though not conclusively, a relationship between IoT technology and the adoption of a 5G business model.

Keywords: Internet of Things · Business models · Fifth generation ports · Port of Barcelona · Belt and Road Initiative

1 Introduction

Seaports are at the crossroads of global supply chain networks, being one of their key nodes [1, 2]. During the second half of the last century, containerization revolutionized international trade, bringing about globalization [3]. Seaports, accordingly, changed their business models, evolving from simple sea-land interfaces, to complex organizations facilitating the seamless and efficient performance of global value chains. Their role as global physical and structural connectivity enablers was emphasized. In the last decade, however, the surge of technological trends such as digitization, data mining, automation and sensorization, have shifted the emphasis from physical to informational connectivity. While interregional spanning government initiatives like China's Belt and Road or the European Union's TEN-T network have mainly focused on infrastructure,

© Springer Nature Switzerland AG 2020
K.-M. Chao et al. (Eds.): ICEBE 2019, LNDECT 41, pp. 302–314, 2020.
https://doi.org/10.1007/978-3-030-34986-8_22

those same plans have given increasing importance to information flows and technological standards.

Under this trend, both scholar and policy discussions have stressed the role that seaports should play in facilitating information flows and data interconnectivity among global supply chain stakeholders. New technologies –is said– require, once more, new business models. However, despite numerous studies about the application to seaports of technologies like electronic data interchange (EDI) or radio-frequency identification (RFID), as well as on seaport's business models, research on how the former influence the latter are difficult to find. The present research presents a case-study about the port of Barcelona, analyzing its policies and initiatives, taking into consideration the context provided by China's Belt and Road Initiative. The aim is to generate explanatory theory [4] on how a specific technology, IoT, functions a driver for the change from one port business model to a more complex one. The main research question is thus presented as follows:

RQ: What is the impact of Internet of Things (IoT) technology on a seaport's business model?

The rest of this paper is organized in the following way: Sect. 2 reviews related works. Section 3 presents a conceptual framework for understanding the relationship between IoT technology and a seaport's business model. Section 4 provides a case study about the Port of Barcelona, as an instance of the interrelationship between technology and business models. Section 5 presents a brief discussion, and Sect. 6 concludes the paper.

2 Related Works

2.1 Business Models for the Internet of Things

Industry 4.0 and the internet of things (IoT) have been closely associated as technological paradigms [5]. Both have been related with the term "Cyber-Physical Systems" (CPS), understood as the connection of information, objects and people in a converging physical/virtual world [6, 7]. Adoption of these paradigms have called for new business models. A business model is broadly defined as the way an organization creates, delivers and captures value [8]. From these general dimensions, [9] and [6] identify a series of "building blocks" and approaches out of which business models can be designed for the particular requirements of IoT and Industry 4.0, among those, the need for real-time capabilities, interoperability, and a services and customer orientation.

Turber et al. [10] present a business model framework for IoT-driven ecosystems, which are said to involve a diverse array of business partners and cross-industry relationships. Following the so-called *Recombination School* of the University of St. Gallen [11], the authors identify 3 dimensions of the IoT business model framework along the questions *who? where? and why? Who* refers to the collaborating partners building the value network; *where*, is the source of value co-creation; and *why*, represents the benefits obtained by the partners. The core idea of the framework is that IoT enhances value co-creation by collaborating partners *(who)*, across a digital ecosystem comprised of four layers: device, network, service and content *(where)*, deriving monetary and non-monetary benefits *(why)*.

IoT-based business models reflect a more general principle about the interrelation between business models and technology. Business models are the channel through which new technologies are transformed into a stream of profits [12], linking technology with performance [13]. The right business model is, therefore, a key factor in developing the potential of IoT technology.

2.2 Business Model Innovation in Seaports

Since the United Nations Conference for Trade and Development (UNCTAD) classified seaports into generations [14], five port generations have been identified [15, 16], with a port falling in them according to how it fares under several categories. Fourth (4G) and Fifth (5G) generation ports imply a globalized e-port, which enhances its role as a node in the global supply network, adopting top ICT systems. 5G ports, however, distinguish themselves from 4G ports by developing customer-centric and community-focused capabilities; constantly evolving to create new value for its stakeholders [17, 18].

This customer and community focus require a significant degree of business model innovation (BMI). BMI in the context of seaports has been associated with the evolution from a "landlord" to a "port developer" role along four levers: organization, management, technology and co-creation; an evolution that renders itself into the development of new value-creation businesses [19]. This way, from a baseline emphasis on costs and efficiency improvement, ports aim to co-create new sources of value, strengthening not only their *structural* connectivity (related to their position in the global transport network), but also their *strategic* connectivity [20]. The latter concept is associated with knowledge-intensive interorganizational exchanges between ports, and was initially proposed in a case-study regarding the ports of Rotterdam and Shanghai [21], noticing the establishment of *Smart and Secure Trade Lanes*, where entire end-to-end supply-chains are certified, rather than the individual firms that take part in them.

2.3 Belt and Road Initiative

The 'Silk Road Economic Belt and the 21st-Century Maritime Silk Road', better known as the 'Belt and Road Initiative' (BRI), was established by the Chinese government in 2013. Its fundamental vision is to enhance connectivity and infrastructure in Asian, African and European countries, along several corridors [22]. Among its objectives are the strengthening of business partnerships and the development of connectivity networks, intergovernmental policy dialogues and convergence of technical standards [23].

Transport interconnectivity is one of the main objectives of the BRI, and while its main focus is on infrastructure development (ports, rail, pipelines, etc.), there is also an interest in technology and communications. As observed in [24], a key aspect for the successful integration aimed through the BRI, is logistics information sharing, which requires international cooperation in logistics, informatization, and technology standards. Among these standards, the ones related to IoT layers are deemed crucial for truly integrated logistic corridors. Success of IoT logistic uses cases like container tracking strongly depend on widely adopted technological standards [25]. To this end the Chinese government has been developing efforts on IoT standardization, mostly through the China Communications Standard Association (CCSA) [26].

3   Conceptual Framework

3.1   Smart Port (5G) Business Model

Following the previously referred *Recombination School*, Table 1 presents differences between 4G and 5G seaports. A slightly modified version of the framework in [10], describes the models along the dimensions provided by the following three questions: *what* (value creation), *how* (value delivery) and *why* (value appropriation).

Table 1. Seaports business models framework

	Fourth generation port (4G)	Fifth generation port (4G)
What Value creation	• Emphasis on the integration role of ports on the global supply chain • IT for cargo clearance and tracking • Gateway port • Efficiency and costs reduction	• Emphasis on environmental sustainability (green port) and high-tech (smart port) • IT not only tracking and tracing, but also for event management anticipation and performance measurement • Competitive transshipment center • New value creation
How Value delivery	• Deliverables related to logistic functionalities and port transport • Logistics as back of port functions • Partial automation, limited sensorization • Structural connectivity	• Deliverables related to port user and community stakeholders • Logistics as part of maritime logistic chain • Full or significant automation and sensorization • IT standards setting • Strategic connectivity
Why Value appropriation	• Landowner port • Revenues from enhanced performance in traditional port functions • Driven by competition with other ports in the logistics chain	• Developer port • Revenues from new value creating activities • Driven by customer-centric community interests • Sustainability as non-monetary benefit

This characterization of fifth generation ports complements the one presented in [18] with traits taken from the "smart port" concept as developed in [27] and [2]. Given the focus of this research on technology, some port characteristics (port cluster, maritime cluster, inland connectivity) have been omitted as categories of generation classification.

This conceptual framework aims to show how some of the "building blocks" associated with IoT technology, are also associated with or play a catalyst role in bringing about the traits that distinguish a fifth generation seaport. In other words, an explanation is offered on how IoT drives business model innovation, from a 4G model to a 5G one. To this end, the likely impact of IoT is analyzed in relation to three seaport areas: operations, strategies and investments.

3.2   IoT as a Driver of Business Model Innovation in Seaports

In this conceptual framework, the use of the term "seaport" is broad, referring not only to port authorities, but also including the private companies and organizations that perform core port related activities, like bulk, container or oil/gas terminals. A seaport is thus understood as an organic system, continuously evolving and adapting to new economic, technological and regulatory patterns [18], which generate new demands from economic actors. This evolution and adaptation to patterns and corresponding demands, has covered the sort of activities and services performed (operations), the policies and strategies adopted, and the kind of investments done.

Regarding port operations, the first sub-research question is posed as follows:

RQ₁: How will IoT impact seaports' operations and functionalities?

IoT technology, with its enhanced sensorization, generates increasing amounts of automatically generated data that make possible to obtain a real-time vision of the state of a supply-chain network. The updated position and state of a specific container in the intermodal transportation chain, for instance, could be visualized. This increased visibility would generate significant benefits in terms of efficiencies, risk management, and security. Given their position as key nodes in the global supply network, seaports' have a special place as sources of aggregated real-time information. Supply-chain stakeholders, therefore, will demand from seaport authorities and private operators the provision of services that would make possible this enhanced visibility. Those services would be performed at several of the four IoT layers: devices, network, applications and content [10, 26].

Regarding policies and strategies, the second sub-research question is defined as follows:

RQ₂: How will IoT impact seaports' strategies and policies?

As IoT-generated data will span wide geographic areas, covering several countries and ports along the global maritime routes, technology standards will be crucial for optimal performance. Coordination between ports authorities, carriers, terminal operators and other players, regarding the adoption of IoT standards, especially regarding device and network layers, will be a factor for competitive advantage of one transportation chain over another. Seaports would therefore adopt policies and strategies that increase standardization, cooperation and information sharing with other ports and stakeholders. These policies would amount to enhanced *strategic connectivity*, as defined in the previous section.

Finally, in relation to port investments, the third sub-research question is presented in the following way:

RQ₃: How will IoT impact seaports' investments?

The development of port infrastructures for enhanced structural connectivity (gantry cranes, automatic vehicles, logistic parks, inland transportation) have required significant capital-intensive investment. However, as seaports have already developed their infrastructures, their focus is shifting to value creation coming from data and information. IoT, with its potential for enhanced information flows, demands from seaports new investments. While these investments would require capital (particularly in the device and network layers), technological expertise and information management capabilities become more important. It is, therefore, expected that new port investments will be more knowledge-intensive than capital-intensive.

Figure 1 presents the conceptual framework containing the constructs of the sub-research questions. The model sketches the impact of IoT on seaports activities, and the reflection of these activities on a seaport's business model in terms of the 5G characteristics.

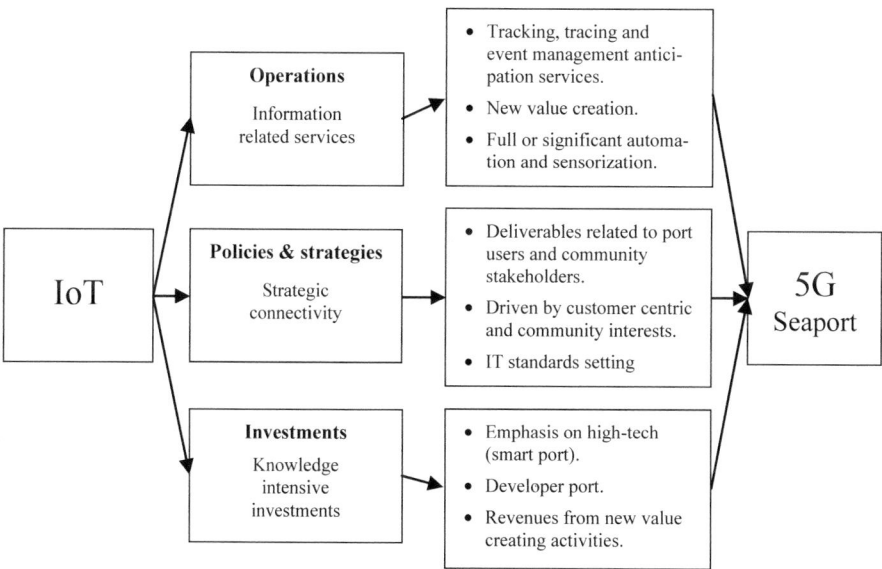

Fig. 1. Conceptual framework

4 Case Study

4.1 Port of Barcelona Strategic Initiatives

The port of Barcelona ranks third in Spain on TEU volume (3.4 M), behind Valencia (5.1 M) and Algeciras (4.8 M) [28]. From an economic perspective, it had in 2018 one of its best years, with €53.7 million in profits (an 8% increase relative to 2017) [29]. It is currently considered the preferred Spanish port for merchandise originating in Chinese ports [30]. Also, there has been important Chinese investment in the port's infrastructure: Hong Kong-based Hutchison Port Holdings Ltd., has invested €465 million in the development of the Barcelona Europe South Terminal (BEST) [31].

The general strategic approach of the Port of Barcelona is contained in the III Strategic Plan (2015–2020), which lays out several strategic axes and objectives, some of which could easily be related with the "Smart Port" or 5G labels. For instance, the 8th Strategic Objective is "to promote innovation in port's processes and services" [32, p. 55]. Topics as automation, sensorization, information flows and port community systems (PCS) are mentioned as R&D objects for the period covered.

This emphasis on technological innovation and information can be noticed in some of the cooperation memoranda recently signed in 2018 by Barcelona Port with two Chinese ports: Shenzhen [33] and Ningbo-Zhoushan [34]. In both cases, the ports agreed to an exchange of information between them, covering several aspects, including sustainability, infrastructure and logistic platforms, innovation and development of technological solutions for the ports. Moreover, both memoranda expressly mention that one of their objectives is to share growth strategies under the Belt and Road Initiative.

4.2   LoRaWAN

During 2018, supply-chain technology companies Kerlink and Datalong16 conducted a pilot on the port of Barcelona, testing LoRaWAN, an IoT technology [35]. LoRaWAN is a wireless platform that gathers data through embedded sensors and antennas. Unlike global positioning system (GPS), it does not rely on satellites, but on the position of the antennas. The pilot consisted in an experiment where both LoRaWAN and GPS were used to monitor five port police vehicles and five vehicles belonging to the port's fleet. Each vehicle was equipped with a GPS and a LoRaWAN device. Also, it was installed a telecommunications network covering the port's areas of influence.

The pilot test showed that LoRaWAN, while having less accuracy than GPS, had also a considerably lower energy consumption. Also, its usage does not require an operator, such that costs are also significantly lower than with GPS.

With LoRaWAN, IoT solutions like container tracking, cold-chain control, water quality, etc., could be implemented with substantially lower environmental and economic costs.

Figure 2, shows a possible port application for LoRaWAN in the context of a seaport's container terminal. It reflects the 4 layers of IoT (device, network, application and content) in the steps.

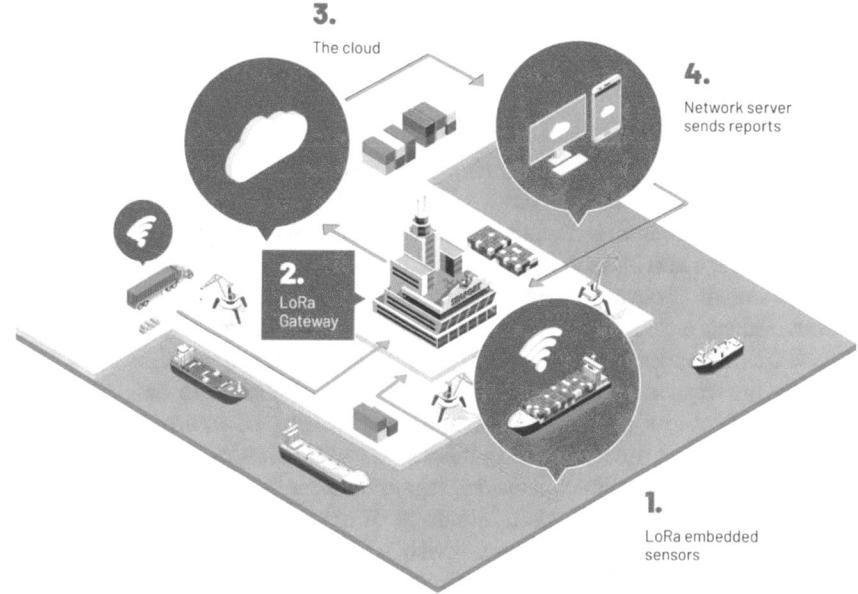

Fig. 2. LoRaWAN technology application for seaports [35].

4.3 PierNext and OpenPort

During the second half of 2018, the Port of Barcelona launched the project PierNext, defined as a *"digital knowledge hub"*, looking to become a space open to collaboration for the port's stakeholders [36]. The general objective is to serve as an innovation ecosystem where trends and initiatives could be shared and debated.

In the same vein, OpenPort was created as an initiative that looks to enhance competitiveness through open innovation and co-creation between the port's community and the startup ecosystem. To this end, OpenPort's three axes are innovative networking, innovative knowledge, and financing of startups and projects [37].

5 Discussion

The strategies and activities developed in the Port of Barcelona, above referred, constitute an example of changes in the way a port approaches value creation. In this section, a preliminary evaluation is done regarding the sub-research questions presented in Sect. 3, as well as the main research question.

The LoRaWAN wireless system, comparatively tested with GPS technology in the port of Barcelona, is an instance of an extension in the kind operations performed in the port (1st sub-research question). While physical activities like cargo handling remain the core service, data and information related activities like container tracking or cold-chain control are also being included as part of the adoption of IoT technology. However, the activities performed were of a testing nature, and did not show specific functionalities that could be developed through IoT. The pilot was limited to test the efficiency of LoRaWAN as a sensorization device in comparison with GPS, and at the moment, the case does not show new ways of value creation, delivery or appropriation, based on IoT technology.

The second sub-research question was related to policies and strategies. Here, the estimated impact of IoT would be shown in cooperation policies and information sharing between stakeholders, in order to allow for optimal performance of services. In this line, the cooperation memoranda signed between the port of Barcelona and the ports of Shenzhen and Ningbo-Zhoushan are examples of steps taken in that direction. The inclusion of innovation, sustainability, and the development of technological solutions reflects a "relational focus on creating added value through knowledge-intensive interorganizational exchanges" [20, p. 25], proper of strategic connectivity. Moreover, the express mention of the Belt and Road Initiative in both memoranda, indicate a willingness to frame cooperation and sharing of information/knowledge under an institutional umbrella. Unfortunately, the generality of the memoranda does not allow to extract specific conclusions regarding a moderating role for BRI on the adoption of IoT-related cooperation initiatives. It represents, however, a first step, as well as a new focus on strategic connectivity.

In relation to a change in the nature of port investments (3rd sub-research question), the OpenPort and PierNext initiative show a clear instance of this. Both aim at enhancing port interrelations with its users and stakeholders, in order to co-create new value, particularly regarding innovation. While capital-intensive, infrastructure investments like the BEST terminal have been the most significant in the last years, knowledge-intensive projects like OpenPort and PierNext show a new approach. The focus changes from developing infrastructure to building enhanced relationships with port stakeholders, where open innovation and co-creation are key for performance. However, a likely impact of IoT in this shift from capital to knowledge intensive investments, cannot be established.

The distinction between capital-intensive and knowledge-intensive investments is also applicable to the Belt and Road Initiative itself. Cooperation agreements like those signed between the port of Barcelona and the ports of Shenzhen and Ningbo-Zhoushan, are explicitly identified as part of initiatives taken under the institutional framework provided by the BRI.

Do these changes in the areas of operations, strategies and investments show an impact of IoT technology on business model innovation? Is it possible to establish a causal relationship between IoT adoption and an evolution towards a 5G seaport business model?

From the port of Barcelona case-study presented, a conclusive answer to these questions cannot be given. While both the adoption of IoT technology and the undertaking of information-related, knowledge-intensive and value co-creating initiatives are present in the port's strategy and projects, it is too early to establish a specific impact of the former over the latter. As these projects develop, additional data and information might offer new light on the interrelationship between IoT technology and seaports business models.

Nonetheless, what can be observed from the case study is an evolution towards a 5G business model. The way that *what*, *how*, and *why* questions of the seaports business model framework would be answered, shows elements of an enhanced focus in high-tech, community stakeholders, co-creation, strategic connectivity, and a port developer role. Table 2 presents a preliminary analysis of the activities and initiatives referred, assigning it to the areas covered by the sub-research questions, and also to aspects of value creation, delivery and appropriation. Below the specific activities and/or initiatives, one or more of the 5G seaport traits are identified.

Table 2. Preliminary analysis

	Operations	Strategies	Investments
What Value creation	The pilot test of LoRaWAN shows an interest in the application of IoT for generating additional value, based on more advanced monitoring of data • *Tracking, tracing and event management anticipation services* • *Automation & sensorization*	The cooperation memoranda signed with the ports of Shenzhen and Ningbo-Zhoushan, expressly mention the exchange of information & technology • *Strategic connectivity* • *IT standards setting*	PierNext aims to enhance stakeholder collaboration, open innovation, and co-creation. OpenPort focuses on innovative networking and innovative knowledge • *Driven by customer centric and community interests* • *Emphasis on high-tech (smart port)*
How Value delivery	The incorporation of the IoT 4 technology layers into port's processes, is seen in the likely application of LoRaWAN • *Full or significant automation and sensorization*	The III Strategic Plan of the port of Barcelona, focuses on innovation in port processes and services, and mentions automation, sensorization and information flows • *Emphasis on high-tech (smart port)*	OpenPort, as an initiative of the port of Barcelona, is a source of financing of innovative startups. The port indirectly becomes an entrepreneurship promoter and investor • *Deliverables related to port users and community stakeholders*
Why Value appropriation	The LoRaWAN pilot testing was performed through private technology companies Kerlink and Datalong16 • *Developer port*	The cooperation memoranda signed with the ports of Shenzhen and Ningbo-Zhoushan, expressly mention the exchange of information & technology • *Strategic connectivity* • *IT standards setting*	Both PierNext, with its stakeholder collaboration, and OpenPort with its startup ecosystem, indicate a form of co-appropriation of the value generated in the port context • *Developer port* • *Revenues from new value creating activities*

6 Conclusions

This paper has presented a case-study about the port of Barcelona, with the objective of researching the relationship between the adoption of internet of things (IoT) technology and innovation of business models in a seaport. The research has taken into consideration the institutional context given by China's Belt and Road Initiative. While no definite conclusion can be given as to whether IoT drives seaports to adopt 5G business model related initiatives, the case shows that both sides take place contemporaneously, and that institutional initiatives like BRI might have an enhancing impact.

This paper has considerable limitations, which arise from the early stage of both the research itself and its object of study. IoT application in the port of Barcelona is being merely tested through pilots; it is too early, also, to assess the impact that the cooperation memoranda signed with the ports of Shenzhen and Ningbo-Zhoushan could have, due to their recency.

However, as these and other inter-port cooperation initiatives are deployed into concrete projects and activities, future research can shed light on the role that strategic connectivity plays on IoT and other technologies applicable to seaports and transport logistics. It can also assess whether the Belt and Road Initiative, in particular, plays a moderating role in business model innovation for the seaports directly or indirectly affected by it.

References

1. Zuidwijk, R.: Ports and global supply chains. In: Geerlings, H., Kuipers, B., Zuidwijk, R. (eds.) Ports and Networks: Strategies, Operations and Perspectives, pp. 26–37. Routledge (2018). https://doi.org/10.4324/9781315601540
2. Chen, J., Huang, T., Xie, X., Lee, P.T.W., Hua, C.: Constructing governance framework of a green and smart port. J. Mar. Sci. Eng. 7(4), 83 (2019). https://doi.org/10.3390/jmse7040083
3. Stopford, M.: Maritime Economics. Routledge, New York (2009)
4. Gregor, S.: The nature of theory in information systems. MIS Q. 30(3), 611–642 (2016)
5. Manavalan, E., Jayakrishna, K.: A review of Internet of Things (IoT) embedded sustainable supply chain for industry 4.0 requirements. Comput. Ind. Eng. 127, 925–953 (2019). https://doi.org/10.1016/j.cie.2018.11.030
6. Ibarra, D., Ganzarain, J., Igartua, J.I.: Business model innovation through Industry 4.0: a review. Procedia Manuf. 22, 4–10 (2018). https://doi.org/10.1016/j.promfg.2018.03.002
7. Müller, J.M., Buliga, O., Voigt, K.-I.: Fortune favors the prepared: how SMEs approach business model innovations in Industry 40. Technol. Forecast. Soc. Chang. 132, 2–17 (2018). https://doi.org/10.1016/j.techfore.2017.12.019
8. Osterwalder, A., Pigneur, Y.: Business Model Generation: A Handbook for Visionaries, Game Changers, and Challengers. Wiley, Hoboken (2010)
9. Dijkman, R.M., Sprenkels, B., Peeters, T., Janssen, A.: Business models for the Internet of Things. Int. J. Inf. Manag. 35(6), 672–678 (2015). https://doi.org/10.1016/j.ijinfomgt.2015.07.008

10. Turber, S., Vom Brocke, J., Gassmann, O., Fleisch, E.: Designing business models in the era of internet of things: towards a reference framework. In: Tremblay, M.C., Van der Meer, D., Rothenberger, M., Gupta, A., Yoon, V. (eds.) Advancing the Impact of Design Science: Moving from Theory to Practice, pp. 17–31. Springer, Heidelberg (2014). https://doi.org/10. 1007/978-3-319-06701-8_2

11. Gassmann, O., Frankenberger, K., Sauer, R.: Exploring the Field of Business Model Innovation: New Theoretical Perspectives. Palgrave Macmillan, London (2016)

12. Teece, D.J.: Business models and dynamic capabilities. Long Range Plan. **51**(1), 40–49 (2018). https://doi.org/10.1016/j.lrp.2017.06.007

13. Baden-Fuller, C., Haeflinger, S.: Business models and technological innovation. Long Range Plan. **46**(6), 419–426 (2013). https://doi.org/10.1016/j.lrp.2013.08.023

14. UNCTAD: Port marketing and the challenge of the third generation port, Geneva (1994)

15. UNCTAD: The fourth generation port. UNCTAD Ports Newsl. **19**, 9–12 (1999)

16. Flynn, M., Lee, P.T.W., Notteboom, T.: The next step on the port generations ladder: customer-centric and community ports. In: Notteboom, T. (ed.) Current Issues in Shipping, Ports and Logistics, pp. 497–510. University Press Antwerp, Brussels (2011)

17. Lee, P.T.W., Lam, J.S.L.: Container port competition and competitiveness analysis: asian major ports. In: Lee, C.Y., Meng, Q. (eds.) Handbook of Ocean Container Transport Logistics - Making Global Supply Chain Effective, pp. 97–136. Springer, New York (2015). https://doi.org/10.1007/978-3-319-11891-8_4

18. Lee, P.T.W., Lam, J.S.L.: Developing the fifth generation ports model. In: Lee, P.T.W., Cullinane, K. (eds.) Dynamic Shipping and Port Development in the Globalized Economy, pp. 186–210. Palgrave Macmillan, New York (2016). https://doi.org/10.1057/97811375 14233

19. Hollen, R.M.A., Van Den Bosch, F.A.J., Volberda, H.W.: Business model innovation in the Port of Rotterdam Authority (2000–2012). In: Zuidwijk, R., Kuipers, B. (eds.) Smart Port Perspectives: Essays in honour of Hans Smits, pp. 29–47. Erasmus University, Rotterdam (2013)

20. Hollen, R.M.A.: Exploratory studies into strategies to enhance innovation-driven international competitiveness in a port context: towards ambidextrous ports. Ph.D. thesis, Erasmus University, Rotterdam (2015)

21. Van Den Bosch, F.A.J., Hollen, R.M.A.: Insights from strategic management research into the Port of Rotterdam for increasing the strategic value of Shanghai Port for China: the levers of strategic connectivity and institutional innovation. In: Asian Economic Transformation: System Design and Strategic Adjustment, pp. 113–126. Shanghai Forum, Shanghai (2015)

22. Lee, P.T.W., Hu, Z.H., Lee, S.J., Choi, K.S., Shin, S.H.: Research trends and agenda on the Belt and Road (B&R) initiative with a focus on maritime transport. Marit. Policy Manag. **45** (3), 282–300 (2018). https://doi.org/10.1080/03088839.2017.1400189

23. Huang, Y.: Understanding China's Belt & Road initiative: motivation, framework and assessment. China Econ. Rev. **40**, 314–321 (2016)

24. Qin, J.: The Belt and Road initiative and the development of China's transport. In: Liu, W., Zhang, H. (eds.) Regional Mutual Benefit and Win-Win Under the Double Circulation of Global Value, pp. 269–283. Springer, Singapore (2019). https://doi.org/10.1007/978-981-13-7656-6

25. Choi, H.R., Moon, Y.S., Kim, J.J., Lee, J.K., Lee, K.B., Shin, J.J.: Development of an IoT-based container tracking system for China's Belt and Road (B&R) initiative. Marit. Policy Manag. **45**(3), 388–402 (2018). https://doi.org/10.1080/03088839.2017.1400190

26. Chen, S., Xu, H., Liu, D., Hu, B., Wang, H.: A vision of IoT: applications, challenges and opportunities with China perspective. IEEE Internet Things J. **1**(4), 349–359 (2014). https://doi.org/10.1109/JIOT.2014.2337336

27. Yang, Y., Zhong, M., Yao, H., Yu, F., Fu, X., Postolache, O.: Internet of things for smart ports: technologies and challenges. IEEE Instrum. Measur. Mag. **21**(1), 34–43 (2018). https://doi.org/10.1109/MIM.2018.8278808
28. ANAVE: Valencia, Algeciras y Barcelona entre los 10 primeros puertos de la UE por movimiento de contenedores en 2018, News. https://www.anave.es/prensa/ultimas-noticias/2087-valencia-algeciras-y-barcelona-entre-los-10-primeros-puertos-de-la-ue-por-movimiento-de-contenedores-en-2018. Accessed 12 July 2019
29. PortSEurope: Barcelona port books €53.7 million profit in 2018, News, 04 February 2019. https://www.portseurope.com/barcelona-port-books-e53-7-million-profit-in-2018. Accessed 12 July 2019
30. La Vanguardia: Cataluña se afianza como puerta de entrada de China en Europa, 10 February 2017. https://www.lavanguardia.com/economia/20170210/413981980751/catalunya-china-inversion-empresas.html. Accessed 12 July 2019
31. Catalan News: Chinese company Hutchison invest an extra €150 million in Barcelona Port terminal extension, 06 June 2014. http://www.catalannews.com/business/item/chinese-company-hutchison-invests-an-extra-150-million-in-barcelona-port-terminal-extension. Accessed 12 July 2019
32. Port de Barcelona: III Pla Estratègic 2015–2020, Barcelona (2015)
33. Europapress: El Puerto de Barcelona y el de Shenzhen aumentan su cooperación, 11 October 2018. https://www.europapress.es/catalunya/noticia-puerto-barcelona-shenzhen-aumentan-cooperacion-20181011145835.html. Accessed 12 July 2019
34. Europapress: El Puerto de Barcelona firma un memorando de colaboración con el puerto chino Ningbo Zhoushan, 07 November 2018. https://www.europapress.es/catalunya/noticia-puerto-barcelona-firma-memorando-colaboracion-puerto-chino-ningbo-zhoushan-20181107132227.html. Accessed 12 July 2019
35. PierNext, LoRaWAN, the IoT network serving ports. https://piernext.portdebarcelona.cat/en/technology/lorawan-the-iot-network-serving-ports-2/. Accessed 12 July 2019
36. PierNext, about us. https://piernext.portdebarcelona.cat/en/about-piernext/. Accessed 12 July 2019
37. PierNext, OpenPort: se buscan empresas innovadoras con soluciones únicas. https://piernext.portdebarcelona.cat/economia/openport-se-buscan-empresas-innovadoras-con-soluciones-unicas/. Accessed 12 July 2019

Agents for Facilitation of e-Business

Acoustic Signal Processing for Acoustic Source Localisation in an Elastic Solid

Hongyu You$^{(\boxtimes)}$ ⓘ, Ming Yang ⓘ, Xiang Fei ⓘ, Kuo-Ming Chao ⓘ, and Hanchao Li ⓘ

Coventry University, Coventry CV1 5FB, UK
{youh2, ab2032}@uni.coventry.ac.uk,
{aa5861, csx240, lih30}@coventry.ac.uk

Abstract. Many research projects have been carried out to achieve close range positioning in the context of indoor localisation using propagation's arrival time difference or multi-path structure with defined features such as Received Signal Strength (RSS). However, most researches used electromagnetic waves as signal carriers, for example, modern Wi-Fi routers, can provide a sufficient coverage for short range positioning and indoor localisation applications. Electromagnetic waves, however, are sensitive to the physical environment and its high traveling speed is prone to resulting into a low accuracy for short range localisation because of the limited bandwidth. In this paper, a probabilistic algorithm using acoustic signal pattern-matching templates is proposed in order to overcome the disadvantages associated with electromagnetic wave localisation approaches, while maintaining the precision. The resolution of the proposed passive acoustic locating method is verified to 3 cm. The structure of the system and relevant experimental parameters in an application of Human Computer Interface (HCI) are described and discussed.

Keywords: Position fingerprint · Acoustic features · Multi-path structure · Acoustic sources pattern matching

1 Introduction

During the past few years, a series of complicated short-range localisation technologies have been developed. They utilize the feature of dimensional quantities and pattern matching methods to solve indoor positioning problems that cannot be solved by RADAR or GPS-based systems [1–3]. Since electromagnetic wave's velocity is close to the speed of light, they are destined for long distance detection and fuzzy control. Ultra-high-speed sampling devices will be likely required to compensate the short time difference. Nowadays, the most commonly used audible acoustic sampling frequency is 44 kHz, but acoustic signal travels faster in resilient elastic solids than in the air, so that 44 kHz is not adequate enough for short-range localisation. For example, sound speed in Gorilla Glass 3 (which is the glass material for LG Nexus 5's screen) [4] is estimated at 4154.44 m/s; while the sampling time-interval is fixed at 0.0226 ms, which means, theoretically, systems equipped with 44 kHz AD converter and Time Difference of Arrival (TDOA) algorithm will have a minimum resolution of

© Springer Nature Switzerland AG 2020
K.-M. Chao et al. (Eds.): ICEBE 2019, LNDECT 41, pp. 317–329, 2020.
https://doi.org/10.1007/978-3-030-34986-8_23

approximately 10 cm in the absence of system delay thus such systems are not suitable for human computer interaction. In that case, how to realise short-range positioning has become a hot research topic. A deterministic algorithm that integrating pattern matching had been proposed by Microsoft in 2000 [5]. It pointed out that it is possible to realise short-range localisation with an RSS templates and the Euclidean distance between sampled vectors and vectors in the RSS template. The key points about position pattern matching techniques are feature extraction which is also referred as location-specified fingerprint extraction, and correlation matching between collected data and real-time received data. In this case, location-specified fingerprint plays a crucial role as it is a unique feature set that is associated with a pre-defined position in a physical space. These features can be obtained by multi-path effect: that emissions from a wave source are subject to reflection (flat surfaces, e.g. walls, floor and table), diffraction (sharp edges, e.g. small apertures) and scattering (small-sized objects, e.g. leaves), such superposition of multi-path waves affects the magnitude, frequency and phase of an acoustical signal thus produces distinctive shape of the received signal. As a result, a wave usually is split into infinite wavelets and each of them has different power density and drift. Typical multipath components are displayed in Fig. 1. Transducers fitted at specific physical positions receive a collection of time-delayed responses (multi-path components), which can be used for various applications. In 2003, Ahonen and Eskelinen [6] pointed out that 3G cellphone networks using multi-path structure achieved a positioning accuracy of 67% within 25 m and 95% within 188 m.

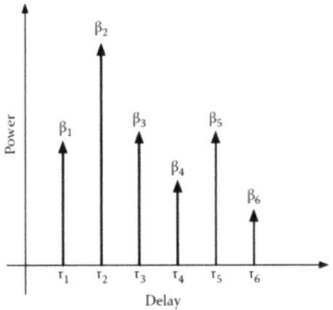

Fig. 1. Power distribution at a transducer [1]. Detectable components are subject to transducer's electrical characteristics. Here are six significant multipath components (corresponding to sets $\{\beta 1, \beta 2, \beta 3, \beta 4, \beta 5, \beta 6\}$ and $\{\tau 1, \tau 2, \tau 3, \tau 4, \tau 5, \tau 6\}$ respectively) which have great amplitudes than the background noise. Components can be dealt independently as the unique fingerprint for pattern matching analysis.

2   Literature Review

Three standard ways that utilize location features and pattern matching were delineated [1]. In the first one, called networks positioning, Radio Frequency (RF) signals are emitted by mobile devices at certain locations then they are received by receivers at fixed locations. Next, the received signals are sent to computer systems for further analysis along with the previous stored data. In the second one, called self-positioning, the transducers provide signals for mobile devices and the mobile devices locate themselves with internal algorithms. The processing steps are in a reversed order in comparison to the former case. The last one is called hybrid positioning which is similar to the first case but uses network's server nodes to handle calculations. All in all, how to make use of the uniqueness of the received signal feature for pattern matching became an essential research theme (Refi in Fig. 2). Apart from the existence of multipath effect, there are other factors that influence the shape of fingerprints such as travelling speed and environmental temperature. In other words, position fingerprints are not restricted to components structure. It could be either unidimensional like signal roundtrip time or multidimensional like multi-path components distribution but thanks to probabilistic analysis and network pattern matching, it is possible to link these independent elements together for further analysis. From the perspective of electromagnetism, due to the severe attenuation and multi-path effect, the signal strength has become the most important feature in positioning because electromagnetic waves are always subject to characters of the carrier, transmit power and shadowing attenuation. Though there are many inevitable shortcomings, researches on electromagnetic wave localisation are still in domination, mainly because the fast propagation speed (radio waves are technically regarded as optic rays when subcarriers' frequencies are greater than 500 MHz) [7] and the widely use of commodity Wi-Fi routers with 802.11n specification. The daily used device is regarded as a universal carrier platform which is capable of providing parameters such as RSS, phase error and multipath structure for indoor localisation.

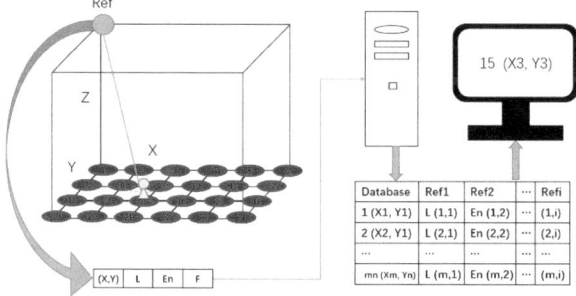

Fig. 2. Two steps of pattern matching: the first stage is called calibration, signals are collected from pre-defined locations and stored with coordinates in a database. In the 2nd stage, acoustic waves from source position are sampled at fixed position then compared with the signals in database, the coefficient indicates the degree of correlation of the source location to a pre-defined location. Therefore, coordinates of the signal with highest value will be the output.

In the context of signal processing, acoustic wave has a number of merits over electromagnetic wave. Electromagnetic wave travels as fast as light in vacuum, but the practical speed varies with frequency in homogeneous medium. In practice, with attenuation effect, the further the wave field propagates away from the source, the more distorted the received signal will be, resulting in a warped reflection of the original signal. The widely used Wi-Fi routers are equipped with 2.4 GHz/5 GHz single transceiver hence even the direction of transceiver antenna will take an effect on localisation features. Besides, practicable electromagnetic waves are usually modulated for long distance loss free transmission. For instance, multiple antennas are used to constrain shadowing loss and large-scale fading. In contrast, acoustic wave is classified as mechanical wave which only transmits energy instead of generating infinite variable electromagnetic fields, thus the speed is totally decided by the medium it propagates within. Though the acoustic propagation velocity is much slower in comparison to electromagnetic waves, time difference of arrival methodology becomes extremely suitable for low sampling speed acoustical localisation and in some cases, the velocity may increase many times in materials with high density and good elasticity. Besides, acoustic waves in low frequency bands have strong penetrability so that they lose less information during the propagation. In summary, if the dynamic response within the spread medium is adequately strong and the time delay is tolerable, acoustic wave can be used to replace electromagnetic wave for short range localisation. Base on the idea, a number of signal feature extraction and processing experiments were designed using shock sensors, microphones and data acquisition board.

The nature of acoustic waves is elastic wave, therefore different states of elastomer have different propagation features. In liquids or gas, acoustical wave transmission is simplified in comparison to solids because there is only longitudinal wave in gas and liquids due to the fact that shear torsion only exists in solids. When acoustic wave penetrates solids, the primary wave (longitudinal) and the secondary wave (transverse) are generated firstly, followed by the surface wave. Accordingly, the coexistence of transverse waves, longitudinal waves, leaky surface acoustic waves, generalized Rayleigh waves, Rayleigh waves, Love waves, Sezawa waves and Stoneley waves, etc. makes the feature extraction more complicated because the waves listed above have different propagation velocities and superposition principles. In addition, the features are changing much rapidly in fluids and gas if disturbances occur internally. Acoustical features in highly resilient elastic solids like glasses, on the other hand, are not changing drastically when forces are applied externally due to the absolute stability (vicinities that centred round the forcepoint recovers from deformation in an extremely short period when the external force is revoked). Therefore, experiments were designed on a smooth-surfaced glass plate firstly.

In this paper, signals from an acoustic source are detected by 4 sensors which were installed on the boundary line of the glass plate. These sensors have different responses in terms of acceleration and frequency for comparison and feasibility verification. This paper aims at verifying the feasibility of acoustical pattern matching localisation and its accuracy. In the context of human computer interaction (HCI), different hand activities will be observed in time and frequency domain for discrimination and creation of pattern matching template (which will be analyzed for corresponding characteristics). The data generated during the tests will be recorded for further offline training and analysis purposes. The paper is organized at follows. Section 2 validates the signal collection system. Section 3 examines the results and the analysis against experimental data. Section 4 concludes the feasibility of building an acoustical Natural User Interface (NUI) on a surface of a block.

3   Experimental Validation

3.1   Experiments Settings

Experiments were conducted on a composite table with a rectangular shape and dimensions of $139.7 \times 80.5 \times 2.3$ cm^3 while the glass plate on the table is rectangular with dimensions of $40 \times 30 \times 0.4$ cm^3. The setups are shown in Fig. 3. Interaction with the glass panel was achieved indirectly; according to Meijer [8], there is a certain relationship among energy consumption of fingers, finger acceleration and movement angular velocity. Bouten [9] have also verified that the absolute integral of acceleration and angular velocity of an object is linear with energy consumption. To acquire the most stable signals and avoid differences caused by different integration angles, a stylus pen with rounded point was used for tapping instead of finger touch and fingernails scratching. Acoustic sources created by the stylus pen was transmitted into the glass plate and collected by a pair of piezoelectric ceramic sensors and a pair of accelerometers [2] (Murata PKS1-4A1, single Y-axis, 4.5 mm thickness, and 34.4 mm diameter. BU-21771, single Y-axis, 7.87 mm $\times$ 5.54 mm $\times$ 4.06 mm) which were glued onto upper corner with a space of 1.5 cm. Theoretically, single sensor is adequate enough to achieve works from sampling to pattern matching but sensor characteristic comparison is the main reason for 4 sensors with 4 differential channels on the Data Acquisition card (DAQ). The closest test points were located 10 cm away from sensors to provide a well-distributed multi-path component. Diffused signals were then discretized by the 2MS/s 16-bit DAQ-2010 card. The sampling rate was originally set to 10 kHz in regard to the centroid frequency of taps. In the tests, the centroid frequency varies from 300 Hz (in the composite table) to 4000 Hz (in the glass plate). Gain of the amplified circuit was set to 40 dB with NE5532P amplifier which frequency response is from 10 Hz to 19.4 kHz. To induce clear frequency domain images at low frequency band, a 2nd order Butterworth lowpass filter is fitted to isolate low frequency components.

3.2   Database Establishment and Methodology Design

Signal collection and data storage were accomplished with ADLINK Application Programming Interface (API). For signal acquisition, the drive code consists of 4 parts: module initialization, channels configuration, buffer pointer assignment and data DMA transfer. Real-time data display and frequency response were presented in a 3×2 grid graph including accumulated input signal surveillance; the refresh rate of real-time data display equals to double buffer mode's single buffer size which is 10000 samples per unit per channel. The signal screening was achieved by a threshold filter (which was also used in real-time signal isolation) firstly (Fig. 4). Then a pre-allocated self-extend matrix was called to save sorted signals from picked channel. The matrix would encode itself to MAT data formats automatically after signal sampling phase completes. In this case, each grid was tapped 10 times randomly. In other words, 10 acoustical signals were collected at an interval of 3 cm on the surface of the glass plate. Signals from A1 and H8 were shown in Fig. 5. Besides, the average of the prior 10 signals were calculated and saved as the 11th signal. Hence, there were 704 signals stored in the database. Every 11 signals were saved in the database subfolder thus the profile path represents grid coordinates. To avoid preload overflow and counter asynchronism, timing functions were used to regulate the time source consumed by each segment functions and consequently call for timeout when any errors occur.

Methodologies used to measure the resolution of the system were designed according to Taguchi methods - methods that advocate the minimal number of experiments which could give the full information of all the factors that affect the performance parameter. In our case, $N_{Taguchi}$ equals to 26, which has been rounded up to 30, a multiple of 2 and 3 [10]. Therefore, each grid would be tested 30 times. The experiment starts with A1, A8, H1, H8 firstly, then indent along the diagonal line and meet in the centre of the plate. Theoretically, the features are less obvious as the distance decreases resulting in an accuracy decline. There are 6 rounds of regression testing corresponding to distance {21 cm, 15 cm, 9 cm, 4.5 cm, 3 cm, 1.5 cm} respectively (Fig. 6).

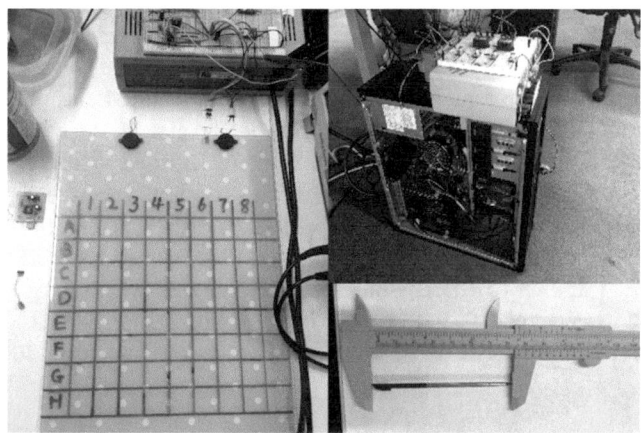

Fig. 3. System building for acoustic wave detection. The glass board has been divided into 8 × 8 grids which is similar to a chess board. Each grid is square with dimensions of 3 cm × 3 cm. The sensors on left were disconnected and removed since the sensor characteristic comparison phase A 4-channel, asynchronous data acquisition card DAQ-2010 with double buffer was installed for sampling. The stylus pen is from HTC touch.

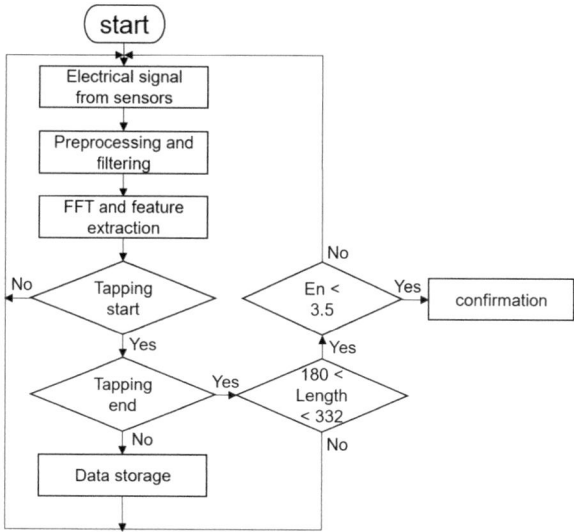

Fig. 4. Signal screening flowchart. En is the energy density acquired from logarithmic processed (10log) power spectrum. The logarithm is used to increase the relative low amplitude components, so that signal characteristics masked in low amplitude noise were revealed.

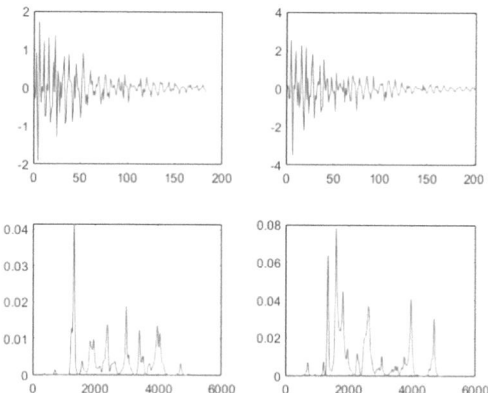

Fig. 5. Signals stored in database. A1 (left) and H8 (right) in time domain and frequency domain. It is clear that acoustic frequency components in frequency domain at location A and B are different. Though their images in time domain look similar in naked eyes, the wavelength, amplitude and form factor of A and B are different as well.

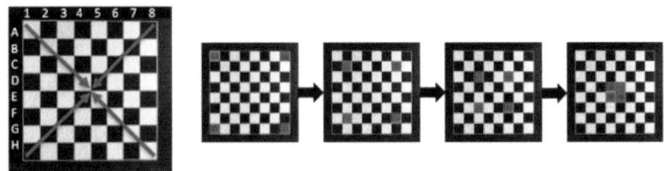

Fig. 6. Resolution tests. The interval was set to 3 cm. Generally speaking, database contains more samples if a shorter interval is set. Hence, the establishment of a database may take hours. For example, if the interval was set to 1 cm, there would be 24 × 24 = 576 points thus the location features are likely to be identified as closest to a measured vector around the correct vector resulting in wrong output.

3.3   Algorithm Design

Kalman filter is not integrated into this experiment because the covariance matrices Q and R in predict and update steps usually were estimated with minimum norm quadratic unbiased estimation. However, iterative procedures (Jacobian matrix) are necessary for the estimation, thus the properties of data and system model will take an act on the estimate procedures which may cause filtering diverging [11]. From the perspective of discrete pulse signal, truncation error may weaken even eliminate features generated by multi-path structure. On the other hand, mathematically, that Kalman filter is optional when signals are isolated as unidirectional vertical acceleration.

From [12] and with no user control input, we have the following Kalman equations:

$$\begin{cases} x_k = F_k x_{k-1} + w_k \\ z_k = H_k x_k + v_k \end{cases} \tag{1}$$

Where

$$\begin{cases} F_k, the\ state - transition\ model \\ H_k : the\ observation\ model \\ w_k, process\ noise, w_k \sim \mathcal{N}(0, Q_k) \\ v_k, observation\ noise, v_k \sim \mathcal{N}(0, R_k) \end{cases} \tag{2}$$

Q_k and R_k are the corresponding covariances, k is the discrete time index of the signal, t is the real time, x_k is the kth true-state signal, and z_k is the kth observation-state signal.

Impulse signals have the following properties:

$$\begin{cases} x_{k-1}(T) = x_k(T),\ at\ T = t \\ \begin{cases} \lim_{\Delta t \to 0} x_k(t + \Delta t) - x_k(t) = 0, \\ impulse\ signal\ lasts\ within\ a\ limit\ time \end{cases} \\ \begin{cases} \lim_{\Delta t \to 0} x_k(t) - x_k(t - \Delta t) = x_k(t), \\ impulse\ signal\ from\ silence \end{cases} \end{cases} \tag{3}$$

By substract the limit difference of the kth signal with $(k-1)^{th}$ signal, the following equation can be obtained:

$$\begin{cases} \begin{cases} x_k(t - \Delta t) = x_{k-1}(t - \Delta t) \\ x_k(t + \Delta t) = x_{k-1}(t + \Delta t) \end{cases}, \quad \forall \Delta t > 0 \\ \qquad or \\ \qquad x_k(T) = x_{k-1}(T), \forall k, T \end{cases} \tag{4}$$

Hence, $x_k(T) = x_{k-1}(T), \forall k, T$, which leads to $x_k = F_k x_k + w_k$ from (1). When impulse signal is not applicable, $x_k = w_k$. In contrast, $(I - F_k)x_k = w_k$, for $x_k \neq 0$. Thus, $\lim_{\Delta t \to 0} w_k = 0$ & $\lim_{\Delta t \to 0} F_k = I$. It indicates the noise, w_k, has negligible to the impulse signal. Similarly, it can be deduced that the noise, v_k, is negligible to the impulse signal, where $H_k \to I$ & $\lim_{\Delta t \to 0} v_k = 0$. Therefore, both noises have negligible impact to the observation signals.

Cross-correlation analysis module was embedded into the main loop of signal acquisition after the establishment of database. It compares the defined features from database with features from real-time sampled signal. Once the most relevant vector in database is decided, it reads the storage path of that vector and outputs its storage path as coordinates. Feature sets that are integrated into the cross-correlation analysis are wavelength, energy density, frequency components. The correlation coefficient was calculated according to (5).

$$R_{ij}(corrcoef) = \frac{C_{ij}}{\sqrt{C_{ii}C_{jj}}} \tag{5}$$

Because of the fact that in the cross-correlation analysis of frequency components, row vector is used rather than point to point comparison, the equation listed above will then be extended to:

$$r = \frac{\sum_m \sum_n (A_{mn} - \bar{A})(B_{mn} - \bar{B})}{\sqrt{\left(\sum_m \sum_n (A_{mn} - \bar{A})^2\right)\left(\sum_m \sum_n (B_{mn} - \bar{B})^2\right)}} \tag{6}$$

where $\bar{A}$ and $\bar{B}$ are average values of A and B. The operator r represents the correlation coefficient [13]. The criteria of correlation coefficient are not academically unified yet, but they are commonly considered as follows (Table 1):

Next is to link all segmentations together to realise the acoustical signal-based pattern matching algorithm. The algorithm outline is as follows (Table 2).

Table 1. Correlation coefficient range.

Correlation coefficient level	Extent of relativity
0.00– ± 0.30	Minor correlation
±0.30– ± 0.50	Positive correlation
±0.50– ± 0.80	Significant correlation
±0.80– ± 1.00	High correlation

Table 2. Algorithm design

Input: Real-time electronic signal, $\overrightarrow{s_t}$

Output: Index of the location, L_I

---

1. System initialisation.
2. Definition of four diagonal axes,
   L_1 ($A1 \sim D4$), L_2 ($A8 \sim D5$), L_3 ($H1 \sim E4$) $and\ L_4$ ($H8 \sim E5$) from Fig. 6.
3. Each point was calibrated with 10 signal samples, $\overrightarrow{s_{1_1}}, ..., \overrightarrow{s_{1_{10}}} \in L_1; \overrightarrow{s_{2_1}}, ..., \overrightarrow{s_{2_{10}}} \in L_2; etc.$
4. Automatically generate 11th 'average' signal for each point such that $\bar{s}_a \in L_1, \bar{s}_b \in L_2, \bar{s}_c \in L_3\ and\ \bar{s}_d \in L_4$.
5. Extract corresponding features, $\overrightarrow{s_{1_1}} \to F_{1_1}, \overrightarrow{s_{1_2}} \to F_{1_2}, ..., \bar{s}_a \to \bar{F}_a, ... etc.$
6. Group the four points into two diagonal subgroups, $G_1 := (L_1, L_4), G_2 := (L_2, L_3)$.
7. Deviation error of G1-G2, ϵ_1.
8. Load database for pattern matching.
9. Real-time electronic sound signal transmission, $\overrightarrow{s_t}$.
10. Apply FFT^2/N on $\overrightarrow{s_t}$.
11. Features extraction of time domain, and of PSD, F_t.
12. For each feature, f_i, inside F_t.
    a. Evaluate the cross-correlation coefficients against $\bar{F}_a, \bar{F}_b, \bar{F}_c$, and $\bar{F}_d$ respectively, $c_{1,i}, c_{2,i}, c_{3,i}, c_{4,i}$.
    b. Based on G_1 and G_2, $m_i := c_{1,i} + c_{4,i}, n_i := c_{2,i} + c_{3,i}$.
    c. If $m_i - n_i > \epsilon_1$
       i. Evaluate under G_1: $c_{1,i} - c_{4,i}$
       ii. Output $L_1|f_i$ or $L_4|f_i$, using L1-L4 error, ϵ_2.
    d. Else
       i. Evaluate under G_2: $c_{2,i} - c_{3,i}$
       ii. Output $L_2|f_i$ or $L_3|f_i$, using L2-L3 error, ϵ_3.
    e. End
13. End
14. Define $\mathcal{F} := \{L_{j_1}|(f_1, \bar{F}_a, \bar{F}_b, \bar{F}_c, \bar{F}_d), ..., L_{j_i}|(f_i, \bar{F}_a, \bar{F}_b, \bar{F}_c, \bar{F}_d)\}$ where $j_i \in 1,2,3,4$ and $i \in \mathbb{N}^+$.
15. If $L_{j_1}|(f_1, \bar{F}_a, \bar{F}_b, \bar{F}_c, \bar{F}_d) \notin mode(\mathcal{F}) \in \{L_1, L_2, L_3, L_4\}$ or $L_{j_1}|(f_1, \bar{F}_a, \bar{F}_b, \bar{F}_c, \bar{F}_d) \in mode(\mathcal{F})$ with $|mode(\mathcal{F})| > 1$, where f_1 is the frequency component.
    a. Evaluate the cross-correlation coefficients against $F_{1_k}, F_{2_k}, F_{3_k}$ and F_{4_k} respectively, for $k \in \{1, ...10\}$ from the actual sample(s).
    b. Repeat 12 – 14 to get $\mathcal{G} := \{L_{j_1}|(f_1, F_{1_k}, F_{2_k}, F_{3_k}, F_{4_k}), ..., L_{j_i}|(f_i, F_{1_k}, F_{2_k}, F_{3_k}, F_{4_k})\}$ where $j_i \in 1,2,3,4$ and $i \in \mathbb{N}^+$.
    c. If $|mode(\mathcal{F} \oplus \mathcal{G})| = 1$
       i. Output $mode(\mathcal{F} \oplus \mathcal{G})$.
    d. Else
       i. Output $L_{j_1}|(f_1, \bar{F}_a, \bar{F}_b, \bar{F}_c, \bar{F}_d)$ or $L_{j_1}|(f_1, F_{1_k}, F_{2_k}, F_{3_k}, F_{4_k})$ against the set $mode(\mathcal{F} \oplus \mathcal{G})$, with preference of $L_{j_1}|(f_1, \bar{F}_a, \bar{F}_b, \bar{F}_c, \bar{F}_d)$.
    e. End If
16. Else
    a. Output $mode(\mathcal{F})$.

As shown in the above algorithm. More specifically, L_1, L_4 and L_2, L_3 were grouped for better performance. After the consolidation, deviation ϵ_1 of G1-G2 cross-correlation coefficient was calibrated using test data. Similar procedure was applied to ϵ_2 and ϵ_3.

4 Results Analysis

From Fig. 7, it is obvious that the accuracy decreases with distance. Moreover, the overall results have shown a declining trend. But at 15 cm, the accuracy has decreased to approximately 80% then it kept going up with distance being shortened to 4 cm. This phenomenon indicates that the feature sets at 15 cm are not that effective in comparison to other indicators of distances. In terms of external conditions, interferences such as door opening, air flow, interaction habits, samples status, interaction angle, interaction energy, friction, temperature, power supply, amp drift and accumulated error will skew the results. There are also internal factors that may take an effect on the performance. e.g. according to test results, the coefficient of L_1 and L_2 are always greater than L_3 and L_4 thus 3 deviation values $\epsilon_1, \epsilon_2, \epsilon_3$ were used to offer preference when the coefficients of 4 directions have successive approximate values. Other factors include signal aliasing, multi-path counteraction, and deliberate changes in waveshape. But in general, the system has reached an accuracy of 3 cm to 4.5 cm.

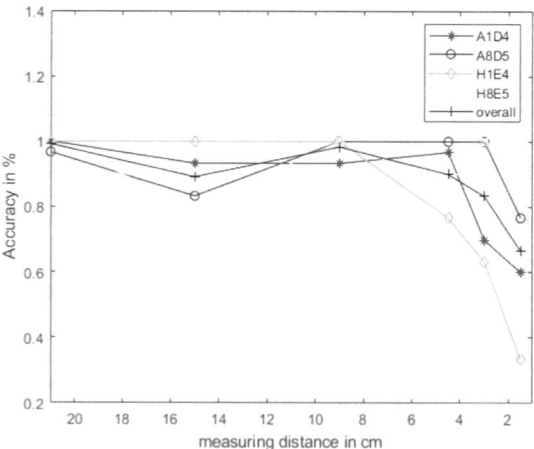

Fig. 7. Test results. There are 6 points distributed under 4 directions in total. The periphery points were settled at an interval of 21 cm while the innermost points were settled at an interval of 1.5 cm. Plot distance on the horizontal x-axis against accuracy on the vertical y-axis.

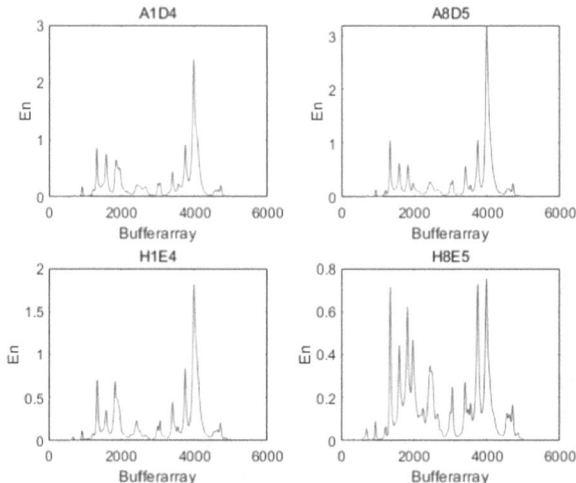

Fig. 8. Averaged signals of the 4 directions. The H8E5 points have some isolated features that can be told by naked eyes. Not only the average energy density is the lowest, but it also has more frequency components between 1500 and 2500 Hz. Images of the other three directions, on the other hand, is not changing dramatically but the energy density contributes more for cross-correlation analysis.

Figure 8 reveals why H8E5 axis remains high accuracy until the end of the tests. Its frequency components and energy density differ from all the other axes. This is probably caused by inhomogeneous medium or a crack inside the glass plate. As for H1E4, it is almost a duplicate of A8D5 with different scale. Thus, in the last two sets of experiments, results gave by the system were biased in favour of D5 (51.7%) to D4 (1.7%). In this case, more features should be explored for accuracy improvement.

5 Conclusion and Future Works

This paper explored the feasibility of integrating acoustics based multi-path components on the surface of a glass plate. The experiment setups were first clarified against radio wave-based pattern matching experiments. Consequently, empirical expressions for acoustical cross-correlation analysis algorithm was designed. The acoustical signal utilisation described in this paper is of interest in a number of NUI applications. With the result from experiments and project TAI-CHI conducted by EU [4, 14], it is possible to transfer solid surfaces into human computer interfaces. In this case, analysis on factors that makes H8E5 axis different should be proposed and further tests need to be designed.

References

1. Karimi, H.: Advanced Location-Based Technologies and Services, 1st edn. Taylor & Francis, Boca Raton (2013)
2. Ji, Z.: Development of Tangible Acoustic Interfaces for Human Computer Interaction, 1st edn. Cardiff University, Cardiff (2007)
3. Hassan-Ali, M., Pahlavan, K.: A new statistical model for site-specific indoor radio propagation prediction based on geometric optics and geometric probability. In: Zhang, J., Lei, Y., Poor, V., Chiang, M. (eds.) IEEE Transactions on Wireless Communications, vol. 1, pp. 112–124. IEEE Communications Society, New York (2002)
4. Shumailov, I., Simon, L., Yan, J., Anderson, R.: Hearing your touch: a new acoustic side channel on smartphones. https://arxiv.org/pdf/1903.11137.pdf. Accessed 29 Aug 2019
5. Bahl, P., Padmanabhan, V.: RADAR: an in-building RF based user location and tracking system. In: Sidi, M., Rom, R., Schulzrinne, H. (eds.) IEEE INFOCOM 2000, pp. 775–784. IEEE (2000)
6. Ahonen, S., Eskelinen, P.: Mobile terminal location for UMTS. IEEE Aerosp. Electron. Syst. Mag. **18**(2), 23–27 (2003)
7. Pahlavan, K., Krishnamurthy, P.: Principles of Wireless Networks, 1st edn. Prentice Hall of India, New Delhi (2002)
8. Meijer, G., Westerterp, K., Verhoeven, F., Koper, H., Hoor, F.: Methods to assess physical activity with special reference to motion sensors and accelerometers. IEEE Trans. Biomed. Eng. **38**(3), 221–229 (1991)
9. Bouten, C., Koekkoek, K., Verduin, M., Kodde, R., Janssen, J.: A triaxial accelerometer and portable data processing unit for the assessment of daily physical activity. IEEE Trans. Biomed. Eng. **44**(3), 136–147 (1997)
10. Grosse, I.: Chapter 2 Introduction to Taguchi Method. http://www.ecs.umass.edu/mie/labs/mda/fea/sankar/chap2.html. Accessed 18 June 2019
11. Almagbile, A., Wang, J., Ding, W.: Evaluating the performances of adaptive Kalman filter methods in GPS/INS integration. J. Glob. Position. Syst. **9**(1), 33–40 (2010)
12. Jin, C., Jia, Z.: Finger tapping recognition algorithm based on wearable input system. http://www.jinoux.com/Seaula/pages/paper/document/124.htm. Accessed 07 June 2019
13. Asuero, A., Sayago, A., González, A.: The correlation coefficient: an overview. Crit. Rev. Anal. Chem. **36**(1), 41–59 (2006)
14. Rolshofen, W., Dietz, P., Schäfer, G., Kruk, R.: TAI-CHI: tangible acoustic interfaces for computer-human interaction. EU (2008)

Design of a Healthy Diet Control System for the Elderly Group by Using Raspberry Pi

Ching-Lung Lin[1(✉)], Shu-Chi Lin[1], Yung-Te Liu[1],
Huang-Liang Lin[1], Yuan-Chang Huang[2], and Po-Hsun Huang[1]

[1] Department of Electrical Engineering, Minghsin University of Science
and Technology, Hsinchu, Taiwan
cll@must.edu.tw
[2] Department of Electrical Engineering, Chung Yuan Christian University,
Taoyuan City, Taiwan

Abstract. In aging society, the number of elderly groups gradually increased, although the awareness of the elderly for health care increased little by little. But many elderly people on diet control or cannot achieve perfect control, improper eating habits, so that the original suffering from diabetes, cardiovascular disease and other symptoms become more serious. Therefore we need a system that can monitor the daily diet. The patients with diseases do not eat food that they cannot eat, get enough nutrition within the appropriate scope, and control the daily diet of the elderly. They can eat healthily and ease their mind to eat. The Healthy Diet Control System is designed for the elderly Group. The system is composed of Raspberry Pi and RFID card with a smart control system.

Keywords: Smart health care · Healthy diet · Raspberry Pi

1 Introduction

1.1 A Subsection Sample

The elderly population is a huge problem of Taiwan. According to the information released by the Government in Taiwan in 2016, the ratio of elderly population (REP) whose older than sixty five years old has exceeded 7% in November 1993, which is called aging population society. In April 2019, the REP has achieved 14%, which is called aged po pulation society. The REP is estimated to 20% in 2026, Taiwan will become a super aged society. The Estimation of the ratio of the elderly population as shown in Table 1 [1, 2].

As the management of Long-Term Care (LTC) is more diverse and different, while the demand added, in order to improve the development of the LTC system and to ensure the quality of services, it also to protect the dignity and rights of LTC service providers. The government has established the "LTC services law" in 15 May 2017, and in 2016 Implementation on the road. After the adoption of the LTC services law, how to formulate follow-up related services and institutions and the preparation of financial resources will be a key factor in the development of LTC in the future [2] (Fig. 1).

© Springer Nature Switzerland AG 2020
K.-M. Chao et al. (Eds.): ICEBE 2019, LNDECT 41, pp. 330–341, 2020.
https://doi.org/10.1007/978-3-030-34986-8_24

Table 1. Estimation of the ratio of the elderly population

Year	REP (%)	Aging index (%)	The population ratio of the senior to the younger	Average age (years)
2016	13.2	98.8	1:1.0	40.4
2019	14.0	117.6	1:0.95	42.0
2021	16.8	130.2	1:0.8	43.1
2031	24.4	204.5	1:0.5	48.4
2041	30.4	298.7	1:0.3	52.4
2051	35.9	383.5	1:0.3	55.5
2061	38.9	406.9	1:0.2	56.9

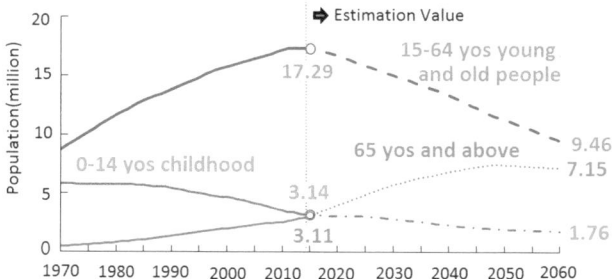

Fig. 1. Three-stage population trend estimation.

Because of the lacking of human resources in the future, the application of digital care, cloud service, big data integration is more important. In the basic care of life must rely on digital and integration system to achieve the best quality and efficiency. So, the "dietary" is more important, "sick from the mouth to the mouth" is very reasonable. The elderly people have to change their dietary, and change in the way of cooking so that dietary from "eat full" to "eat well", but also let modern people's nutrition is too much, many diseases also come from this reason. It is very important to control your diet to maintain your health.

Diet and chronic diseases are quite closely related, and studies of the association between diet and chronic diseases have long focused on the association with large amounts or micronutrients such as fat, saturated fat, protein and alcohol, and the availability and absorption of nutrients are often influenced by food cooking methods and dietary patterns. As a result, the correlation between overall dietary quality and health and chronic diseases has been gradually emphasized.

2   Relationship Between Diseases and Food in the Elderly

Most diseases of the elderly is related the dietary closely. The elderly body began to deteriorate, such as resistance, metabolism, motion. There are four diseases can be used to control the of diet and control the disease effectively. In the process of aging, there are many health problems. Most of the elderly over 65 have more than two of chronic disease. According to the Survey of 2014, the Elderly in the Taiwan, 81.1% of the elderly suffer from chronic diseases which are "diabetes (25%)", "hypertension (54%)", "heart disease (21%)", "arthritis (18%)" and "osteoporosis (33%)".

2.1 Diabetes

Diabetes is a chronic disease caused by congenital deficiencies, relative deficiencies in insulin hormones secreted by beta cells in islets (also known as Islands of Langerhans), or due to metabolic abnormalities in sugar (or carbohydrates) caused by obesity. Diabetics in addition to basic high-sugar food cannot eat, cannot eat easy to absorb high blood sugar, as well as lipid metabolism abnormal high cholesterol-type foods, and finally, will make protein metabolism may be difficult, causing urea nitrogen retention and aggravation of the disease of soybean products [2].

2.2 Cardiovascular Disease

The World Health Organization defines cardiovascular disease as a group of heart and vascular diseases, including hypertension, coronary heart disease (heart attack), cerebrovascular disease (stroke), peripheral vascular disease, heart failure, rheumatism, congenital heart disease, cardiomyopathy, etc., and heart attacks and strokes are usually as acute. Mainly due to blockage caused blood not to flow to the heart or brain, and cardiovascular disease risk factors are unhealthy diet, lack of exercise, the use of cigarettes and alcohol, 80% of cardiovascular disease is caused by these risk factors, the impact on the body may be blood lipids, blood pressure, blood sugar rise.

2.3 Degenerative Disease

Degenerative knee arthritis can be divided into four levels according to severity: the first level is already the phenomenon of joint cartilage softening, and the occurrence of mild pain. Finally, the fourth stage cartilage can be severely worn, not only in pain but also with difficulty standing [1].

2.4 Osteoporosis

Osteoporosis is due to low bone mass or bone loss leading to the reduction of bone mass and bone microstructure damage, so that bone brittleness caused fractures, adult bone mass in the 20s to 30 years old to peak, after 30 years of age will gradually decline, the main factors affecting peak bone mass are genes, hormones, exercise and nutritional intake. As age increases the rate of bone loss, bone density decreases and the bones become hollowed out gradually, making the bones lighter, more fragile, and easier to break [1].

3 Structure of the Healthy Diet Control System

3.1 What Are Raspberry and RFID Card Reader?

The Raspberry Pi 3 is designed for the educational purpose. In this paper, we made an application for taking health care of elderly group. The Pi 3 is in small size like a credit card, it has different programming at different applications. Although Pi 3 is slower than the laptop or desktop, but it availability of price in the market is low. It comes with a **64 bit** quad core processor, GPU, UART, board Wi-Fi, Bluetooth, 4 USB ports and i/o pins and can be connected with external peripherals and helps in running number of operations like regular computer [3, 4] (Fig. 2).

Fig. 2. Each module monitoring different characteristic of the food.

Specification of Raspberry Pi 3

- The Pi 3 comes with **GPIO** (General Purpose Input Output) pins that are essential to maintain connection with other electronic devices. These input output pins receive commands and work based on the programming of the device.
- The UART pins are the serial input output pins that are used for serial communication for data and for the conversion of debugging code.
- It comes with **64 bit quad core processor**, on board Wi-Fi and Bluetooth and USB features.
- It has a processing speed ranging from 700 MHz to 1.4 GHz where RAM memory ranges from 256 to 1 GB.
- The **CPU** is considered as the brain of the device which is responsible for executing numbers of instructions based on mathematical and logical operation.
- The **GPU** is another advanced chip incorporated in the board that carries out function of image calculation.
- The **Ethernet port** is incorporated on this device that sets a pathway for communicating with other devices.

- The Board has four **USB ports** that are used for communication and **SD card** is added for storing the operating system.
- **Power source connector** is a basic part of the board that is used to provide 5 V power to the board.
- The Pi 3 supports two connection options including **HDMI** and **composite**.

The RFID card reader is used in this project. It can be directly loaded into the variety of reader molds. RFID card reader uses voltage of 3.3 V, simple few lines through the SPI interface directly with any user CPU board is connected to the communication module can guarantee stable and reliable work, reader distance [3, 4].

Specification of RFID Card Reader
- Module Name: MF522-ED
- Working current: 13–26 mA/DC 3.3 V
- Standby current: 10–13 mA/DC 3.3 V
- sleeping current: <80 μA
- peak current: <30 mA
- Working frequency: 13.56 MHz
- Card reading distance: 0–60 mm(mifare1 card)
- Protocol: SPI
- data communication speed: Maximum 10 Mbit/s
- Card types supported: mifare1 S50, mifare1 S70, mifare Ultra-Light, mifare Pro, mifare Desfire
- Dimension: 40 mm × 60 mm

Environment
- Working temperature: −20–80°
- Storage temperature: −40–85°
- Humidity: relevant humidity 5%–95%
- Max SPI speed: 10 Mbit/s

3.2   Hardware Structure of Healthy Diet Control System

The demand of LTC service industry continues to rise, but it also appears the lack of service demand and human resources. It is necessary to build a high technology management and monitoring system for LTC to achieve high quality service and high efficiency is very important. To achieve management and monitoring dietary of elderly group, the healthy diet control system (HDCS) is designed by suing Raspberry pi and RFID card, linking to the network, applying the cloud platform, database to make real-time monitoring of the elderly dietary, to ensure that the elderly get appropriate services, improve the quality of life, food safety, access to the highest quality of life. The structure of the HDCS is shown in Fig. 3 [4].

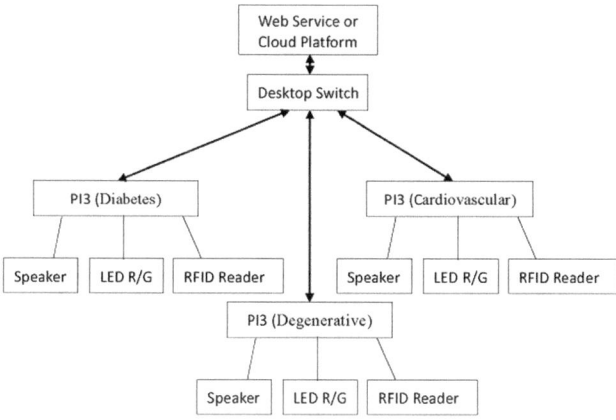

Fig. 3. The structure of the healthy diet control system.

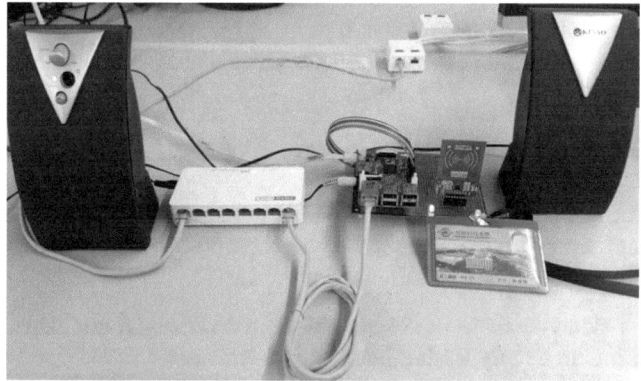

Fig. 4. The module of the HDCS.

There are three modules (is shown as Fig. 4) in the HDCS. Each module is compose of Raspberry Pi 3, RFID reader, 2 LEDs and a speaker. The Raspberry P3 is a main controller of the HDCS. The Pi 3 receives the chronic diseases data from RFID card. The RFID reader reads data from RFID card of the elderly people who have. 2 LEDs and a speaker are the alarm devices to advice the dietary of elderly people.

In this case, we discriminates all foods in 3 areas in the dining room of a elderly group. Each area disposed one set of a Pi 3, as shown in Fig. 4. The Pi 3 receives the chronic diseases data from RFID card and to determine the card owner dietary. The First area food is unsuitable for the patient of the diabetes, the second area food is unsuitable for the patient of the hypertension or heart disease, the third area food is unsuitable for the patient of the degenerative disease (Fig. 5).

Fig. 5. Each module monitoring different characteristic of the food.

When a elderly with a personal RFID card order dishes in the restaurant, his/her RFID card (with the diabetes) will be read at the entrance, and he/she takes food in area 1, the HDCS module of the area 1 will turn on red LED to warm this elderly that the food of this area is unsuitable for the patient of the diabetes.

4   Establish the Health Diet Control System

Ensuring that the elderly are getting proper services, improving their quality of life, eating safe and worry-free. To get the highest quality of life. For this purpose, the HDCS must be established 4 function as following:

- Establish a cloud database of diseases of the elderly in the diet table.
- Implementing an elderly health diet control system.
- Monitor and record the diet of elderly.
- Analyze the recorded data to understand the eating habits of the elderly and improve.

4.1   Database Established

Firstly, we must establish a cloud database, and buildup a database to link disease and food related. After the database is established, then the disease and food are linked, patient data is entered, patient RFID card data is read, patient disease is associated, and then the food is associated to write to the form. So the basic database set-up is very important, the database platform is the choice of PHP and MySQL, a commonly used network cloud database [5–7].

The database form has information on personal data, diseases data, and food data, and codes the data of this three item to facilitate the correlation of late data when it is read. When we create a MySQL form, we select the form name, domain name, field pattern, encoding, and so on, and when the selection is complete, to start creating a disease data form, personal data form and food data form, shown as Figs. 6, 7 and 8.

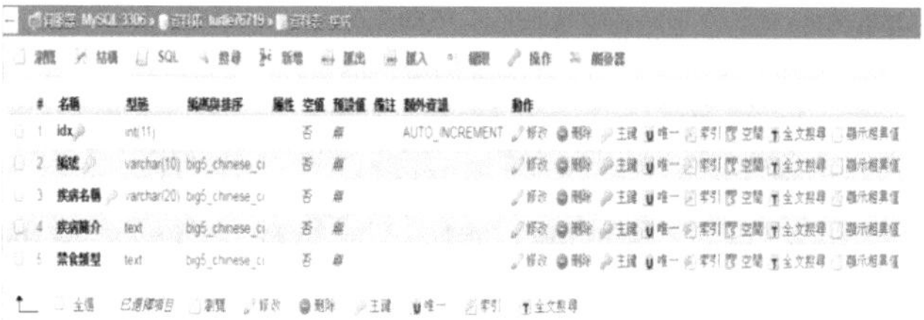

#	名稱	型態	編碼與排序	屬性	空值	預設值	備註	額外資訊	動作
1	idx	int(11)			否	無		AUTO_INCREMENT	修改 刪除 主鍵 唯一 索引 空間 全文搜尋 顯示相異值
2	編號	varchar(10)	big5_chinese_ci		否	無			修改 刪除 主鍵 唯一 索引 空間 全文搜尋 顯示相異值
3	疾病名稱	varchar(20)	big5_chinese_ci		否	無			修改 刪除 主鍵 唯一 索引 空間 全文搜尋 顯示相異值
4	疾病簡介	text	big5_chinese_ci		否	無			修改 刪除 主鍵 唯一 索引 空間 全文搜尋 顯示相異值
5	禁食類型	text	big5_chinese_ci		否	無			修改 刪除 主鍵 唯一 索引 空間 全文搜尋 顯示相異值

Fig. 6. Created a disease data form.

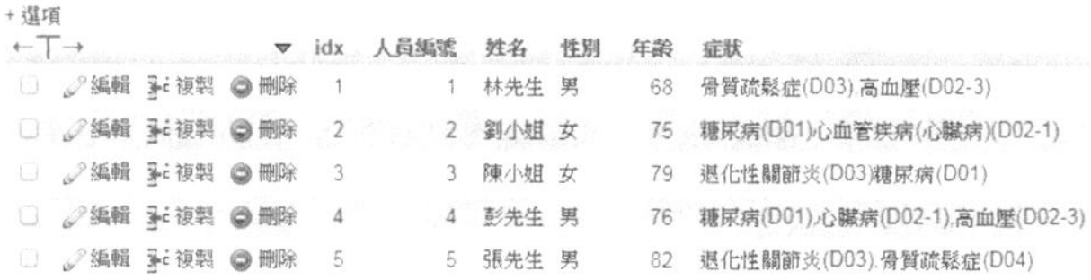

+ 選項

	idx	人員編號	姓名	性別	年齡	症狀
編輯 複製 刪除	1	1	林先生	男	68	骨質疏鬆症(D03),高血壓(D02-3)
編輯 複製 刪除	2	2	劉小姐	女	75	糖尿病(D01),心血管疾病(心臟病)(D02-1)
編輯 複製 刪除	3	3	陳小姐	女	79	退化性關節炎(D03)糖尿病(D01)
編輯 複製 刪除	4	4	彭先生	男	76	糖尿病(D01),心臟病(D02-1),高血壓(D02-3)
編輯 複製 刪除	5	5	張先生	男	82	退化性關節炎(D03),骨質疏鬆症(D04)

Fig. 7. Personal data form.

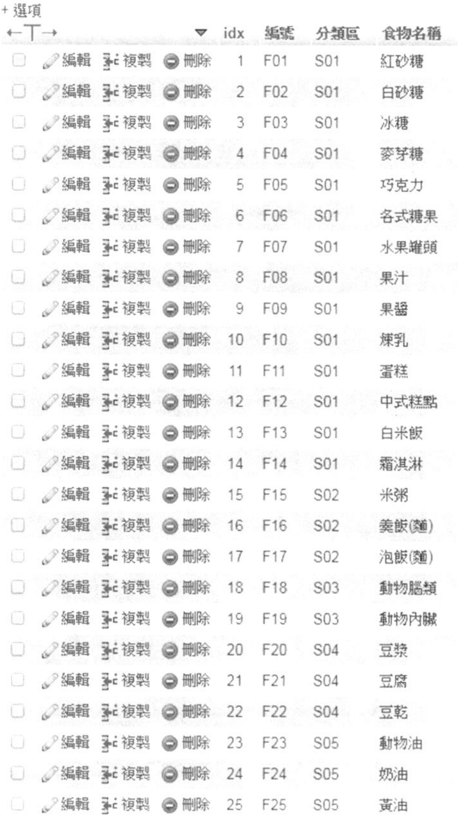

+ 選項

	idx	編號	分類區	食物名稱
編輯 複製 刪除	1	F01	S01	紅砂糖
編輯 複製 刪除	2	F02	S01	白砂糖
編輯 複製 刪除	3	F03	S01	冰糖
編輯 複製 刪除	4	F04	S01	麥芽糖
編輯 複製 刪除	5	F05	S01	巧克力
編輯 複製 刪除	6	F06	S01	各式糖果
編輯 複製 刪除	7	F07	S01	水果罐頭
編輯 複製 刪除	8	F08	S01	果汁
編輯 複製 刪除	9	F09	S01	果醬
編輯 複製 刪除	10	F10	S01	煉乳
編輯 複製 刪除	11	F11	S01	蛋糕
編輯 複製 刪除	12	F12	S01	中式糕點
編輯 複製 刪除	13	F13	S01	白米飯
編輯 複製 刪除	14	F14	S01	霜淇淋
編輯 複製 刪除	15	F15	S02	米粥
編輯 複製 刪除	16	F16	S02	羹飯(麵)
編輯 複製 刪除	17	F17	S02	泡飯(麵)
編輯 複製 刪除	18	F18	S03	動物腦類
編輯 複製 刪除	19	F19	S03	動物內臟
編輯 複製 刪除	20	F20	S04	豆漿
編輯 複製 刪除	21	F21	S04	豆腐
編輯 複製 刪除	22	F22	S04	豆乾
編輯 複製 刪除	23	F23	S05	動物油
編輯 複製 刪除	24	F24	S05	奶油
編輯 複製 刪除	25	F25	S05	黃油

Fig. 8. Food data form.

4.2 Data Linking

MySQL databases must be linked to PHP web pages and made into form formats for future use. At this point, php and MySQL need a linking program to link all data in the MySQL databases, to complete the basic elderly data, it also links the personals' RFID UID code with the personal data codes, and then connected to the disease data codes. Finally, the linker links disease data codes to food data codes, links personal data codes to food data codes, a basic databases association diagram is shown as Fig. 9.

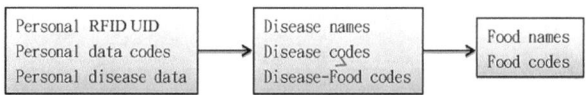

Fig. 9. A basic databases association diagram.

4.3 The Flowchart of HDCS

The elderly eats a meal in the restaurant, his/her RFID card (with the diabetes) will be read at the entrance, and he/she takes food in an area, the HDCS module of the area will turn on red LED and speaker alarm to warm this elderly that the food of this area is unsuitable for the patient of the diabetes. If the foods of this area are suitable for the elderly, the HDCS module of the area will turn green LED and Speaker saying welcome for the elderly. The system flowchart of HDCS is shown as Fig. 10.

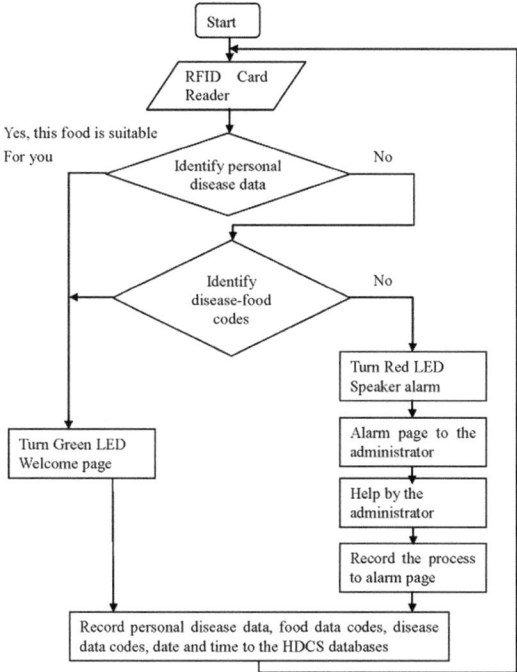

Fig. 10. The system flowchart of HDCS.

5   Testing of HDCS for the Elderly Group

When the elderly are ready to pick up a meal, the RFID card must be sensing before the meal is desirable. The card has been written into the personal data and disease data of each elderly, which corresponds to the item stowaway to the meal collection item. The elderly choose the food in an area which they cannot eat, the system records it to the MySQL database in real time and warns that in each recording form, the time and date of the violation will appear, and the horn output transmitted to the food area will be warned to discourage the elderly from eating. All information will be recorded in the recording form. The forms are shown as Figs. 11 and 12. The recording form of Ms Liu is shown as Fig. 13.

高齡者姓名	疾病狀況	感應時間	感應食物類別
林先生	骨質疏鬆症(D03),高血壓(D02-3)		
劉小姐	心臟病(D02-1)		
陳小姐	糖尿病(D01)		
彭先生	心臟病(D02-1),高血壓(D02-3)		
張先生	退化性關節炎(D03),骨質疏鬆症(D04)		

人員表單
疾病表單
食物表單

Fig. 11.   The recording form of HDCS.

高齡者姓名	疾病狀況	感應時間	感應食物類別
林先生	骨質疏鬆症(D03),高血壓(D02-3)	2018 05 21 07.31	S11
劉小姐	心臟病(D02-1)	2018 05 21 07.07	S01
陳小姐	糖尿病(D01)	2018 05 21 08.02	S10
彭先生	心臟病(D02-1),高血壓(D02-3)	2018 05 21 07.21	S11
張先生	退化性關節炎(D03),骨質疏鬆症(D04)	2018 05 21 08.05	S02

人員表單
疾病表單
食物表單

Fig. 12.   All information will be recorded in the recording form.

Of course, the elderly are not easy to discourage, so the administrator of the control center founds that there is a warning page appeared, the administrator must go to the scene to discourage, advise the elderly not to take food from the area. The warning page is shown as Fig. 14. And elderly people because of age or their personality habit. Sometimes forget to eat unsuitable food, sometimes deliberately and insist on taking food that cannot be eaten. HDCS will the record can be used to do future analysis in

1	劉小姐	女	75

心臟病	S01高糖份,S05油類,S06高油脂,S07高鹽份,S08辛辣食物,S13堅果類食物,S14碳酸飲料

飲食紀錄

日期: ▾ 年 ▾ 月 ▾ 日 查詢

時間: ▾ 時 - ▾ 時 查詢

食物類別: ◉ S01 ○ S02 ○ S03 ○ S04 ◉ S05 ◉ S06 ◉ S07 ◉ S08 ○ S09 ○ S10
○ S11 ○ S12 ◉ S13 ◉ S14 ○ S15 查詢

日期	時間	食物類別	人員附註
5/21	7:07	S01	經勸導後離去
5/21	12:13	S08	經勸導後離去
5/21	12:22	S08	人員已離去
5/21	18:39	S13	經勸導後離去
5/21	18:46	S08	經勸導後執意食用
5/22	7:12	S01	經勸導後離去
5/22	7:22	S01	人員已離去
5/22	12:21	S08	經勸導後離去
5/22	12:28	S08	人員已離去
5/22	18:31	S08	人員已離去
5/23	12:17	S08	人員已離去
5/23	18:36	S08	人員已離去
5/24	12:25	S08	人員已離去
5/25	12:19	S08	人員已離去
5/26	12:30	S01	人員已離去
5/26	18:51	S08	人員已離去
5/27	12:27	S01	人員已離去

Fig. 13. The recording form of Ms Liu.

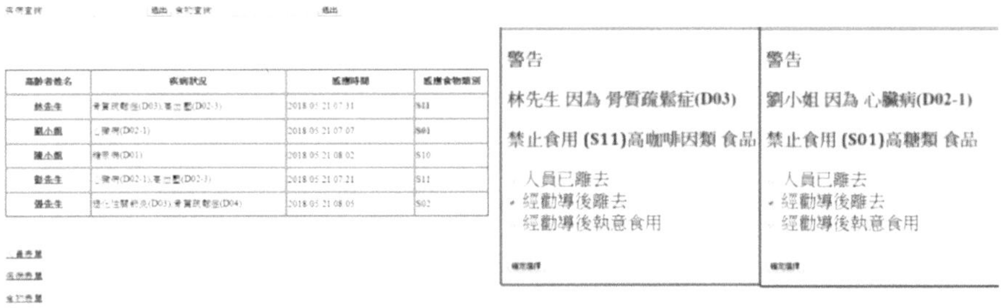

Fig. 14. The warning page is shown at administrator's computer.

each time, and in conjunction with the body's regular health check-up, to observe the relationship between each elderly person's physical condition and diet, it can be seen that after warning and change into a healthy diet of the elderly body and condition gradually improved, without listening to advice is the condition has not changed or worse, All the data are clear, as the records show.

6 Conclusions

In this time, the fertility decline and the lack of manpower, the HDCS in the elderly group is to achieve the purpose of control and management, the useage of RFID card features, with Raspberry Pi 3, linked to the cloud. The system created, system modified, data recorded, and analysis of MySQL Database, the management of the health diet control system becomes very convenient and easier to query.

The test results found that in the first phase of the warning, the system discouraged 27% of the elderly. In the second phase, about 63% of the elderly were discouraged. The results showed that the system performed very well in terms of control, reducing the risk of eating by mistake by 90% in the elderly with systematic reminders and making the diet more health for the elderly. This will help improve the disease, which is the main purpose of this study, using a healthy eating control system to achieve healthy eating habits to maintain physical health.

References

1. Lin, C.-L., Weng, L.-S., Chang, H.-H., Lin, C.-F.: Telecare system using RF communication technology in elderly center. In: Proceedings of the 2009 13th International Conference on Computer Supported Cooperative Work in Design, Santiago, Chile, pp. 444–449 (2009)
2. Hsieh, C.-Y.: Study of a healthy diet control system for elderly groups. Thesis of The Master degree (2018)
3. Kamal, N., Ghosal, P.: Three tier architecture for IoT driven health monitoring system using Raspberry Pi. In: 2018 IEEE International Symposium on Smart Electronic Systems (iSES) (Formerly iNiS), pp. 167–170 (2018)
4. Lin, C.-L., Chuang, S.-Y., Lin, S.-H., Lin, C.-F.: To create a health-care on demand platform by using internet of things for elderly center. In: 2018 IEEE International Symposium on Smart Electronic Systems (iSES) (Formerly iNiS), pp. 167–170 (2018)
5. Wardi, Achmad, A., Hasanuddin, Z.B., Asrun, D., Lutfi, M.S.: Portable IP-based communication system using Raspberry Pi as exchange. In: 2017 International Seminar on Application for Technology of Information and Communication (iSemantic), pp. 198–204 (2017)
6. Navya, K., Murthy, M.B.R.: A zigbee based patient health monitoring system. Int. J. Eng. Res. Appl. **3**(5), 483–486 (2013)
7. Pardeshi, V., Sagar, S., Murmurwar, S., Hage, P.: Health monitoring systems using IoT and Raspberry Pi—a review. In: 2017 IEEE International Conference on Innovative Mechanisms for Industry Applications (ICIMIA) (2017)

A Web Service Composition Method Based on OpenAPI Semantic Annotations

Andrei Netedu(✉), Sabin C. Buraga(✉), Paul Diac(✉), and Liana Ţucăr(✉)

Alexandru Ioan Cuza University of Iaşi, Iaşi, Romania
{mircea.netedu,busaco,paul.diac,stefania.tucar}@info.uaic.ro

Abstract. Automatic Web service composition is a research direction aimed to improve the process of aggregating multiple Web services to create some new, specific functionality. The use of semantics is required as the proper semantic model with annotation standards is enabling the automation of reasoning required to solve non-trivial cases. Most previous models are limited in describing service parameters as concepts of a simple hierarchy. Our proposal is increasing the expressiveness at the parameter level, using inherited concept properties that define attributes The paper also describes how parameters are matched to create, in an automatic manner, valid compositions. The composition algorithm is practically used on descriptions of Web services implemented by REST APIs expressed by OpenAPI specifications. Our proposal uses knowledge models to enhance these OpenAPI constructs with JSON-LD annotations in order to obtain better compositions for involved services.

Keywords: Web service composition · Semantics · JSON-LD · OpenAPI

1 Introduction

In the current software landscape, *Web services* are the core elements of Service-Oriented Architectures (SOAs) [6] that has been already used for many years with vast popularity. A Web Service provides a straightforward functionality defined through a public interface. In the enterprise context, this interface was traditionally expressed in WSDL (Web Service Description Language) – a standardized XML (Extensible Markup Language) dialect. Nowadays, Web services are usually built according to the REST (REpresentational State Transfer) architectural style [8]. There are several pragmatic solutions able to describe their APIs (Application Programming Interfaces) by using lightweight formats such as JSON (JavaScript Object Notation), a well-known data-interchange format based on ECMAScript programming language. In order to compose REST-based Web services, the paper proposes a novel method able to select suitable services automatically by using a knowledge-based approach described in Sect. 2. Without actually executing services, valid compositions are automatically generated depending on concepts from an ontology that denotes the semantics of input

© Springer Nature Switzerland AG 2020
K.-M. Chao et al. (Eds.): ICEBE 2019, LNDECT 41, pp. 342–357, 2020.
https://doi.org/10.1007/978-3-030-34986-8_25

parameters, properties, and output (the result). The proposed algorithm – producing certain encouraging results – is presented, evaluated, and discussed in Sect. 3.

We think that our proposal improves productivity and reduces costs within a complex (micro)service-based system. Additionally, the involved knowledge could be easily described, in a pragmatic way, by JSON-LD semantic constructs augmenting the OpenAPI description of each involved service, in order to offer the proper support for intelligent business operations – see Sect. 4.

For a motivating case study, several experiments were also conducted by using the proposed algorithm. From a pragmatic point of view, the popular schema.org model was chosen to convey Web services conceptual descriptions augmenting input parameters and expected results. Our approach added a suitable conceptualization that was necessary to discover compositions on cases where it could not be possible before, such as in the study presented in Sect. 5.

The paper enumerates various related approaches (Sect. 6), and ends with conclusions and further directions of research.

2   Problem Definition

2.1   Preliminaries

A *Web service* represents a software system designed to support inter-operable machine-to-machine interaction over a network and can be viewed as an abstract resource performing tasks that form *a coherent functionality* from the point of view of provider's entities and requester's entities.[1]

From a computational point of view, a Web service is a set of related methods described by the same interface. This interface could be declared by adopting various specifications: WSDL 2.0 – a classical XML-based Web standard[2] – and OpenAPI 3.0 Specification – a modern solution declaring the public interface of a Web service (REST API) in different formats like JSON or YAML (Yet Another Markup Language).

We refer to a Web service as a *single method* or the minimal endpoint that can be accessed or invoked at a time using some values as parameters – i.e. having some prior knowledge. In our context, we also work only with "information providing" Web services, or *stateless services* that do not alter their state and are not sensitive to the outside world states, time or any external factors.

The main focus of the proposed method is on improving the semantic description of services motivated by the lack of means to express several composition techniques on previous models and software solutions in Sect. 6.

We enhanced the service composition problem by modeling semantics in the manner described below. It was inspired by our previous experience and the

---

[1] Web Services Glossary, W3C Working Group Note, 2004 – https://www.w3.org/TR/ws-gloss/.
[2] Web Services Description Language (WSDL) Version 2.0, W3C Recommendation, 2007 – https://www.w3.org/TR/wsdl20/.

shortcomings we found on expressing certain natural cases of composition. More precisely, it was not possible to describe service parameters with properties or any relations/interaction between these parameters. Adding the new elements to the problem definition is also done with inspiration from the data model used by popular ontologies such as schema.org [10].

Using this approach, we managed to fix the issues that appeared in examples where the previous model failed. Our main addition is consisting of concept properties and how they are used for composition, allowing interaction between concepts and their properties based on simple constructions that increase the expressiveness.

2.2 Proposed Formal Model

Service parameters are defined over a set of *concepts*. Let $\mathbb{C}$ be the set of all concepts that appear in a repository of services, that are all possible concepts or the problem universe. As in the previous modeling, the concepts are first organized by using the *isA* or *subsumes* relation. This is a binary relation between concepts and can be considered somewhat similar to the inheritance in object-oriented programming. If a concept c_a *isA* c_b then c_a can substitute when there is need of a c_b. Also, for any concept c in $\mathbb{C}$, there can be only one direct, more generic concept then c, i.e. we do not allow multiple inheritance. Obviously, *isA* is transitive and, for convenience, reflexive: c_a *isA* c_a. This implies that $\mathbb{C}$ together with the *isA* relation form a tree (a taxonomy) or, more generally, a forest (a set of taxonomies or a complex ontology).

However, the new element added to the problem are concept *properties*. Any concept has a set of properties, possibly empty. Each property p is a pair $\langle name, type \rangle$. The name is just an identifier of the property, and for simplicity, it can be seen as a string – for concrete cases, this identifier is actually an IRI (Internationalized Resource Identifiers)[3], a superset of URIs (Uniform Resource Identifiers). The type is also a concept from the same set of concepts $\mathbb{C}$ and can be considered as the range of a property.

From an ontological point of view [1], a property p is defined as a relation between Web resources. The values of a property are instances of one or more concepts (classes) – expressed by *range* property. Any resource having a given property is an instance of one or more concepts denoted by *domain* property.

Properties are inherited: if c_a *isA* c_b, and c_b has some property $\langle name_x, type_x \rangle$ then c_a also has property $\langle name_x, type_x \rangle$. For example, if an *apple* is a *fruit*, and *fruit* has property $\langle hasColor, Color \rangle$ expressing that *fruit* instances have a color, then *apples* also must have a color.

It is important that property names do not repeat for any unrelated concepts. For any concepts that are in a *isA* relation, all properties are passed to the specialization one by inheritance. This restriction is only imposed to avoid confusion and does not reduce the expressiveness, as properties can be renamed.

---

[3] IRI (Internationalized Resource Identifiers – https://tools.ietf.org/html/rfc3987.

Syntactically, to define a property, the following are to be known: its name, its type, and the most general concept that the property can describe. For example, consider that the *hasColor* property can describe the *apple* concept, but also the more general *fruit* concept. If concept *fruit isA physicalObject*, the next more general concept than *fruit* – i.e., its parent in the concepts tree –, under the assumption that not all physical objects are colored, then we can say that *hasColor* can most generally describe *fruit*, but not any *physicalObject* or any other more general concept. However, it can describe other concepts in the tree, together with all their descendants. For simplicity, we consider further that all the properties are defined within $\mathbb{C}$, thus the concepts, *isA* relation and properties structure are in $\mathbb{C}$ – the particular ontological model.

A *partially defined concept* denotes a pair *(c, propSet)*, where c is a concept from $\mathbb{C}$ and *propSet* is a subset of the properties that c has defined directly or through inheritance from more generic concepts. At some moment of time (or in some stage of a workflow), a partially defined concept describes what is currently known about a concept. It does not refer to a specific concept instance, but rather generally to the information that could potentially be found for any instance of that concept.

A Web Service $\mathbf{w}$ is defined by a pair of input and output parameters: $(\mathbf{w}_{in}, \mathbf{w}_{out})$. Both are sets of partially defined concepts. All are defined over the same structure $\mathbb{C}$, so all service providers must adhere to $\mathbb{C}$, thus adding the requirement that $\mathbb{C}$ is publicly available and defined ahead of time. In order to be able to validly call a service, all input parameters in $\mathbf{w}_{in}$ must be known together with their specified required properties. After calling the service, all output parameters $\mathbf{w}_{out}$ will be learned with the properties specified at output.

Parameter Matching. Let $\mathbf{P}$ be a set of partially defined concepts, and $\mathbf{w} = (\mathbf{w}_{in}, \mathbf{w}_{out})$ a Web service. The set $\mathbf{P}$ matches service $\mathbf{w}$ (or, equivalently, $\mathbf{w}$ is *callable* if $\mathbf{P}$ is known) if and only if $\forall$ partially defined concept $\mathbf{pdc} = (c, propSet) \in \mathbf{w}_{in}$, $\exists\ \mathbf{p} = (c_{spec}, propSuperSet) \in \mathbf{P}$ such that c_{spec} *isA* c and $propSet \subseteq propSuperSet$. We define the addition of $\mathbf{w}_{out}$ to $\mathbf{P} : \mathbf{P} \oplus \mathbf{w}$ as:

$$\left\{ \left(c, \left\{ p \middle| \begin{matrix} c\ has\ p \\ \exists (c', propSet') \in w_{out}\ and\ p \in propSet' \\ c'\ isA\ c \end{matrix} \right\} \right) \middle| \nexists (c, pSet) \in P \right\} \bigcup$$

$$\left\{ \left(c, pSet \cup \left\{ p \middle| \begin{matrix} c\ has\ p \\ \exists (c', propSet') \in w_{out}\ and\ p \in propSet' \\ c'\ isA\ c \end{matrix} \right\} \right) \middle| \exists (c, pSet) \in P \right\}$$

or the union of $\mathbf{w}_{out}$ with $\mathbf{P}$ under the constraint of $\mathbf{P}$ matching $\mathbf{w}$ (defined as parameter matching above). Also, by *c has p* we refer to the fact that property p is stated for c directly or by inheritance. $\mathbf{P} \oplus \mathbf{w}$ contains (1) new concepts that are in $\mathbf{w}_{out}$ and (2) concepts already in $\mathbf{P}$ possibly with new properties from $\mathbf{w}_{out}$ specified for corresponding concepts or their specializations.

In words, after a call to a service, all its output parameters are selected, and for each concept together with its selected properties *(c, propSet)* in $\mathbf{w}_{out}$, *propSet* is added to c, c's parent in the concepts tree or the ascendants until we reach the first node that gains no new information or the root. More precisely, for each p in *propSet* we add p to our knowledge base for c, for the parent of c, and so on until p is no longer defined for the node we reached. The node where this process stops can differ from one p property to another p' property, but once the process stops for all properties in *propSet* there is no need to go further.

Chained Matching. Let $\mathbf{P}$ be a set of partially defined concepts and $(w_1, w_2, \ldots, w_k)$ an ordered list of services. We say that $\mathbf{P} \oplus w_1 \oplus w_2 \oplus \cdots \oplus w_k$ is a chain of matching services iff w_i matches $\mathbf{P} \oplus w_1 \oplus w_2 \oplus \cdots \oplus w_{i-1}; \forall i = 1 \ldots k$. This is the rather primitive model for multiple service calls, that is a requirement for defining the composition. For simplicity, we avoid for now more complex workflows that could handle parallel and sequential service execution constructs.

Web Service Composition Problem. Given an ontology having a set of concepts $\mathbb{C}$ and a repository of Web services $W = (w_1, w_2, \ldots, w_n)$, and two sets of partially defined concepts **Init** and **Goal**, all defined over $\mathbb{C}$, find a chain of matching services $(w_{c1}, w_{c2}, \ldots w_{ck})$ such that $(\emptyset, \mathbf{Init}) \oplus w_{c1} \oplus w_{c2} \oplus \cdots \oplus w_{ck} \oplus (\mathbf{Goal}, \emptyset)$.

The $(\emptyset, \mathbf{Init})$ and $(\mathbf{Goal}, \emptyset)$ are just short ways of writing the initially known and finally required parameters, by using mock services.

We can also imagine $(\mathbf{Init}, \mathbf{Goal})$ as a Web (micro-)service – in this context, the problem requires finding an "implementation" of a Web (micro-)service using the services available in a certain development environment (e.g., a public or private repository).

3   Automatic Service Composition

3.1   Algorithm Description

The proposed algorithm is intended to describe a generic solution that generates a valid composition. Tough it considers some basic optimizations like special data structures and indexes, there are many ways in which it can be improved, so we shortly describe some of them after the basic algorithm description.

In a simplified form, considered main entities have the following structure:

```
class Concept {      // full or partial type
    String name;     // a label
    Concept parent;  // isA relation
    Set<Property> properties; } // proper and inherited
class Property {
    String name;     // a label
    Concept type; } // property's range
class WebService {
    String name;
    Set<Concept> in, out; } // I/O parameters
```

Global data structures that are most important and often used by the algorithm are presented below:

```
Set<Concept> C; // knowledge: concepts, isA, properties
Set <WebService> webServices; // service repository
WebService Init, Goal; // user's query as two fictive services
Map <Concept,Set<Property>> known;
// partial concepts: known.get(c) = concept's known properties
Map <Concept,Map<Property, Set<WebService>>> required;
// .get(c).get(p) services that have property p of C in input
Map <WebService, Map<Concept, Set<Property>>> remaining;
// .get(w).get(c) = properties of C necessary to call W
Set<WebService> callableServices; // with all input known
```

The algorithm described next uses the above structures, and is composed of three methods: initialization, the main composition search method which calls the last (utility) method, that updates the knowledge base with the new outputs learned form a service call.

Several obvious instructions are skipped for simplicity like parsing input data and initializing empty containers.

```
void initialize() { // read problem instance
 Init.in = Goal.out = ∅; webServices.add(Goal);
 for (WebService ws : webServices) {
  for (Concept c : ws.in) {
   for (Property p : c.properties) {
    required.get(c).get(p).add(ws);
    remaining.get(ws).get(c).add(p);
}}}} // container creation skipped
```

After reading the problem instance, the described data structures have to be *loaded*. **Init** and **Goal** can be used as web services to reduce the implementation size, if they are initialized as above. Then, for each parameter in service's input, we add the corresponding concepts with their specified properties to the *indexes* (maps) that efficiently get the services that have those properties at input and the properties that remain yet unknown but required to validly call a service.

```
List<WebService> findComp(WebService Init, Goal) {
 List<WebService> composition; //result
 callService(Init); // learn initial

 while (!(required.get(Goal).isEmpty() ||
         callableServices.isEmpty())) {
  WebService ws = callableServices.first();
  callableServices.remove(ws);
  composition.add(ws);
  callWebService(ws);
 }
 if (remaining.get(Goal).isEmpty()) { return composition; }
 } else { return null; } // no solution
}
```

The main method that searches for a valid composition satisfying user's query is *findComp()*. The result is simplified for now as an ordered list of services. As long as the **Goal** service is not yet callable, but we can call any other new service, we pick the *first* service from the *callableServices* set. Then we simulate its call, basically by learning all its output parameter information, with the help of *callWebService()* method below. We add the selected service to the composition and remove it from *callableServices* so it won't get called again. If *callableServices* empties before reaching the **Goal**, then the query is unsolvable.

```
void callWebService(WebService ws) {
 for (Concept c : ws.out) {
  Concept cp = c; // concept that goes up in tree
  boolean added = true; // if anything new was learned
  while (added && cp != null) {
   added = false;
   for (Property p : c.properties) {
    if (cp.properties.contains(p)&&!known.get(cp).contains(p)) {
     added = true; known.get(cp).add(p); // learn p at cp level
     for (WebService ws: required.get(cp).get(p)) {
      remaining.get(ws).get(cp).remove(p);
      if (remaining.get(ws).get(cp) .isEmpty()) {
       // all properties of cp in ws.in are known
       remaining.get(ws).remove(cp); }
      if (remaining.get(ws). isEmpty()) {
       // all concepts in ws.in known
       callableServices.add(ws); }
   }}}
   cp = cp.parent;
}}}
```

When calling a Web service, its output is learned and also expanded (we mark as learned properties for more generic concepts). This improves the algorithm's complexity, as it is better to prepare detection of newly callable services than to search for them by iteration after any call. This is possible by marking in the tree the known properties for each level and iterating only to levels that get any new information, as higher current service's output would be already known.

The optimization also comes from the fact that all services with inputs with these properties are hence updated only once (for each learned concept and property). As it gets learned, information is removed from the *remaining* data structure first at the property level and then at the concept level. When there's no property of any concept left to learn, the service gets callable. This can happen only once per service. The main loop might stop before reaching the tree root if at some generalization level, all current services' output was already known, and this is determined by the *added* flag variable.

3.2 Possible Improvements

One important metric that the algorithm does not consider is the size of the produced composition. As can be seen from the overview description above, the solution is both deterministic and of polynomial time complexity. This is possible because the length of the composition is not a necessary minimum. Finding the shortest composition is `NP-Hard` even for problem definitions that do not model semantics. The proposed model introduces *properties*; this addition does not significantly increase the computational problem complexity. Even if the shortest composition is hard to find, there are at least two simple ways to favor finding shorter compositions with good results.

One is based on the observation that when a service is called, it is chosen from possibly multiple callable services. This choice is not guided by any criteria. It is possible to add at least a simple *heuristic score* to each service that would estimate how useful is the information gained by that service call. Also, this score can be updated for the remaining services when information is learned.

Another improvement is based on the observation that services can be added to the composition even if they might produce no useful information – there is no condition check that they add anything new, or the information produced could also be found from services added later to the composition. To mitigate this, another algorithm can be implemented that would search the resulting composition *backward* and remove services that proved useless in the end. Both of the above improvements can have an impact on the running time as well, as the algorithm stops when the goal is reached.

3.3 Empiric Evaluation

To assess the performance on larger instances, a random test generator was built.

The generator first creates a conceptual model with random concepts and properties and then a service repository based on the generated ontology. Service parameters are chosen from direct and inherited properties. In the last step, an ordered list of services is altered by rebuilding each service input. The first one gets its input from **Init**, which is randomly generated at the beginning. Each other service in order gets its input rebuilt from all previous services outputs, or from valid properties of generalizations of those concepts. Finally, **Goal** has assigned a subset of the set of all outputs of the services in the list. The total number of services and the list size are input parameters for the generator. The intended dependency between services in the list is not guaranteed, so shorter compositions could potentially exist.

Table 1 shows the algorithm performance. The first two columns briefly describe the input size, by the number of concepts and properties in the generated ontology and total number of services in the repository. The column *result composition size* measures the length of the composition found by the algorithm. The last column, *dependency list size*, measures the length of the composition generated by the tests generator algorithm. The *dependency list* constitutes a valid composition, hidden within the repository and may contain useless services as the *dependency* is not guaranteed.

Table 1. Algorithm run times and resulting composition size on random generated instances.

Ontology size (#classes + #props.)	Repository size: #services	Run time in seconds	Result composition size: #services	Dependency list size: #services
10 (5 + 5)	10	0.002	**3**	5
20 (10 + 10)	20	0.003	**4**	10
50 (30 + 20)	20	0.007	**12**	20
20 (10 + 10)	50	0.011	**6**	20

4   Extending OpenAPI with JSON-LD

Another aim of this research is to show how current OpenAPI specification[4] and our proposed extension to the JSON-LD model can be used for automatic Web service composition.

4.1   From Formalism to Semantic Descriptions

As a first step, we applied the above mathematical model and algorithm for a set of Web services defined with the help of OpenAPI specification expressed by JSON constructs.

OpenAPI specification is used to describe Web services (APIs) aligned to the REST (REpresentational State Transfer) architectural style [8] and defines in a standardized manner a meta-model to declare the interfaces to RESTful APIs in a programming language-agnostic manner.

These APIs correspond to a set of Web services that forms a repository. For our conducted experiments, we considered an API as a collection of services where each different URI path of the API denoted a different service – this can also be useful in the context of microservices. Thus, we can group services based on information related to the location of the group of services. In practice, this is useful as generally a Web server would likely host a multitude of Web services.

We used OpenAPI specification to describe the input and output parameters of each different service. Those parameters are Web resources that represent, in our mathematical model, partially defined concepts. OpenAPI also helps us match parameters on a syntactic level by specifying the data types of the parameters.

Example. As a real-life case, the public REST API (Web service) provided by Lyft[5] is described. For the operation of getting details about an authenticated user (`GET/profile`), the result – i.e., *w.out* in our mathematical model and, in fact, output data as a JSON object composed by various properties such as

---

[4] OpenAPI Spec – https://github.com/OAI/OpenAPI-Specification.
[5] Lyft API – https://developer.lyft.com/docs.

id: string (authenticated user's identifier), *first_name*: string, *last_name*: string, *has_taken_a_ride*: boolean (indicates if the user has taken at least one Lyft ride).

Each property can have as a result – the *range* mentioned in Sect. 2 – a datatype denoted by a concept, according to the proposed mathematical model. The Web resources processed by a service are then linked to assertions embedded into JSON-LD descriptions. JSON-LD model supports machine-readable annotations of involved properties for a given JSON object. All JSON-LD statements could be converted into RDF (Resource Description Framework) or OWL (Web Ontology Language) [1] to be further processed, including automated reasoning.

For this example, the *first_name* property of the returned JSON object could be mapped – via JSON-LD annotations – to the *givenName* property of *Person* concept (class) defined by the schema.org conceptual model or by FOAF vocabulary[6]. Additionally, the output parameter – i.e., the returned JSON object – could be seen as an instance of *Person* class. This approach could be further refined, if a *Customer* and *Driver* class are defined as a subclass of *Person* concept. The *subclass* relation from the (onto)logical model – formally written as *Customer* $\sqsubseteq$ *Person* – is equivalent to the *isA* relation of the mathematical formalism presented in Sect. 2. Using this taxonomy of classes, the algorithm could select the Web service as a candidate solution for Web service composition.

Secondly, we chose the JSON-LD model to describe the entities used by a group of services. For each different Web resource specified in the OpenAPI description of considered Web services, a corresponding semantic description could be stated. This semantic description attaches to each Web resource a URI specifying a concept in an ontology. Similarly, for each property of the corresponding Web resource.

4.2   Generic Approach

Generally, considering an OpenAPI specification stored in a JSON document:

```
{ "paths": {
    "/resource": {
      "get": {
        "operationId": "service",
        "parameters": [ {
          "name": "parameter",
          "in": "query", "schema": { "type": "string" } } ],
        "responses": {
          "200": {
            "description": "Success",
            "schema": { "$ref":"#/definitions/Response"
      }}} /* a response object composed by properties and
           values, e.g. { response: string } */
}}}}
```

---

[6] FOAF (Friend Of A Friend) Specification – http://xmlns.com/foaf/spec/.

an abstract JSON-LD annotation has the following form, where the context is a knowledge model used to denote Web service entities and the *Concept1* and *Concept2* classes are used for each parameter instead of a generic JSON datatype (in this case, a string).

```
{ "@context": "http://ontology.info/",
  "@id": "parameter", "@type": "Concept1" // w.in
  "response": { "@type": "Concept2" } // w.out
}
```

Also, a convenient mapping (a-priori given or automatically generated) between JSON datatypes (string, number, boolean, object, array) and ontological concepts could be attached as meta-data for the considered set of Web services in order to facilitate the matching process. This enhancement is inspired by ontology alignment strategies [22]. The mapping itself could be directly expressed in JSON-LD via @context construct used to map terms (in our case, datatype names and/or property names) to concepts denoted by URIs. The concepts, properties, restrictions, and related entities (such as individuals and annotations) form the knowledge base – usually, specified by using OWL and RDF [1].

5   Case Study: A Transport Agency

To illustrate the benefits of our approach, we have considered the following example, according to the problem definition in Sect. 2. Also, we have used concepts from schema.org. The services' interfaces are stored in an OpenAPI compliant document. The resources (service parameters) are described via JSON-LD by using the method exposed in Sect. 4.

The case study specifies the car company operating processes via web services. The car company services are simplistic and perform small tasks that can significantly benefit from a web service composition solution. Describing the services using OpenAPI and JSON-LD, the approach showcases how easy it is to represent complex relations between resources by combining object-oriented concepts in both structure and semantics. On top of this, because the REST services are similar on an architectural level, the solution guarantees that the situation presented in the case study can easily be extended and applied in real-world scenarios. Our study contains six services, several resources, but also the query to be solved by a composition of all six services in the correct order. We are supposing that a customer needs a *vehicle* to transport a given *payload*. This person knows his/her current *GeoLocation(latitude, longitude)* and a time frame *Action(startTime, endTime)* in which the transport should arrive. The **Goal** is to obtain the *Action(location)* where the *vehicle* will arrive. The six services – each of them implementing a single operation – are specified below:

```
getCountryFromLocation   in  = GeoLocation(lat,lon)
                         out = Country(name)
getTransportCompany      in  = AdministrativeArea(name)
                         out = Organization(name)
```

```
getClosestCity   in  = GeoLocation(lat,lon)
                 out = City(name)
getLocalSubsidiary  in  = Organization(name), City(name)
                    out = LocalBusiness(email)
getVehicle  in = Vehicle(payload), LocalBusiness(email)
            out= Vehicle(vehicleIdentificationNumber)
makeArrangements
   in = Vehicle(vehicleIdentificationNumber),
        Organization(name,email), Action(startTime,endTime)
   out = Action(location)
```

In OpenAPI terms, a HTTP GET method is defined to obtain a JSON representation of the desired Web resource, for each *getResource* operation – i.e. using `GET/country` with *GeoLocation* as input parameter and *Country* as output. Without JSON-LD constructs, these parameters have regular JSON datatypes like string or number. As defined by schema.org, *LocalBusiness*[7] *isA Organization* – or, equivalent, in (onto)logic terms: *LocalBusiness* $\sqsubseteq$ *Organization*. Similarly, *Country isA AdministrativeArea*. A valid composition satisfying the user request can consist of the services in the following order: **Init** $\rightarrow$ *get-CountryFromLocation* $\rightarrow$ *getTransportCompany* $\rightarrow$ *getClosestCity* $\rightarrow$ *getLocalSubsidiary* $\rightarrow$ *getVehicle* $\rightarrow$ *makeArrangements* $\rightarrow$ **Goal**. The order is relevant, but not unique in this case. This can be verified by considering all resources added by each service and also by the use of the *isA* relation. For example, *LocalBusiness(email)* can be used as *Organization(email)*.

The Java implementation of the algorithm from Sect. 3 reads the OpenAPI specification document describing the above scenario and finds the highlighted composition based on schema.org conceptual model. We also validated the OpenAPI and JSON-LD files using available public software tools included in Swagger Editor[8] and JSON-LD Playground[9].

6   Related Work

Several approaches [3, 16, 18] were considered to tackle the problem of Web service composition in the case of "classical" specification of Web services by using the old SOAP, WSDL (Web Service Description Language), and additional WS-* standards and initiatives [6]. Also, several semantic-based formal, conceptual, and software solutions are proposed by using DAML-S and its successor OWL-S[10] service ontologies (in the present, they are almost neglected by the current communities of practice) and language extensions to Web service descriptions.

---

[7] LocalBusiness, a particular physical business or branch of an organization – https://schema.org/LocalBusiness.
[8] Swagger Editor – http://editor.swagger.io/.
[9] JSON-LD Playground – https://json-ld.org/playground/.
[10] OWL-S: Semantic Markup for Web Services – https://www.w3.org/Submission/OWL-S/.

Various initiatives, methods, and software solutions could be mentioned:

- A mathematical model of the semantic Web service composition problem considering AI planning and causal link matrices [12].
- A linear programming strategy for service selection scheme considering non-functional QoS (Quality of Service) attributes [4].
- Automated discovery, interaction, and composition of the Web services that are semantically described by DAML-S ontological constructs [23].
- A process ontology (called OWL-S) to describe various aspects of Web services using SOAP protocol [15].
- A service composition planner by using hybrid artificial intelligence techniques and OWL-S model [11].
- A hybrid framework which achieves the semantic web service matchmaking by using fuzzy logic and OWL-S statements [7].
- An automated software tool using MDA (Model-Driven Architecture) techniques to generate OWL-S descriptions from UML models [24].
- Various efforts for annotating SOAP-based Web service descriptions by using semantic approaches – e.g., SAWSDL (Semantic Annotations for WSDL and XML Schema) – within the METEOR-S system [21].

In contrast, there are relatively few recent proposals focused on resolving the problem of automatic service composition in the context of the new pragmatic way of describing the public REST APIs by using OpenAPI specification. Several solutions and tools are covered in [9] and [13]. From the modeling service compositions perspective, our formal model presents several similarities to the solutions proposed by [2] and [5].

Concerning enhancing Web services with semantic descriptions [26], several recent initiatives considering OpenAPI-based approaches are focused on:

- Extending the OpenAPI specification with a meta-model giving developers proper abstractions to write reusable code and allowing support for design-time consistency verification [20]. This approach is not using any semantic descriptions and does not provide JSON-LD-based support for reasoning.
- Adopting fuzzy inference methods to match services considering QoS metrics [27], or denoted by OpenAPI specifications [17].
- Generating, in an automatic manner, the suitable GraphQL-based wrappers for REST APIs described by OpenAPI documents [28].
- Using the RESTdesc[11] for service composition and invocation processes in the context of the Internet of Things – e.g., smart sensors [25].
- Capturing the semantics and relationships of REST APIs by using a simplified description model (Linked REST APIs), to automatically discover, compose, and orchestrate Web services [19].

---

[11] RESTdesc – http://restdesc.org/.

7   Conclusion and Future Work

The paper focused on REST-based Web services composition by using a straightforward method that adopts a conceptual approach for modeling semantics of (micro)service parameters. Starting with a formal model described in Sect. 2, an automatic service composition algorithm was proposed and evaluated – see Sect. 3.

To prove the feasibility of the proposed formalism, we extended OpenAPI description of stateless services with JSON-LD constructs, in order to conceptually explain the involved entities (resources) of the service's interface. This new method was detailed in Sect. 4.

For practical reasons, we adopted the popular schema.org model to quickly determine taxonomic relationships between concepts according to the described algorithm. This was exemplified by the case study presented in Sect. 5. Our approach is general enough to choose convenient ontologies for each specific set of composable (micro-)services. We consider that our proposal is also a suitable solution for service-based business Web applications deployed in various cloud computing platforms.

An alternative approach is to use the SHACL (Shapes Constraint Language) model[12] to specify certain restrictions on RDF and, equally, on JSON-LD data – this direction of research is to be investigated in the near future in order to provide support for accessing semantically enriched digital content [14].

We are aware that the proposed method presents several shortcomings – e.g., lack of support regarding service meta-data such as quality of service, various restrictions, work and data-flows, and others. These aspects will be considered, formalized, and implemented in the next stages of our research.

References

1. Allemang, D., Hendler, J.: Semantic Web for the Working Ontologist: Effective Modeling in RDFS and OWL. Elsevier, Waltham (2011)
2. Baccar, S., Rouached, M., Verborgh, R., Abid, M.: Declarative web services composition using proofs. Serv. Oriented Comput. Appl. 1–19 (2018)
3. Blake, M.B., Cheung, W., Jaeger, M.C., Wombacher, A.: WSC-06: the web service challenge. In: The 8th IEEE International Conference on E-Commerce Technology and The 3rd IEEE International Conference on Enterprise Computing, E-Commerce, and E-Services, CEC/EEE 2006, p. 62. IEEE (2006)
4. Cardellini, V., Casalicchio, E., Grassi, V., Presti, F.L.: Flow-based service selection for web service composition supporting multiple QoS classes. In: 2007 IEEE International Conference on Web Services, ICWS 2007, pp. 743–750. IEEE (2007)
5. Cremaschi, M., De Paoli, F.: A practical approach to services composition through light semantic descriptions. In: European Conference on Service-Oriented and Cloud Computing, pp. 130–145. Springer (2018)

---

[12] Shapes Constraint Language (SHACL), W3C Recommendation, 2017 – https:// www.w3.org/TR/shacl/.

6. Erl, T.: SOA: Principles of Service Design. Prentice Hall, Upper Saddle River (2007)
7. Fenza, G., Loia, V., Senatore, S.: A hybrid approach to semantic web services matchmaking. Int. J. Approx. Reason. **48**(3), 808–828 (2008)
8. Fielding, R.T.: Chapter 5: RE presentational state transfer (REST). Architectural styles and the design of network-based software architectures, Ph. D. thesis (2000)
9. Garriga, M., Mateos, C., Flores, A., Cechich, A., Zunino, A.: RESTful service composition at a glance: a survey. J. Netw. Comput. Appl. **60**, 32–53 (2016)
10. Guha, R.V., Brickley, D., Macbeth, S.: Schema.org: evolution of structured data on the web. Commun. ACM **59**(2), 44–51 (2016)
11. Klusch, M., Gerber, A., Schmidt, M.: Semantic web service composition planning with OWLS-xplan. In: Proceedings of the 1st International AAAI Fall Symposium on Agents and the Semantic Web, pp. 55–62. sn (2005)
12. Lécué, F., Léger, A.: A formal model for semantic web service composition. In: International Semantic Web Conference, pp. 385–398. Springer (2006)
13. Lemos, A.L., Daniel, F., Benatallah, B.: Web service composition: a survey of techniques and tools. ACM Comput. Surv. (CSUR) **48**(3), 33 (2016)
14. Levina, O.: Towards a platform architecture for digital content. In: Proceedings of the 15th International Joint Conference on e-Business and Telecommunications, ICETE 2018, Volume 1: DCNET, ICE-B, OPTICS, SIGMAP and WINSYS, pp. 340–347. SciTePress (2018)
15. Martin, D., Burstein, M., Mcdermott, D., Mcilraith, S., Paolucci, M., Sycara, K., Mcguinness, D.L., Sirin, E., Srinivasan, N.: Bringing semantics to web services with OWL-S. World Wide Web **10**(3), 243–277 (2007)
16. Milanovic, N., Malek, M.: Current solutions for web service composition. IEEE Internet Comput. **8**(6), 51–59 (2004)
17. Peng, C., Goswami, P., Bai, G.: Fuzzy matching of OpenAPI described REST services. Proc. Comput. Sci. **126**, 1313–1322 (2018)
18. Rao, J., Su, X.: A survey of automated web service composition methods. In: International Workshop on Semantic Web Services and Web Process Composition, pp. 43–54. Springer (2004)
19. Serrano Suarez, D.F.: Automated API discovery, composition, and orchestration with linked metadata. Ph.D. thesis (2018)
20. Sferruzza, D., Rocheteau, J., Attiogbé, C., Lanoix, A.: Extending OpenAPI 3.0 to build web services from their specification. In: International Conference on Web Information Systems and Technologies (2018)
21. Sheth, A.P., Gomadam, K., Ranabahu, A.H.: Semantics enhanced services: Meteor-s, SAWSDL and SA-REST. Bull. Tech. Comm. Data Eng. **31**(3), 8 (2008)
22. Shvaiko, P., Euzenat, J.: Ontology matching: state of the art and future challenges. IEEE Trans. Knowl. Data Eng. **25**(1), 158–176 (2013)
23. Sycara, K., Paolucci, M., Ankolekar, A., Srinivasan, N.: Automated discovery, interaction and composition of semantic web services. Web Semant. Sci. Serv. Agents World Wide Web **1**(1), 27–46 (2003)
24. Timm, J.T., Gannod, G.C.: A model-driven approach for specifying semantic web services. In: IEEE International Conference on Web Services, pp. 313–320. IEEE (2005)
25. Ventura, D., Verborgh, R., Catania, V., Mannens, E.: Autonomous composition and execution of REST APIs for smart sensors. In: SSN-TC/OrdRing@ ISWC, pp. 13–24 (2015)

26. Verborgh, R., Harth, A., Maleshkova, M., Stadtmüller, S., Steiner, T., Taheriyan, M., Van de Walle, R.: Survey of semantic description of REST APIs. In: REST: Advanced Research Topics and Practical Applications, pp. 69–89. Springer (2014)
27. Wang, P., Chao, K.M., Lo, C.C., Huang, C.L., Li, Y.: A fuzzy model for selection of QoS-aware web services. In: 2006 IEEE International Conference on e-Business Engineering, ICEBE 2006, pp. 585–593. IEEE (2006)
28. Wittern, E., Cha, A., Laredo, J.A.: Generating GraphQL-wrappers for REST (-like) APIs. In: International Conference on Web Engineering, pp. 65–83. Springer (2018)

GWMA Algorithm for Host Overloading Detection in Cloud Computing Environment

Jen-Hsiang Chen[1(✉)] and Shin-Li Lu[2]

[1] Department of Information Management, Shih Chien University,
Kaohsiung Campus, Taipei, Taiwan
jhchen@g2.usc.edu.tw
[2] Department of Industrial Management and Enterprise Information,
Aletheia University, Taipei, Taiwan
shinlilu@mail.au.edu.tw

Abstract. Energy consumption and Service Level Agreement (SLA) in Cloud computing environment are important issues in cloud management. Dynamic consolidation of the Virtual Machines (VMs) need effective and efficient distribution for VMs migration to hosts in data center. The process of VMs migration needs to evaluate host capability, VM placement and reallocation, which satisfy SLA criterions under a flexible service plan. Therefore, the plan is to select effective resource allocation to achieve cost minimization, reduce energy consumption and avoid SLA violation. We *propose* Generally Weighted Moving Average (GWMA) algorithm to detect overloaded hosts, which deals with dynamic consolidation of VMs based on an analysis of historical data of the resource usage by VMs. It increases the accuracy in calculation of the upper threshold for host overloading and consequently increases accuracy in identification to deal with VMs migration issue.

Keywords: Generally Weighted Moving Average (GWMA) · Service Level Agreement (SLA) · VM migration

1 Introduction

According to Gartner Forecast survey the worldwide public cloud services market is projected to grow 17.5% in 2019 to total $214.3 billion, and Worldwide Public Cloud Service Revenue Forecast in 2022 will reach $331.2 billion [1]. Increasing demand of infrastructure consumes more electric energy. The importance of reduction in energy consumption and CO_2 is acknowledged by various research efforts [2]. Cloud providers may try to utilize their infrastructure at least up to 70%, where traditional on-premise server utilization is 5%–10% on average [3]. It is imperative to reduce the number of active servers, and increase the number of users per server while processing more workloads at optimal levels [4]. An Infrastructure as a Service (IaaS) is the optimization architecture of cloud computing infrastructure and it ensures better resource management technologies of virtual machine (VMs) to be utilized in cloud management [5].

© Springer Nature Switzerland AG 2020
K.-M. Chao et al. (Eds.): ICEBE 2019, LNDECT 41, pp. 358–370, 2020.
https://doi.org/10.1007/978-3-030-34986-8_26

Cloud providers manage optimize resource usage and reduce energy usage by dynamic consolidation of virtual machines (VMs), while using live migration and switching idle nodes to the sleep mode [6]. VMs can be migrated from hosts to hosts depending on different applications needs, the VM can be moved automatically or manually to a set of specified hosts. To ensure VMs can be moved freely and selects a host that meets specifications. Operational efficiency of servers, operating costs for data centers have direct impact on the host server where as too much load affects the performance. Hence as a result service quality may be reduced, on the other hand less load will lead to low resource utilization and results in increased operating costs. To address this problem and drive Green Cloud computing, cloud resources need to be allocated not only to satisfy Quality of Service (QoS) requirements, but also reduce energy consumption.

The presented study recognizes the problem of dynamic virtual machine consolidation using virtualization, live migration of VMs based on threshold of hosts and switching idle nodes to the sleep mode. Therefore, the approach is effective for utilizing resources and achieving energy efficiency in green cloud computing [7–9]. This paper also proposes GWMA algorithm to improve the problem of dynamic virtual machine consolidation based on energy efficiency. The paper is organized as follows: Sect. 2 presents optimization policy for resource allocation. Section 3 presents GWMA algorithm for Host Overloading Detection. Experiments and their results are described in Sect. 4. Finally, conclusion and future is presented in Sect. 5.

2 Optimization Algorithm for Resource Allocation

The main objective of resource allocation is to optimize the response time, task completion time, network bandwidth, memory provisioning and energy saving [10]. The distribution of resource allocation includes user's requirement for VMs, and placement of VM into host (Fig. 1).

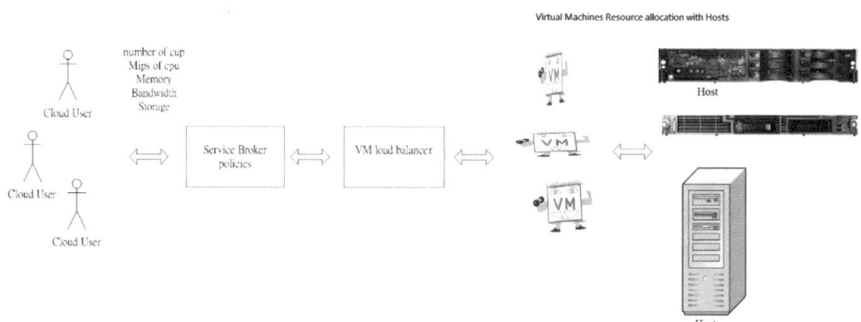

Fig. 1. Policies with resource allocation

Dynamic consolidation of the Virtual Machines and hosts involves three steps. Firstly, detection of a host load should be normal working threshold based on Qos requirement. Secondly, selecting VMs in an overloaded host or all VMs in a underload host are listed to be migrated VMs. Finally, reallocating the listed VMs to other active or reactivated hosts (see Fig. 2). In addition, the related algorithm for each step are investigated in the subsection below.

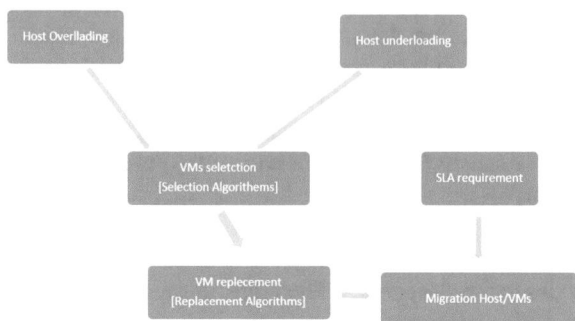

Fig. 2. VMs Replacement process

2.1   Algorithms for Host Overloading Detection

The overloading detection algorithms apply the predicate mechanism by historical data, there are some algorithms are below [11].

Median Absolute Deviation (MAD) is a measure of statistical dispersion. It is a more robust estimator of scale than the sample variance or standard deviation.

Interquartile Range (IR) is called the mid spread or middle fifty is a measure of statistical dispersion, being equal to the difference between the third and first quartiles.

Local Regression (LR) is a fitting simple models to localized subsets of data to build up a curve that approximates the original data.

Robust Local Regression (RLR) is modified transforms Local Regression into an iterative method. The initial fit is carried out with weights defined using the tricube weight function.

2.2   Algorithms for VM Selection

Once it has been decided that a host is overloaded, the next step is to select particular VMs to migrate from this host. There are three algorithms for VM selection, which are applied iteratively. After selection of a VM to migrate, the host is checked again for being overloaded. If it is still considered as being overloaded, the VM selection policy is applied again to select another VM to migrate from the host. This is repeated until the host is considered as being not overloaded. There are three algorithms for VM selection [11].

The Minimum Migration Time (MMT) algorithm migrates a VM that requires the minimum time to complete a migration relatively to the other VMs allocated to the host.

The Random Choice (RC) algorithm selects a VM to be migrated according to a uniformly distributed discrete random variable.

The Maximum Correlation (MC) algorithm shows the higher the correlation between the resource usages by applications running on an oversubscribed server, the higher the probability of the server overloading. According to this idea, we select those VMs to be migrated that have the highest correlation of the CPU utilization with other VMs.

2.3 VM Replacement

VM replacement like an NP-hard problem to allocate VMs and hosts by minimizing the cost. VMs has requited capacities and host offer different resource, which solution seems to VMs reallocation concern optimal energy consumption. The algorithms employed are of First Fit (FF), First Fit Decreasing (FFD), Worst Fit Decreasing (WFD) and Second Worst Fit Decreasing (SWFD), Best Fit Decreasing (BFD) to solve NP-hard problem.

In addition, heuristic approaches have been implement for VM consolidation problem, such as genetic algorithm (GA) [12–14], Multi-Objective Ant Colony Optimization (MOCO) algorithm [15], the Honey Bee algorithm [16]. These kind of approaches are based on minimize resource that can reach near optimal solution for VM reallocation problems.

3 GWMA Algorithm for Host Overloading Detection

Shin and Sohn [17] proposed that forecasting exchange rates using exponentially weighted moving average (EWMA) combination technology resulted in better forecasting than the GARCH model, neural network model, and random walk model. The EWMA method is a simple and easy method to understand and use in forecasting future stock prices or exchange rate volatility. The popular Google machine learning software TensorFlow, which adopt EWMA as one of mechanism that indicates "*When training a model, it is often beneficial to maintain moving averages of the trained parameters. Evaluations that use averaged parameters sometimes produce significantly better results than the final trained values*" [18]. Therefore, the follow up research the generally weighted moving average (GWMA) extend to exponentially weighted moving average (EWMA) by adding design and adjustment parameters algorithm [19]. Afterward, the GWMA algorithm has successfully used in monitoring small process shifts in various data type such as the multivariate GWMA chart [20], the Poisson GWMA chart [21], the autocorrelated GWMA chart [22], and the nonparametric GWMA chart [23]. Recently, Lin et al. [24] and Sheu et al. [25] applied the GWMA method in forecasting field to predict the volatility of a combined multi-country stock index.

The EWMA method was first introduced by Roberts [26] and defined as follows:

$$y_t = \lambda x_t + (1 - \lambda)y_{t-1} \tag{1}$$

where λ is a smoothing constant satisfying $0 < \lambda \leq 1$, y_t is EWMA value, and x_t is observation at time t. Substituting the value y_{t-1} into Eq. (1) yields:

$$y_t = \lambda \sum_{j=1}^{t} (1 - \lambda)^{j-1} x_{t-j+1} + (1 - \lambda)^t y_0 \tag{2}$$

Sheu and Lin [19] first developed the GWMA model, which was motivated by Sheu and Griffith [27], to enhance the detection ability of EWMA models. The design q and adjustment α parameters are adopted in the GWMA model to be shown it is sensitive in small fluctuations. The GWMA method is described briefly as follows:

$$y_t = \sum_{j=1}^{t} (q^{(j-1)^\alpha} - q^{j^\alpha}) x_{t-j+1} + q^{t^\alpha} y_0 \tag{3}$$

When $\alpha = 1$, and $q = 1 - \lambda$, then Eq. (3) can be transformed to be consistent with the foregoing EWMA Eq. (2). Therefore, the EWMA model is a special case of the GWMA model when $\alpha = 1$.

Where y_0 is the starting value, usually set equal to the target value. This paper uses the GWMA method, combined with the all past observations at time t to predict the next observation $\hat{y}_{t+1}$ at time $t + 1$. Hence,

$$\min \sum_{j=1}^{t} (y_t - \hat{y}_t)^2 \tag{4}$$

$$\text{s.t. } s \cdot \hat{y}_{t+1} \geq 1, \quad x_{t+1} - x_t \leq t_m$$

where $s \in R^+$ is the safety parameter; and t_m is the maximum time required for a migration of any of the VMs allocated to the host.

4   Experiment and Evaluation

This section provides a simulation of overloading host detection with EWMA algorithm and its assessment. Then it compares proposed algorithm with algorithm in [11] and the results were analyzed and examined. CloudSime framework [28] is a popular cloud toolkit and simulator, many researches are evaluated proposal novel approaches based on the platform [9, 11, 29], we also use it to simulate the proposed algorithm.

4.1   Experimental Setting

We evaluate the proposed algorithms by extensive simulation using the CloudSim toolkit and the workload data from 10 days of the resource usage by more than a thousand PlanetLab VMs provisioned for multiple users. The datacenter in this simulation give 800 heterogeneous physical hosts to examine and evaluate in Clouds framework. The both types of hosts are:

HP ProLiant ML110 G4 (Intel Xeon 3040, two cores × 1860 MHz, 4 GB)
HP ProLiant ML110 G5 (Intel Xeon 3075, two cores × 2660 MHz, 4 GB).

The datacenter offer 5 instance of single-core virtual machines, which are:

High-Memory Extra Large Instance: 3.25 EC2 Compute Units, 8.55 GB
High-CPU Medium Instance: 2.5 EC2 Compute Units, 0.85 GB
Extra Large Instance: 2 EC2 Compute Units, 3.75 GB
Small Instance: 1 EC2 Compute Unit, 1.7 GB
Micro Instance: 0.5 EC2 Compute Unit, 0.633 GB

4.2   Workload Data

The simulation of Workload Data of Cloudsim toolkit used the workload data from the CoMon project, which is a monitoring infrastructure associated with PlanetLab workloads [11]. This Cloudsim used a workload data collected for 10 days in March and April 2011, which carries out a reasonable and appropriate comparison. For more realistic results, a CPU utilization dataset was used, the data were collected from more than thousands of operational VMs in over 500 locations around the world. The utilization measurements is 5-min time interval. In the experiment, we also used the same data in our study for assessing the proposed GWMA algorithm as detection mechanism to compare it with its counterpart approaches.

4.3   Performance Metrics

There are six parameters to be assessed and compared the proposed algorithm with those of other studies. These metrics are described below [11]:

- Number of VM migrations metric measure the total value VM migration from overloaded and underloaded hosts for replacement.
- Energy Consumption metric measure the total energy consumption for physical resources.
- PDM metric measure the performance degradation due to VM Migration (PDM).
- SLATAH metric is SLA Violation Time per Active Host that can be defined as the percentage of the period when the host experiences a CPU utilization of 100%.
- SLAV metric encompasses both performance degradation due to host overloading and SLA Violation (SLAV). It indicates the duration in which the allocated resources to the host is lower than the required amount. SLAV is calculated as SLAV = SLATAH * PDM.
- ESV metric measure the total energy consumption with SLA violation, which is used to measure the simultaneous improvement in both metrics. ESV is calculated as $ESV = Energy * SLAV$
- Energy-SLA-Migration (ESM) measure the simultaneous minimization of energy, SLA violation, and number of VMs migrations. ESM and is calculated as $ESM = E \times SLAV \times Number\ of\ VM\ migrations$ [30].

4.4   Simulation Results

Local Regression (LR) VM allocation policy with Minimum Migration Time (MMT) VM selection policy reached the best performance in previous work [11]. The authors used the latest 10 historical data from the resource usage by VMs to make the local regression, therefore, we have evaluated the various training historical number data. 30 historical data will have the better performance than others. We set the 30 length of training data and safety parameter $S = 1.2$ to evaluate our proposal GWMA to compare LR.

In addition, we experiment with four values of parameter λ, which $\lambda = 1\%$, 2.5%, 5% and 10% will be used in evaluation for GWMA performance comparison. The seven metrics will be analyzed using following experiments.

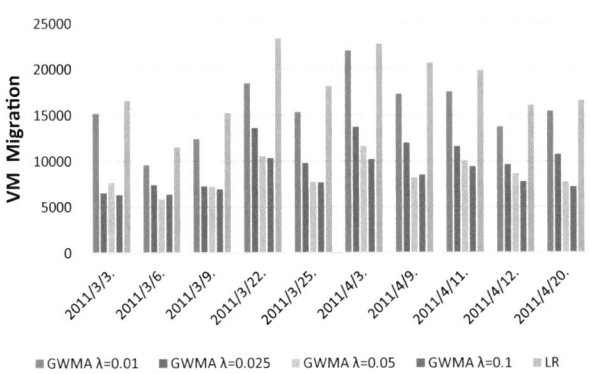

Fig. 3. The number of migration metric

Figure 3 presents number of Migration, number of migration in proposed GWMA is much less than Local regression. The GWMA consequently increased accuracy in identification of overloaded hosts among the reasons for the significant reduction in the number of migrations in GWMA algorithm as compared with the others. Compare the result for various values of λ in GWMA, which are 15214, 6538, 7717 and 6331 of number of migration for $\lambda = 1\%$, 2.5%, 5% and 10% respectively, and 16615 is in LR on 2011/3/3. As for other days, the result seem to be same.

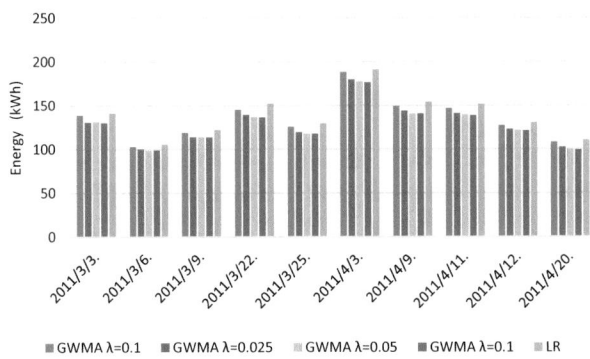

Fig. 4. The energy consumption metric

Figure 4 demonstrates energy consumption, generally LR consumed higher than four GWMA parameters, and $\lambda = 5$ and 10% save more energy than the other two.

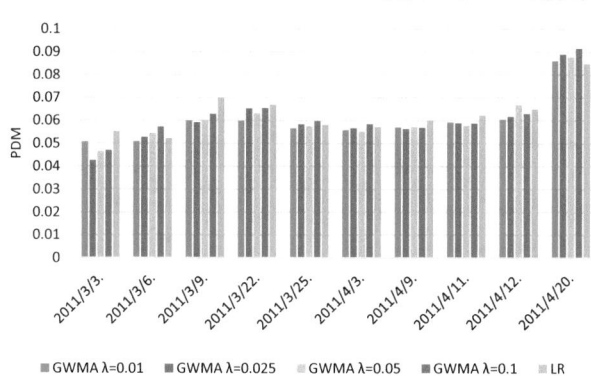

Fig. 5. The PDM metric

Figure 5 demonstrates SLA performance degradation (PDM) due to migration, which show LR had higher, lower or similar performance degradation due to more VMs migration in 10 days period.

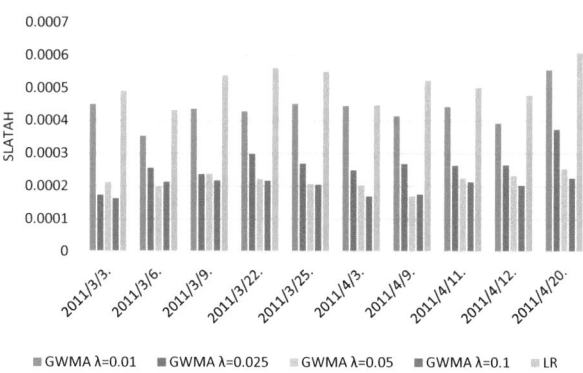

Fig. 6. The SLA violation Time per Active Host metric (SLATAH).

Figure 6 shows general view that SLA Violation Time per Active Host (SLATAH) in LR is high than GWMA. $\lambda = 5\%$, 10% in GWMA are much lower than other two parameters.

Figure 7 show the number of host switch off. The number of LR algorithm switched more times than GWMA, and $\lambda = 5\%$, 10% have lower number of showdown.

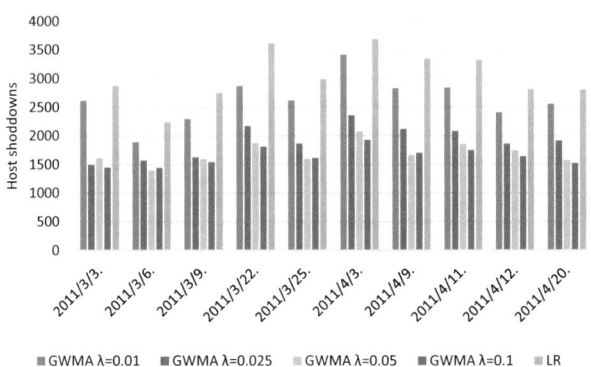

Fig. 7. Number of Host showdowns

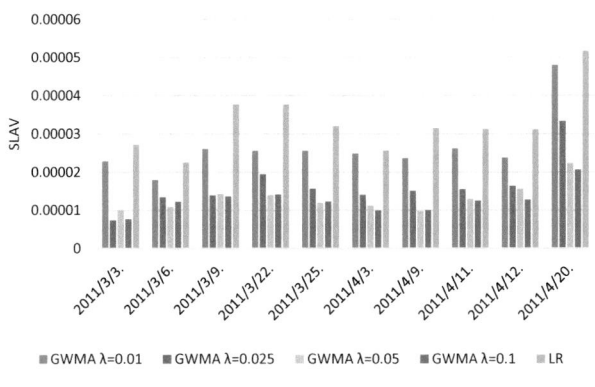

Fig. 8. The SLAV metric

SLAV metric is calculated from SLATAH metric multiple PDM metric. According to result of Fig. 5 for PDM and 6 for SLAV performance. Figure 8 presents GWMA performed well than LR, and the value $\lambda = 5\%$ or 10% in GWMA parameter seems to better than two.

The formula of metric ESV = Energy * SLAV. The ESV metric in Fig. 9 demonstrated GWMA performed better than LR. $\lambda = 5\%$ and 10% present better result than $\lambda = 1\%$.

Figure 10 show ESM for minimization energy, SLA violation, and number of VMs migrations, where $\lambda = 5\%$ and 10% looks they have the better performance than the other two. In additionally, the Table 1 show sum of ESM in 10 days period that $\lambda = 10\%$ has better performance in ESM metric.

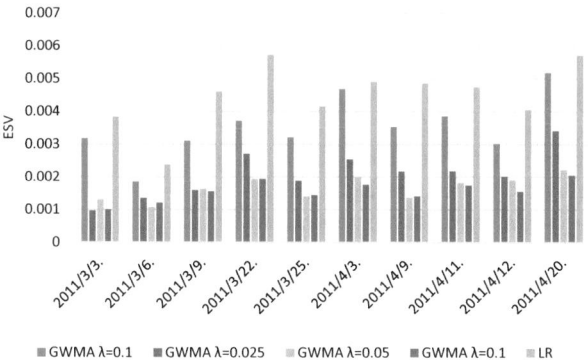

Fig. 9. The ESV metric

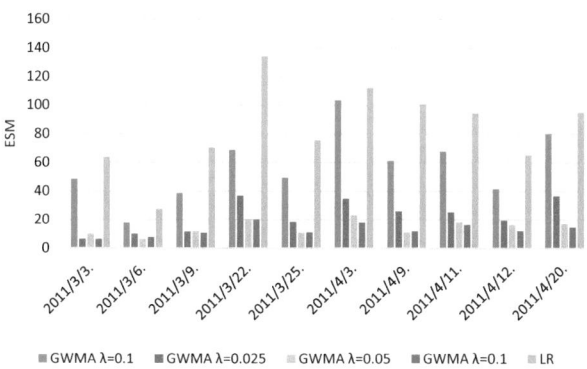

Fig. 10. ESM metric

Table 1. Total value of ESM in 10 days

$\lambda = 0.01$	$\lambda = 0.025$	$\lambda = 0.05$	$\lambda = 0.1$	LR
577.08	225.80	146.26	129.33	839.20

Minimizing energy and SLA violations are main metrics of the resource management system [5]. The results of ESV and ESM are significant decrease in energy and improvements in SLAV metrics. The proposed GWMA has proof and shown increase of performance than Local Regression.

5 Conclusion and Future Work

The recent research for host overloading detection, El-Moursy et al. [31] adopt Multi-Dimensional Regression Utilization algorithm (MDRHU) for Host Overload Detection and it simulated in PlanetLab workloads to evaluate his performance. They implement the proposal mechanism to compare with local regression, which improve over 20% in ESM metric. However, we provided GWMA can improve much more than local regression in the same workloads environment.

As for performance for range of parameter λ values setting, while $\lambda = 5\%$ and 10% provides better performance in dynamic VMs migration. The weight for past period training is high effect. Therefore, GWMA algorithm improve the using of an optimized upper threshold that leads to more efficient and effective utilization of processing resources on the hosts.

As for future work, the novel heuristic approaches or combined algorithms for dynamic consolidation of the Virtual Machines will be further investigated. Furthermore, the resource of cpu and memory and network BW utilization could be considered as the parameters in cloud environment. Additionally, new algorithms as the maximum General Weighted Moving Average (Maximum-GWMA) method or Auxiliary Information General Weighted Moving Average (AIB-GWMA) as an extension of the GWMA model will be explored.

References

1. Gartner Forecasts Worldwide Public Cloud Revenue to Grow 17.5 Percent (2019). https://www.gartner.com/en/newsroom/press-releases/2019-04-02-gartner-forecasts-worldwide-public-cloud-revenue-to-g. Accessed 10 Jun 2019
2. Wajid, U., Cappiello, C.: On achieving energy efficiency and reducing CO_2 footprint in cloud computing. IEEE Trans. Cloud Comput. **4**(2), 152–165 (2016)
3. Garg, S., Buyya, R.: Green cloud computing and environmental sustainability. In: Murugesan, S., Gangadharan, G. (eds.) Harnessing Green IT: Principles and Practices, pp. 315–340. Wiley Press, London, October 2012. ISBN: 978-1-1199-7005-7
4. Balasooriya, P.N., Wibowo, S., Wells, M.: Green cloud computing and economics of the cloud: moving towards sustainable future. J. Comput. (JOC) **5**(1), 15–20 (2016). ISSN: 2251-3043
5. Pietri, I., Sakellariou, R.: Mapping virtual machines onto physical machines in cloud computing: a survey. ACM Comput. Surv. **49**(3), 49 (2016)
6. Beloglazov, A., Abawajy, J., Buyya, R.: Energy-aware resource allocation heuristics for efficient management of data centers for cloud computing. Future Gener. Comput. Syst. **28**(5), 755–768 (2012). ISSN: 0167-739X
7. Ashraf, A., Porres, I.: Machine consolidation in the cloud using ant colony system. Int. J. Parallel Emergent Distrib. Syst. **33**(1), 103–120 (2018)
8. Esfandiarpoor, S., Pahlavan, A., Goudarzi, M.: Structure-aware online virtual machine consolidation for datacenter energy improvement in cloud computing. Comput. Electr. Eng. **42**, 74–89 (2015)
9. Alavi, S.E., Noorimehr, M.R.: Optimization of dynamic virtual machine consolidation in cloud computing data centers. Int. J. Adv. Comput. Sci. Appl. (IJACSA) **7**(9) (2016)

10. Chen, J.H., Tsai, C.F., Lu, S.L., Abedin, F.: Resource reallocation based on SLA requirement in cloud environment. In: 2015 IEEE 12th International Conference on e-Business Engineering (2015)
11. Beloglazov, A., Buyya, R.: Optimal online deterministic algorithms and adaptive heuristics for energy and performance efficient dynamic consolidation of virtual machines in cloud data centers. Concurrency Comput.: Pract. Exp. **24**(13), 1397–1420 (2012)
12. Wu, G., Tang, M., Tian, Y.-C., Li, W.: Energy-efficient virtual machine placement in data centres by genetic algorithm. In: International Conference on Neural Information Processing (ICONIP 2012), vol. 7665, pp. 315–323 (2012)
13. Joseph, C.T., Chandrasekaran, K., Cyriac, R.: A novel family genetic approach for virtual machine allocation. Procedia Comput. Sci. **46**, 558–565 (2015)
14. Janani, N., Shiva Jegan, R.D., Prakash, P.: Optimization of virtual machine placement in cloud environment using genetic algorithm. Res. J. Appl. Sci. Eng. Technol. **10**(3), 274–287 (2015)
15. Gao, Y., Guan, H., Qi, Z., Hou, Y., Liu, L.: A multi-objective ant colony system algorithm for virtual machine placement in cloud computing. J. Comput. Syst. Sci. **79**(8), 1230–1242 (2013)
16. Singh, A., Hemalatha, N.M.: Cluster based bee algorithm for virtual machine placement in cloud data centre. J. Theor. Appl. Inf. Technol. **57**(3) (2013)
17. Shin, H.W., Sohn, S.Y.: Application of an EWMA combining technique to the prediction of currency exchange rates. IIE Trans. **39**, 639–644 (2007)
18. Shin, H.W., Sohn, S.Y.: Application of an EWMA combining technique to the prediction of currency exchange rates. IIE Trans. **39**, 639–644 (2007). Exponential Moving Average, ThesorFlow (2019). https://www.tensorflow.org/api_docs/python/tf/train/Exponential MovingAverage. Accessed 1 Feb 2009
19. Sheu, S.H., Lin, T.C.: The generally weighted moving average control chart for detecting small shifts in the process mean. Qual. Eng. **16**, 209–231 (2003)
20. Yang, L., Sheu, S.H.: Integrating multivariate engineering process control and multivariate statistical process control. Int. J. Adv. Manufact. Technol. **29**, 129–136 (2006)
21. Sheu, S.H., Chiu, W.C.: Poisson GWMA control chart. Commun. Stat. Simul. Comput. **36**, 1099–1114 (2007)
22. Sheu, S.H., Lu, S.L.: Monitoring the mean of autocorrelated observations with one generally weighted moving average control chart. J. Stat. Comput. Simul. **79**, 1393–1406 (2009)
23. Lu, S.L.: An extended nonparametric exponentially weighted moving average sign control chart. Qual. Reliab. Eng. Int. **31**, 3–13 (2015)
24. Lin, C.Y., Sheu, S.H., Hsu, T.S., Chen, Y.C.: Application of generally weighted moving average method to tracking signal state space model. Expert Syst. **30**, 429–435 (2013)
25. Sheu, S.H., Lin, C.Y., Lu, S.L., Tsai, H.N., Chen, Y.C.: Forecasting the volatility of a combined multi-country stock index using GWMA algorithms. Expert Syst. **35**(3), e12248 (2018)
26. Roberts, S.W.: Control chart tests based on geometric moving average. Technometrics **42**, 97–102 (1959)
27. Sheu, S.H., Griffith, W.S.: Optimal number of minimal repairs before replacement of a system subject to shocks. Naval Res. Logistics **43**, 319–333 (1996)
28. Calheiros, R.N., Ranjan, R., Beloglazov, A., De Rose, C.A.F., Buyya, R.: CloudSim: a toolkit for modeling and simulation of cloud computing environments and evaluation of resource provisioning algorithms. Softw.: Pract. Exp. **41**(1), 23–50 (2011). ISSN: 0038-0644

29. Mosa, A., Sakellariou, R.: Virtual machine consolidation for cloud data centers using parameter-based adaptive allocation. In: ECBS 2017, 5th European Conference on the Engineering of Computer Based Systems, Larnaca, Cyprus, 31 August–1 September 2017
30. Arianyan, E., Taheri, H., Sharifian, S.: J. Inf. Sci. Eng. **32**, 1575–1593 (2016)
31. El-Moursy, A.A., Abdelsamea, A., Kamran, R., Saad, M.: Multi-dimensional regression host utilization algorithm (MDRHU) for host overload detection in cloud computing. J. Cloud Comput.: Adv. Syst. Appl. **8**(1), 8 (2019)

Security and Privacy Driven Techniques for e-Business Management

Privacy Preserving Government Data Sharing Based on Hyperledger Blockchain

Yurong Hao[1], Chunhui Piao[2(✉)], Yongbin Zhao[1],
and Xuehong Jiang[2]

[1] Shijiazhuang Tiedao University, Shijiazhuang 050043, Hebei, China
[2] Construction Information Center of Hebei Province,
Shijiazhuang 050043, Hebei, China
pchls2011@126.com

Abstract. With the further advancement of the government information resource sharing mechanism, the realization of data sharing between departments under the support of safe and effective technology has become focus of common concern for government departments and the public. Blockchain provides new ideas and ways for government departments to share data securely and efficiently. On the basis of analyzing the risk of disclosure of private information of citizens and enterprises in government data sharing, this paper discusses the data sharing framework of privacy preserving among government departments based on hyperledger blockchain and describes the corresponding data sharing process. Through an example of data sharing application between government departments, the feasibility and advantages of the proposed privacy preserving government data sharing framework based on blockchain technology are presented.

Keywords: Data sharing · Privacy preserving · Blockchain · Hyperledger

1 Introduction

In recent years, with the development of China's digital government, the government data have been accumulated continuously. The growing cooperation of departments results in the need to share and exchange such data [1]. However, there is a contradiction between security and efficiency in realizing data sharing in e-government. Government is a major collector and sharer of data [2]. Its data are widely stored in different units, departments, systems and even different network environments. These data are digital records of the whole social activities and non-material wealth that can be reused. They play a key supporting role in cross-department business collaboration and cross-department service collaboration and are the core content of government data sharing and application.

For a long time, due to the division of government administrative departments, different government data are controlled by their respective departments, resulting in mutual blockade and barriers between different departments, making it difficult to share data resources. However, government data sharing can transfer government information from one department to another, greatly improving the phenomenon of "data

K.-M. Chao et al. (Eds.): ICEBE 2019, LNDECT 41, pp. 373–388, 2020.
https://doi.org/10.1007/978-3-030-34986-8_27

barrier", making the data play the greatest value, promoting collaborative work. In addition, privacy preserving is also widely concerned before data sharing. On the one hand, the government service platform involves a large number of sensitive information of personnel and enterprises [3]. Meanwhile, the data sharing process flows through more personnel and processing links, which may lead to more threats to citizens' privacy. It's not only prone to man-made error, but vulnerable to hacker attacks, resulting in information leakage [3]. Therefore, in the process of deepening the sharing of government data, the safe and efficient sharing of data between departments has become a challenging problem that needs to be urgently studied and solved.

The characteristics of blockchain, such as traceability, decentralization and tamper resistance, provide a secure and trusted environment for cross-departmental data interconnection. Blockchain allows government departments to independently authorize visitors and access data, record data calling behavior and accurately take responsibility in case of data leakage events, thus greatly reducing the security risk of e-government data sharing and improving law enforcement efficiency [3]. In view of the problem of privacy disclosure in government data sharing, this paper attempts to start with the implementation mechanism of blockchain, discusses the applicability and enforceability of blockchain technology in government data sharing, constructs a government data sharing framework between government departments based on blockchain and describes its corresponding data sharing process. On the one hand, it provides ideas for the theoretical research of government data sharing; on the other hand, it further promotes the construction of government data sharing through model construction and reduces the risk of privacy disclosure.

2   Related Works of Data Sharing Using Blockchain

2.1   Privacy Preserving Government Data Sharing Requirements

Government data sharing refers to the process in which government departments transfer information or data from one department to another by means of information sharing platform or other technical means in accordance with laws and regulations in order to give full play to data value. However, due to the lack of unified planning and design of business information systems by government departments based on their own needs, many departments repeatedly collect and study information that has been collected and processed by other departments. The real-time, consistency and accuracy of information are not only no guaranteed, but cost a lot of administrative costs, which brings troubles to government departments at all levels in making administrative decisions.

The sharing and cooperation of government data has also attracted increasing attention from governments around the world. As the leader of the global open data, the United States has established three open sharing modes of government data among government departments: hierarchical, parallel and regional. It also has been in the forefront of the world in terms of government-enterprise cooperation, government-citizen cooperation in opening up government data resources and improving the value of government data utilization; Many Southern African Development Community (SADC) countries are adopting and implementing e-government systems to improve the efficiency and effectiveness of their service delivery systems. Zimbabwe and Zambia aren't an exception [4]. In June 2018, the "Implementation Plan for Further Deepening the Internet + Government Service and Promoting the 'One Network, One Door, One Time' Reform of Government Service" also put forward further clear requirements for the integration and sharing of government information systems. China has made some progress in promoting the sharing of government data. By December 2018, the implementation leading group had broken through more than 2,800 data barriers and formed 520,000 data catalogues, of which 94.5% could be shared [5].

However, in the process of sharing government data, there are still the following problems. One department can't determine which data other departments need and how the data should be provided, nor can it know what shareable data other departments can provide. Data sharing between government departments mainly depends on the requirements of interest-driven or higher-level departments, that is, department can't fully trust the shared data obtained. Therefore, the establishment of a reasonable and feasible government data supply-demand correlation mechanism and trust mechanism is one of the key issues to be discussed in the open sharing of government data.

2.2   Privacy Issues in Government Data Sharing

With the emergence of the Internet, data sharing among government departments is also increasing. While realizing the benefits of data sharing, we need to solve the privacy problems caused by the use of the Internet [6]. In the process of sharing government data, as long as it involves the transmission and processing of government data, it may cause the disclosure of private information. Generally, government data contain a large number of personal and enterprise-related data, which have great economic utility value. However, these valuable data are often threatened by accidental or malicious infringement [7]. These threats mainly come from both internal and external threats. Internal threats generally refer to the disclosure of privacy by government personnel due to improper operation, illegal use or profit-driven. External threats generally originate from attackers using system vulnerabilities to invade the background to steal information and intercept information during data sharing. Therefore, how to protect the privacy of government data is the key problem that the government faces while implementing government data sharing and it's also the core capability that needs to be strengthened urgently.

In September 2017, Equifax, a well-known credit agency in the United States, was attacked by hackers, affecting more than 143 million Americans. It's reported that besides the United States, users in Britain and Canada were also affected to varying degrees. In October of the same year, 41 Hyatt hotels' payment systems in 11 countries were hacked, resulting in a large amount of data leakage, of which 18 hotels in China were affected. The leaked information included the cardholder's name, card number, expiration date and internal verification code. In June 2018, 1 billion YTO express data were sold. In April 2019, millions of user records of Facebook, which had not yet completely pulled itself out of the "whirlpool" of leaked data, were posted on Amazon's cloud server. The risk of privacy disclosure in government data sharing has attracted great attention from various government departments. Government departments of various countries ensure data security and sensitive information by formulating corresponding policies and regulations [8]. For example, Australia's "Privacy Act", Germany's "Data Protection Act" and the European Parliament's "General Data Protection Act". These bills emphasize that data privacy is a basic right of citizens and clearly regulate the content of rights such as data permission and default privacy preserving. Faced with the challenges of network and information security, the Chinese government has issued a series of relevant laws and regulations in recent years, such as the "Implementation Plan for Integration and Sharing of Government Information Systems" issued on May 3, 2017. "The Cyber Security Law of the People's Republic of Government Information Systems", which came into force on June 1, 2017. These laws protect privacy from disclosure and invasion [9]. It provides legal basis for information protection in government data sharing.

2.3   Research on Data Sharing Based on Blockchain

Blockchain is an innovative application mode of distributed data storage, point-to-point transmission, consensus mechanism, encryption algorithm and other computer technologies in the Internet era. The State Council's "Notice of 'the 13th Five-Year Plan' for National Informatization" issued in 2016 emphasizes the necessity of research and development of new technologies such as blockchains and their strengthening strategies. For this reason, domestic and foreign researchers have carried out a large number of researches on blockchain technology under various application backgrounds. Tan et al. [11] have realized the protection, verification, recovery and sharing of digital files through technologies such as blockchain + digital signature + IPFS. By summarizing the advantages and characteristics of blockchain technology, Xu et al. [12] designed a library resource sharing service model based on blockchain technology on the basis of analyzing the disadvantages of traditional library resource sharing service and elaborated the specific implementation process of this service on blockchain in detail.

At the same time, researchers at home and abroad have also conducted a series of studies on the application of blockchain technology in government data sharing. Taylor [13] believes that the use of blockchain technology by government departments isn't only safe and reliable, but can reduce administrative costs and improve government transparency. Xia et al. [14] studied the access control management in the medical data sharing system. It mainly designed a data sharing scheme based on blockchain, allowing data owners to access electronic medical records from the shared database after authentication. Jin et al. [15] put forward a blockchain-based e-government security sharing demand model and a guarantee scheme. Through in-depth analysis of the key factors and operation mechanism of the current government information resources sharing, Yu et al. [16] have given the overall framework of the model based on blockchain technology and constructed the sharing and exchange system of government information resources. Through in-depth research on 13 e-government blockchain technology projects carried out in 5 countries, Adegoega et al. [17] believe that blockchain technology can be applied to the infrastructure construction of government manage, public services and government information resource sharing. The pilot project of e-government blockchain infrastructure carried out by various governments has fully demonstrated the prospect and practical application value of blockchain technology in e-government and government information resource sharing research.

2.4   Research on Data Sharing Based on Blockchain

The application modes of blockchain are divided into public blockchain, consortium blockchain and private blockchain. The public blockchain is open to all the public and anyone can participate. It needs to realize completely decentralized trust through "mining", which greatly reduces efficiency. The read and write permissions of the private blockchain are completely owned by an organization. The write permissions are limited to the internal peers of the organization and are more applicable to the internal of the organization. The consortium blockchain is between the public blockchain and the private blockchain. It has strict restrictions on the members. The peers restrict each other and have high trust strength. It's mainly applicable to cooperation within or between organizations and institutions. This paper holds that among the above three blockchain types, consortium blockchain is more controllable and more conform to the application requirements of government data sharing. Although consortium blockchain is not as efficient as application programs with centralized architecture in data storage speed, it's still applicable to government data sharing among government departments from the perspectives of data security, data traceability, scalability, etc. The specific performance is as follows:

(1)   Data are not disclosed by default. The data in the consortium blockchain is only accessible to members and users within the alliance, which to a large extent ensures the privacy of government data transfer between departments.
(2)   Partial decentralization. The consortium blockchain only belongs to the within the alliance and is easy to reach consensus.
(3)   Strong controllability. As long as most departments in the alliance reach a consensus, the block data can be changed.
(4)   Fast trading speed. Due to the small number of department members, it's easy to reach consensus and the transaction speed is faster.

The characteristics of "ledger sharing" and "information sharing" in the consortium blockchain system can change many key areas of public services, such as data storage, sharing and traceability, which are highly consistent with the government's goal of becoming increasingly open and transparent [18]. For government, blockchain technology not only means paperless office, optimization of efficiency and security of data sharing, but also means a series of changes from optimization of data management process to governance thinking [19]. Therefore, this paper will focus on the construction of privacy-preserving government data sharing framework based on consortium blockchain to realize the constraint, trust and consensus mechanism generated by data sharing among departments, namely consortium blockchain sharing mechanism.

Hyperledger is one of the basic frameworks for consortium blockchain implementation. It's an open source project initiated by the Linux Foundation in 2015, attracting the participation of many companies including Huawei, Tencent Cloud, IBM, Intel, Cisco, Oracle and others [3]. Fabric is one of several hyperledger projects. It released version 1.0GA in July 2017, officially entering an active state. Hyperledger is different from Bitcoin and Ethernet in the past. It realizes the separation of codes and distributed ledgers and is more suitable for hot plug program development. This paper makes a brief comparison between Ethereum and Fabric, which are currently relatively mature blockchain platforms, as shown in Table 1.

Table 1. Comparison between ethereum and fabric

	Ethereum	Fabric
Consensus	POW	PBFT
Decentralization	Completely decentralization	Partial de-centralization
Throughput	≥ 25 TPS	≥ 1000 TPS
Fault tolerance rate	50%	33%
Scalability	No	Yes
Pluggability	No	Yes

Secondly, hyperledger isn't a completely decentralized framework. It does not generate blocks when there is no transaction. The PBFT consensus protocol recommended by hyperledger also does not require a large amount of computational power access. It has higher operation efficiency and is more suitable for practical application of government data sharing. The main features are as follows:

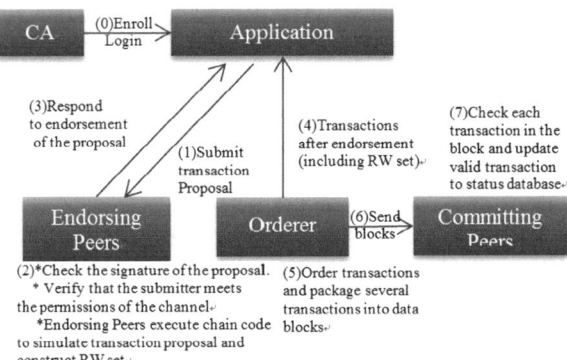

Fig. 1. Hyperledger fabric communication transaction flow

(1) Trust cost between peers is low. Due to the adoption of a distributed logical processing architecture, hyperledger separates each step. When a peer processes the current service, it only needs to process according to the constraints of the current step, thus greatly saving the trust cost between peers and improving the efficiency of the transmission and execution of transaction data. In the process of government data sharing, department members can only share data after obtaining permission.

(2) Members must be authenticated before joining. MSP (Membership Service Provider) mechanism is used in hyperledger to describe the identity of members in the network. Membership service is the concrete realization of MSP. In the process of government data sharing, the identity certificates held by department personnel are issued by specific CA. Department personnel need get permission to upload or download data, which is consistent with hyperledger's member permission mechanism.

(3) Modularization enhances the expansibility of the department. Hyperledger adopts modular architecture and supports hot plug and is more suitable for the transformation of the existing government information system. In order to access the blockchain, if the original mature government information system is torn down and replaced, a lot of costs will be wasted. However, hyperledger's modular architecture can complete the transformation of the software blockchain with less cost. Therefore, using modular processing methods can expand the access range of the data sharing alliance faster and reduce the cost of data sharing.

(4) Multi-channel characteristics. Hyperledger uses the concept of channel to isolate the information of the channel. This channel is dedicated to blockchain privacy preserving and allows the data on the channel to be isolated separately. In the sharing of government data, the same department can add different channels according to its own data need to facilitate data flow. Using Channel can solve the problems in the process of data request and call between different departments and ensures the confidentiality of direct participants. On the other hand, due to the close communication and high degree of data sharing among multiple departments participating in the same channel, the sharing efficiency of data among departments is greatly improved.

As shown in Fig. 1 above, the fabric contains the main components such as client, CA, peer and orderer. The CA can generate or cancel the identity certificate of the member. Any operation on the blockchain needs to be signed with the certificate associated with the user. Peers are mainly divided into endorsing peers and committing peer according to functions. Among them, the endorsing peers are used to execute proposal. It has not only the ability to record books, but to endorse in transactions. After obtaining the application's request, the endorsing peer will check the legality of the requested message. After the verification is passed, the endorsing peers execute a chaincode to simulate the transaction to realize a Proposal Response and return it to the application program. In order to improve the operation efficiency of the system, the endorsement policy is executed asynchronously and the application node will be in a waiting state after the request is issued. Only when the endorsement policy is met, that is, the endorsement quantity meeting the threshold is returned, then this endorsement process is finished and the broadcast transaction is considered valid. The ordering service node sorts the received transactions, packages the transactions sorted by the same channel into blocks and broadcasts the blocks to members in the channel. The function of the committing peers is to perform a series of checks on the transactions after the ordering service. Their status can always be consistent, because all the committing peers check the transactions in the same order and write the legal transactions into the account book in turn. In fabric, all peer nodes are accounting nodes. After receiving the block message, the accounting node verifies the validity of the block and submits it to the local ledger to update the blockchain data.

3 Government Data Sharing Framework for Privacy Preserving Based on Hyperledger Blockchain

The administrative functions of government departments have their own division of labor and business information systems are usually managed and operated separately. A large amount of business data collected or generated is owned by a specific department and stored in a department database or data warehouse. Data sharing among government departments involves the following issues:

(1) The data holder needs to let other data requester know what shared data it can provide and how to provide it;
(2) The data requester can make a query request according to the shareable data index of the data holder to obtain the required shared data;
(3) Privacy preserving processing of shared data, that is, assuming that the data requester is not completely trusted, the data holder desensitizes the source data in advance and then provides it to the data requester;
(4) Share data index and data sharing records to meet requirements such as decentralization, traceability and non-repudiation.

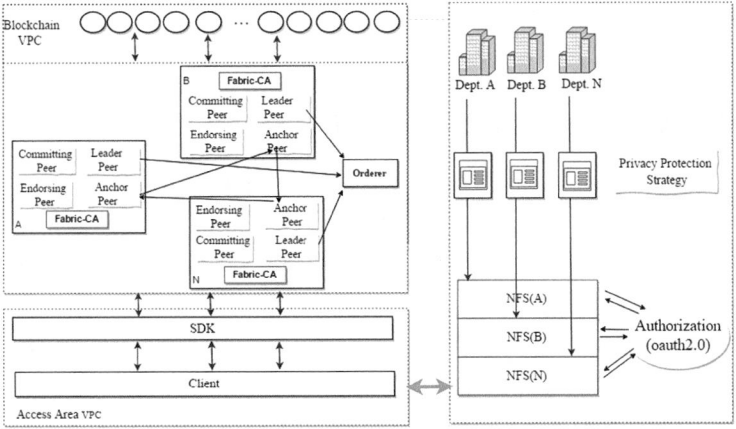

Fig. 2. Government data sharing framework for privacy preserving based on hyperledger

This paper mainly discusses the data sharing privacy preserving solution between government departments. In view of the above problems, combined with the idea of hyperledger and considering the privacy-preserving requirements, this paper constructs a privacy preserving government data sharing framework to strengthen mutual cooperation and promote government departments. The privacy preserving government data sharing framework is shown in Fig. 2. The blockchain is deployed in an independent VPC (Virtual Private Cloud), which does not occupy the VPC quota of users. At the same time, the VPC can be quickly connected with VPCs of other users, thus facilitating users to directly access their own organizations and nodes through their VPC without worrying about user network interconnection and subsequent expansion.

The client is an interactive interface that processes data interaction related operations. It consists of data index publishing, data query request, data sharing request, data download request and other modules. The client uses a centralized database for data storage and is centralized platform architecture. But in the data sharing phase, it's necessary to use the decentralized hyperledger architecture to share some of the data. Therefore, each user who logs in from the client is connected to a peer in the blockchain and uses the SDK method to interact with the blockchain. Members may invite new members to join the alliance. In each service channel in the network, participating members need to go through an admission mechanism to obtain read and write permissions and participate in accounting permissions on the blockchain.

In order to make the government data more adaptable and highly reliable in the context of interdepartmental sharing, based on the idea of "service on the chain, data under the chain", this paper publishes the data index to the blockchain for other departments to retrieve, while keeping sensitive data outside the blockchain. For sensitive data, the data holder may perform different forms of privacy processing on the requested data according to the requirements of the data requirement department and then upload it to NFS. "Privacy processing" is the privacy preserving process performed by the data providing department to protect sensitive information before sharing data, such as anonymization, de-identification and noise disturbance. NFS is a network file system that allows resources to be shared between computers on the network, which can greatly save local storage space. In order to further ensure the security of the data, the data requester needs to pass the OAUTH2.0 protocol authentication when downloading data to NFS and the data can be obtained only after the authentication succeeds.

The privacy preserving government data sharing framework aims to establish a data supply-demand correlation mechanism between government departments, Trust mechanism and privacy preserving mechanism. The blockchain records the detailed information such as the index file of the original data owned by each department, as well as the information of data sharing events with other departments, such as sharing the information of both sides, sharing the information of data description, sharing the time of events and so on. This information defines the ownership and shared use of data resources. When disputes arise over the existence of shared data, the information can act as a "witness" to resolve data accountability issues.

In general, we assume that there is no privacy leak in the data collection process. After the new data collection is completed, the collection department will expand the index file of the collected data to the blockchain. When data sharing is required, first, the data requester will query based on the blockchain information to clarify the department in which the required information is located. The data sharing request is then made to the data holder. When a data holder receives a data sharing request, According to the sensitivity of the requested data, a certain amount of privacy processing is carried out. The privacy-processed data is stored in the NFS and shared with the data requester. Blockchain's use of data cannot be tampered with, while ensuring the identity of each peer, speed up the information query and improve the efficiency. Moreover, the use of privacy-preserving technology to ensure the security of data can reduce the risk of privacy disclosure, indicating that the framework design has a certain degree of rationality.

4 Government Data Sharing Process for Privacy Preserving Based on Hyperledger Blockchain

In view of the existing problems of government data sharing, this paper constructs a privacy preserving government data sharing framework based on Hyperledger. Meanwhile, the corresponding privacy preserving government data sharing process has also changed. Before introducing the government data sharing privacy preserving process, it is first necessary to clarify that when the government department collects a new dataset, the metadata, acquisition time, collection department and other information of the data set will be packaged into the corresponding block in the blockchain. The sharing process between government departments is shown in Fig. 3.

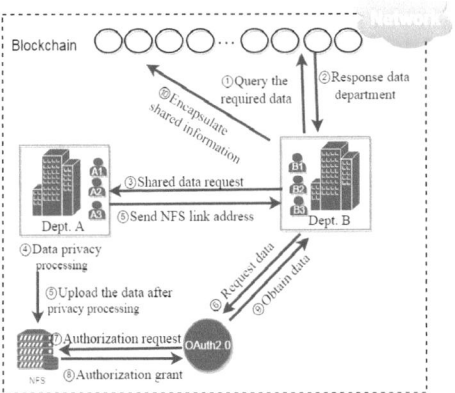

Fig. 3. Government data sharing process for privacy preserving based on hyperledger

This paper assumes that there is no privacy leakage in the data collection and storage process. Data sharing between departments is implemented using channels. Because different members in the department have different functions, the department can join different channels according to their different functions. Members only need to maintain the ledger information of the channels they join. For members in the channel, the communication is transparent, but other peers in the network aren't aware of the data communication within the channel, which can not only ensure the identity of users in the channel but also enable each user to trust each other, at the same time, it can also ensure the structural characteristics of the data to alleviate the problems of data structure chaos, thus solving the "data barrier" phenomenon. Based on this, the framework can also query according to the needs of the data requestor, accurately obtain the department where the required data is located and make requests for shared data.

It is assumed that members A1, A2, A3 in government department A and members B1, B2 and B3 in government department B are added to channel C due to work requirements. In order to share the information transparently, the data request will generate a data record in the initial stage, which will be continuously updated until the end of the event transaction. The recorded data items include the event transaction number (TxNum), the shared data information description (Metadata) and the data requester (Requester), the data holder (Holder), the current status of the event (CurStatus) and the time corresponding to the status.

Assuming that government department A needs to share the data of department B, the privacy preserving process for sharing government data between government departments based on the blockchain can be described as follows:

(1) The data requirement department A is clear about the data that it wants to request. A1, a member of the department A, logs on to the client and initiates the data query request in blockchain through SDK, at which time the record items include: {TxNum, Metadata, Requester, T_{query}, CurStatus}; with T_{query} is the query request time; CurStatus: A1 Queried.

(2) If the request initiated by A1 is legal and the information of querying data exists in the index directory, the consortium blockchain will return to the data providing department of the data required by A1; At this time, the record item is updated to {TxNum, Metadata, Requester, Holder, T_{query}, CurStatus}; with Holder is the data holder side; CurStatus: A1 requested.

(3) Member A1 in the data demand department A sends a data sharing request to the data providing department B. At this time, the record item is updated to {TxNum, Metadata, Requester, Holder, T_{query}, $T_{request}$, CurStatus}; with $T_{request}$ is the data sharing request time; CurStatus: A1 requested.

(4) Member B1 in the data providing department B selects appropriate privacy preserving technology to process the data to be provided according to the data sharing request sent by A1 and the privacy preserving policy of the department, uploads the privacy data to NFS and sends the NFS link address to A1 to ensure the availability and privacy of the shared data. At this time, the record item is updated to{TxNum, Metadata, Requester, Holder, T_{query}, $T_{request}$, $T_{B\_response}$, CurStatus}; with $T_{B\_response}$ is the time when the data holder responds to the data sharing request; CurStatus: B1 sent NFS link address.

(5) A1 makes a data request to NFS through the link address. After authorization passes, A1 obtains the requested data. At this time, the record item is updated to {TxNum, Metadata, Requester, Holder, T_{query}, $T_{request}$, $T_{B\_response}$, T_{get}, CurStatus}; with T_{get} is the time at which the data requestor gets the data; CurStatus: A1 acquired.

(6) When the data state becomes acquired, the event will be triggered automatically and the data requester will encapsulate the shared event information (e.g. data sharing information, shared data description information, shared event occurrence time, etc.) and send it to the blockchain. At this time, the record item is updated to {TxNum, Metadata, Requester, Holder, T_{query}, $T_{request}$, $T_{B\_response}$, T_{get}, CurTime, CurStatus}; with CurTime is the time it takes to encapsulate the information; CurStatus: A1 acquired.

Each step executed above can be further refined. For the request of query class, its concrete details are as follows:

(1) Data request A1 initiates a data query request;
(2) Endorsing peers in the same channel (endorsement strategy designation) will carry out relevant verification after receiving A1 data query request. After verification, the endorsing peer executes the chaincode to obtain the data query result and signs the read-write set (data query result) and then sends it back to A1.

For query class requests, there is no need to submit them to the ordering service. If it is a request to update a ledger, taking publishing a data index as an example, the specific implementation steps are as follows:

(1) Member A1 in Data Demand Department A initiates a request to publish a data index;
(2) Endorsement members in the same channel carry out relevant verification after receiving A1's request for publishing data index; After verification, the endorsement member (endorsement policy designation) calls and executes the chaincode to obtain the read-write set, then signs the read-write set and sends it back to A1.
(3) After collecting the number of endorsement member responses satisfying the endorsement policy, A1 sends the read-write set obtained in the endorsement request, the signatures and channel numbers of all endorsement members to the orderer;
(4) The ordering services sort the request transactions according to the channel number and then pack the request transactions in the same channel into a data block and broadcast it to the members in the channel;
(5) After receiving the data blocks sent by the orderer, committing peers (non-endorsing peers) check the requested transactions one by one and mark "valid" and "invalid", then package them into blocks and write them into the blockchain ledger.
(6) Accounting members (all members in the channel, namely A1, A2, A3, B1, B2, B3) verify the validity of the block and submit it to the local ledger to complete the account function and update the blockchain data.

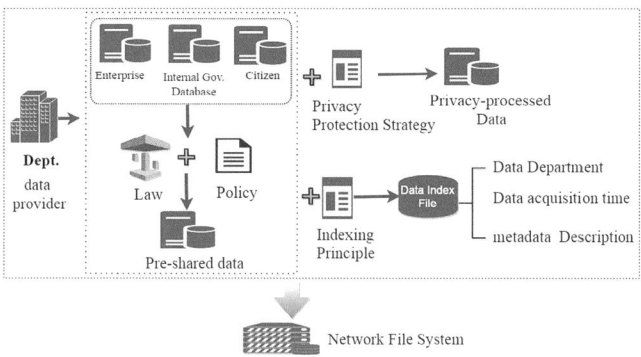

Fig. 4. Data index processing and privacy preserving processing flow

In order to process data efficiently, each government department stores the original data in the local database and the shared data after privacy processing is stored in NFS. The data indexing process and privacy preserving process are shown in Fig. 4. This process mainly introduces the privacy and indexing process of data. Among them, the index file after data indexing is uploaded to the blockchain through the client, so that the blockchain records the data directories stored in various departments for other departments to query. The data index file mainly contains detailed information such as the department to which the data belongs, the data collection time, the available forms of data and metadata description. According to the existing laws, regulations and policies, the data providing department pre-screens the collected data to obtain pre-shared data. After receiving the data sharing requests from other departments, the data providing department can process different types of data in different degrees and forms according to the requirements of the data demand department. For example, for table data to be shared, privacy preserving technologies such as greedy clustering anonymity, multidimensional sensitive k-anonymity and the like can be adopted. For statistical data to be shared, differential privacy [20] or local differential privacy technology can be used for processing. The processed data has high security, thus speeding up information circulation and reducing risks caused by information asymmetry.

5   Superiority Analysis of Government Data Sharing Framework for Privacy Preserving Based on Hyperledger

The framework established has certain advantages in solving the problems existing in government data sharing. It can solve some problems faced by traditional government data sharing, such as disjunction between supply and demand, lack of trust mechanism, low sharing efficiency and privacy disclosure risk. It provides new ideas for solving the problems. Its frame advantages are shown in Table 2:

(1) It is easier for participants to share data. Since the setting of government departments is predetermined and the number is limited, it is suitable to establish a departmental consortium blockchain. Each government department can participate in one or more channels manage one or more peers and don't need a reliable third-party central peer and takes non-repudiation responsibility for the shared data provided. At the same time, the modular design of the framework provides better scalability.

(2) Data sharing scope and efficiency are higher. The government department consortium blockchain is only owned by each government department within the alliance and has the characteristics of strong controllability. It can not only realize point-to-point data information sharing and exchange among multi-departments, but also realize backup and traceability management of data information, thus greatly increasing the scope and efficiency of government information sharing.

(3) The sharing of personalized privacy preserving for data is realized and the data has higher security. Each government department formulates its own privacy preserving policy according to the data sharing regulations. According to the type, sensitivity and difference of the data requestor to be shared, appropriate privacy

preserving technology is selected to desensitize the data to be shared. When attacked or data leaked, the data security can still be guaranteed and the data security is higher.

(4) The data are traceable. The data directories of governments departments are stored on the blockchain and when the data requester conducts information inquiry, the departments where the corresponding information is located can be quickly inquired. The tracing characteristics of the blockchain and the continuous updating of the records mentioned in this paper make the basis of government decision more clear, so that there is evidence to follow. Each request is recorded on the chain and seen by other members, so it's not feasible to make data requests at will.

(5) Further safety protection. Hyperledger is used to reinforce the known blockchain attacks, such as witch attack, overflow attack and eclipse attack.

Through the introduction of the government data sharing framework for privacy preserving based on hyperledger, combined with the characteristics of hyperledger and the advantages of privacy preserving processing, it clearly shows that the use of hyperledger can prevent data from being tampered with, ensure the identity of each peer, improve the reliability and authenticity of data, speed up information query and improve work efficiency. At the same time, the use of privacy preserving technology ensures the safety of data and reduces the risk of privacy disclosure. It further shows that the design of the framework is reasonable.

Table 2. Comparative analysis between the framework and traditional blockchain

	Traditional blockchain	This framework and its advantages
Credit environment Between Government Departments	It is more suitable for the scene with no credit relationship at all	This model can share data in a weak credit environment; Modular design makes participation more convenient; Personalized privacy preserving can be realized in data sharing
Data security	① Only one-to-one transmission can be carried out and the demand cannot be met; ② The data on the chain is transparent and all data can be seen when entering the system; ③ There are hidden dangers in the design of block chain technology, such as eclipse attack, witch attack	① For the complicated departments and agencies within the government, visit control is indirectly realized through channels; ② All the shared data are handled go through privacy preserving, which can still ensure data security in case of attack or data leakage; ③ The solar eclipse attack and witch attack existing in the traditional block chain are solved
Data storage	① It is impossible to prevent malicious nodes or requires a large amount of computational maintenance to keep nodes from doing evil; ② All data are recorded on the chain	① Each request is recorded on the chain and seen by other members, so it is not feasible to make data requests at will; ② Only the data index is uploaded to the block chain and sensitive data is still under the chain

6   Conclusion

The characteristics of blockchain technology, such as decentralization, non-tampering and information desensitization conform to the needs of information sharing and trust building in e-government. Therefore, the blockchain technology is applied to e-government to provide trust endorsement through data traceability and full records. Using hyperledger's blockchain architecture is safer than public blockchain and the decentralized alliance mechanism ensures the rights and interests of all users on the platform. The application of channels also makes data sharing more efficient and data docking clearer, which promotes the rapid development of data sharing. Although the government data sharing service based on hyperledger meets various needs of government, many practical problems will be encountered in the implementation process. For example, the current fabric has a low transaction per second and there may be bottlenecks for highly concurrent event processing. This needs to be improved in future research and practice to promote its application in government data sharing.

References

1. Otjacques, B., Hitzelberger, P., Feltz, F.: Interoperability of E-government information systems: issues of identification and data sharing. J. Manag. Inf. Syst. **23**(4), 29–51 (2007)
2. Welch, E.W., Feeney, M.K., Park, C.H.: Determinants of data sharing in U.S. city governments. Gov. Inf. Q. **33**(3), 393–403 (2016)
3. Ministry of Industry and Information Technology Information Center: 2018 White Paper on China's Block Chain Industry [J/OL], 15 2018. http://www.miit.gov.cn/n1146290/n1146402/n1146445
4. Willard, M., Maharaj, M.S.: Privacy, security, trust, risk and optimism bias in e-government use: the case of two Southern African development community countries. S. Afr. J. Inf. Manag. **21**, 1–9 (2019)
5. Huang, R., Chen, C.: Cooperation model of US government data opening and sharing. Libr. Inf. Work. **60**(19), 6–14 (2016)
6. Sarathy, R., Muralidhar, K.: Secure and useful data sharing. Decis. Support Syst. **42**(1), 204–220 (2004)
7. Clarke, S.: Reducing the impact of cyberthreats with robust data governance. Comput. Fraud. Secur. **2016**(07), 12–15 (2016)
8. Liu, Y.: Research on government affairs openness in big data environment. Inf. Syst. Eng. **5**, 146–147 (2018)
9. Sun, L., Li, G.J.: The research based on OGD application analysis model. Libr. Inf. Serv. **3**, 97–108 (2017)
10. Nakamoto, S.: Bitcoin: a peer-to-peer electronic cash system (2008). Maxwell, J.C.: A Treatise on Electricity and Magnetism, 3rd edn, vol. 2, pp. 68–73. Clarendon, Oxford (1892)
11. Tan, H., Zhou, T., Zhao, H., Zhao, Z., Wang, W., Zhang, Z., et al. File data protection and sharing method based on block chain [J/OL]. J. Softw. 1–15 (2019). https://doi.org/10.13328/j.cnki.jos.005770
12. Chuan, X., An, G., Liu, J.: Research on the construction of resource sharing service model of university libraries based on block chain theory. Libr. Res. **49**(03), 56–62 (2019)
13. Taylor, S.: Distributed ledger technology: beyond blockchain, 41. Government Office for Science, UK (2016)

14. Xia, Q., Sifa, E.B., Asamoah, K.O., et al.: MeDShare: trust-less medical data sharing among cloud service providers via blockchain [J/OL]. IEEE Access **5**(99), 14757–14767 (2017)
15. Jin, Y., Danny, X., Wang, G., Zeng, Z.: Research on e-government big data security sharing based on block chain. Inf. Secur. Res. 11 (2018)
16. Yimin, Yu., Chen, T., Duan, Z., et al.: Research on sharing model of government information resources based on block chain. E-Government **04**, 58–67 (2019)
17. Ojo, A., Adebayo, S.: Blockchain as a next generation government information infrastructure: a review of initiatives in D5 countries. In: Ojo, A., Millard, J. (eds.) Government 3.0 – Next Generation Government Technology Infrastructure and Services, pp. 283–298. Springer (2017)
18. Wang, M.L., Lu, J.Y.: Research on block chain technology and its application in government governance. e-Government **02**, 2–14 (2018)
19. Zhang yi, Xiao congli, ning xiaojing. the influence of block chain technology on government governance innovation [J]. e-government. 2016(12)
20. Piao, Ch.H., Shi, Y.J., Yan, J.Q., Zhang, Ch.Y., Liu, L.P.: Privacy-preserving governmental data publishing: a fog-computing-based differential privacy approach. Futur. Gener. Comput. Syst. **90**, 158–174 (2019)

Deep Learning for Trust-Related Attacks Detection in Social Internet of Things

Mariam Masmoudi[1](✉), Wafa Abdelghani[2], Ikram Amous[1], and Florence Sèdes[2]

[1] MIRACL Laboratory, Sfax University, Sfax, Tunisia
mariam.masmoudi19@gmail.com, ikram.amous@enetcom.usf.tn
[2] IRIT Laboratory, Paul Sabatier University, Toulouse, France
{abdelghani.wafa,sedes}@irit.fr

Abstract. Social Internet of Things (SIoT) is a new paradigm where the Internet of Things (IoT) is merged with social networks, allowing objects to establish autonomous social relationships. However, face to this new paradigm, users remain suspicious. They fear the violation of their privacy and revelation of their personal information. Without reliable mechanisms to enhance trustworthy communications between nodes, SIoT will not reach sufficient popularity to be considered as a leading technology. Hence, trust management becomes a major challenge to ensure qualified services and guaranteed security.

Several works in the literature have tried to diagnose this problem. They proposed various trust evaluation models based on different features and aggregation methods, aiming to classify benign nodes of the SIoT network. However, related works did not allow to detect malicious nodes and couldn't identify their types of attacks.

As a result, we suggest a new trust-evaluation model in a deep learning framework. This model permits to find out the type of trust-related attacks performed by malicious nodes, which will be isolated from the network in order to achieve a reliable environment. Based on authentic data, experimentation is able to prove our system performance.

Keywords: Internet of Things (IoT) · Social Internet of Things (SIoT) · Trust-evaluation model · Trust-related attacks · Deep learning · Multi-Layer Perceptron (MLP)

1 Introduction

Internet with its multiple assets and capabilities have given rise to the Internet of Objects, also known as Internet of Things (IoT); a true revolution in the world of information, communication and technology. IoT corresponds to a set of interrelated objects, which communicate with each other via the internet network in which individuals will be able to remotely control their data and devices.

© Springer Nature Switzerland AG 2020
K.-M. Chao et al. (Eds.): ICEBE 2019, LNDECT 41, pp. 389–404, 2020.
https://doi.org/10.1007/978-3-030-34986-8_28

Integrating the social component into the IoT gives birth to a new paradigm called Social Internet of Things (SIoT), which offers a variety of attractive services and applications. However, face to this emerging paradigm, users remain cautious and wary. They fear the revelation of their personal data and the violation of their privacy.

As a result, trust becomes one of the main obstacles to the adoption of this new model.

In SIoT, it is very important to know to whom we should trust and whom we should not trust [2]. Indeed, a malicious node can perform a different trust-related attack in a system, which can transmit incorrect or malicious information leading to an inappropriate change in the system.

Several works in the literature have tried to deal with this problem. The proposed trust-evaluation models neither allow to detect malicious nodes nor identify their types of attacks. Indeed, detecting malicious nodes by specifying the type of attack is a complex problem, which requires an in-depth analysis. We propose to apply a deep learning technique. This choice is based on the fact that using deep learning methods have yielded satisfactory results in some sub-areas related to Social IoT, such as Recommendation Systems and Social Networks.

The major contributions of this article are as follows:

- In this paper, we propose a new system designed to detect malicious nodes in order to isolate them and to obtain a reliable system. Indeed, some types of attacks may be more dangerous than others, depending on the context of application.
- It is therefore crucial to detect not only the malicious nodes but also the types of attacks they have made.
- This work lies in the continuation of our previous works [1]. Our goal is to improve it based on deep learning techniques.

The remainder of the paper is organized as follows. The key concepts of our work, such as SIoT, trust attacks and trust management mechanism are defined in Sect. 2. Then, we survey related works in Sect. 3. In Sect. 4, we detail the proposed trust-evaluation model. The experimentation results are discussed in Sect. 5. Finally, Sect. 6 concludes the paper and outlines our future works.

2   Background

In this section, we begin with the definition of the SIoT paradigm. Later, we define the trust concept and the notions that revolve around this term.

2.1   Trust in SIoT

Trust is a concept used in myriad disciplines. It is a complex term that has no common definition. Its definition depends on the field of study of the authors. There is therefore a rich and a growing literature on this concept.

The Oxford dictionary defines trust as "a firm belief in the reliability, truth, ability, or strength of someone or something" [2]. It can be also seen as a basic fact in human life [14]. Some authors view trust as a decision in relation to an individual perception while others perceive it as the prediction of behaviors [13]. Indeed, trust is a complex concept comprising not only security but also goodness, reliability etc.

According to [5], apart from its importance in everyday life, trust is a useful concept for applications and services in a networking system, like SIoT.

Trust is a relationship including at least two entities: a "trustor" entity and a "trustee" entity [11]. The former represents an entity that is supposed to initiate an interaction with another entity, while the latter is the second entity that provides the necessary information to the trustor at its request [12].

We adopt the following definition "Trust is defined as a belief of a trustor in a trustee that the trustee will provide or accomplish a trust goal as trustors expectation" [1].

2.2 Trust Attacks in SIoT

An attack is a malicious behavior performed by a malicious node to reach a variety of malevolent purposes. In this work, we are interested in trust-based attacks, in which a malicious node can disturb the system to reduce its effectiveness.

A malicious IoT node can execute the following attacks [11]:

- **Self-Promoting Attack (SPA):** A node can promote its reputation (by providing good recommendations to itself) to be chosen as a service provider.
- **Bad-Mouthing Attack (BMA):** A node can ruin the reputation of other benevolent nodes to decrease their chances of being selected as service providers.
- **Ballot-Stuffing Attack (BSA):** A node can boost the reputation of other malicious nodes to increase their chances of being selected as service providers.
- **Discriminatory Attack (DA):** A malicious node can discriminatorily attack nodes with which it does not maintain a social relationship because of human nature or propensity toward strangers.
- **Opportunistic Service Attack (OSA):** A malicious node can provide good services to gain a high reputation, then start providing bad services or collaborate with other malicious nodes to perform BSA or BMA attacks. We do not focus on the OSA attack in this work because at the beginning, nodes have a good behavior and cannot be detected. That's to say, they will be detected as soon as they do one of the mentioned attacks.

2.3 Trust Management Mechanism

Some works in the literature focused on the development of trust-management mechanisms to manage the establishment, propagation, storage and update of trust. According to [1], the architecture of the trust management mechanism is shown in Fig. 1.

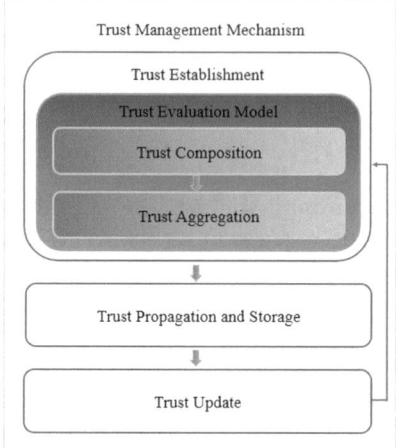

Fig. 1. Trust management mechanism

In this work, we concentrate only on the trust establishment step. The latter, the main component of trust management mechanisms, develops a trust evaluation model. In future works, we will focus on the other steps.

A trust evaluation model consists of two major phases.

- **The composition phase:** consists in selecting the features to be considered in trust values computing, such as honesty, cooperativeness, similarity of interests or intelligence, etc.
- **The aggregation phase:** consists in choosing a method to combine the values of different considered features to come up with the final trust value. For this, several methods are used, like the weighted mean, the fuzzy logic, machine learning, etc.

3   Related Work

Several works in the literature have recently addressed the trust management problem. In this section, we give a glance at the different trust-evaluation models that have been proposed in the literature. These models are classified according to the used aggregation method.

3.1   Aggregation Based on Combinatorial Logic

[10] proposed a trust management protocol that is composed of three factors: honesty, cooperativeness, and community interests. This protocol allows new nodes to quickly establish trust relationships with other nodes and to survive in insecure environments. The authors did not propose a specific method for aggregating these factors, but they used the combinatorial logic.

3.2   Aggregation Based on the Weighted Mean

[6] presented two trust management models. A subjective trust model that is similar to the social network, and an objective trust model that is inspired by peer-to-peer networks. The subjective model factors are associated with centrality in the social network, direct experience and indirect experience. Objective factors however are centrality in the peer-to-peer network, short-term observations, and long-term observations. All these factors are aggregated by a weighted mean. Both models help to establish trust between objects and isolate the malicious nodes from the network.

[8] offered a trust management model composed of a set of factors that are distributed in four dimensions: The QoS dimension refers to the quality of the provided service by the device. The security dimension points to security settings, such as the security communication protocol, the range of the network, or the device intelligence. The reputation dimension refers to the recommendations (scores or opinions) given by the other devices. The social dimension is calculated based on the type of relationship between the nodes (parental relation, proximity relation,). These factors are aggregated with a weighted average. This model allows reliable communication between the nodes.

3.3   Aggregation Based on the Multi-criteria Utility Theory

[7] suggested a trust model that incorporates five trust metrics, namely recommendation, Reputation and Knowledge, user preferences, and the context of the service. All these factors are aggregated by the multi-criteria utility theory. This model is used to evaluate trust between two SIoT entities before exchanging data and offering services to improve service quality and experience.

3.4   Aggregation Based on D-S Theory

[9] came up with a trust model that is composed of a variety of trust factors related to node behaviors, such as packet forwarding capacity, repetition rate, consistency of the packet content, delay, and integrity. They used the D-S theory to aggregate all these factors. This method helps to face attacks and inappropriate behaviors and to ensure all nodes reliability.

3.5   Aggregation Based on Machine Learning

[12] relied on a trust calculation model made up of five factors: (i) Co-work relationship, (ii) Cooperativeness-Frequency-Duration, (iii) Reward, (iv) Mutuality and centrality and finally (v) Co-location relationship. They used SVM, a supervised machine learning algorithm to combine these factors. This work classifies the interactions between objects in terms of either reliable or unreliable.

[1] recommended a trust model composed of six factors: (i) Reputation, (ii) Honesty, (iii) Quality of provider, (iv) Users similarity (v) Direct experience and (vi) Rating frequency. They used perceptron, a supervised machine learning algorithm, to aggregate these factors. This model is able to detect the malicious nodes.

3.6   Synthesis

Different trust models were suggested in the literature to provide reliable services and interactions in SIoT environments. In this section, we try to compare and analyze these different models in Table 1 according to two criteria:

(i) the proposed trust-evaluation model.
(ii) the goal of the proposed models.

We notice that the majority of the proposed models aim to classify benign nodes according to the trust degree of services. They did not allow neither to detect the malicious node nor to identify their attacks types. The models presented by [10] and [1] were able to detect attacks without identifying its types. Whereas, other works like [7] could not detect any type of trust-related attacks.

We know that an attack may be more dangerous than another depending on the context in which it is applied. The detection of malicious nodes by specifying the types of trust-related attacks is an intricate problem that requires a deep analysis of node behaviors.

Deep learning process has yielded satisfactory results in the analysis of classical social networks and other measure trust works. Since SIoT is a combination of social networks and IoT, we suggest to use deep learning technique to aggregate the proposed features in previous works.

4   Trust Evaluation Model

In SIoT, it is very important to know to whom we should trust and whom we should not trust [2]. Indeed, a malicious node can transmit incorrect or malicious information causing an inappropriate modification in the system.

A trust evaluation model is registered in the system in which it must fulfill the role of the guarantor of worthiness. This reliability is compromised by the different types of trust-related attacks. For all these reasons, we propose a new system that aims not only to detect whether a node has made an attack or not, but also to determine what type of attack it is.

Thus, we will present our trust evaluation model and as we know that a trust evaluation model is composed of two phases: (1) Composition Phase, (2) Aggregation Phase. In Fig. 2, we expose the architecture of our system.

We have as input a SIoT network composed by users with their social relationships, devices, services and interactions.

In the first composition phase, we select the features that allow us to detect the types of interactions between users, then we generate a data-set based on features as attributes and types of interactions as classification class.

The second aggregation phase consists of applying a deep learning algorithm on the data-set to generate a model that classifies the interactions between users in five classes, either non-attack or one of the types of attacks: BMA, BSA, SPA and DA.

Table 1. Comparison of related works

	Trust evaluation model		Objective	
	Composition phase: Features	Aggregation phase	Attack detection	Node ranking
[10]	Honesty, Cooperativeness, Community-interest	Combinatory logic	×	
[6]	Centrality, Direct experience, Indirect Experience	Weighted mean		×
[8]	Quality of service, Security, Reputation, Social relationship	Weighted mean		×
[7]	Recommendation, Reputation, Knowledge, User preferences, Context of the service	Multi-criteria utility theory		×
[9]	Packet forwarding capacity, Repetition rate, The consistency of the packet content, The delay, The integrity	D-S theory		×
[12]	Co-work relationship, Cooperativeness-Frequency-Duration, Reward, Mutuality and centrality, Co-location relationship	Machine learning		×
[1]	Reputation, Honesty, Quality of provider, Users similarity, Direct experience, Rating frequency	Machine learning	×	

4.1 Composition Phase

This phase consists in defining the features that allow to measure the trust value. These features are derived from the informal description of each type of attack and permit to describe and quantify the behaviors of malicious nodes. We briefly present each feature as follows:

- **Reputation:** stands for the global reputation of a user ui in the network.
- **Honesty:** this feature represents whether a user is honest or not. An honest user is someone whose rates reflect his real opinion.
- **Quality of provider:** Quality of provider shows whether the services provided by the user ui present a good or bad Quality of Service.
- **Similarity:** refers to the similarity between the user ui and the user uj.
- **Rating-Frequency:** points to the frequency of rating attributed by the user ui to the user uj.

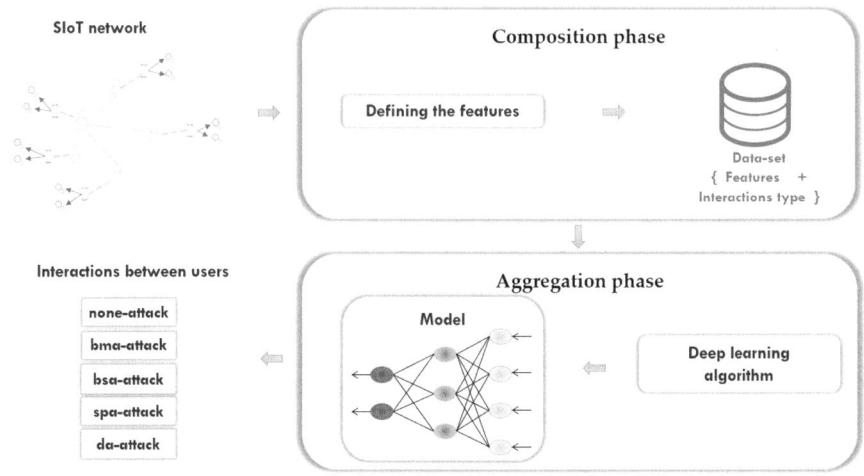

Fig. 2. System architecture

– **Direct-Experience:** is concerned with the opinion of a node i about its past interactions with a node j.

These features are detailed in the work proposed by [1].

4.2   Aggregation Phase

Once we have chosen the different features that describe the behavior of different nodes in the network, the next step consists in selecting a method to aggregate them. Our goal is to detect whether the user is either malicious or benign. Unlike the benign user, a malicious user is the one who performs one of the following attacks BMA, BSA, SPA or DA.

For this purpose, attack-detection systems are considered as a key technique to identify malicious nodes. We view our system as a classification problem. We establish a supervised learning process.

We will classify user's interactions into the following five classes:

(i) The bma-attack class, when a user performs a BMA attack.
(ii) The bsa-attack class, when a user performs a BSA-type attack.
(iii) The spa-attack class, when a user performs a SPA type attack.
(iv) The da-attack class, when a user performs a DA-type attack.
(v) The none-attack class, when a user has not performed any of the attacks listed.

The attributes are the trust features namely: Reputation, honesty, Quality of provider, similarity, Direct-Experience, Rating-Frequency and the classification class.

Each instance is as follows: [rep (Ui), rep (Uj), hon (Ui), hon (Uj), qop (Ui), qop (Uj), rateT (Ui), rateT (Uj), dExp (Ui; Uj), rateF (Ui; Uj), sim (Ui; Uj), class]; with Ui and Uj are users.

The proposed attack-detection system requires an in-depth analysis of node behaviors. We propose to use the deep learning technique to aggregate the proposed features as we mentioned in Synthesis Sect. 3.6.

Deep learning is the next-generation machine learning. It drives machine learning to the next level, through the use of artificial neural networks. Indeed, the latter simulate the way a human brain perceives, organizes and makes decisions from a set of data, such as images, videos, texts, speeches, etc.

We distinguish several types of neural networks, namely Multi-Layer Perceptron (MLP), Convolutional Neural Network (CNN) and Recurrent Neural Network (RNN).

The Convolutional Neural Network (CNN) is the most suitable type for image and video recognition. It can extract features automatically during learning. The Recurrent Neural Network (RNN) is used for tasks that involve sequential inputs such as speech and text [5]. He is able to learn and use the context for the prediction of the next word in a sequence. Whereas, the Multi-Layer Perceptron (MLP) is for classification and forecasting models. Moreover, the MLP is the most used for numerical data.

Since our learning data are neither images nor texts but are numerical data that are factors weights, we chose to apply Multi-Layer Perceptron (MLP).

The MLP[1] is structured in several successive layers related to each other. The output of one layer is the input of the next. Three different types make up these layers, namely the input layer, the output layer, and the hidden layer.

The Perceptron[2] computes the weighted average of its inputs that will be transmitted to an activation function to produce the output. For this, it is important to choose the best activation function.

The MLP entries are associated with the synaptic weights. To determine optimal weights, an algorithm named back-propagation (or Optimizer) is utilized. The latter is repeated thousands of times for each neuron of the neural network from the last layer to the first layer. In addition, the back-propagation algorithm calculates the error between the actual output and the output given by the network using an error calculation function in the first place. In the next step, it propagates this error backward by adjusting the weights in order to decrease it.

Figure 3 is a summary of what was mentioned in the paragraph above.

Thus, a deep learning model has a combination of values of the parameters to be fixed to create it and train it. These parameters are configured according to a series of experiments on all the data. All these parameters will be described in the next section Learning method.

---

[1] http://www.statsoft.fr/concepts-statistiques/reseaux-de-neurones-automatises/reseaux-de-neurones-automatises.htm#.XApVjGhKjIU.
[2] http://www.grappa.univ-lille3.fr/polys/apprentissage/sortie005.html.

The steps for creating the model are as follows:

– Model setting
– Generation of the corresponding model
– Training
– Save the model with the different parameters used.

The saved model will be used in the test phase to evaluate the performance of our system which classifies interactions between users to detect trust-related attacks. And we know that an attack may be more dangerous than another depending on the context in which it is applied. It would be attractive, in this case, to be able to figure out the type of the attack executed by a malicious nodes.

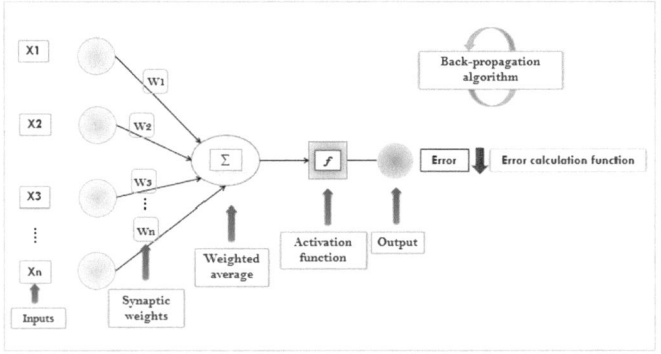

Fig. 3. Running of the multi-layer perceptron

5   Results and Evaluation

5.1   Experimental Setup

Data-set Description: Since real data are unavailable, several works in the literature offered experiments based on simulations. In our work, we evaluated the performance of our model based on simulations applied to a real data-set named "Sigcomm".

Sigcomm[3] data-set contains users, their profiles and their lists of interests. It also involves social relations between users, the interactions between them and the frequency of proximity of each couple of users.

We generate for each user one or more devices and we divide interactions of a user by his devices. Table 2 shows statistics and Fig. 4 shows description of the resulting data-set.

Based on the dataset, we performed simulations, whose purpose is to generate the different types of attacks according to their descriptions, namely BMA, BSA, SPA and DA.

---

[3] https://crawdad.org/thlab/sigcomm2009/20120715/.

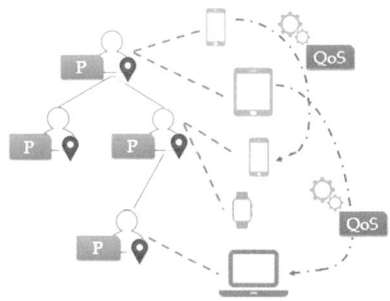

Fig. 4. Data-set description

Table 2. Statistics about Sigcomm data-set

Users	75
Users interests	711
Social relationships between users	531
Interactions between users	32000
Devices	300
Services	364
Proximities	285788

For example, in the case of a BMA attack:

- A malicious user Ui who has a bad reputation (Rep)Ui $= 0.2$ and provides two services Si,1 and Si,2 of poor quality (QoS)Si,1 $= 0.2$ and (QoS)Si,2 $= 0.1$.
- Uj has a good reputation (Rep)Uj $= 0.8$ and who provides three services Sj,1; Sj,2 and Sj,3 of good quality (QoS)Sj,1 $= 0.7$; (QoS)Sj,2 $= 0.8$ and (QoS)Sj,3 $= 0.9$.
- Ui will try to ruin the reputation of the legitimate user Uj by performing the BMA attack.
- Ui will interact with Uj. He will invoke different services provided by Uj and assign bad votes for each interaction.

Figure 5 shows description of BMA attack.

After simulations we have generated a CSV file, which includes 1613 instances. The following Table 3 shows statistics for each type of attack.

Evaluation Metric: To test the relevance of our classification model, we will apply the most commonly used evaluation measures in the classification problems, namely Recall, Precision and F-measure.

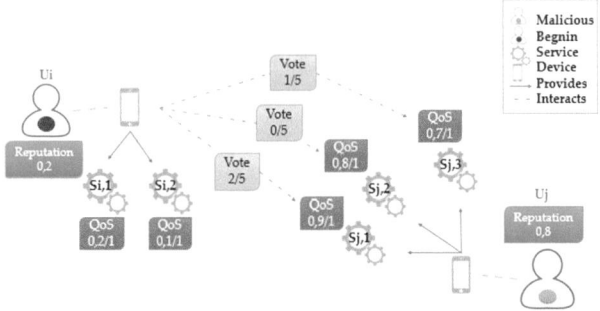

Fig. 5. Bad-mouthing attack (BMA)

Table 3. Statistics about the type of trust-related attack

Class	Number of instances
Attack-bma	150
Attack-bsa	154
Attack-da	149
Attack-spa	160
Attack-none	1000

Learning Method: We utilized Multi-Layer Perceptron (MLP) algorithm implemented in the PyCharm[4] tool based on Python, in order to build our multi-class classification function. To parameterize the MLP, we tried different values for each parameter. Table 4 reports the settings that give us the best performance.

5.2   Experimental Results

Evaluation for Each Type of Trust-Related Attack: These experiments are done to assess the resiliency of our system according to the type of trust-related attacks. Figure 6 proves that our system has the best performance in detecting both BSA and DA attacks.

Evaluation of All Types of Trust-Related Attack: This experimentation mixes all types of trust-related attacks. We evaluate the ability of our system to detect all types of trust-related attacks. Figure 7 indicates that our system gives satisfactory results both in terms of recall and precision.

---

[4] PyCharm is an integrated development environment (IDE). It is specific to the Python language and was developed by JetBrains. It is a multi platform that includes a built-in text editor, compiler and Python debugger.

Table 4. Multi layer perceptron settings

Name		Value
Input layer	*Number of neurons*	11
	Activation function	*Hyperbolic tangent*
Hidden layer	*Number of hidden layers*	10
	Number of neurons	*80*
	Activation function	*Rectified Linear Units (ReLu)*
Output layer	*Number of neurons*	*5*
	Activation function	*Sigmoïd*
Optimizer		Adam
Error calculation function		Categorical_crossEntropy
Epoch		25
Batch-size		10

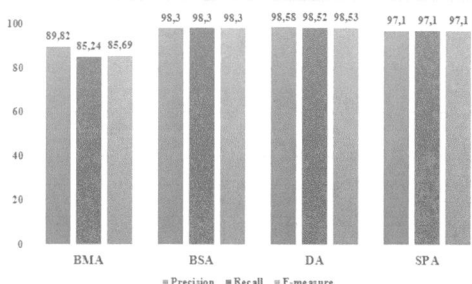

Fig. 6. Evaluation according to the type of attacks

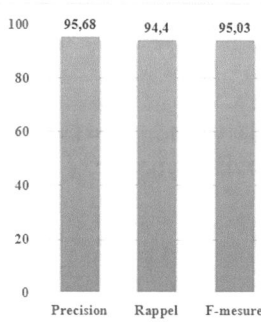

Fig. 7. Evaluation of the performance of our system facing all types of trust-related attacks

5.3   Comparison

In the work proposed by [1], the authors proved the relevance of the aggregation method they proposed (machine learning) compared to the most used method in the literature (the Weighted mean). The method as proposed does not allow to determine the type of performed attack, but only to detect whether there was an attack or not. So, they have two possible classes, namely (i) malicious user class, (ii) benign user class. While, our work allow not only to figure out the malicious or the benign user but to specify the type of attack performed by a malicious user too.

Hence, to prove the performance of our proposed aggregation method (Deep Learning more precisely, the Multi-Layer Perceptron (MLP)) we compare the obtained results with those attained from the Machine Learning technique.

Figure 8 displays the obtained results. Our proposed method gives better results in terms of Precision, Recall and F-measure than those given by [1].

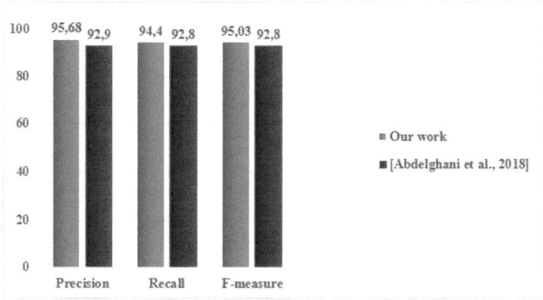

Fig. 8. Comparison with related works

According to the aforementioned results, we can notice that our solution improves the work proposed by [1] with 1.6% in terms of Recall and 2.78% in terms of Precision. Therefore, we can conclude that experimental outcomes indicate that the proposed model will improve the trust degree and the detection of malicious users depending on the type of the performed attack.

6   Conclusion and Perspectives

Social Internet of Things (SIoT) is a new paradigm where the Internet of Things (IoT) is merged with social networks, allowing objects to make autonomous social relationships. However, facing this new paradigm, users stay careful. They fear the revelation of their personal information. Hence, trust management becomes a major challenge to obtain a reliable environment.

To deal with this challenge, we proposed a new method based on deep learning technique designed to detect the type of trust-related attacks performed by malevolent nodes in order to isolate them and to achieve a trustworthy environment. This method is based on supervised learning.

According to the evaluation phase, we showed the performance of our attack detection system with an accuracy value of 95.63% and 94.4% as a recall value.

However, in this work we have discussed the level of trust related to user. The extension of this work would be the definition of a trust management system for device-level. This device level allows to measure trust for device node.

Another perspective as continuity to this work would be to treat the other stages of trust management mechanism, such as propagation, storage and updation of trust based on the proposed trust-evaluation model.

Acknowledgment. This work was financially supported by the PHC Utique program of the French Ministry of Foreign Affairs and Ministry of higher education and research and the Tunisian Ministry of higher education and scientific research in the CMCU project number 18G1431.

References

1. Abdelghani, W., Zayani, C.A., Amous, I., Sèdes, F.: Trust evaluation model for attack detection in social internet of things. In: International Conference on Risks and Security of Internet and Systems. Springer (2018)
2. Ramanathan, A.: A Multi-level Trust Management Scheme for the Internet of Things. University of Nevada, Las Vegas (2015)
3. Chen, R., Bao, F., Guo, J.: Trust-based service management for social internet of things systems. IEEE Trans. Dependable Secur. Comput. **13**(6), 684–696 (2016)
4. LeCun, Y., Bengio, Y., Hinton, G.: Deep learning. Nature **521**(7553), 436 (2015)
5. Lin, Z., Dong, L.: Clarifying trust in social internet of things. IEEE Trans. Knowl. Data Eng. **30**(2), 234–248 (2018)
6. Nitti, M., Girau, R., Atzori, L.: Trustworthiness management in the social internet of things. IEEE Trans. Knowl. Data Eng. **26**(5), 1253–1266 (2014)
7. Truong, N.B., Um, T.W., Lee, G.M.: A reputation and knowledge based trust service platform for trustworthy social internet of things. In: Innovations in Clouds, Internet and Networks (ICIN), Paris, France (2016)
8. Bernabe, J.B., Ramos, J.L.H., Gomez, A.F.S.: TACIoT: multidimensional trust-aware access control system for the internet of things. Soft. Comput. **20**(5), 1763–1779 (2016)
9. Yu, Y., Jia, Z., Tao, W., Xue, B., Lee, C.: An efficient trust evaluation scheme for node behavior detection in the internet of things. Wirel. Pers. Commun. **93**(2), 571–587 (2017)
10. Bao, F., Chen, R., Guo, J.: Scalable, adaptive and survivable trust management for community of interest based internet of things systems. In: IEEE Eleventh International Symposium on Autonomous Decentralized Systems (ISADS). IEEE (2013)
11. Abdelghani, W., Zayani, C.A., Amous, I., Sèdes, F.: Trust management in social internet of things: a survey. In: Conference on e-Business, e-Services and e-Society, pp. 430–441. Springer, Cham (2016)
12. Jayasinghe, U., Lee, G.M., Um, T.W., Shi, Q.: Machine learning based trust computational model for IoT services. IEEE Trans. Sustain. Comput. **4**, 39–52 (2018)

13. Gambetta, D.: Can we trust. Trust.: Mak. Break. Coop. Relat. **13**, 213–237 (2000)
14. Luhmann, N.: Trust and Power Chichester. Wiley, Hoboken (1979)
15. Kalaï, A., Zayani, C.A., Amous, I., Abdelghani, W., Sèdes, F.: Social collaborative service recommendation approach based on user's trust and domain-specific expertise. Future Gener. Comput. Syst. **80**, 355–367 (2018)

Clustering-Anonymity Method for Privacy Preserving Table Data Sharing

Liping Liu, Chunhui Piao[⊠], and Huirui Cao

Shijiazhuang Tiedao University, Shijiazhuang 050043, Hebei, China
pchls2011@126.com

Abstract. In the era of big data, the open sharing of government data has increasingly attracted the attention of governments. However, there is privacy leakage risk in the government's data sharing. For the scene of sharing the table data, this paper proposes a approach for privacy-preserving data sharing in this paper based on anonymity clustering. Firstly, we preprocess the data table, and the records in the table are clustered by k-mediods clustering algorithm. The data table is divided into multiple sub-tables according to the distance between records. Then, the data records in the sub-table are divided based on the information loss parameter value, and the anonymous table data is adjusted so that the sensitive attribute values in the equivalence class are different. Last, Laplace noises are added to the value of sensitive attribute to ensure the privacy of the shared data. Compared with the classical k-anonymous MDAV algorithm in execute time, information loss and information entropy, the experimental results show that the proposed algorithm can reduce the operating time, improve the privacy protection to some extent, and has certain availability from the three aspects.

Keywords: Data sharing · Table data · Privacy preserving · K-mediods clustering · K-anonymity

1 Introduction

With the construction and development of digital government, government data are increasing gradually, with larger scale and more types, showing the characteristics of diversification and complexity [1]. For a long time, the value of data cannot be make the best used owing to the widespread phenomenon of "information island" and "data barrier". The Government data sharing can transfer government information from one department to another, thus improving the phenomenon of "data island", making the full use of data and improving the quality of government services [2].

The main goal of privacy protection research in data sharing is to design related application models or tools so that the shared data can effectively protect individual private information in a hostile environment, and the data utility is not much loss, thus achieving the balance between privacy preserving and data availability [3]. And the table data sharing is an important way of data sharing [4]. A table contains multiple attribute information, and some attributes are sensitive needed to be processed to ensure personal protection privacy. Data anonymization is a general privacy protection

© Springer Nature Switzerland AG 2020
K.-M. Chao et al. (Eds.): ICEBE 2019, LNDECT 41, pp. 405–420, 2020.
https://doi.org/10.1007/978-3-030-34986-8_29

technology. The core idea is to use a certain method (generalization/concealment technology, clustering anonymity technology, replacement technology, etc.) to anonymity transform the original data set, and generate a new ones. That table makes the attacker unable to distinguish from the data perspective to ensure the privacy of the data.

This paper proposes a privacy preserving anonymity method based k-mediods clustering algorithm for table data sharing oriented to ensure data privacy while ensuring data privacy. Compared with the traditional anonymous algorithm MDAV [5], the experimental analysis shows that the proposed can improve the efficiency of the algorithm and provide effective privacy protection.

2   Introduction to Basic Technology

2.1   Analysis the Attention of Privacy Protection

Privacy is the information of being protected from unwanted access by others [6]. However, the development of information technology will inevitably enhance the possibility of data information disclosure, which in turn will limit the development of information technology [7]. Therefore, the privacy has draw a widely public attention. We take the number of the papers published related to privacy in a year as a measure to show the concern of people from all walks of life to the issues of "privacy" and "privacy protection" more intuitively. In this paper, "privacy" as the topic keyword, and then searched in the search results with "privacy protection" as subject keywords to search in HowNet. Figure 1 depicts the changes of the number of articles published each year since 1990, according to the results. Has increasingly concerned about the privacy issues and privacy protection issues, since2003, can be seen from Fig. 1. We can also conclude people focused much attention on the issue of privacy protection, which accounts for about half.

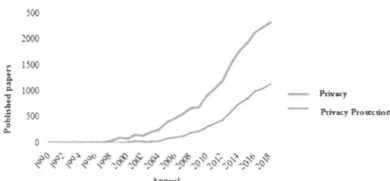

Fig. 1. "Privacy" and "privacy protection" concerns

2.2   Anonymous Technology

Anonymity technology [8] is an important method in the field of privacy protection. The data table divided into sensitive attributes and non-sensitive attributes.

The traditional privacy protection model makes too many assumptions about the attack model and the attacker's background knowledge, which leads to some defects

[9]. This problem was not well solved until the emergence of differential privacy technology.

If the sensitive attribute values in the equivalence class are identical in the anonymous data table, it is impossible to resist homogeneous attacks on anonymous tables by attackers based on background knowledge [10]. Moreover, when the attacker's background knowledge is large enough, there is also the risk of privacy disclosure [11].

2.3 Differential Privacy Method

Dwork, a Microsoft scholar, proposed a new privacy protection model-differential privacy firstly in 2006 [12]. The basic idea is to process data information by adding noise to the original data or statistical data, converting the original data, so that when adding or deleting a record, it will not affect the overall statistical attribute value, thus achieving the effect of privacy protection. The model can mitigate the maximum background attack risk, and defines a quantitative evaluation method for privacy protection level.

Definition 1 [13]. Let A be a randomized algorithm. Let D, D' be two datasets that differ in at most one entry (we call these database neighbors). Let $\epsilon > 0$. Define A to be ϵ-differentially private if for all neighboring databases D, D'. For final result O, we have

$$\Pr[A(D) = O] \leq e^{\varepsilon} \times \Pr[A(D') = O] \tag{1}$$

To achieve differential privacy protection, we need to add noise to the results of the query. Common noise adding mechanism are Laplace mechanism and exponential mechanism [14].

2.4 The K-Mediods Algorithm

K-mediods algorithm [15] is a classic algorithm to solve clustering problems. It is simple, fast, capable of processing large data sets, compact within clusters and distinct differences between clusters. The process is as follows:

(1) K cluster samples are randomly selected as initial cluster centers.
(2) Calculating the distance from the remaining sample points to each initial clustering center, then dividing the sample points into clusters with the shortest distance from the clustering center to form k clusters.
(3) Calculate the distance sum from the sample point to other points except the center point in each cluster. When the distance sum is the smallest, this point is selected as the new cluster center.
(4) If the new cluster center set is different from the original cluster center set, Step 2 is returned; if it is the same, the clustering algorithm ends.

3   Data Table Privacy Protection Clustering Anonymous Algorithm

3.1   Analysis the Attention of Privacy Protection

It is known the most widely used model is k- Anonymous model, simple structure and few restrictions. Further research is based on the k-Anonymous model for design and implementation [16], which is suitable for government departments to use. However, the k-Anonymous model unable to resist the maximum background knowledge attack, homogeneous attack, and re-identification attack. Based on the differential privacy can resist the biggest background knowledge attack and provides poor data availability [17]. The combination of the two can theoretically enhance the degree of privacy protection, also reduce the risk of privacy disclosure. In addition, because clustering algorithm has the characteristic of grouping data records, data records with high similarity can be assigned to a group to prepare for anonymous processing.

This research studies a privacy protection table data sharing method based on clustering anonymity to cope with the homogeneous attacks and background knowledge attacks. The method firstly divides data records according to the distance between table records, then constructs anonymous groups according to the amount of information loss, and generalizes anonymous groups so that they meet the k- anonymity condition, then performs differential privacy processing on sensitive attributes in anonymous table data. It should be noted that the algorithm proposed in this chapter is aimed at static data sharing scenarios.

(1) Partition of Table Data Records Based on k-mediods Clustering. According to the distance between records in the data table, k-mediods clustering algorithm is used to cluster the data records to obtain K clusters;

(2) Anonymous processing of table data. Each cluster obtained through Step 1 processing is processed separately. The data in the cluster is divided according to the loss of information, so that the number of records in each cluster after division is between k_2 and $2k_2$. Then adjusted so that the sensitive attribute values are not completely equal in each cluster. Finally, generalization processing is carried out on each cluster;

(3) Carry out differential privacy treatment on sensitive attribute values in table data. If the sensitive attribute is classified, add "Num" column and record the noisy statistical data value. If the sensitive attribute is numeric, cluster it and replace the attribute value with average value.

3.2   Initial Partition of Table Data Records Based on K-Mediods Clustering

Data table attributes can be divided into classified attributes and numerical attributes according to the representation of attribute values, and classified attributes can be divided into ordered classified attributes and unordered classified attributes. For example, the grades of "excellent", "good", "pass" and "fail" are classified in an orderly way. Gender "male" and "female" are unordered classification attributes.

During clustering, in order to better reflect the distance of data, the values of ordered classification attributes are quantized into numerical values in order and then treated as numerical attributes. Meanwhile, due to the different value ranges of each attribute, it has relatively great influence on the calculation of the table record distance, so it is necessary to normalize the numerical attribute values and the quantized ordered classification attribute values. The normalization formula is as follows:

$$x_i' = \frac{x_i - x_{min}}{x_{max} - x_{min}} \tag{2}$$

Where x_i is the original value of a numeric attribute, x_{min} is the minimum value of the attribute, and x_{max} is the maximum value of the attribute. This table data record division is to use clustering technology to divide n records in a data table into multiple clusters, so that records with high similarity are divided into one group. And in order to meet the following k-anonymity requirements, clusters do not meet the anonymity requirements need to be adjusted after clustering is completed. So, combined with k-mediods clustering algorithm, the specific flow of table data record division is as follows:

(1) Normalization. Quantifying the non-sensitive ordered classification attribute, then treating the attribute as a numerical attribute, and normalizing the numerical attribute data in all non-sensitive attributes in the data table;
(2) Division of clusters. According to the distance of non-sensitive attributes between table data records, k-mediods clustering algorithm is used to cluster the table data and divide the records into k_1 clusters;
(3) Adjust cluster record according to k- anonymous parameter k_2. If the number of data records in the divided clusters is greater than k_2, no adjustment will be made; If the number of data records in the obtained cluster C_i is less than k_2, the record closest to the center point of the cluster C_i is added to the cluster C_i while ensuring that the data records in the cluster where the record is located are still greater than K_2.
(4) Repeat Step 3 until the records in each cluster are greater than or equal to K_2;
(5) Divide the data into different subdatasheets according to different clusters.

The data table contains two types of attributes, classified attribute and numerical attribute. So when using clustering algorithm to divide records, different data distance calculation methods need to be adopted when calculating the distance between records. In addition, the problem of optimal clustering results, i.e. the selection of the optimal number of divided clusters k1, needs to be considered when carrying out clustering algorithms.

(1) Calculation formula of distance between data table records

Different attributes need to be calculated separately because there are many attributes in the data table. The formula for calculating the distance between numerical attributes is as follows:

Algorithm 1: Table Data record partition Algorithm

Input: Government Data Table T, Number of divided clusters $k1$, Anonymous

   budget k_2

Output: cluster

1.**Begin**
2. //Quantification the ordered classification attributes
3. **for** (i=1; i<=number ordered classification attributes; i++)
4. quantTD =quantify (ordered classification attribute[i])
5. **End for**
6. //standardization the numeric attributes
7. **for** (i=1; i<= Numeric attribute in non-sensitive attributes; i++)
8. initTD =init (the value of numeric attribute[i])
9. **End for**
10.clusters = getClustersByK-medoids(initTD, $k1$)
11. // adjustment the each cluster
12.**for** (i=1; i<=$k1$; i++)
13. **if** (cluster[i].size<k_2)
14. record.needNum=k_2-cluster[i].size;
15. **for** (j=1; j<=record.needNum; j++)
16. **if** (m != i & cluster[m].size > k_2)
17. NearRecordmin←min(dist (cluster[i].medoids, cluster[m].EachRecords))
18. **End if**
19. NearestRecord=min(NearRecord)
20. getNewCluster[i]=add(NearestRecord, Cluster[i])
21. delete the NearestRecord in the original cluster
22. **End for**
23. **End if**
24. **End for**
25. return NewCluster
26. **for each** cluster
27. create a table
28. **End for**
29. **End**

$$dist(x_i, x_j) = |x_i - x_j| \qquad (3)$$

The calculation formula for classified attributes is as follows:

$$\delta(x_i, x_j) = \begin{cases} 0 & (x_i = x_j) \\ 1 & (x_i \neq x_j) \end{cases} \qquad (4)$$

Assuming that there are m numeric attributes and n typed attributes in the data table, the distance calculation formula for any two records X_i and X_j in the data table is as follows:

$$d(X_i, X_j) = \sqrt{\sum_{p=1}^{m} (dist(x_{ip}, x_{jp}))^2} + \sqrt{\sum_{q=1}^{n} \delta(x_{iq}, x_{jq})} \tag{5}$$

Where x_{ip} and x_{jp} are the p-th numerical attribute values of record X_i and record X_j respectively, x_{ip} and x_{jp} are the q-th sub-type attribute values of record X_i and record X_j respectively.

Algorithm 2: Table Data Anonymity Algorithm
Input: Government Data **clusters**, Anonymous budget k_2
Output: Anonymous Data Table T

1. **Begin**
2. **Function (clusters,k_2)**
3. **foreach** cluster
4. **If** cluster.size>2k_2-1
5. **//Perform the Step 2**
6. r_1, r_2=max_loss(cluster)
7. **foreach** record in cluster except r_1 and r_2
8. loss1=calculate(cluster1,record)
9. loss2=calculate (cluster2,record)
10. distribute record ←min(loss1,loss2)
11. **End for**
12. get two Newclusters
13. adjustment the each cluster make each   Newcluster. Size>=k_2
14. **End if**
15. **Function (Newclusters,k_2)**
16. **End for**
17. **foreach** cluster
18. **If** each.Sensitive attribute.valueType==1
19. SenValue=Sensitive attribute.value
20. **foreach** cluster expect itself and cluster. Size=k_2
21. **If** cluster. Size>k_2 and   Num(cluster.sensitive.value!=SenValue)>2
22. select record ←different from SenValue
23. **End if**
24. **End for**
25. Add(cluster, record) ←min(loss(cluster, each record))
26. **End if**
27. **End for**
28. Anonymizing each cluster
29. Form an anonymous data table
30. **End**

---

(2) Determination of the number k1 of data record partition clusters

K-mediods clustering algorithm is used to divide similar records into groups, to prepare for anonymization and to minimize information loss caused by anonymization. When determining the number of clusters in a cluster, the similarity within the cluster is mainly considered. This method determine the number k_1 of data record partition clusters by the intra-group square error and SSE. However, with the increase of k_1, the data records in each cluster will gradually decrease, and the distance between records in the cluster should be smaller and smaller. The SSE value should decrease with the decrease of k_1. Besides, when determining the k_1 value through SSE, we should mainly pay more attention to its change. When SSE decreases relatively slowly with the increase of k_1, it is considered that there is little change in the clustering effect of further increasing k_1, then the k_1 value is the best clustering number. Therefore, if each k_1 value and the corresponding SSE value are expressed in a line chart, the corresponding k_1 value at the inflection point is the optimal clustering number.

3.3 Design of Table Data Anonymity Algorithm Based on Information Loss

Processing each sub-table in turn which are obtained after the table data record division process. Its core idea is to divide the data records in the sub-table so that the number of records in each cluster is between k2 and 2k2 − 1, while ensuring that the values of sensitive attributes in each cluster are not unique. Therefore, the specific process of table data anonymity algorithm is as follows:

(1) Determine whether the number of data records in the dataset is greater than $2k_2 - 1$, and if it is greater, then execute Step 2;

(2) Select two records such as r_1 and r_2 as the two initial clusters in the dataset. Which two records are combined into one cluster has the largest information loss among all the records two-two combinations. Then execute Step 3;

(3) Calculate the change of information loss separately after each record is divided into two clusters to divide the record into the cluster with less information loss, and adjust the data records so that the data records in each cluster are at least k_2. Then let two clusters as the new datasets and go back to Step 1.

(4) When the number of records in all datasets is between k_2 and $2k_2 - 1$, it is determined in turn whether there is a unique value of the sensitive attribute in each dataset. And if there exists, execute Step 5.

(5) Select data records in other datasets that are different from sensitive attribute values in the dataset Q. At the same time, it is must ensure that if the data record is deleted, the number of data records in the dataset is still big or equal to k_2, and the value of sensitive attributes is not unique.

(6) Calculate the change of information loss if the selected data record is divided into corresponding dataset Q, and divide the data record which makes the information loss smaller into dataset Q.

(7) Get each datasets with the number of records between $[k_2, 2k_2 - 1]$ and the value of sensitive attributes is not unique. Then each dataset is generalized to get anonymous data table.

> **Algorithm 3: Differential Privacy Algorithm for Table Data**
> Input: Anonymous Data Table T
> Output: Differential Privacy Anonymous Data Table
> 1. **Begin**
> 2. laplaceNoise ← getLaplaceNoise(Lapdatas)
> 3. **foreach** Classification Sensitive Attribute Value Combination
> 4. Frequency statistics ← different Classification Sensitive Attribute Value Combination
> 5. lapstatistics ← add(Frequency statistics, laplaceNoise)
> 6. **End for**
> 7. **foreach** Numerical Attribute[i]
> 8. clusters = getClustersByK-medoids(Attribute[i], k1)
> 9. **foreach** cluster
> 10. record.attribute[i]=mean.cluster.attribute[i]
> 11.**End for**
> 12. delete the duplicate records
> 13.**End for**
> 14.**Share**
> 15.**End**

We use generalization to achieve anonymity. The rule is that if a non-sensitive attributes is numerical attribute, it will be generalized to the attribute's range, such as if the dataset is {1 2 3 3}, it can be generalized to [1, 3]. And if it is a classified attribute, it can be generalized to a complement set of values in its set, such as if the dataset is {working in private enterprises, working in state-owned enterprises}, it can be generalized to {working in private enterprises/working in state-owned enterprises}. Formula (4)–(9) when anonymizing table data. The amount of information loss is also the basis when adjusting data records, that can minimize the loss of information and increase the availability of data.

3.4   Design of Privacy Protection Algorithms for Sensitive Attributes Based on Differential Privacy

After anonymous processing, many equivalent classes are generated, but because the value of sensitive attributes has not been processed, attackers can still infer the value of individual sensitive attributes through background knowledge attacks, thus causing privacy disclosure. Therefore, in order to better protect personal privacy, the sensitive attribute values are processed by differential privacy and noise, and then data sharing is performed. The specific steps are as follows:

(1) If there are classified attributes insensitive attributes, the frequency statistics are performed according to different combinations of classified sensitive attributes. Then added the "Num" column at the corresponding location to record the noised data;

(2) If there is a numerical attribute in the sensitive attribute, each numerical attribute is clustered separately, and the value is replaced by its average value. The number of clusters is $\lfloor n/k_3 \rfloor$, n is the number of records in the data table, k_3 is the number of data records in the cluster which can be determined according to the availability requirements.

4 Design and Analysis of Experiment

4.1 Data Set

In order to verify the availability and effectiveness of the data sharing algorithm based on clustering anonymity privacy protection, the real data set of Philippine family income and expenditure provided by Kaggle is used as an experimental data set in this paper [18]. The data set contains 41,544 pieces of citizen information, covering a variety of private data such as family information, family income and expenditure information, and property information. The table consisting of attributes serves as data table T to be shared. There are 5 attributes selected in this paper, such as Household Head Sex, the Household Head Marital Status, Household Head Age, Total Number of Family members and the Household Head Class of Worker. Among them, the gender of household head and the marital status of household head are classified attributes, the number of family members and the age of household head are numerical attributes, and the type of work of household head is treated as a sensitive attribute. The proposed algorithm is compared with the classical anonymous MDAV algorithm, and the performance, availability and privacy protection of the proposed method are analyzed.

4.2 Evaluation of Privacy Protection Algorithm

(1) Privacy measurement [19]

In the anonymous privacy protection, information entropy is usually used to measure the degree of privacy protection [20]. The degree of privacy protection is reflected according to the probability distribution of data records in the data table. The larger the entropy value, the more uniform the probability distribution of the data, the lower the probability of successful attack and the higher the degree of privacy protection. Conversely, the smaller the entropy value, the lower the degree of privacy protection [21]. The information entropy of the j-th sensitive attribute in the equivalence class C_i of anonymous data table is defined as follows:

$$H(C_i(S_j)) = -\sum_t (n_t/m) \log(n_t/m) \tag{6}$$

In the formula, m is the number of data records in the equivalent class C_i, and n_t is the number of t-th possible values of the j-th sensitive attribute in the equivalent class C_i. Therefore, the average information entropy of equivalence classes in anonymous data tables is defined as follows:

$$H(T) = \frac{\sum\limits_{i=1}^{k}\sum\limits_{j=1}^{n_1}\left[-\sum\limits_{t}(n_t/m)\log(n_t/m)\right]}{k} \tag{7}$$

(2) Availability Measurement

The amount of information loss is often used in anonymous privacy protection to measure data availability [22]. In quasi-identifying attributes, assuming that the number of numeric attributes is n_2 and the number of categorical attributes is n_3. The information loss of numeric attributes is defined as follows in equivalent class C_i.

$$Loss(C_i(A)) = |C_i| \times \sum_{j=1}^{n_2} \frac{\max(C_i(A_j)) - \min(C_i(A_j))}{\max(A_j) - \min(A_j)} \tag{8}$$

In the formula, $\max(C_i(A_j))$ represents the maximum value of the j-th numeric attribute in the equivalent class C_i, $\min(C_i(A_j))$ represents the minimum value of the j-th numeric attribute in the equivalent class C_i, $\max(A_j)$ represents the maximum value of the j-th numeric attribute in the data table, $\min(A_j)$ represents the minimum value of the j-th numeric attribute in the data table, $|C_i|$ represents the number of items recorded in the equivalent class C_i. Therefore, in data table T, the amount of information loss after all numeric attributes are anonymized can be written as:

$$Loss(T, A) = \sum_{i=1}^{k} Loss(C_i(A)) \tag{9}$$

In the formula, k is the number of equivalence classes, and the information loss amount of the classified attribute in the equivalence class C_i is defined as follows:

$$Loss(C_i(B)) = |C_i| \times \sum_{j=1}^{n_3} \frac{h(C_i(B_j))}{h(B_j)} \tag{10}$$

In the formula, 1 represents the number of different values of the j-th classified attribute in the equivalent class C_i, 2 represents the number of different values of the j-th classified attribute in the data table T. Therefore, in data table T, the amount of information loss after all numeric attributes are anonymized can be written as:

$$Loss(T, B) = \sum_{i=1}^{k} Loss(C_i(B)) \tag{11}$$

In summary, when the number of data records in the data table is N, the average information loss of the anonymous table T is:

$$Loss(T) = \frac{Loss(T,A) + Loss(T,B)}{(n_2 + n_3) \times N} \tag{12}$$

4.3 Analysis of Experimental Results

(1) The Effect of Different Conditions on Anonymous Processing Time

The time used in this method mainly comes from anonymity processing, and the average anonymity time can reflect the algorithm's performance. The data size can affect the anonymity processing time. And anonymous processing time mainly comes from the process of anonymous group construction and anonymous group adjustment. As shown in Fig. 2, it can be seen that with the increase of the data size, the average anonymous processing time increases gradually. This is because with the increase of data size, the number of iterations and adjustments may increase in the process of anonymous processing, which makes the processing time increase gradually. And compared with MDAV method, the method proposed in this paper has smaller processing time under different data size.

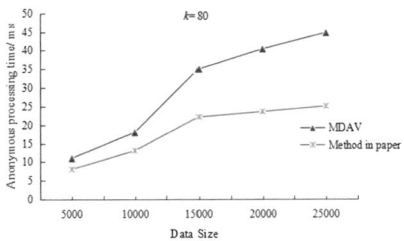

Fig. 2. Anonymous processing time under different data size

(2) The Effect of K Value on Information Availability

The information loss is a criterion to measure data availability. The greater information loss, the lower data availability. Figure 3 shows an graph that the change of information loss in different K values. It can be seen from the figure that with the increase of K value, the amount of information loss increases gradually, that is, the availability of data decreases gradually. This is because the data set has a large amount of data, and when k takes a small value, the difference between multiple equivalent classes is reduced. When the value of K increases gradually, the equivalents with smaller differences merge gradually, so the amount of information loss changes little. However, when the gap between equivalent classes is large, the value of K increases again, which makes the amount of information loss change greatly. At this time, as can be seen from Fig. 3, the amount of information loss of MDAV method is not much different from that of the method proposed in this chapter. It is because, the method proposed, although the value of sensitive attributes in equivalent classes is constrained in the process of anonymity processing, so that the value of sensitive attributes in equivalent classes does not all have the same situation, which increases the information loss, but in the process of anonymous group construction. When dealing with the loss of information, the loss of information in anonymous group is smaller, so the difference between the two methods is not significant.

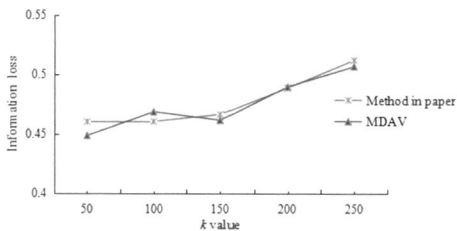

Fig. 3. Change of information loss under different K values

(3) The influence of K value on privacy protection

Information entropy is a standard to measure the degree of privacy protection. As shown in Fig. 4, the information entropy increases gradually with the increase of K value, that is, the degree of data privacy protection increases gradually. Compared with the MDAV method, the proposed method has larger information entropy, because the sensitive attributes in the equivalent class are constrained, so that the values of the sensitive attributes in each equivalent class are not equal, thus increasing the information entropy of the data table.

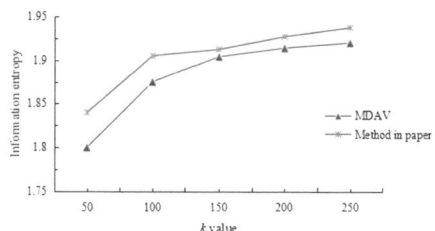

Fig. 4. Change of information entropy under different k values

(4) The analysis of the availability of statistical value

The information loss is a criterion to measure data availability. n order to illustrate the availability of data after differential privacy processing, the original data, the data after differential privacy denoising and the data after original frequency differential privacy denoising are compared. Let the $\varepsilon = 1$ [23]. From Table 1 and Fig. 5, it can be seen that the gap between the three is small, and the data is available to some extent.

However, when the gap between equivalent classes is large, the value of K increases again, which makes the amount of information loss change greatly. At this time, as can be seen from Fig. 3, the amount of information loss of MDAV method is

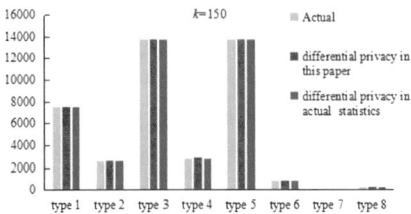

Fig. 5. Comparison of different noise addition methods and actual values

Table 1. Statistic comparison

Sensitive attribute values	Data type		
	Actual	Differential privacy in this paper	Differential privacy in actual statistic
Type 1	7536	7505.281413	7535.217489
Type 2	2581	2581.54122	2581.276248
Type 3	13766	13772.94839	13765.77608
Type 4	2820	2844.389056	2819.640393
Type 5	13731	13760.22544	13731.17357
Type 6	811	810.925564	810.7238023
Type 7	14	10.00770442	15.70966186
Type 8	285	307.1096449	281.9542926

not much different from that of the method proposed in this chapter. It is because, the method proposed, although the value of sensitive attributes in equivalent classes is constrained in the process of anonymity processing, so that the value of sensitive attributes in equivalent classes does not all have the same situation, which increases the information loss, but in the process of anonymous group construction. When dealing with the loss of information, the loss of information in anonymous group is smaller, so the difference between the two methods is not significant.

5   Conclusion

In this paper, a governmental table data sharing method based on clustering anonymity is proposed aiming at the scenario of sharing governmental table data. In the method, it can be divided into three steps. Firstly, the k-mediods clustering algorithm is used to cluster the tables' records, and obtains several data tables. Then, each data table is anonymously processed according to the amount of information loss, and generates anonymous data tables. Finally, noise is added to the sensitive attribute values. The usability and privacy of the proposed algorithm are proved by example analysis and comparison with the classical k-anonymity algorithm of MDAV.

In the future, we will carry out further research work by using different data sets of different types (numerical type, string type) and different data volumes to carry out specific experiments and analyze their impaction on information loss.

References

1. Xiaolin, X., Ming, C.: Research on government service data sharing in the environment of digital government. Adm. Trib. **01**, 50–59 (2018)
2. Susanto, H., Almunawar, M.N.: Security and privacy issues in cloud-based e-government. In: Cloud Computing Technologies for Connected Government, pp. 292–321. IGI Global (2016)
3. Yuan, Y.: Privacy Protection Method and Research in Data Sharing. Harbin Engineering University, (2014)
4. Zhang, Y.: Research on Sensitive Information Protection in Data Sharing. Dalian Maritime University (2012)
5. Xia, Z., Han, J., Juan, Yu., Guo, T.: MDAV algorithm for implementing (k, e) -anonymity model. Comput. Eng. **36**(15), 159–161 (2010)
6. Ronglei, H., Yanqiong, H., Ping, Z., Xiaohong, F.: Design and implementation of medical privacy protection scheme in big data environment. Netinfo Secur. **9**, 48–54 (2018)
7. Li, C.: Analysis of the research status of privacy protection under the environment of big data. Comput. Knowl. Technol. **12**(18), 29–31 (2016)
8. Wang, B., Yang, J.: Research on anonymity technique for personalization privacy-preserving data publishing. Comput. Sci. **39**(4), 168–171+200 (2012)
9. Zhu, T., He, M., Zou, D.: Differential privacy and applications on big data. J. Inf. Secur. Res. **1**(3), 224–229 (2015)
10. Wu, L.: Research on Clustering Algorithm of Data Table Anonymity. Xidian University (2017)
11. Ren, W.: Association rules based background knowledge attack and privacy protection. Shandong University (2011)
12. Dwork, C.: Differential privacy. In: Proceedings of the 33rd International Conference on Automata, Languages and Programming, pp. 1–12 (2006)
13. Dwork, C.: Differential privacy. In: Encyclopedia of Cryptography and Security, pp. 338–340. Springer, New York (2011)
14. Dwork, C., Mcsherry, F., Nissim, K., Smith, A.: Calibrating noise to sensitivity inprivate data analysis. In: Proceedings of the 3th Theory of Cryptography Conference (TCC), pp. 363–385 (2006)
15. Xie, J., Guo, W., Xie, W.: A neighborhood-based K-medoids clustering algorithm. J. Shaanxi Norm. Univ. (Nat. Sci. Ed.) **40**(4), 1672–4291 (2012)
16. Lu, Y., Sinnott, R.O., Verspoor, K.: A semantic-based k-anonymity scheme for health record linkage. Stud. Health Technol. Inform. **239**, 84–90 (2017)
17. Liu, X., Li, Q.: Differentially private data release based on clustering anonymization. J. Commun. **37**(5), 125–129 (2016)
18. Chunhui, P., Yajuan, S., Jiaqi, Y., et al.: Privacy-preserving governmental data publishing: A fog-computing-based differential privacy approach. Future Generation Computer Systems, S0167739X18300773 (2018)
19. Liu, H.: Clustering-based data publishing for differential data anonymization. J. Hainan Norm. Univ. Nat. Sci. **27**(01), 23–26 (2014)

20. Xiong, J.B., Wang, M.S., Tian, Y.L., Ma, R., Yao, Z.Q., Lin, M.W.: Research progress on privacy measurement for cloud data. Ruan Jian Xue Bao/Journal of Software 29(7), 1963–1980 (2018). (in Chinese). http://www.jos.org.cn/1000-9825/5363.htm
21. Shi, Y., Zhou, W., Zang, S., et al.: A comprehensive evaluation model of privacy protection based on probability statistics and del-entropy. Chin. J. Comput. **4**, 786–799 (2019)
22. Chen, X.: Research and Implementation of Data Anonymized Privacy Protection Method. Jiangsu University of Science and Technology (2018)
23. Blum, A., Ligett, K., Roth, A.: A learning theory approach to non-interactive database privacy. In: Fortieth annual ACM Symposium on Theory of Computing (STOC 2008). ACM (2008)

Blockchain-Based Food Traceability: A Dataflow Perspective

Xingchen Liu, Jiaqi Yan[⊠], and Jinbei Song

School of Information Management,
Nanjing University, Nanjing, Jiangsu, China
jiaqiyan@nju.edu.cn

Abstract. Blockchain is a promising technology to address the problems of information asymmetry and information tampering in food traceability. Food data items can be tracked with tamper-resistant blockchain data tagging with the food processing information along the food supply chain. The blockchain dataflow, which identifies the input and output data of the food processing workflow, plays a vital role to design the blockchain-based food traceability systems. In this paper, we provide a framework to identify the blockchain dataflow items with the dimensions of data properties (i.e., static or dynamic) and data requirements (i.e., mandatory or optional). Based on this framework, we propose a dataflow reference model for stakeholders along food supply chains, and provide a dataflow matrix to analyze the stakeholders' access rights to the traceable food data. The dataflow reference model and matrix can serve as tools to standardize data formats in the design of blockchain-based food traceable systems.

Keywords: Dataflow · Food traceability · Blockchain

1 Introduction

With the improvement of quality of people's life, the issues of food traceability are also receiving increasing attention from society. Many organizations in the world are actively conducting research of food traceability. Despite the continuous development of traceability technology, food traceability and quality supervision are not fully realized today. The existing food traceability systems adopt either of the two architectures: centralized architecture or distributed architecture [1]. And among them, most organizations adopt a centralized architecture. However, the fundamental reason why food safety is difficult to solve is that the traditional centralized trust mechanism brings three challenges. The first one is self-interest of supply chain members [2]. The second one is information asymmetry among participants in supply chain. The third one is limitations of traceability cost and quality monitoring [3]. Because of the information gap between food producers and consumers, the former is exempted from liability risk based on economic interests, which is the primary reason for food safety.

Blockchain has the characteristics of distributed collaboration, transparency, openness, anonymity and untouchability. It can solve the problem of lack of trust and information asymmetry in traditional food traceability. The rise of blockchain

© Springer Nature Switzerland AG 2020
K.-M. Chao et al. (Eds.): ICEBE 2019, LNDECT 41, pp. 421–431, 2020.
https://doi.org/10.1007/978-3-030-34986-8_30

technology has provided new ideas for solving problems such as information asymmetry and information tampering in food traceability [4].

Dataflow is a key factor in ensuring food traceability. Dataflow refers to the management of input and output of the food traceability process. The food traceability system based on blockchain combines traceability data with quality management system, the key of which lies in management of dataflow. Although blockchain-based food traceability can ensure decentralized management and tamper-resistant information, it does not completely solve the problem of dataflow. There are two main problems in blockchain-based food traceability: first of all, whether all the nodes in the blockchain can access food traceability data, and it may reveal enterprises' privacy. Secondly, which food data items should be traced and uploaded to the blockchain.

To address two issues above, this paper provides a framework to identify the data items which should be traced. The framework will help the participants in the supply chain better define the food traceability data. In addition, this paper also proposes a dataflow reference model along the supply chain and a data matrix analyzing the stakeholders' access right to the food traceability data. Research on traceability dataflow based on blockchain in the field of food safety is of great strategic importance for improving food traceability. It not only ensures data security and tamper-resistant, but also strengthens government regulation of the food industry.

The paper is organized in the following manner. After the introduction and literature review, we analyze the traceability dataflow and classify the traceability data items. Then, we provide a framework for food traceability dataflow in Sect. 3. We propose a dataflow reference model, and discuss the access rights of enterprises to traceability data in Sect. 4, after which some suggestions are made for enterprises and government to carry out blockchain-based food traceability in Sect. 5. Finally, we conclude in Sect. 6 with some closing thoughts and comments.

2   Literature Review

2.1   Food Traceability

For the past few years, food traceability has become an outstanding problem around the world. For example, some researchers have described the dilemma faced by food traceability. In 2015, Tang [5] has listed major food safety incidents in China from 2003 to 2014. And through these major events, he pointed out that there are big loopholes in food traceability. In 2017, Yanqing [6] has pointed out that the successful implementation of food traceability systems faces many challenges due to the scale, diversity and complexity of the food supply chains.

Moreover, some researchers also proposed new frameworks or models for food traceability. In 2015, Pizzuti [7] has generated a general framework for the global track and trace system for food. In 2017, Ali [8] has proposed a food supply chain (SC) integrity framework in the context of halal food. Similarly, Vukatana [9] has developed a data model for wine traceability. Some researchers also attempted to apply advanced technology in food traceability, especially RFID technology, in supply chain management. In 2016, Tian [10] has analyzed the advantages and disadvantages of using RFID and blockchain technology in building the agri-food supply chain traceability system.

2.2   Blockchain Adoption in Food Traceability

One way to solve the problem of food traceability is by using blockchain technology to store traceability data in chronological order, so as to ensure that traceability data cannot be tampered with [11]. Blockchain technology offers many benefits to food traceability because it provides a distributed, transparent and secure way for untrusted parties to execute transactions. This is a key element of food traceability because many participants are involved in the food supply chain, from raw material production to final sales [12]. Moreover, blockchain technology can also capture traceability data, not only the traceable attributes of food (e.g. production time and location), but also those attributes that improve transparency (e.g., the whole life cycle of food, the way food is produced, etc.) [13].

The blockchain technology can solve the food safety problems caused by information asymmetry. The food traceability system based on blockchain can collect, process and distribute the input and output data of each link in the supply chain. All stakeholders in the supply chain can obtain valuable information through many ways, which greatly reduces the information asymmetry caused by information fragmentation, while providing support for consumers' correct decision-making. Tracking the sources of all the raw materials that make up the final product is particularly important for traceability, because every region of the world has different approved testing methods and is challenged by different potential pollutants. Therefore, food manufacturers must manage the traceability data they receive from different sources, including raw material information, relevant documents (authentic certificates, country of origin, etc.), other validated test data, and all factory production and testing data [14].

Each entity in the food supply chain is a node in the blockchain network, including suppliers, processors, transporters, consumers, and regulators. In order to achieve the life cycle of food traceability, every enterprise must broadcast key information to all the nodes in the blockchain, and the node must obtain the consensus before uploading the traceability data to the blockchain. The process of uploading food traceability data to the blockchain is shown in Fig. 1 below (see Fig. 1). The left side of the red dotted line represents the entities involved in food traceability. The right side of the red dotted line represents storage. Blockchain and databases are storage technologies. Regulators regularly conduct spot checks on the process of food production, and return to the certificate of inspection if the inspection is qualified. Otherwise, enterprises should re-standardize the process of food production. Then the enterprise packages the hash value of the traceability data into a block and broadcasts it to other nodes in the blockchain. If consensus is reached, the block is added to the blockchain, otherwise it cannot be added. In addition to storing the hash value on the blockchain, enterprises also store the original traceability data in the database.

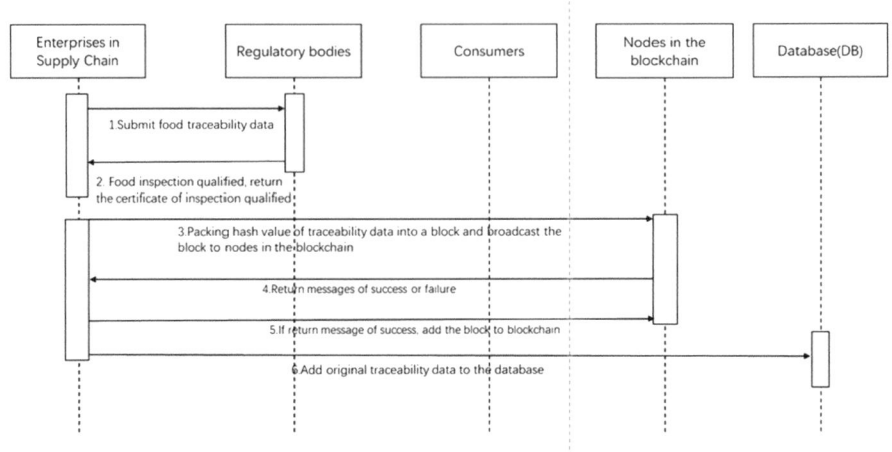

Fig. 1. The process of uploading traceability data to the blockchain

3   A Framework for Food Traceability Dataflow

In order to help enterprises to identify traceability data items, we propose a framework to classify these data items in this section. Although the length and complexity of the supply chain are different, the logistics and dataflow are based on the same principle. In every link of the supply chain, raw materials, ingredients and products are assigned identifiers according to batches. When products move across the supply chain, they are often referred to as trade units (TUs) or logistics units (LUs) and give identifiers, which may be different from batch identifiers [15]. Traceability data is generated from various processes in the supply chain and keeps the information of the smallest product unit. There are many data items related to a single trade unit and logistics unit in the food supply chain, but enterprises often do not know which data items need to be traced, or how to judge the authenticity of traceability data. Kumar and Diesel [16] believe that in order to share data among supply chain entities, it is necessary to clearly identify, define and standardize the data items, otherwise there will be many problems in accessing and using data. Therefore, a management framework is proposed to identify and classify these traceability data items.

In Fig. 2, traceability data items can be divided into static data and dynamic data. Static data are data that will not change over time. Dynamic data is the data that products will change over time. Because the ownership of products will change when products move on the supply chain, and many data items will change [17]. So, we can judge the accuracy of data according to the static data or dynamic data. If the data belongs to static data, it will not change in the block chain, on the contrary, it will change with the position of the product in the supply chain. According to whether data items need to be traced, traceability data can also be divided into mandatory data and optional data. Mandatory data refers to the data that describes the key elements of food traceability (which may vary according to different supply chains) and must be stored in the blockchain and archived for all nodes. Compared with mandatory data, optional data has little effect on food traceability. As shown in Fig. 2, we identify the data items from two dimensions. According to the category of data item, we can judge whether to trace the data item and whether the value of the data item will change. In the following framework, we also provide data sample for four types (see Fig. 2).

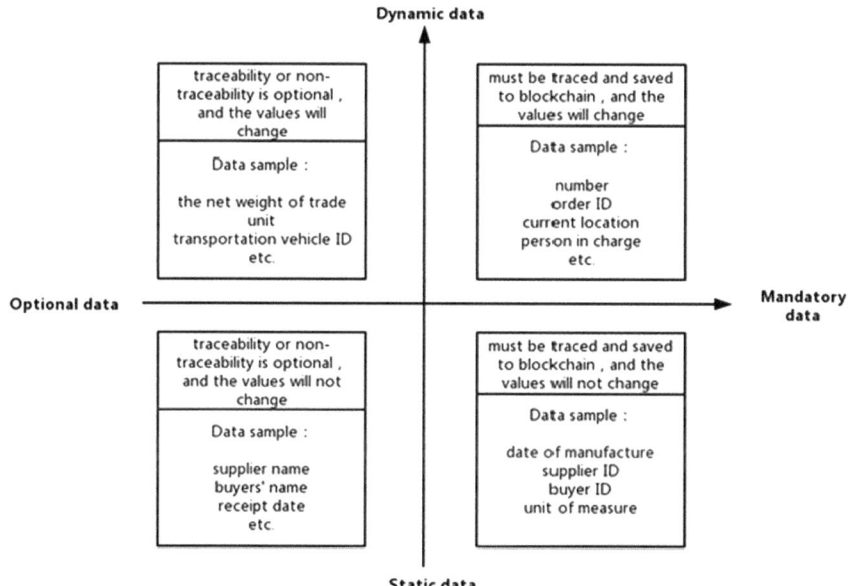

Fig. 2. The framework for food traceability dataflow

Food traceability dataflow is also closely related to product identification. Product identification of food involves two attributes, one is content attribute and the other is process attribute. Content attribute is also the physical attributes of a product. It is difficult for consumers to perceive the content attributes of food. The process attributes of a product refer to the characteristics of the production process, such as the growth environment of food, processing equipment and so on. Basically, most content attributes belong to static data, while process attributes belong to dynamic data. Therefore, when analyzing the dataflow of food traceability, we should first summarize the content and process attributes of food, and then identify which data items should be traced with the framework proposed above.

4 Data-Flow Model and Data Matrix for Food Traceability

Data sharing in the supply chain can improve the effectiveness of the supply chain and food traceability. It is also conducive to inventory level, forecast demand, monitor the status of sales and orders and production planning [18]. In the traditional supply chain network, every entity lacks trust with each other, so the willingness to share data is weak, which cannot guarantee the transparency and authenticity of the traceability data. Blockchain technology can ensure that data is difficult to tamper with and data sharing is transparent, so it is an effective means to improve data sharing and the management of traceability data in supply chain. It can also achieve decentralized management, so it can greatly improve the trust among entities in the supply chain.

So, this paper develops a blockchain-based dataflow model for food traceability, including logistics, data flow and capital flow. The extended view in Fig. 3 below describes the key data requirements among entities in the supply chain and the direction of dataflow. Because traceability includes forward-traceability and backward-traceability. Forward-traceability refers to the direction of production process, which can predict potential quality hazards and food sales inventory. Backward-traceability refers to the opposite direction of production process, which can find the source of defective food and recall defective food in time. So, the direction of dataflow in Fig. 3 is bidirectional. In addition, the dataflow model in Fig. 3 also clarifies the logistics and capital flows among entities (see Fig. 3). Financial institutions mainly include banks, lending institutions and insurance companies. Financial institutions provide financial services for enterprises in the supply chain. The Food and Drug Administration and the Quality and Technical Supervision provide authoritative certification for regular qualified products. The model also includes the role of consumers. The collection of consumer's behavior data can further verify traceability data and avoid cross-regional sales of food in violation of agreements.

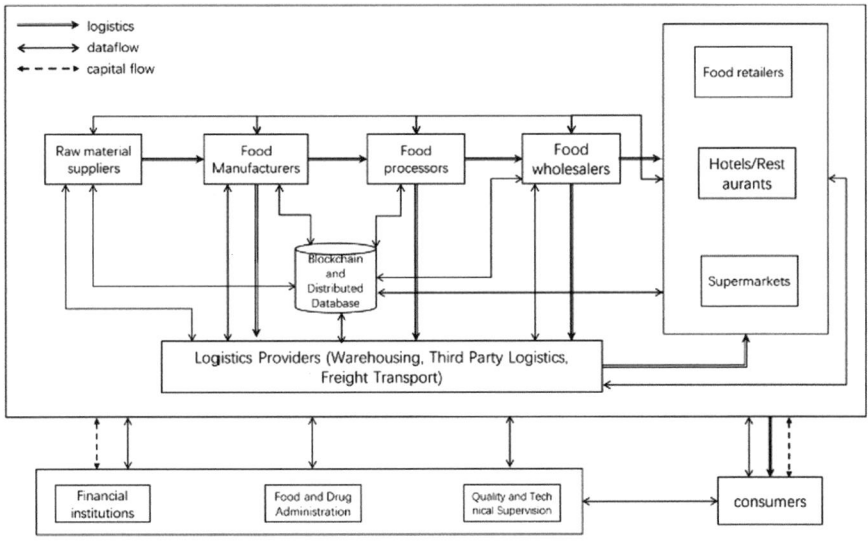

Fig. 3. Dataflow model of food traceability based on business scenario

The development of food traceability requires data sharing among partners to facilitate cooperation, and food traceability is a complex process involving more

Table 1. The entity's access rights to individual data items

the life cycle of supply chain	entities in the supply chain	raw material suppliers	food manufactures	Food processors	food wholesalers	food retailers	logistics providers
raw material link	supplier's information (name . place of origin)	w	r	r	r	r	r
	raw material quality inspection	w	r	r	r	r	r
	composition of raw materials	w	r	r	r	r	r
	the batch number of raw materials	w	r	r	r	r	r
	quantity	w	r	r	r	r	r
	price	w	r	—	—	—	—
	geographical data	w	r	r	r	r	r
	order contract	w	r/w	—	—	—	—
production link	manufacturer's information	r	w	r	r	r	r
	environmental Information (temperature, humidity)	—	w	r	r	r	r
	production equipment	—	w	r	r	r	r
	geographical data	r	w	r	r	r	r
	quantity	—	w	r	r	r	r
	price	r	w	r	r	r	r
	date of manufacture	—	w	r	r	r	r
	Inventory information	—	w	r	r	r	r
	processing order contract	—	w	r/w	—	—	—
	sales order contract	—	w	—	—	r/w	—
	logistics contract	—	w	—	—	—	r/w

processing link	processor's information	r	r	w	r	r	r
	packaging material	—	r	w	r	r	r
	quantity	—	r	w	r	r	r
	discharge date	—	r	w	r	r	r
	geographical data	r	r	w	r	r	r
circulation link	logistics provider's information	r	r	r	r	r	w
	geographical data	r	r	r	r	r	w
	input information	—	r	r	r	r	w
	output information	—	r	r	r	r	w
	warehousing information	—	r	r	r	r	w
sales link	wholesaler's information	r	r	r	w	r	r
	retailer's information	r	r	r	r	w	r
	wholesale price	r	r	r	w	r	r
	retail price	r	r	r	r	w	r
	wholesale quantity	r	r	r	w	r	r
	retail quantity	r	r	r	r	w	r
	geographical data	r	r	r	w	w	r

dynamic data. On the premise of ensuring privacy and confidentiality, data sharing in the supply chain can promote transparent and standardized management of the whole food production process, reduce the uncertainty of supply chain management, reduce inventory costs and improve the operation efficiency of the supply chain. The data matrix listed in Table 1 defines the entity's access rights to individual data items in the life cycle of supply chain. By setting read and write permissions, it can ensure that useful data can be effectively shared among entities, and confidential data will not be disclosed.

5   Management Implications

In our study, there are several important points for government and enterprises to carry out food traceability based on blockchain.

First, all stakeholders in the supply chain should strengthen the data exchange and circulation among enterprises. At present, governments and many enterprises have built food traceability systems. These food traceability systems are independent and disconnected among them, thus forming an information island. Blockchain-based food traceability can connect food enterprises. Because the blockchain technology has the characteristics of decentralization and privacy protection, it can enable these distrustful enterprises to share data safely in the network without worrying about leaking enterprises' privacy.

Second, the enterprise's original ERP system should be combined with the food traceability system based on blockchain. And enterprises should pay more attention to the management of dynamic data. Enterprise's original ERP system contains many different types of data, such as trade, logistics, delivery, warehousing information, etc. These data items constitute the attributes of multiple dimensions of food [19]. These data items include dynamic data and static data, among which dynamic data is more important for enterprises and governments because effective sharing of dynamic data and knowledge is a prerequisite for food traceability. Dynamic data evolves over time, allowing partners to leverage the latest information to achieve greater competitiveness and agility [18]. In addition, the application of blockchain technology can strengthen the automatic management of traceability data. At present, food traceability data rely heavily on manual records, which greatly reduces the efficiency of traceability and the quality of data. Through smart contract, risk identification and judgment of data can be carried out automatically, which can greatly improve the work efficiency of supply chain, and also improve the accuracy of data.

Third, the standard of traceability data should be preset in advance before the traceability system is built. The standard of food traceability should specify which data items need to be traced, and the degree of data sharing, whether it should be shared to all nodes, or only to the relevant nodes. The significance of establishing the standard of traceability data lies in the fact that traceability data is the core and foundation of food traceability.

6   Conclusions

In order to solve the problem of information asymmetry and information tampering in food supply chain, this paper analyses the data requirements of all stakeholders in the supply chain from the perspective of dataflow management, classifies and identifies traceability data, and proposes a dataflow model based on blockchain technology. Strengthening dataflow management is of great significance to the development and construction of food traceability.

Blockchain, as the underlying technology of distributed storage, has solved some problems in the field of food safety, such as data tampering, information asymmetry, and mutual distrust among enterprises in the supply chain. It also provides a new solution for the management of food safety in China. With the advent of the Internet of Things era, the collection methods of traceability data will be more diversified in the future. How to manage these data from different sources and different structures is a problem that food enterprises and governments need to think about in the future, and also a key way to enhance consumer confidence in the future.

Acknowledgement. This work was supported by Natural Science Foundation of China (NSFC No. 71701091, 71701043, and 71704078) and the Chinese Ministry of Education Project of Humanities and Social Science (No. 17YJC870020).

References

1. Lin, Q., Wang, H., Pei, X., et al.: Food safety traceability system based on blockchain and epcis. IEEE Access **7**, 20698–20707 (2019)
2. Yan, J., Li, X., Sun, S.X., et al.: A BDI modeling approach for decision support in supply chain quality inspection. IEEE Tran. Syst. Man Cybern. Syst. (2017)
3. Chen, S., Shi, R., Ren, Z., et al.: A blockchain-based supply chain quality management framework. In: The Fourteenth IEEE International Conference on e-Business Engineering, IEEE (2017)
4. Mattila, J., Seppala, T., Holmstrom, J.: Product-centric information management – a case study of a shared platform with blockchain technology. In: Industry Studies Association Conference (2016)
5. Tang, Q., Li, J., Sun, M., et al.: Food traceability systems in China: the current status of and future perspectives on food supply chain databases, legal support, and technological research and support for food safety regulation. Biosci. Trends **9**(1), 7–15 (2015)
6. Duan, Y., Miao, M., Wang, R., et al.: A framework for the successful implementation of food traceability systems in China. Inf. Soc. **33**(4), 226–242 (2017)
7. Pizzuti, T., Mirabelli, G.: The global track&trace system for food: general framework and functioning principles. J. Food Eng. **159**, 16–35 (2015)
8. Ali, M.H., Tan, K.H., Ismail, M.D.: A supply chain integrity framework for halal food. Br. Food J. **119**(1), 20–38 (2017)
9. Vukatana, K., Sevrani, K., Hoxha, E.: Wine traceability: a data model and prototype in Albanian context. Foods **5**(1), 11 (2016)
10. Tian, F.: An agri-food supply chain traceability system for China based on RFID & blockchain technology. In: 13th International Conference on Service Systems and Service Management (ICSSSM), pp. 1–6. IEEE (2016)
11. Galvez, J.F., Mejuto, J.C., Simal-Gandara, J.: Future challenges on the use of blockchain for food traceability analysis. TrAC Trends Anal. Chem. (2018)
12. Lin, Y.P., Petway, J.R., Anthony, J., Mukhtar, H., Liao, S.W., Chou, C.F., Ho, Y.F.: Blockchain: the evolutionary next step for ICT e-agriculture. Environments **4**(3), 50 (2017)
13. Yiannas, F.: A new era of food transparency powered by blockchain. Innov. Technol. Gov. Globalization **12**(1–2), 46–56 (2018)
14. Gemesi, H.G.: Food traceability information modeling and data exchange and GIS based farm traceability model design and application (2010)
15. Olsen, P., Aschan, M.: Reference method for analyzing material flow, information flow and information loss in food supply chains. Trends Food Sci. Technol. **21**(6), 313–320 (2010)
16. Kumar, K., van Diesel, H.G.: Managing conflict and cooperation in interorganizational systems. MIS Q. **20**(3), 279–300 (1996)

17. Folinas, D., Manikas, I., Manos, B.: Traceability data management for food chains. Br. Food J. **108**(8), 622–633 (2006)
18. Du, T.C., Lai, V.S., Cheung, W., et al.: Willingness to share information in a supply chain: a partnership-data-process perspective. Inf. Manag. **49**(2), 89–98 (2012)
19. Lin, J., Shen, Z., Zhang, A., et al.: Blockchain and IoT based food traceability for smart agriculture. In: Proceedings of the 3rd International Conference on Crowd Science and Engineering, p. 3. ACM (2018)

Domain Knowledge Synthesis
for e-Business Management

An Artifact for Learning the TOGAF Architecture Development Method

Pierre Hadaya[1], Abderrahmane Leshob[1,2]([✉]), and Julien Nicolas de Verteuil[1]

[1] UQAM School of Management (ESG UQAM), Center of Expertise in Digital
Business Transformation, Montreal, Canada
{hadaya.pierre,leshob.abderrahmane}@uqam.ca
[2] LATECE Laboratory, University of Quebec at Montreal, Montreal, Canada

Abstract. In order to achieve and maintain a competitive advantage
in today's global economy, organizations need to align their information
technology (IT) with their strategy while lowering the cost of ownership
of their IT. To do so, an increasing number of organizations are adopting The Open Group Architecture Framework Architecture Development
Method (TOGAF ADM). Indeed, this method seems to be the most
widely adopted, complete and promising EA method. However, understanding and applying this method well is not as simple as it seems.
The objective of the present research is to design and develop a tutoring
artifact that implements the core concepts of intelligent tutoring systems
to facilitate and accelerate the learning of the fundamental concepts of
the TOGAF ADM. To attain this objective, we adopt a design science
approach (DSA). This research contributes to the literature by synthesizing and clarifying key writings on the ADM. Furthermore, the proposed
artifact, in addition to addressing an unmet need in the market, also
helps users be rapidly more proficient in their work which in turn should
enable their organizations to be more effective, agile and efficient.

1 Introduction

In order to achieve and maintain a competitive advantage in today's global
economy, organizations need to align their information technology (IT) with
their strategy, while lowering the cost of ownership of their IT [1,2]. This, however, is not an easy task. Indeed, many organizations tend to implement new
IT solutions in piece meal fashion without paying enough attention on how to
integrate them into their current ecosystem. In addition to limiting the benefits
that a good alignment between IT and strategy can provide, this can lead to
inefficiencies or even significant problems for the organization, today or in the
future [3]. In order to address these challenges, more and more organizations
are now starting to adopt an enterprise architecture (EA) practice [4]. EA is a
high-level definition of the components of an organization (e.g., business processes, IT systems and actors) and a representation of the interactions that exist
between these components, in order to enable the implementation and execution of the organization's strategy [5–7]. An EA practice, in turn, includes all the

© Springer Nature Switzerland AG 2020
K.-M. Chao et al. (Eds.): ICEBE 2019, LNDECT 41, pp. 435–449, 2020.
https://doi.org/10.1007/978-3-030-34986-8_31

resources, activities, methodologies, techniques, frameworks, tools and principles that enable an organization to define its target EA and the roadmap required to implement it [8,9].

Several methodologies (e.g., Enterprise Architecture Planning and Activity-Based Methodology) have been proposed by researchers and practitioners to help organizations define their target EA and to align their IT with their business strategies in an efficient manner [2–4,7,9–15]. An EA methodology identifies and describes each step that must be completed to document the current and target EA and govern the resulting transformation projects [9]. Among all the methodologies available on the market today, The Open Group Architecture Framework Architecture Development Method (TOGAF ADM) seems to be the most widely adopted, complete and promising [7,15]. TOGAF ADM (in short the ADM) comprises 10 phases whose central activity is business requirements management [8]. In addition to be used in an iterative manner, the ADM has been designed and developed to be modular so that its use and benefits grow in parallel with the degree of maturity of the EA in place within the organization. Lastly, the ADM is an open and flexible methodology, which allowed it to grow in popularity for it now has more than 40,000 certified professionals [16].

Despite its popularity and the benefits it provides to organizations, learning the fundamental concepts at the heart of the ADM is not without its difficulties. The methodology employs several concepts that are intrinsically interconnected, which complicates the learning of the fundamental concepts of the ADM. Indeed, although each phase of the methodology is documented in great detail, implementing and using the ADM requires a deep understanding of the different concepts at its heart [8]. For example, it is necessary to understand the distinction between three fundamental concepts of the methodology - building blocks, artifacts and deliverables - and to correctly identify the interrelationship that exists between these concepts. That is why many experts and authors warn that any organization trying to apply the ADM by assuming that it is sufficient to follow the steps without understanding the fundamental concepts at the heart of the ADM is more likely to fail [17–20]. In addition, it is difficult for a new ADM user to develop an expertise on the methodology since carrying out a complete cycle of the ADM may require an important investment in time and effort, depending on the level of detail desired, the complexity of the ecosystem and the available resources of the organization [9,10].

The objective of the present research is to design and develop a tutoring artifact that implements the core concepts of intelligent tutoring systems to facilitate and accelerate the learning of the fundamental concepts of the ADM in order to help new users/EA practitioners rapidly get hands on experience with the methodology. To attain this objective, we adopt a design science approach (DSA) following the six steps proposed by [21]. This research contributes to the literature by synthesizing and clarifying key writings on the ADM. Furthermore, the proposed artifact, in addition to addressing an unmet need in the market, also helps users to be rapidly more proficient in their work which in turn should enable their organizations align their information technology (IT) with their strategy in a more effective and agile manner while lowering the cost of ownership of their IT.

The rest of the document is structured as follows. Section 2 justifies the relevance and originality of this research contribution. Section 3 details the adopted research methodology. Section 4 synthesizes key writings on the ADM. The first sketch of the artifact is then detailed in Sect. 5 while Sect. 6 presents the results of the evaluation of the artifact. Finally, the conclusion exposes the contributions as well as future research avenues.

2 Relevance and Originality of the Contribution

Before proceeding with the development of the proposed artifact, we conducted a literature review to ensure its relevance and originality. Among the various types of literature reviews that exist, we choose the scoping review [22] as the objective of our review was to examine the extent, range and nature of research activities on the subject [23] while focusing more on the breadth than the depth of coverage of the literature.

Our scoping review allowed us to identify eighteen existing resources that practitioners can use to learn the fundamental concepts of the ADM. These resources are of four types: static documents (e.g., books, white papers, and scientific articles), training, hypermedia and tools. In the context of this research, a *tool* refers to the ability of a software solution to support modeling languages for defining, developing, generating, editing, and/or managing architectural views [24]. All but one of the tools identified (ABACUS, ADOIT, Alfabet, ARIS, BiZ-Zdesign Enterprise Studio, Enterprise Architect, HOPEX Enterprise Architecture Suite, iServer, Troux, and Visual Paradigm) were certified by The Open Group.

Although each type of resource identified is relevant to learn the basic concepts of the ADM, each of them has weaknesses that limit the learning of the methodology. Indeed, *static documents* and *hypermedia* are linear. It is therefore not possible to adapt them. *Training*, in turn, is expensive and generally intended for individuals who already have previous EA experience and some knowledge of the ADM. Therefore, in the case of a new user, the pace of the training imposed on the students is often considered too expeditious. Finally, EA *tools*, in addition to being generally expensive, focus on the use of the TOGAF framework rather than on the learning of the ADM concepts.

In addition to uncovering the resources available today, our scoping review allowed us to identity the following functionalities that a resource must possess to facilitate and accelerate the learning of the fundamental concepts of the ADM:

– A *tutor component* accompanying ADM learners in the realization of their first EA projects. This feature will enable new ADM users to be effective in creating her/his first EA projects. Indeed, because it is generally required to adapt the ADM according to the context in which it is used in the workplace [8,9], it is necessary to provide new users with a functionality/strategy that accompanies them in the selection of the different architectural products that will have to be made during their first EA projects. This feature allows to better control the choice of different building blocks, artifacts and deliverables that will be used for each phase of the ADM.

- *TOGAF ADM knowledge base* which provides functions to manage the ADM. This feature includes: (i) *Templates* to facilitate the standardization and reuse of EA products arising from the ADM [8,25], (ii) *Hypermedia documentation* which facilitates the understanding of complex topics via an interactive environment [26,27], and (iii) A *glossary* that supports the use of a common vocabulary [9,25] which facilitates and ensures the proper use of terms and concepts commonly used by the methodology [28].
- *An intuitive interface* to carry out the information about the learning process of the ADM. This feature allows the user to control access to content from the TOGAF ADM knowledge base, the tutor component and the state of the ADM learner at a pace that suits him [27].
- A *repository manager* which makes it possible to classify, order and make available, for the purpose of reuse, the various architectural products resulting from the ADM cycles [8,9]. In this context, any EA project benefits greatly from the reuse of the products that have been realized in previous projects while the accompanying tool becomes a highly complementary functionality to the repository. The absence of a repository leads necessarily to a loss of efficiency [8].

Having the existing resources available as well as the functionalities that an artifact should have to facilitate and accelerate the learning and implementation of the fundamental concepts of the ADM, we then compared the 18 resources according to the evaluation criteria. To do this, we used an evaluation scale ranging from 1 to 4 (1 = very unsatisfactory, 2 = unsatisfactory, 3 = satisfactory and 4 = very satisfactory) to measure our level of satisfaction with the integration of each criteria within each of the eighteen resources. Results of this thorough analysis clearly showed that each of the resource available today is unsatisfactory on at least one criteria and hence demonstrated the relevance and originality of our contribution as the artifact we propose will be more appropriate than all the resources available to date, to facilitate new users' learning of the ADM.

3   Research Methodology

This research adopts the design science (DS) approach. The goal of DS is to rely on existing scientific literature to solve important issues that have no known solution [29]. More precisely, this study adopts the six-step methodology proposed by Peffers *et al.* in [21]. Sections 1 and 2 of this manuscript exposed the result of the first two steps of the methodology: *Problem identification and motivation* and *Define the objectives for a solution*.

The objective of the third step of DS methodology, *Design and Development*, is to design and develop the artifact to address the problem identified during the first phase [21,30]. This phase comprises two activities: the literature review and the design and development of the artifact. The results of this step are presented in Sects. 4 and 5.

The objective of the fourth step, *Demonstration*, is to establish that the proposed artifact can be used to solve the identified problem [21,30].[1]

The objective of the fifth step, *Evaluation*, is to measure the degree of usefulness, quality and effectiveness of the artifact by referring to the objectives of the solution that were defined, to determine if the artifact is relevant to address the problem previously identified [21,30]. The results of this step are presented in Sect. 6.

The objective of the sixth phase, *Communication*, is to inform the scientific community of the results of the research [21,30]. For this purpose, the results and conclusion of this research are summarized in the present manuscript.

4   Literature Review

To put the ADM into practice, it is important to have an understanding of the different concepts that are at the heart of this methodology [8]. This section summarizes the results of our literature review of the two main ADM writings: the *TOGAF® Version 9.1* standard available on The Open Group website and the *Modeling Enterprise Architecture with TOGAF* book from [8]. It is divided into two subsections. The first subsection describes the types of architectural products created when using the ADM. The second subsection, in turn, presents each of the ten phases of the ADM and identifies the different iterations that can occur when applying the ADM.

4.1   Building Blocks, Artifacts and Deliverables

Three types of architectural products are created when using TOGAF ADM: building blocks, artifacts and deliverables. A buildings block is an architectural element with a clearly defined perimeter that meets the business needs of an organization [9]. A building block can be used to define an actor, a service, a technological or application component, an organizational unit or any other element or set of architectural elements. There are two types of building blocks: Architecture Building Blocks (ABB) and Solution Building Blocks (SBB). An ABB consists of a logical specification of an architectural element while an SBB represents technical or physical components of a candidate solution for an architecture building block [8].

For an enterprise architect to complete his project, the building blocks must be documented using artifacts [9]. To this end, the TOGAF framework offers a range of fifty-six artifacts that can be created while using the ADM [9]. These artifacts are of three different types: catalogs, matrices, and diagrams. Catalogs are structured lists of building blocks [9]. For example, an application catalog

---

[1] As part of this research project, two ADM experts took part in the demonstration to validate the artifact in the process of ADM learning. Both experts had more than ten years of experience in EA in the financial sector as well as theoretical and practical knowledge of the ADM. Because of space constraints, we will not present the results of this phase.

is an artifact that lists the different applications that are part of an EA project application portfolio. This application catalog clearly reveals to stakeholders the various applications that are part of the context of the architectural project. Matrices, on the other hand, consist of tables showing the relationships that exist between two or more building blocks [9]. Thus, a stakeholder matrix is an artifact that identifies the key stakeholders in the realization of the EA project, as well as each of these stakeholders main concerns. As for diagrams, they graphically show building blocks and their relationships [9].

Finally, the deliverables are architectural products that have been contractually defined and approved by all stakeholders of the architectural project [8]. For example, an architecture work request is a defined document, approved and signed by all stakeholders, that is used to trigger an architectural project. This deliverable includes the project's mission, the project sponsor, the business objectives, the internal and external constraints, as well as the resources available for the project. Deliverables typically take the form of documents consisting of building blocks and artifacts [8]. The ADM proposes twenty-one deliverables typically initiated, refined and finalized during an ADM cycle. Once the architectural project is complete, the building blocks, artifacts and deliverables are archived for reuse as needed in future iterations of ADM.

4.2   The Ten Phases of the ADM

The ADM includes ten phases [9]. The first phase, *Preliminary*, is executed only during the first EA project. Its purpose is to prepare the organization to undertake architectural works by establishing the structure, governance, principles, methods and tools that will be used in subsequent phases of the ADM [8]. The second phase, *Requirements Management*, is essential to the methodology and is performed during each project. The purpose of this phase is to centrally identify the requirements that will need to be met throughout each subsequent phase of the ADM to meet the needs of the EA project [8]. Centralized requirements management facilitates the documentation, prioritization and assessment of the impacts of each requirement. In addition, centralized requirements provides enterprise architects with the vision to better guide architectural decisions throughout the ADM cycle while aligning these requirements with business objectives [8]. The other eight phases of ADM are also carried out during each EA project. However, some may be omitted if they are considered irrelevant to completing the architectural project or if there is, for example, a time constraint. These remaining eight phases of ADM are subdivided into four parts as shown in Fig. 1: Business, IT, Planning and Change.

The *business* part aims to establish the scope of the EA project, identify key stakeholders, define commitments as well as describe the current and target business architectures and assess the gaps and impacts between the two [8]. This part of the ADM has two phases: *Phase A - Architecture Vision* and *Phase B - Business Architecture*. Phase A aims to start an EA project, as well as determine the stakeholders, principles, objectives, requirements, scope and risks associated

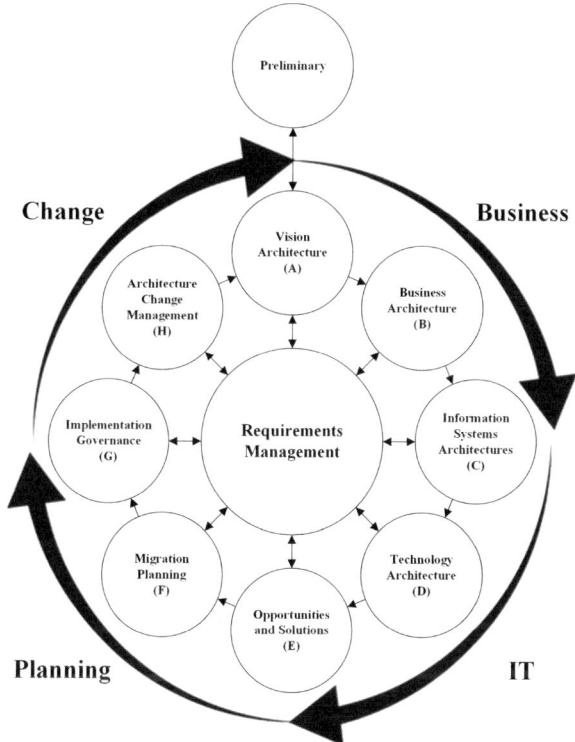

Fig. 1. The four parts of the ADM.

with the architectural work [8]. Phase B expands on Phase A, by specifying business objectives, organizational units, functions and services, processes, roles and business entities of the organization. During this phase, the current and target architectures as well as the requirements are specified and documented. When the current and target architectures are established, a gap analysis between them is performed. The purpose of the gap analysis is to identify candidate building blocks for the EA project roadmap and the impacts that these will have on the current architecture of the organization [9].

The *IT* part aims to describe current and target IT architectures of the organization as well as to assess the gaps and impacts between the two [8]. This part of ADM has two phases: *Phase C - Information Systems Architectures* and *Phase D - Technology Architecture*. In addition, Phase C is subdivided into two sub-phases: *Phase C1 - Data Architecture* and *Phase C2 - Application Architecture*. The first sub-phase, Data Architecture (C1), aims to identify the data used by the different actors, as well as the functions, the services of the organization and the data structure [16]. As for the Application Architecture (C2) sub-phase, it seeks to define the applications and their functionalities, the stakeholders in relation to these applications and the data used by these applications

[16]. Finally, Phase D concerns all the infrastructure components and technical platforms that will support the architectural elements mentioned in the previous phases. As with the Business Architecture (Phase B), these phases also generate current and target architectures, gap analyses and roadmaps to plan the transitions.

The *Planning* part aims to finalize the architecture roadmap, define the projects to be carried out and plan their execution [8]. This part of the ADM contains two phases: *Phase E - Opportunities and Solutions* and *Phase F - Migration Planning*. The previous phases described the current architecture, target architecture and roadmaps for each perspective of the architecture in silos. In this part, the operational planning of the roadmap is global [8]. Phase E therefore aims to prepare a consolidated version of the architecture roadmap based on the gap analyses between the current and target business, information systems and technology architectures, and to determine if transition architectures will be required [9]. A transition architecture is an intermediate architecture between the current and target architectures [9]. As for Phase F, its goal is to carry out an implementation and migration plan, in addition to ensuring that the architectural work offers added value for the organization [9].

The *Change* part seeks to put in place the governance for the implementation and monitoring of the execution of EA projects [8]. This part of the ADM includes two phases: *Phase G - Implementation Governance* and *Phase H - Architecture Change Management*. Phase G concerns the final approval of architectural contracts by an architectural committee. This architecture committee is a multidisciplinary horizontal unit within the organization and its mandate is to oversee the implementation of the architectural work [9]. Finally, Phase H manages the change requests that will have an impact on the architecture and the trigger of a new ADM cycle [8].

The ADM thus comprises ten phases for which a goal has been identified. For each of these phases, the methodology presents a series of steps allowing their execution as well as the realization of the architectural products that accompany them [9]. These different architectural products will serve both as outputs for certain phases and as inputs for other phases. For example, the stakeholder matrix (an artifact) is an output of Phase A, but it will also be used as an input in the subsequent phase (Phase B). Indeed, before you can proceed to define a business architecture, it is important to identify in advance the stakeholders of the EA project. Each phase of the ADM will have objectives, inputs, steps, outputs, and milestones that are well documented on The Open Group website.

It is important to mention that the ADM is an iterative methodology containing the ten phases producing building blocks, artifacts and deliverables. Thus, the ADM identifies four cycles of iterations: context iteration, architecture definition iteration, transition and planning iteration and governance iteration [8]. The first, *Context Iteration*, includes the Preliminary and Architecture Vision (A) phases. This iteration aims to establish or adjust the approach, principles, scope, vision and governance of the architecture [9]. The second, *Architecture Definition Iteration*, includes the Business Architecture (B), Information Systems Architectures (C), and Technical Architecture (D) phases. This iteration confirms that the entire architecture has been suitably defined [9]. The third, *Transition Iteration*, focuses on the Opportunities and Solutions (E) Migration Planning (F) phases. The objective of this iteration is to support the creation of a global and formal roadmap for the architecture that has been defined [9]. Finally, Governance Iteration, covers the Implementation Governance (G) and Architecture Change Management (H) phases. This iteration aims to support the governance of change in the transition to the target architecture. The ADM is therefore a methodology that ensures that the target EA is implemented in a progressive and scalable manner, taking into account changes occurring along the way [9].

5   First Sketch of the Artifact

Based on the objectives for our solution as well as the results of our literature review in the previous section, we proceeded to design and develop a first sketch of the proposed artifact to facilitate the learning of the ADM.

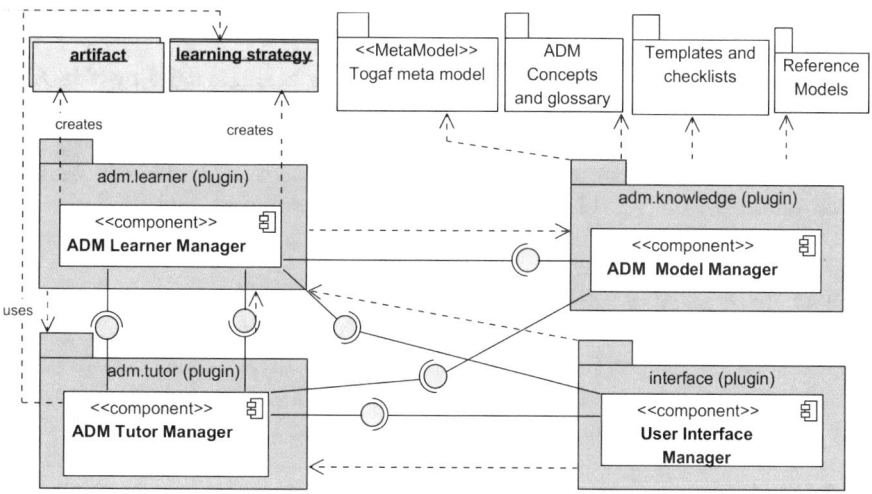

Fig. 2. Artifact architecture.

5.1   Core Architecture

We designed the core components of the artifact with the Eclipse Modeling Framework^{TM} (EMF). EMF is a Java-based modeling framework that implements EMOF (Essential Meta-Object Facility). As shown in Fig. 2, the artifact is based on four Eclipse plugins representing ITS components.

The first plugin is *adm.knowledge*. It represents the core of the artifact. This plugin is based on the ADM_MODEL_MANAGER component that provides functions to manage the ADM knowledge base. It uses the TOGAF reference models and the TOGAF Resource Base that support the ADM such as ADM templates, guidelines, checklists, ADM concepts and glossary.

The second plugin is *adm.learner*. It is based on the ADM_LEARNER_MANAGER component that is responsible for managing the learner's (e.g., student, EA practitioner) state and its evolution in the learning process.

The third plugin is *adm.tutor*. The core of the tutor plugin is based on the ADM_TUTOR_MANAGER component that determines what to do next, relative to the user (learner) current location in the process. This component uses *when-then* rules that manipulate ADM objects to guide the learner and help him/her complete successfully EA projects.

The fourth plugin is *interface*. It is responsible for carrying out the information from the domain model (*adm.knowledge*) and the learner model (*adm.learner*).

5.2   User Interfaces

This subsection presents two interfaces of the proposed artifact: The *Deliverables Matrix* interface and the *Artifacts Map* interface.

The *Deliverables Matrix* interface shown in Fig. 3 is managed by The ADM MODEL MANAGER (AMM). It provides the user with the opportunity to view and interact with the list of deliverables. These deliverables are presented under the deliverable life cycle criterion. The life cycle of a deliverable indicates in which phase the deliverable was partially created, updated, and defined. Based on Deliverables Matrix, the ADM TUTOR MANAGER (ATM) guides the user to create/complete other required deliverables.

The *Artifacts Map* interface shown in Fig. 4 allows the learner to view and interact with the artifact mapping for each phase. These artifacts are grouped according to their types (catalog, matrix and diagram) for each of these phases. This interface is managed by ADM LEARNER MANAGER (ALM) component that: (i) displays the map using the service provided by the AMM, (ii) allows the user to select an artifact, and (iii) lists the corresponding artifacts that were produced during the realization of the EA project from the learner context.

6   Evaluation

The evaluation of the proposed artifact included the following four steps: the recruitment of participants, the preparation of a deliverable on the ADM by the participants (in groups), the presentation of the deliverable by the groups to the researchers, and the analysis of the data collected by the researchers. Students enrolled in an introductory EA course taught at the master's level by one of the researchers at his University were recruited to participate in the study. Students were invited to participate in an experiment that asked them to complete and

Deliverables Matrix

Add a Deliverable to My Repository

Filter by Projects : **See all Projects** ▼

Deliverables	Pre	A	B	C	D	E	F	G	H	Reqs
Architecture Building Blocks		I	I	I	I	I	D			
Architecture Contract								D	U	
Architecture Definition Document			P	P	P	P	D	U	U	
Architecture Principles		P	P	P	P	I	I	I	U	
Architecture Repository	P	D	I	I	I	I	I	D	I	I
Architecture Requirements Specification			P	P	P	P	D			U
Architecture Roadmap			P	P	P	D	D	I	I	
Architecture Vision		D	I	I	I	U	I	U	I	I
Business Principles, Goals, and Drivers	P	P	D							
Capability Assessment		D	I	I	I	D	I			
Change Request							D	D	D	
Communications Plan		D	I	I	I	I	I			
Compliance Assessment								D	U	
Implementation and Migration Plan						P	D			
Implementation Governance Model							D	I	I	
Organizational Model for Enterprise Architecture	D	I	I	I	I	I	I	I	I	I
Request for Architecture Work	D	I					D	I	D	
Requirements Impact Assessment										D
Solution Building Blocks		I	I	I	I	I	I	D		
Statement of Architecture Work		D	U	U	U	U	I	I	U	I
Tailored Architecture Framework	D	D	I	I	I	I	I	I	I	I

Legend	
Input	I
Partly made	P
Updated	U
Defined	D

Fig. 3. Deliverables matrix.

submit an ADM deliverable. The 17 students who volunteered to participate in the study were divided into four groups. The first two groups were designated as control groups and the other two groups as test groups. Both control groups produced the deliverable only using resources and information available in the public domain. In addition to having at their disposal the same resources and public information, the two test groups also had access to the proposed artifact when preparing their deliverable. Each of the four teams had one week to prepare its deliverable. The following week, each of the four teams had 20 min to present its deliverable to the researchers. Following each closed presentation, the researchers assessed the quality of the deliverable and presentation and asked each of the participants of the group to complete a questionnaire that assessed the extent to which the artifact facilitated his/her learning of the ADM concepts. Prior to completing this questionnaire, the researchers gave a presentation of the artifact to the members of the two control groups exposing to them, one by one, its key features.

Once each of the four groups were met, we determined the quality and effectiveness of the artifact with respect to the research objectives. To do this, we conducted two analyses. The first analysis was based on researchers' evaluations of the deliverables and presentations of the teams. For this analysis, we considered the data collected from both control groups but only from one of the test groups. Indeed, we realized at the end of the evaluation process that the members of one of the two test groups had not followed the training instruction which was to use the

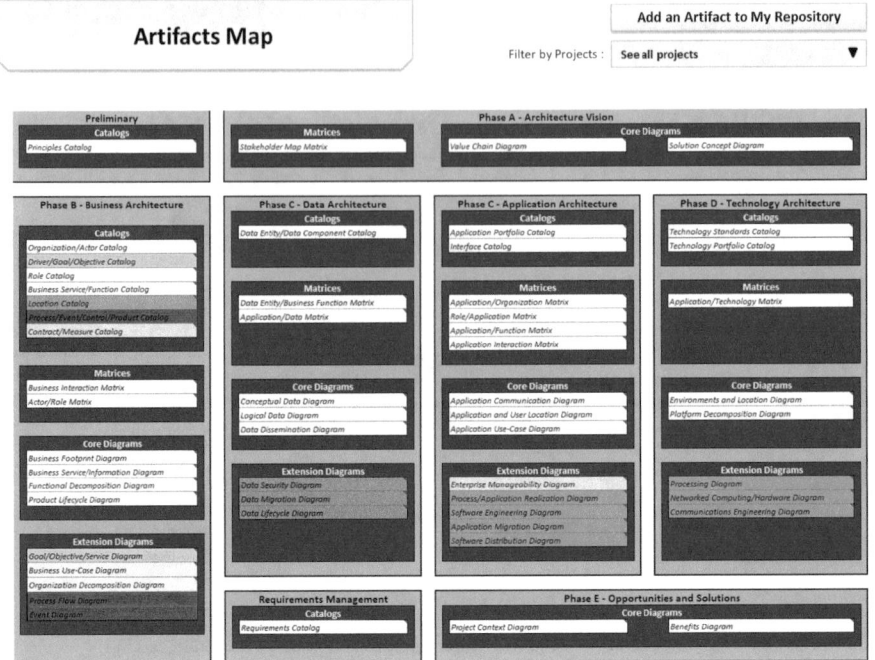

Fig. 4. Artifacts map.

proposed artifact while preparing their deliverable. We were thus forced to discard this group in our analysis. The consolidated results, based on the 7-point Likert scale [31], of the three groups are summarized in Table 1. Results show, without question, that the participants of the test group who used the proposed artifact gave better answers to the questions of the deliverable than the participants in the two control groups who did not have access to the proposed artifact while preparing the deliverable. We used our evaluations of the deliverables and presentations, aggregated the results obtained from the questionnaires and compared the results of the control groups to those of the test groups.

Two analyses were conducted. The first analysis was based on researchers' evaluation of the deliverables and presentations of the teams. For this analysis, we only considered the data from both control groups but only one of the test groups. Indeed, we realized at the end of the evaluation process that the members of one of the two test groups had not followed the training instruction which was to use the proposed artifact while preparing their deliverable. We were thus forced to discard this group in our analysis. The consolidated results of researchers' evaluations, based on the 7-point Likert scale [31], of the three groups are summarized in Table 1. Results show, without question, that the participants of the test group who used the proposed artifact gave better answers to the questions of the deliverable than the participants in the two control groups who did not have access to the proposed artifact while preparing the deliverable.

Table 1. Consolidated results from researchers' evaluations.

Statement	Control	Test
The group gave the right answers to the deliverable questions	4.5	6
The group responded clearly to the deliverable questions	4.25	6
Considering the answers to the questions of the deliverable, as well as our questions during the presentation, the group mastered its subject	4	6
Considering the answers to the questions of the deliverable, as well as our questions during the presentation, the group did quality work	4	6.5
Considering my knowledge of the artifact and participant responses, I believe the artifact helped the group in achieving its deliverable	–	6

The second analysis was based on the data collected through participants self-assessment questionnaires. The consolidated results, based on the 7-point Likert scale [31], of this self-assessment by the 13 participants are noted in Table 2. Again, these include only the inputs from the members of the test group that actually used the proposed artifact to prepare their deliverable. The results demonstrate that participants in the test group felt that they had learned more and understood the fundamental concepts of ADM better than participants in the control groups. Moreover, all participants - both those in the test group that used the proposed artifact to prepare the deliverable and those in the control groups who had a demonstration of the artifact following their presentation - concluded that this artifact helped facilitate the learning of ADM.

Table 2. Consolidated results from participants self-assessment questionnaires.

Statement	Control	Test
Through this deliverable, I understand the fundamentals of TOGAF'S ADM	3.64	5.8
I learned a lot from doing this work	4.94	6
The artifact helped (would have helped) us find right answers to the deliverable questions	5.07	5
The artifact helped (would have helped) us understand the fundamental concepts of TOGAF ADM	4.97	4.4
The artifact helped (would have helped) us put into practice TOGAF ADM	5.44	5.6
The artifact guided (would have guided) us in learning the fundamental concepts of TOGAF ADM and putting the methodology into practice	5.2	5.6
If we had used the artifact, I would have better understood the fundamentals of TOGAF ADM	5.34	-
If we had used the artifact, I would have learned even more things doing this work.*	5.1	-

* Statement present only on the control group survey

7   Conclusion and Future Work

To remain competitive in today's global market, organizations must constantly adapt to changing customer needs. To achieve this, an increasing number of organizations are starting to adopt an EA practice to try to align their IT with their business objectives in an efficient manner [4]. Of all the EA methodology available today, TOGAF ADM is the most widely adopted, complete and promising [7,13,15]. However, despite its popularity, ADM is difficult to learn, since the fundamental concepts of the methodology are not only complex, but also intrinsically interconnected.

The objective of this research was thus to design and develop an artifact that will facilitate the learning of the fundamental concepts of the ADM (version 9.1) for new users of the methodology. To do this, our study used the design science approach in order to design and develop a tutoring system offering a set of specially selected features to enable us to attain our research objective.

The proposed artifact, in addition to addressing an unmet need in the market, will also help EA practitioners be rapidly more proficient in their work which in turn should enable their organizations to align their information technology (IT) with their strategy in a more effective and agile manner while lowering the cost of ownership of their IT. The next challenges we face are: (i) improve the artifact's core components to improve users ability to follow the evolution of their learning of the ADM, (ii) develop a web version of the artifact, (iii) evaluate the artifact in a larger experimental data set, and (iv) extend the method to support full concepts of intelligent tutoring systems.

References

1. Goodhue, D.L., Thompson, R.L.: Task-technology fit and individual performance. MIS Q. **19**(2), 213 (1995). http://www.jstor.org/stable/249689?origin=crossref
2. Lapalme, J.: Three schools of thought on enterprise architecture. IT Prof. **14**(6), 37–43 (2012)
3. Bradley, R.V., Pratt, R.M.E., Byrd, T.A., Simmons, L.: The role of enterprise architecture in the quest for it value. MIS Q. Executive **10**(2), 19–27 (2011)
4. Boh, W.F., Yellin, D.: Using enterprise architecture standards in managing information technology. J. Manage. Inf. Syst. **23**(3), 163–207 (2006)
5. Hadaya, P., Gagnon, B.: Business Architecture: The Missing Link in Strategy Formulation, Implementation and Execution. ASATE Publishing Inc. (2017)
6. Ross, J.W., Weill, P., Robertson, D.: Enterprise Architecture As Strategy: Creating a Foundation for Business Execution. Harvard Business Press, Boston (2013)
7. Tamm, T., Seddon, P.B., Shanks, G., Reynolds, P.: How does enterprise architecture add value to organisations? Commun. Assoc. Inf. Syst. **28**(1), 10 (2011)
8. Desfray, P., Raymond, G.: Modeling Enterprise Architecture with TOGAF: A Practical Guide Using UML and BPMN, 1st edn. Morgan Kaufmann Publishers Inc., San Francisco (2014)
9. The Open Group, TOGAF® Version 9.1 (2011). https://pubs.opengroup.org/architecture/togaf91-doc/arch/

10. Bernaert, M., Poels, G., Snoeck, M., De Backer, M.: Enterprise architecture for small and medium-sized enterprises: a starting point for bringing EA to SMEs, based on adoption models. In: Information Systems for Small and Medium-Sized Enterprises, pp. 67–96. Springer, Heidelberg (2014)

11. DoD, A.S.D.: DoD Architecture Framework Version 2.0 (DoDAF V2. 0). Department of Defense, Washington DC (2009)

12. Schekkerman, J.: How to Survive in the Jungle of Enterprise Architecture Frameworks: Creating Or Choosing an Enterprise Architecture Framework. Trafford Publishing, Victoria (2004)

13. Urbaczewski, L., Mrdalj, S.: A comparison of enterprise architecture frameworks. Issues Inf. Syst. **7**(2), 18–23 (2006)

14. Zachman, J.A.: A framework for information systems architecture. IBM Syst. J. **26**(3), 276–292 (1987). http://ieeexplore.ieee.org/document/5387671/

15. Cameron, B.H., McMillan, E.: Analyzing the current trends in enterprise architecture frameworks. J. Enterp. Archit. **9**(1), 60–71 (2013)

16. The Open Group, TOGAF® 9 Certification - Directory of Certified People (2016). https://togaf9-cert.opengroup.org/certified-individuals

17. Bloomberg, J.: Enterprise Architecture: Don't Be a Fool with a Tool (2014)

18. Kotusev, S.: The critical scrutiny of TOGAF (2016)

19. Macgregor, S.: Keys to Enterprise Architecture Success (2016)

20. McLeod, J.: TOGAF is Agile! What? You Didn't Know This? (2017)

21. Peffers, K., Tuunanen, T., Rothenberger, M.A., Chatterjee, S.: A design science research methodology for information systems research. J. Manage. Inf. Syst. **24**(3), 45–77 (2007)

22. Paré, G., Trudel, M.-C., Jaana, M., Kitsiou, S.: Synthesizing information systems knowledge: a typology of literature reviews. Inf. Manage. **52**(2), 183–199 (2015)

23. Rumrill, P.D., Fitzgerald, S.M., Merchant, W.R.: Using scoping literature reviews as a means of understanding and interpreting existing literature. Work (Reading, Mass.) **35**(3), 399–404 (2010)

24. Steenkamp, A., Alawdah, A., Almasri, O., Gai, K., Khattab, N., Swaby, C., Abaas, R.: Teaching case enterprise architecture specification case study. J. Inf. Syst. Educ. **24**(2), 105 (2013)

25. Winter, R., Fischer, R.: Essential layers, artifacts, and dependencies of enterprise architecture. J. Enterprise Archit. **3**(2), 1–12 (2007)

26. Cairncross, S., Mannion, M.: Interactive multimedia and learning: realizing the benefits. Innovations Educ. Teach. Int. **38**(2), 156–164 (2001)

27. Scheiter, K., Gerjets, P.: Learner control in hypermedia environments. Educ. Psychol. Rev. **19**(3), 285–307 (2007)

28. Jonkers, H., Van Burren, R., Arbab, F., De Boer, F., Bonsangue, M., Bosma, H., Ter Doest, H., Groenewegen, L., Scholten, J.G., Hoppenbrouwers, S., et al.: Towards a language for coherent enterprise architecture descriptions. In: Enterprise Distributed Object Computing Conference: Proceedings. Seventh IEEE International, pp. 28–37. IEEE (2003)

29. Hevner, A., March, S.T., Park, J., Ram, S., von Alan, R.H., March, S.T., Park, J., Ram, S.: Design science in information systems research. MIS Q. **28**(1), 75–105 (2004)

30. Johannesson, P., Perjons, E.: An Introduction to Design Science. Springer, Heidelberg (2014). https://books.google.com/books?id=ovvFBAAAQBAJ&pgis=1

31. Joshi, A., Kale, S., Chandel, S., Pal, D.: Likert scale: explored and explained. Br. J. Appl. Sci. Technol. **7**(4), 396 (2015)

OCBLA: A Storage Space Allocation Method for Outbound Containers

Xinyu Chen[1,2(✉)] and Wenbin Hu[1,2(✉)]

[1] School of Computer Science, Wuhan University, Wuhan, China
{chenll,hwb}@whu.edu.cn
[2] Shenzhen Research Institute, Wuhan University, Wuhan, China

Abstract. In container yard, the way of allocating space for outbound containers determines the transport efficiency of containers and the management cost. The existing space allocation methods usually have problems such as low computational efficiency, large number of shifts etc. This paper proposes an outbound containers' block-location allocation (OCBLA) method to solve this problem, it has two steps, including allocating a block for containers which arrive at the same time and transported by the same vessel and allocating the specific location for each container. The first step aims to balance the load among blocks and reduce the cost of transportation, and last step uses ITO algorithm to reduce the number of shifts. The ITO algorithm regards every optional place as a moving particle and uses drift operator to control particle moving in the direction of the optimal place, wave operator to control particle exploring around. Having a tendency to explore makes the algorithm to find the optimal result in a limited number of tests, thus improving the computational efficiency. Through experiments it can be seen that this method can reduce the number of shifts during the process of container stacking, and get better results while having good real-time performance.

Keywords: OCBLA · ITO algorithm · Shifts · Container stacking

1 Introduction

The port is usually located at the mouth of a river, providing cargo handling services for vessels. It plays an important role in the global logistics transportation. The cargo transported by vessels is usually stored in containers. Containers are the carriers to hold the cargo, so that they can be placed neatly. International organizations have formulated a unified standard for the size of containers, which greatly facilitates the worldwide transportation of goods. Outbound containers are transported to the port by trucks and stored temporarily in the port yard, waiting for vessels. When vessels dock at the berth, the containers in the yard are transported to the loading area by trucks. The transportation process of inbound containers is the opposite. The main function of port logistics system is to ensure that the operations above are carried out in an orderly manner. However, the port logistics system is complex. For arriving vessels, it is necessary to

© Springer Nature Switzerland AG 2020
K.-M. Chao et al. (Eds.): ICEBE 2019, LNDECT 41, pp. 450–463, 2020.
https://doi.org/10.1007/978-3-030-34986-8_32

determine the loading time according to the vessel's departure time, and make sure they can enter the berth on time. When containers enter the port, the port logistics system should solve these problems like allocating space, dispatching transport vehicles, and formulating transport routes.

The operation mechanism of port logistics system needs to be efficient. Otherwise, it will cause congestion in the yard. All kinds of congestion will cause a series of impacts, such as the prolongation of vessel's berthing time and the increase of transportation cost. So the algorithm of allocating space for containers needs to be real-time.

2   Related Work

In order to reduce the operating cost and improve the efficiency of the port, many scholars have done a series of research on port logistics system. Their emphases are different, some focus on the specific optimization of a link in the port logistics system, and some focus on the comprehensive optimization of the whole system.

For [1], a heuristic algorithm was proposed to minimize the number of shifts during the process of container transportation. Lee et al. [2] made a comprehensive analysis of the truck scheduling problem and space allocation problem, they took the cost of transportation and the operation time of containers as objective functions and used heuristic algorithm to solve the problem. Dekker et al. [3] simulated different rules of container storage, and one of the strategies is based on the classification of containers, the same kind of containers place together, the classification is usually based on vessels, container size, the number of containers, etc. In this way, there will be fewer shifts. Another strategy is to analyze the loading time of containers. The containers with earlier loading time are placed on the containers with later loading time. In this way, the shifts caused by different loading times will be fewer. The simulation shows that storage containers by classification has better performance. Min et al. [4] proposed an outer-inner cellular automaton algorithm (CAOI) to solve storage allocating problem. And got a better performance than other state-of-the-art methods, which synchronously optimized the container transportation distance, the number of allocated yard-bays and shifts. Chang et al. [5] established the model with the main objective of reducing the operation time of gantry crane, and solved it with heuristic algorithm and genetic algorithm. Chen et al. [6] analysed the interaction between container crane and container truck. They used three-stage algorithm to schedule the crane. Tang et al. [7] also focused on the coordination of container crane and trucks, aiming at reducing the idle time of equipment between tasks, and used an improved particle swarm optimization algorithm to solve the problem and got better results. Boysen et al. [8] focused on dividing the working areas of container cranes to suit their workload.

Peng et al. [9] took the gantry crane as main research object, and used discrete artificial bee colony algorithm to solve crane scheduling problem [10]. Jiang [11] analysed the influence of crane's initial position on operating time.

Chu et al. [12] set up a two-cycle optimization model, aiming at reducing the waiting time of vessels and keeping balance. Zheng et al. [13] set up a non-linear programming model to improve the efficiency of container management and reduce the waiting time of trucks. Li et al. [14] established two models in the study, respectively aiming to improve the utilization rate of space and reduce the number of shifts. Wu et al. [15] proposed a mixed integer programming model to reduce berthing time of vessels and operation time of containers, and proposed a non-linear mixed integer programming model to reduce calculation time. Lee's [16] research pointed out that blocks were the basic unit of container storage, and put forward the method of optimizing block area. In the research of Yu et al. [17], a spatial allocation model for reducing waiting time of container trucks was established to improve the operation efficiency of the terminal. Bazzazi et al. [18] used genetic algorithm to solve the problem of space allocation in container yard. The algorithm can balance the workload of each block and reduce the time of container transportation and management. In the model established by Sharif et al. [19], two competing objectives were considered, include whether the distribution of containers in each block was balanced or not and the transport distance of containers. This model balances the workload of each block and effectively improves the congestion situation of transport roads.

The remainder of the paper is organized as follows. Section 3 describes the problem we solved. Section 4 introduces the content of outbound containers' block-location allocation (OCBLA) method. Section 5 shows the results of experiments based on the method.

3   Problem Description

Port logistics includes many links, and there are many kinds of containers. This paper mainly considers the allocation of storage space for outbound containers. With the increase of port business, in the case of limited storage space, how to allocate the appropriate location for containers in real time is a key concern. Allocating proper locations for outbound containers can improve the management efficiency and space utilization. Suitability refers to the shorter distance between container's location and container's loading port to reduce the cost of transportation, and the smaller number of shifts to reduce the cost of container's movement, while controlling the distribution of containers in order to prevent congestion. The structure of yard and related concepts are introduced below.

The left part of Fig. 1 shows the diagram of a container port. Outbound containers enter the container yard, and stored in the yard temporarily waiting for vessels. When the vessels arrive, the containers are transported to the dock together. The right part of the figure shows the structure of block and bay, a block contains many bays. Allocating a location for a container involves two steps, including allocating a specific block of the container yard for the container and allocating the specific location of the block for the container. The main optimizing objectives of OCBLA method include reducing the cost of transportation, load balancing and reducing the number of shifts. The container yard covers a

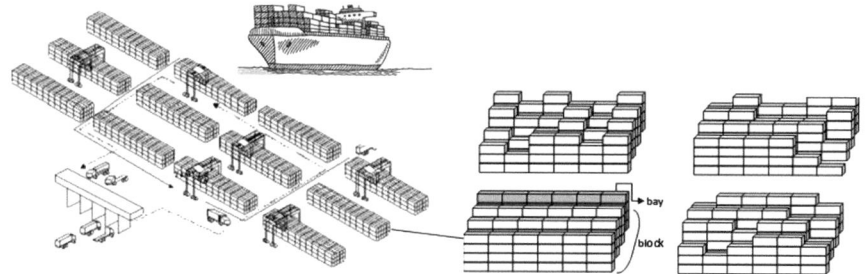

Fig. 1. The diagram of container port.

	A1	A4	C3	A3	A6
	B1	B3	A2	D3	E4
	B2	C2	D2	A5	D4
	C1	D1	E1	B4	C4
time	0	0	1	2	3
weight	2	0	0	4	0

Fig. 2. Shifts caused by time and weight in one bay.

large area, reduce the time of container transportation and improve the transport efficiency can reduce the cost of management. The containers transported by the same vessel are evenly placed in the yard. During loading, multiple routes can be used for transportation to improve transport efficiency and prevent blockage.

The shifts in this method are mainly caused by two factors, including time and weight. Containers with later loading times are placed under containers with earlier loading times. Containers stored on vessels follow the principle that place heavier containers under lighter containers to maintain balance, so in the container yard, heavier containers are placed on lighter containers. Containers that do not follow these two rules will cause shifts which will cause containers' movement. Moving containers needs special equipment and that will have certain expenses. Figure 2 shows the shifts caused by time and weight in one bay, each rectangle represents a container, the letter represents the vessel to which the container belongs, from A to E, the loading time of vessels is getting later and later. The number represents the weight of the container. It can be seen from the figure that the number of shifts caused by time and weight in this case.

Several assumptions are made in this paper.

1. The research aims at outbound containers, and outbound containers are not mixed with inbound containers.
2. All outbound containers are of the same size.

3. Containers have ten heavyweights and the numerical range in the experiment is [1,10].
4. Shift in the experiment only considers the following two cases: Containers with later loading times are placed on containers with earlier loading times and heavier containers are placed under lighter containers.

4   OCBLA Method

This paper proposes an OCBLA method to allocate space for containers. Figure 3 shows the framework of this method, it has two steps, including allocating a block for containers which arrive at the same time and transported by the same vessel and allocating the specific location for each container. The first step aims to balance the load among blocks and reduce the cost of transportation, and last step aims to reduce the number of shifts.

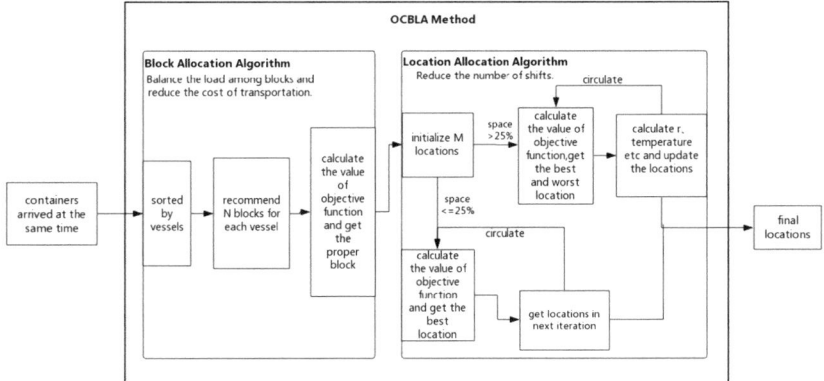

Fig. 3. The framework of OCBLA.

4.1   Block Allocation Algorithm

Block allocation is the first step of the method. It is allocating a certain block for containers that arrive at the same time and transported by the same vessel. The main idea is to categorize containers first, the classification is based on transport vessels, then recommend several blocks for containers according to the value of objective function, the calculation factors in the objective function include the number of vessels which load containers in the same block at the same time and the transport distance of vehicles in order to balance the load among blocks and reduce the cost of transportation. Figure 4 describes the process of block allocation abstractly.

Algorithm 1. Block Allocation

Input: containers
Output: map⟨*vesselId, blockId*⟩ result
  for containers **do**
    if map.containsKey(vesselId) **then**
      map.get(vesselId).add(containerId);
    else
      map.put(VesselId,List⟨⟩(containerId));
    end if
  end for
  for vessels **do**
    List ⟨*block*⟩ blockList=selectBlocks();
    if blockList.isEmpty() **then**
      result.put(vesselId,0);
      break;
    end if
    for block **do**
      temp=$f^i_{k_1} + f^i_{k_2}$;
      min=min≤temp?min:temp;
    end for
    result.put(vesselId,blockId);
  end for

---

1. Categorize containers that arrive at the same time. The standard of classification is the transport vessel, containers transported by the same vessel belong to one category.
2. Every block in container yard has the first objective function $f^i_{k_1}$ stored in database. Recommend N blocks sorted by the value of the first objective function for each vessel. The first objective function is

$$f^i_{k_1} = u_i / (c_i + u_i) \tag{1}$$

In the formula, u_i stands for the number of locations occupied in block i, c_i stands for the number of free locations in block i. The first objective function takes available space as the main evaluation index.
3. This step is mainly to allocate a certain block for each vessel. Each vessel has N blocks to test. For each block, compute the sum of the first objective function $f^i_{k_1}$ and the second objective function $f^i_{k_2}$ synthetically, the block with the smallest value is the selected one. The second objective function is

$$f^i_{k_2} = \theta * d^i_j + \lambda * m^i_j + \gamma * n^i_j \tag{2}$$

In the formula, d^i_j stands for the distance from block i to the berth of vessel j, m^i_j stands for the number of vessels in block i gathering at the same time as vessel j, n^i_j stands for the number of vessels in block i loading at the same time as vessel j.

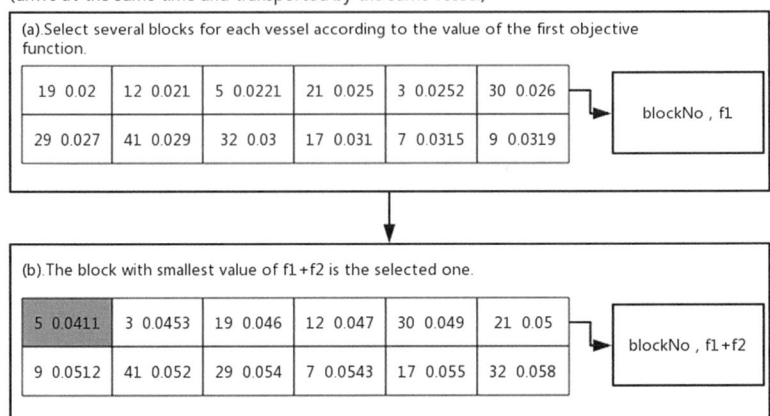

Fig. 4. The process of block allocation.

4.2   Location Allocation Algorithm

After allocating block for each vessel, next step is to allocate specific locations for containers transported by each vessel. This algorithm uses different storage allocation strategies according to the amount of rest space in container yard. When the space of the yard is sufficient, ITO algorithm is used to allocate space. ITO algorithm is based on Brownian motion of particles. Brownian motion is irregular motion of particles. ITO algorithm uses drift operator and wave operator to control the direction of particle's movement. Drift operator can control particle moving in the direction of the optimal place, wave operator can control particle exploring around. Having a tendency to explore makes the algorithm to find the optimal result in a limited number of tests. When the available space is insufficient, the effect of Ito algorithm to reduce the number of shifts is not ideal, so by testing the remaining space to find the proper location. It is found during the experiment that when the space threshold is 25%, the effect is better.

Figure 5 describes the process of location allocation abstractly. It shows the vertical view of the block. Step (a) is initializing locations for container x, step (b) gets the best and worst location in this iteration, red square represents the best position and black square represents the worst position. According to them update the locations in next iteration, the locations in next iteration will be close to the best position. Circulate (b) (c) two steps, and get better results.

1. When the free space is greater than 25%, it uses ITO algorithm to allocate proper locations for containers. The steps are as follows:
2. Initialize M candidate locations for container, regard them as particles in ITO algorithm. The candidate locations are evenly distributed in the block, define the iterations as T.

Algorithm 2. Location Allocation

Input: container
Output: map$\langle containerId, locationId \rangle$ result
  if space$\geq$25% **then**
    List $\langle location \rangle$ locationList=selectLocations();
    for iterations **do**
      for location **do**
        temp=f_b^i
        min=min$\leq$temp?min:temp;
        max=max$\geq$temp?max:temp;
      end for
      Calculate r α β ambient temperature;
      get locations in next iteration;
    end for
    result.put(containerId,locationId);
  else
    List $\langle location \rangle$ locationList=selectLocations();
    for iterations **do**
      for location **do**
        temp=f_b^i
        min=min$\leq$temp?min:temp;
      end for
      get locations in next iteration;
    end for
    result.put(containerId,locationId);
  end if

3. Calculate the value of objective function f_b^i in each location. The formula is

$$f_b^i = \epsilon * (w_{i_0} + w_{i_1}) + \mu * (t_{i_0} + t_{i_1}) \tag{3}$$

w_{i_0} and t_{i_0} stand for the number of shifts caused by weight and time before putting container on location i. w_{i_1} and t_{i_1} stand for the number of shifts caused by weight and time after putting container on location i. ϵ and μ stand for the weight of shifts caused by weight and time. By comparing the values of M objective functions, the best and worst locations in this iteration are obtained.

4. The radius of each particle in next iteration is calculated according to the smallest value of the objective function and the biggest value of the objective function in this iteration. From [20] we can see that the formulas for calculating radius, the ambient temperature, the drift operator and the wave operator are as follows:

$$r = r_{min} + (f_w - f_0) / (f_w - f_b) * (r_{max} - r_{min}) \tag{4}$$

f_w stands for the biggest value of the objective function, f_b stands for the smallest value of the objective function, f_0 stands for the value of test location's objective function. r_{max} and r_{min} stand for the biggest and the

Allocate space for container x.

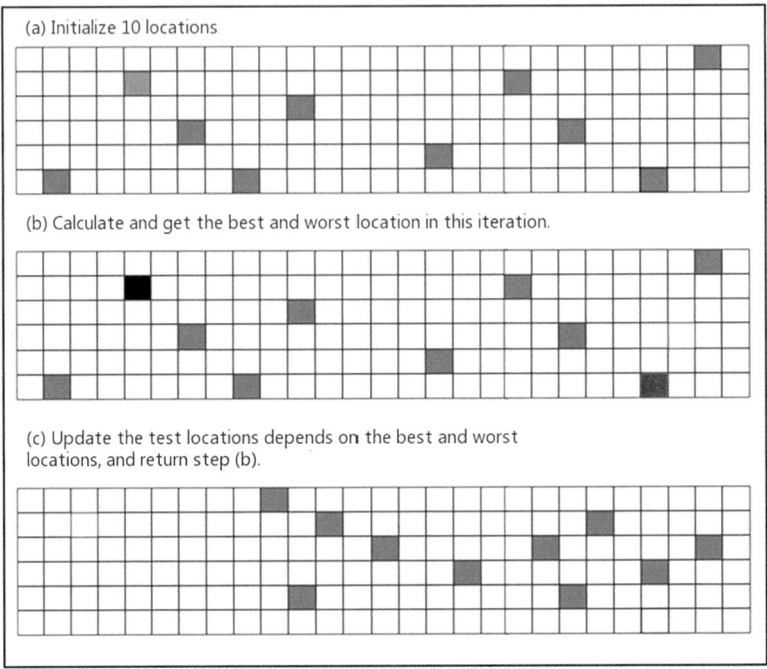

Fig. 5. The process of location allocation.

smallest value of particle's radius. During the experiment, the range of radius is [0,1]. Calculate the ambient temperature of the next iteration, the formula is

$$t = t_0 * speed \tag{5}$$

t_0 stands for the ambient temperature in this iteration, and speed is the annealing speed. The ambient temperature keeps falling in order to keep the convergence of the algorithm. Calculate the drift operator of each particle in next iteration, the formula is

$$\alpha = \alpha_{min} + f_1(r) * f_2(t) * (\alpha_{max} - \alpha_{min}) \tag{6}$$

Calculate the wave operator of each particle in next iteration, the formula is

$$\beta = \beta_{min} + f_1(r) * f_2(t) * (\beta_{max} - \beta_{min}) \tag{7}$$

α_{max} and α_{min} stand for the maximum and minimum of drift operator, β_{max} and β_{min} stand for the maximum and minimum of wave operator. $f_1(r)$ is a function about particle's radius and $f_2(t)$ is a function about ambient temperature. The formula is

$$f_1(r) = \left(e^{-\lambda * r} - e^{-\lambda * r_{max}}\right) / \left(e^{-\lambda * r_{min}} - e^{-\lambda * r_{max}}\right) \tag{8}$$

$$f_2(t) = e^{-1/t} \tag{9}$$

The calculation of drift operator and wave operator are related to the environmental temperature and the radius of particles, and according to the drift operator and wave operator update particles, the updating rules are as follows:

$$B_1 = (B_0 + \alpha + \beta + B_{num})\, \% B_{num} \tag{10}$$

$$R_1 = (R_0 + \alpha + \beta + R_{num})\, \% R_{num} \tag{11}$$

B_1 is the updated bay number, B_0 is the test location's bay number, B_{num} is the number of bays in one block. R_1 is the updated row number, R_0 is the test location's row number, R_{num} is the number of rows in one bay.

5. If the number of iterations is less than T, circulate the steps above, else the optimal location in all iterations is the proper location allocated for container.
6. When the rest space is less than 25%, test the rest locations several times to find a proper location for the container, The steps are as follows:
7. Initialize M candidate locations for container, candidate locations are selected in the order of id, and define the iterations as T.
8. Calculate the value of objective function f_b^i in each location. The formula is

$$f_b^i = \epsilon * (w_{i_0} + w_{i_1}) + \mu * (t_{i_0} + t_{i_1}) \tag{12}$$

w_{i_0} and t_{i_0} stand for the number of shifts caused by weight and time before putting container on location i. w_{i_1} and t_{i_1} stand for the number of shifts caused by weight and time after putting container on location i. ϵ and μ stand for the weight of shifts caused by weight and time. By comparing the values of M objective functions, the best location in this iteration is obtained.

9. Get next M locations selected in the order of id, if the number of iterations is less than T, circulate the steps above.
10. After T iterations, the location which has the smallest value of objective function is the suitable location.

5   Experiment and Analysis

In order to verify the effectiveness of the OCBLA method in improving the number of shifts and computational efficiency, two data sets with different data volumes are used in this experiment.

During experiment, we find that the better results can be obtained when iterations is 50. When the iterations is 50, a series of experiments were carried out to determine the number of locations (particles) initialized for each container in the location allocation process. Table 1 shows the effect of particle number on the algorithm. The experiment was completed on the basis of Data Set 1. The first column is the number of recommended locations for container during each iteration, and the second and third columns are the number of shifts caused by weight and time after handling 30,000 containers. From the table, we can see the proper number of recommended locations for each container is 50. Table 2 shows the value of parameters in this experiment, in practice, the values of each parameter can be changed according to different standards of measurement.

5.1 Data Set 1

The test object in Data Set 1 is a yard that can hold 86400 containers. It shows the number of shifts caused by time and weight when locating 83000 containers. The first column is the number of containers that need to be assigned locations. The second and third columns are the number of shifts caused by weight and time. The experimental data are as follows: There are 15 berths, the distance between each block and berths is randomly generated, and the numerical range is [10,14]; There are 100 vessels; The weight of containers ranges from 1 to 10, the number of containers of each vessel conforms to the normal distribution.

Table 1. The effect of particle number on the algorithm.

Number of particles	Weight	Time
20	1569	2722
30	61	194
40	1	3
50	0	0
60	0	0

Table 2. Experimental parameters

Parameter	Value
N	20
M	50
T	50
θ	0.01
λ	0.02
γ	0.02
ϵ	1.1
μ	1.1
t_0	100
Speed	0.99
α_{max}	3
α_{min}	-3
β_{max}	1
β_{min}	-1

Table 3. The results about Data Set 1.

Number of containers (1e4)	Weight	Time
1	0	0
6	0	0
7	2	7
8	136	1760
8.3	628	4995

The running results about Data set 1 is shown in Table 3. From the results, it can be seen that the number of shifts is very small when the space is sufficient. After storing 70,000 containers, the number of shifts caused by time and weight is almost zero. In the case of limited space, the number of shifts will increase correspondingly, after storing 83,000 containers, the number of shifts caused by time and weight is 5623, but its value is acceptable in practical operation. The number of shifts caused by time is more than that caused by weight is because the classification of time is much more than that of weight.

5.2 Data Set 2

The test object in Data Set 2 is a yard that can hold 900000 containers. It shows the number of shifts caused by time and weight when locating 860000 containers. The experimental data are as follows: There are 30 berths, the distance between each block and berth is randomly generated, and the numerical range is [10,14]; There are 150 vessels; The weight of containers ranges from 1 to 10, the number of containers of each vessel conforms to the normal distribution.

The running results about Data set 2 is shown in Table 4. From the results, it can be seen that the number of shifts is very small when the space is sufficient. After storing 840,000 containers, the number of shifts caused by time and weight is about 1500. In the case of limited space, the number of shifts will increase correspondingly, after storing 860,000 containers, the number of shifts caused by

Table 4. The results about Data Set 2.

Number of containers (1e4)	Weight	Time
1	0	0
79	0	0
80	73	251
82	77	385
84	86	1560
86	209	6589

time and weight is about 6500, but its value is acceptable in practical operation. At the same time, the calculation speed can meet the requirements of production.

Conclusion and Future Work

With the development of economic globalization, trade between countries has become more frequent, and various resources are dispatched around the world. Water carriage has become a popular mode of transportation because of its low cost and large carrying capacity, which leads to the busy port. As the loading and unloading area of cargo, container yard receives and stores containers, dispatches vehicles to transport containers to the dock, and the operational efficiency of each link determines whether vessels can sail on time. The purpose of this paper is to allocate suitable space for outbound containers, which can ensure that the cost of container storage and transportation is low and the load of the whole yard is balanced.

In this paper, we propose an OCBLA method to allocate suitable locations for outbound containers. Allocate a suitable block for each container to balance the load among the blocks and reduce the transportation cost. Allocate a certain location in block for each container to reduce the number of shifts caused by time and weight. Our next work is: Improving the performance of the method in reducing the number of shifts by analyzing the law of container's arrival.

Acknowledgement. This work was supported in part by the Key Projects of Guangdong Natural Science Foundation (No. 2018B030311003).

References

1. Exposito-Izquierdo, C., Melian-Batista, B., Moreno-Vega, M.: Pre-marshalling problem: heuristic solution method and instances generator. Expert Syst. Appl. **39**(9), 8337–8349 (2012)
2. Lee, D.H., Cao, J.X., Shi, Q., et al.: A heuristic algorithm for yard truck scheduling and storage allocation problems. Transp. Res. Part E Logistics Transp. Rev. **45**(5), 810–820 (2009)
3. Dekker, R., Voogd, P.: Container Terminals and Cargo Systems, pp. 131–154. Springer, Heidelberg (2007)
4. Hu, W., Wang, H., Min, Z.: A storage allocation algorithm for outbound containers based on the outer-inner cellular automaton. Inf. Sci. **281**, 147–171 (2014)
5. Chang, D., Jiang, Z., Yan, W., et al.: Developing a dynamic rolling-horizon decision strategy for yard crane scheduling. Adv. Eng. Inform. **25**(3), 485–494 (2011)
6. Chen, L., Langevin, A., Lu, Z.: Integrated scheduling of crane handling and truck transportation in a maritime container terminal. Eur. J. Oper. Res. **225**(1), 142–152 (2013)
7. Tang, L., Zhao, J., Liu, J.: Modeling and solution of the joint quay crane and truck scheduling problem. Eur. J. Oper. Res. **236**(3), 978–990 (2014)
8. Boysen, N.: Determining crane areas in intermodal transshipment yards: the yard partition problem. Eur. J. Oper. Res. **204**(2), 336–342 (2010)

9. Peng, G., Cheng, W., Yi, W., et al.: Gantry crane scheduling in intermodal railroad container terminals. Int. J. Prod. Res. 1–18 (2018)
10. Peng, G., Cheng, W., Zhang, Z., et al.: Gantry crane scheduling with interference constraints in railway container terminals. Int. J. Comput. Intell. Syst. 6(2), 244–260 (2013)
11. Jiang, H.C.: A note on: a flexible crane scheduling methodology for container terminals. Flex. Serv. Manufact. J. 1–7 (2018)
12. Chu, Y., Zhang, X., Yang, Z.: Multiple quay cranes scheduling for double cycling in container terminals. PLoS ONE 12(7), e0180370 (2017)
13. Zheng, H., Liu, B., Dong, Y., et al.: Multi-yard cranes scheduling optimization of inbound full container block with timely relocation. Syst. Eng. Theory Pract. 37(10), 2700–2714 (2017)
14. Li, Y., Zhu, X., Li, W., et al.: Stowage plan based slot optimal allocation in railwater container terminal. J. Control Sci. Eng. 2, 1–10 (2017)
15. Wu, Y., Luo, J., Zhang, D., et al.: An integrated programming model for storage management and vehicle scheduling at container terminals. Res. Transp. Econ. 42(1), 13–27 (2013)
16. Lee, B.K., Kim, K.H.: Optimizing the block size in container yards. Transp. Res. Part E Logistics Transp. Rev. 46(1), 120–135 (2010)
17. Yu, M., Qi, X.: Storage space allocation models for inbound containers in an automatic container terminal. Eur. J. Oper. Res. 226(1), 32–45 (2013)
18. Bazzazi, M., Safaei, N., Javadian, N.: A genetic algorithm to solve the storage space allocation problem in a container terminal. Comput. Ind. Eng. 56(1), 44–52 (2009)
19. Sharif, O., Huynh, N.: Storage space allocation at marine container terminals using ant-based control. Expert Syst. Appl. 40(6), 2323–2330 (2013)
20. Dong, W., Zhang, W., Yu, R.: Convergence and runtime analysis of ITO algorithm for one class of combinatorial optimization. Chin. J. Comput. 34(4), 636–646 (2011)

The Construction of a Domain Knowledge Graph and Its Application in Supply Chain Risk Analysis

Wanyue Zhang[1], Yonggang Liu[2], Lihong Jiang[1], Nazaraf Shah[3],
Xiang Fei[3], and Hongming Cai[1(✉)]

[1] Shanghai Jiao Tong University, Shanghai 200240, China
{wanyuez,jianglh,hmcai}@sjtu.edu.cn
[2] Nanjing Runchain Technology Co. Ltd., Nanjing 210012, China
liu_yonggang@hoperun.com
[3] Coventry University, Conventry, UK
{aa0699,aa5861}@coventry.ac.uk

Abstract. Domain knowledge graphs, which compose scattered information about domain entities, are expressive when organizing information for enterprise systems in the decision-making process. Such knowledge graphs can give us semantically-rich information which can later be applied to fuel different graph mining services to conduct analytical work. In this paper, we discuss a subject-oriented domain knowledge graph based on multi-source heterogenous data consisting of dynamic data generated from daily transactions among companies in interlacing supply-chains and relatively static data demonstrating the basic properties of these enterprises to assist with analytical work. Such high-dimensional graph with strong heterogeneity is rich in semantics and is casted into lower dimensions to be used as inputs for graph mining services, giving us various enterprise correlation chains, aiming to support upper-level application like credit risk assessment. The framework has been testified in real-life information systems.

Keywords: Domain ontology · Knowledge graph construction · Community detection · Data as a service · Supply chain risk

1 Introduction

Full-fledged information about enterprises can be of various use for information systems. Knowledge graphs which capture the relationship among domain entities as well as their built-in features are expressive and rich in semantics, making them perfect for organizing transactional data, so as to reveal the trade volume and business credibility of enterprises and fusing them with relatively static information about enterprises from open data bases. But the heterogeneity of multi-source data leads to the demand of domain ontology construction for data extraction and fusion [1]. Also, information captured by printed contracts and other paper forms recording daily transactions among companies in interlacing supply chains is of large-scale and has temporal relation, which results in the need of dynamic updating as well as batched reading in the

© Springer Nature Switzerland AG 2020
K.-M. Chao et al. (Eds.): ICEBE 2019, LNDECT 41, pp. 464–478, 2020.
https://doi.org/10.1007/978-3-030-34986-8_33

construction of domain knowledge graph to capture the entity relationships in a timely and expressive manner. Another topic that's worth considering is dimension transition, which aims to cast the knowledge graph with high heterogeneity into graphs of lower dimension while fulfilling the input requirement of upper level mining services. Supply chain credit risk assessment can be a significant use case of the mined results of the enterprise knowledge graph. The problem of credit risk contagion is partially triggered by the trading process, and need to be captured and addressed when (1) commercial banks need to make decisions of whether to give loan to an enterprise (2) enterprises need to choose possible business partners. Such features make knowledge graph a suitable tool for analyzing and supply chain credit risks.

In this paper, a framework that constructs a domain specific knowledge graph while addressing the above-mentioned problems is demonstrated. On the basis of the resulting knowledge graph, community detection mechanisms are applied to formulate enterprise correlation chains and give us semantically rich mining results to support decision making process of information systems. These graph mining services are organized according to the business process and the input resources of a series of services are composed using data modeling techniques, making it possible to realize dimension transformation in a schema-on-read manner.

The rest of this paper is organized as follows: Sect. 2 discusses the related works in domain knowledge graph construction, dimension transformation and upper application. Section 3 proposes a complete framework for collecting and processing multi-source heterogeneous data and applying them to construct entity correlation chains which can be of various use for business analysis. The intermediate steps are demonstrated, and detailed description of the data preprocessing, knowledge graph construction and correlation chain generation processes are given. Section 4 is a case study carried out in the context of multiple cooperating enterprises in 2 industrial parks. Finally, we conclude the work and gives some future direction to be further considered.

2   Related Work

2.1   Domain Knowledge Graph Construction

Knowledge graph-based information systems can be of essential use in the times of big data. The mechanisms to be taken into consideration when constructing such information systems include (i) Knowledge representation and reasoning, which touches areas of language processing and demands the extraction of schema graph with the assistance of standard vocabularies [2]. (ii) Knowledge storage, which usually includes graph databases and repositories, and raises the problem of D2R matching. (iii) Knowledge engineering, which can be based on patterns or other mining techniques. (iv) (Automatic) Knowledge learning including schema learning and population [3]. Ontology instance linking, ontology alignment and disambiguation needs to be considered in this phase [4].

Tong et al. [5] attempts to build a nation-wide enterprise KG with information extracted from the Internet so as to facilitate investment analysis. But this work does not take into consideration the valuable information which is generated continuously and extensively when daily transactions take place among enterprises.

2.2 Enterprise Resource Management

Traditional enterprise data warehouses apply the ETL process to the original data generated from transaction processing systems. Data have to be loaded according to a specific schema before they are made available for use in reporting and analytics (schema on write), resulting in the time-consuming schema design process and extra effort for relational joins when needing to accomplish other analytic objectives [6, 7]. Yet another way of accessing data resources is through service governance. Transactional data can be stored and accessed in any structure while no ETL jobs are required, and schema can be determined long after data are loaded by means of composing data (resources) into RESTful services [8]. Basic resource is built as basic structure connected to its data source. It is also called independent resource related to single table in database and is identified with *URI* to store the unique address of a resource, *ObjectID* to be the unique object identifier, *DBConfig* to record the connection information of a data source and *AS* to organize the attribute set with concepts from domain ontology. *BR* s have one-to-one mapping relation with tables of databases. Composed Resource is composed of *BR* s or *CR* s, and can be defined with *RS* to represent different resources relations, *ACS* to be the attribute set of *CR*, C_i to be independent attributes, and r_i to be the relationship between target *CR* and referred resources. Such mechanism of resource configuration can be applied in transforming resources into RESTful services for business purposes, making it possible for storage service provider to offer query in a pay-as-you-go manner. When conducting the recomposition of data to form lower-dimension views, one possible solution is to load the related resources into Spark dataFrames, which facilitates the schema on read process of specifying schema when required attributes are determined by upper level mining services. Also, applying Spark to the data set makes available the GraphX library which brings easy to use graph mining packages.

2.3 Community Detection in Social Networks

The process of discovering cohesive groups or clusters in networks is known as community detection. It is one of the key tasks of social network analysis [9]. A community can be defined as a set of entities related more closely with each other in some perspective than those outside the group. The closeness between entities of a group can be measured via similarity or other distance measures [10].

Basic mechanisms of community discovery include clustering based on modularity. Algorithms utilizing this mechanism include CNM, NewmanFastGN and spectral clustering, which obtains cluster partition by establishing Laplacian matrix and then inputting the clustering method. Another category of community discovery algorithm is done by label propagation. The original Label Propagation Algorithm (LPA) was proposed in [11], in which every node in the graph is initialized with a unique label (i.e. its id), and assigned the label of the highest frequency of occurrence in its adjacent nodes. This original form of label propagation algorithm is simple and shows linear time complexity. However, this algorithm cannot solve problems in which one individual belongs to multiple communities. In other words, communities cannot overlap with each other. To address this problem, SLPA [12] opens up a memory space for the storage of a label set for each node; Speaker nodes choose the label to be transmitted by a customizable rule, listener nodes then select or aggregate the labels sent by its adjacent speaker nodes based on another customizable rule. After multiple iterations, community labels of the listener node are filtered according to some threshold r and the final community attribution is determined. The termination condition of the algorithm is that the label set of each node does not change significantly, in other words, the algorithm converges.

On this basis, there exist COPRA [13] and BMLPA [14] which refine the initialization and updating process of node tags. Firstly, the Rough Core function is used to initialize the network to form several non-overlapping communities, and each node is given a belonging coefficient with regard to the possibility of this node's being a member of one community and its initial value is set to be one. Then, community labels are propagated and updated iteratively. When updating the community label of a central node, the belonging coefficients of nodes neighboring to this central node are summed up and then normalized by community. The labels with relatively small belonging coefficients are removed. This updating process end until the network converges. Finally, if there are any small groups that are included in larger groups, those properly included groups are removed.

In this paper, we propose a knowledge-graph-based information system with domain-ontology-assisted multi-source heterogeneous data extraction and fusion. Graph mining services, typically community detection mechanisms, take this rich-sematic knowledge graph as input, and attempt to support decision-making process of end users with entity correlation chains as an intermediate mining result.

3   Framework

There are altogether 3 major steps in the proposed framework. First of all, data payload is extracted from open knowledge bases and printed contracts or other paper forms recording transactions [15]. Such multi-source data are fused with the assistance of domain ontologies, thus finishing the data preprocessing and knowledge graph construction phase. Then, the information captured by the knowledge graph is reorganized according to different mining objectives and the original knowledge graph is casted into different views. In the next step, diverse graph mining services, which demand different patterns of resource combination as input, construct different entity correlation chains to facilitate different application objectives including risk analysis, link prediction and business partner recommendation. The data organized as a knowledge graph are stored in graph data bases and temporal databases, to facilitate fast query speed and reveal time series to adjust to the interconnectivity and extensiveness of daily transaction data. The composition process is also assisted with service ontology, which organizes a series of services and determines the inputs required by this series of services. In this paper, we generate various entity correlation chains to be the intermediate results to support different business objectives. Such framework is demonstrated in the Fig. 1:

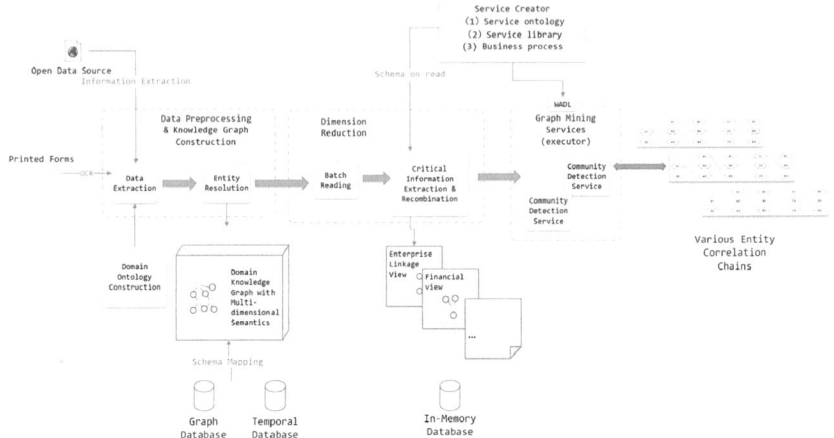

Fig. 1. System framework.

4   Method

4.1   Data Preprocessing Assisted with Domain Ontology

Data utilized to reveal various trading processes are extracted from printed contracts with the assistance of OCR, and every transaction is stored in a semi-structured format so that it can be retrieved. Aside from the printed contracts, we also have to attain information revealing some relatively static features of companies participating in supply-chain activities from online enterprise knowledge bases. Such diversity and complexity of data sources results in the heterogeneity and redundancy of data, and therefore raises the demand of domain ontology construction. Instead of striving to be comprehensive as open knowledge bases, in this paper we attempt to construct domain ontologies according to business objectives. As the schema of entities, attributes and entity relationships stay relatively static in the business context, the topic of ontology evolution is not taken into consideration in this framework.

We apply a top-down knowledge graph construction process to obtain 5 domain ontologies. Instances of different types are linked accordingly. After collecting domain-related vocabularies, developing the domain concept hierarchy and defining the properties and relationships between concepts, the resulting ontology graph is shown in the figure below (Fig. 2):

We consider 5 classes in the schema layer of knowledge graph, namely, *Person*, *Company*, *Contract* and *Activity*. Properties of the *Company* node are designed according to generalized user portrait of supply-chain enterprises, and has static attributes of a company (name, capital, type, etc.), as well as properties that changes when this company participates in trading activities (contractID, amount, location, etc.) within certain supply chain.

Bilateral or tripartite contracts are aggregated into *Activity* nodes so as to reveal the interconnection between companies without too much redundancy in the resulting graph. However, information of original contracts is also stored by *Contract* nodes in

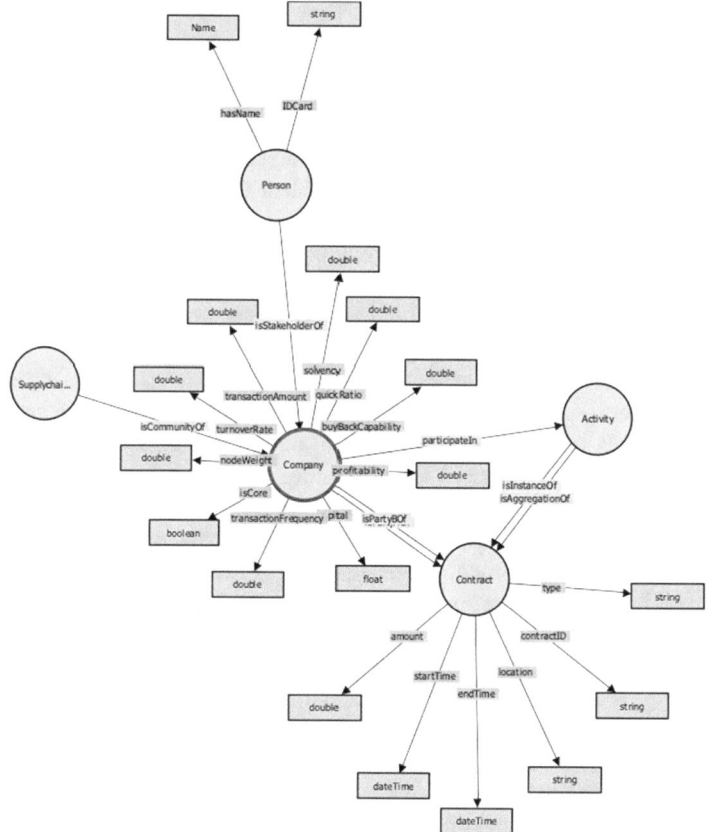

Fig. 2. Schema of corresponding domain knowledge graph.

the knowledge graph so that single transaction remains retrievable. *Supply Chain* entities are the result of higher-level graph mining services.

4.2   Knowledge Graph Construction

Knowledge graphs are used to logically integrate multi-source raw data. The construction process of domain knowledge graphs is usually a combination of the top-down method (designing ontology first and then conduct instance linking, the traditional method used by RDF graph models) and the bottom-up method (extracting data from logs or web tables and utilize critical patterns to enrich the semantics of knowledge graphs).

In this system, we utilize the top-down method as it better fits the context of supply chain credit risk assessment for enterprises. During this process, 3 problems need to be addressed properly: (1) Different entities with the same name need to be distinguished based on their semantic context, (2) Different names which coreference to the same entity need to be linked together, (3) Some references may not have been recorded and such relation need to be inferenced and stored. Taking these into consideration, large-scale heterogeneous data of domain entities form a semantically interlinked graph.

4.3   Dimension Reduction

Critical information describing logistics, finance, and supply chain status is extracted from the domain knowledge graph, loaded into in-memory databases, and recombined into new nodes with new properties to facilitate graph mining services and support business-oriented applications including risk analysis, trade partner recommendation for new contenders and influence prediction. Most of these applications require the formulation of supply chains.

It is worth noticing that in the business context, trading information need to be updated in a batched format. Outdated information is excluded from the knowledge graph in use, but the original data in the semi-structured format are stored persistently for possible use and credit traceability.

The critical information extraction and recombination process is consistent with the resource allocation process. After the dimension reduction process, information from the domain knowledge graph are abstracted to be basic or composed resources to be transformed into RESTful services for the calling of graph mining services. With the assistant of domain ontology, basic resources are composed to support higher level aggregation of calculation. Also, such model can adapt to changes as business objectives or the ontology structure changes [15] as more transactional data are generated or new business applications are required by end users.

4.4   Service and Resource Combination Subjecting to Trading Process

When conducting supply chain risk analysis, the trading process need to be captured from multiple perspectives and the consistency in goods and money need to be assessed. To accomplish such business goal, services are invoked, taking the up-to-date information from the database recording transaction data as input.

Since services can be composed according to business process and objectives [8], it is necessary to formulate service models to describe the interdependency of various services and the required resources as inputs. In this paper, we utilize the service ontology [17] to match candidate services from the service repository and organize the resources required by the resulting series of services.

The service ontology contains five classes, which are *Service, TaskSet, ServiceProvider, ServiceConsumer, ResourceSet* and *ServiceLimitation*. The related properties include *conducts, provides, providedBy, consumes, consumedBy, utilizes, limits, limitedBy, delay* and *queue*. Here, we compose microservices with different objectives with the assist of such ontology, based on the task a service can conduct and determines the required resources of the composed service groups based on the union of ResourceSets of a series of services composed in the previous step. Semantic relations of *Association* are utilized to indicate that one resource is consumed by a service or a task. It is also worth noticing that such service modeling mechanism can also be applied in cloud service discovery.

Since knowledge graph data are acquired with the mechanism described in Sect. 4.2, the composed resource can respond to the calling of composed services to realize certain business goal.

4.5   Supply Chain as Mined Results of Community Detection Mechanisms

Community detection mechanisms are practical in grouping multi-dimensionally correlated companies based on their static properties or trading partnership over a certain period, providing a medium for business personnel to do data analysis from various viewpoints. Therefore, graph-based pattern matching as well as a modified balanced multi-label propagation algorithm is utilized by community detection services to find groups of cooperating companies. The original form of BMLPA is adjusted to consume enterprise nodes and their trading relationships as inputs and gives the output of groups of closely cooperating comprises in the form of entity correlation chains. These groups are later stored in the knowledge graph as supply chains, to support various types of analysis. The modified BMLPA is consistent with the original form of BMLPA in the three stages of core initializing, label propagation and the elimination of properly included small communities, as is described in Sect. 2.3. The adjustment is mainly made in the ranking criteria of nodes and link weights are considered in the propagation process. The algorithm is described in detail as follows:

Algorithm 1: Initialize community labels for companies

Input: Set of company nodes

Output: List of cores

1: Sort all nodes in decreased order of the selected property P. and add to vList sequentially

2: foreach vertex i in sorted vList:

3:   if $ki \geqq 3$ and i.free = true:

4:     find vertex j which has the largest P in N(i)

5:     and j.free is true;

6:     add vertices i and j to a new core;

7:     commNeiber $\leftarrow$ N(i) $\cap$ N(j);

8:     sort commNeiber in the order of vertex degree from small to large;

9:     while commNeiber $\neq$ null:

10:       foreach vertex h in sorted commNeiber://add common neighbor of i, j to core

11:         add h to the core;

12:         delete vertices not in N(h) from commNeiber;

13:         delete vertex h from commNeiber;

14:     if sizeof(core) $\geqq$ 3:

15:       add core to cores;

16:   return cores;

First of all, we use RC (Rough Cores) to extract non- overlapping cores, which are cliques of companies that should be categorized into the same community. We use node property P as the classification criteria. P can be selected to be company's registered capital or sum of contracts' amount a company is involved in, which gives us communities of lowest risk and highest potential return on investment sequentially. The Rough Core algorithm is described in detail in Algorithm 1.

i.free is used to denote whether vertex i belongs to any core, and is initialized to be true for all vertices, thus this first stage of algorithm outputs non-overlapping cores. ki represents the degree of vertex i.

The second stage is the iterative update of community identifier and belonging coefficient of each vertex. This stage corresponds to that of the original BMPLA except that link weights are considered in the propagation process. This stage is demonstrated in Fig. 3:

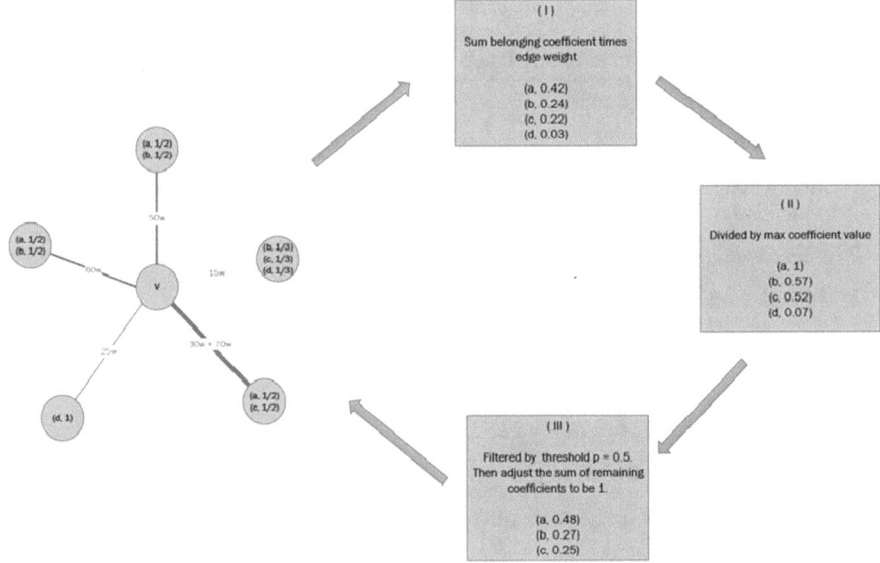

Fig. 3. Propagation of multiple labels for the central node v considering link weights

The links (*activities*) are the aggregation of contracts. We give weights to the links correspond to the amounts of contracts making up an association. To calculate the labels of the central vertex V, first we sum the belonging coefficient, time the link weight, for each label of all neighbors. After finding the maximum value in the outputs of the stage (I), we divide each belonging coefficient by the max value in stage (II). The labels with ratio larger than threshold p are kept as the labels of the central vertex V in this iteration. The belonging coefficients are then normalized to sum up to 1 in stage (III). The difference between the propagating process in this paper and that in the original BMLPA is that we take the link weights into consideration and do coefficient propagation correspondingly.

The above-mentioned stages give us communities of enterprise entities within multiple interlacing supply-chains. As is shown in Fig. 4.

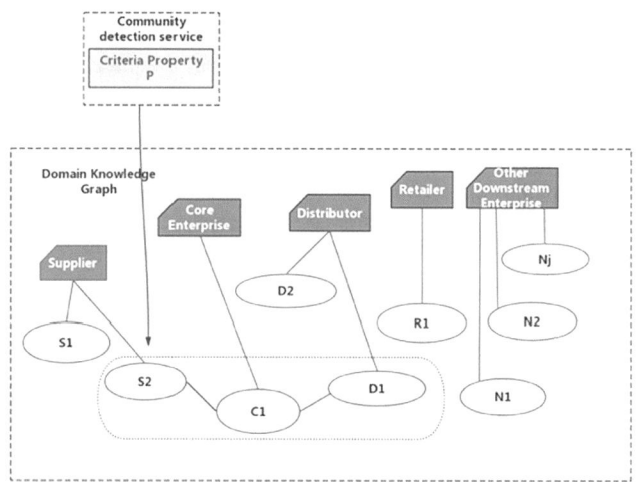

Fig. 4. Community of enterprises based on given criteria property P

Since enterprises are likely to cooperate in a sequential order, forming correlations in the format of chains (supply-chains). The mined communities are also likely to have a chain format, giving us an enterprise correlation chain. Risk assessment can be conducted on such mining result and gives us ranking of supply-chain enterprises in the decreasing order of credit risk. Finally, end users like commercial banks can make loan decisions accordingly.

5   Case Study and Discussion

5.1   Case Study

Small and medium-sized enterprises (SMEs) face the problem of excessive internal financing. It is difficult for commercial banks to assess their credit status because of the information asymmetry between borrowers and lenders and the potential unreliability of their financial information. Supply chain finance, however, enables coordinating enterprises to be credited as a whole, thus opens up new sources of profits for banks and reduces the financing costs of SMEs. However, in the complex process of trading among participants, the dynamic changes and interactions of capital flow, logistics and information are difficult to capture and applied to credit granting. Existing risk assessment systems often rely on large core enterprises' credit guarantee because it is difficult to control the risks caused by active or passive delivery delay and double-spending of digital currencies which happens dynamically in trading activities.

Enterprise knowledge graphs put forward a new way of organizing and utilizing the relatively dynamic information generated in trading activities. Different dimensions of information can be integrated and analyses can be done based on such fused information. Aside from the task of credit risk assessment mentioned above, other possible use cases include influence assessment, business partner recommendation and supply chain stability assessment, all of which can benefit from the intermediate mining result of a supply chain.

In total, we adopt information of 201 coordinating companies and mock 1046 trading activities among them based on some given contracts as templates. These companies perform different roles like supplier, distributor and core enterprise. Revoked companies and companies with critical missing properties are ignored. Figure 5 shows 1000 nodes of the resulting knowledge graph, with 799 contracts (red nodes) and 201 enterprises (blue nodes) from the original dataset.

This study takes contract data extracted to semi-structured databases as input, imitating data extracted from printed forms recording various transactions, to form instances of contracts, and further aggregates contracts between same participating companies to be links in knowledge graphs. More information revealing relatively statistic information about supply-chain enterprises is extracted from open knowledge bases and fused with the assistance of domain ontology shown in Fig. 2.

Fig. 5. The resulting knowledge graph with enterprises and activities as nodes

The original data are casted into lower dimension based on the required format of input of upper level application. One example of such dimension reduction is shown in Fig. 6.

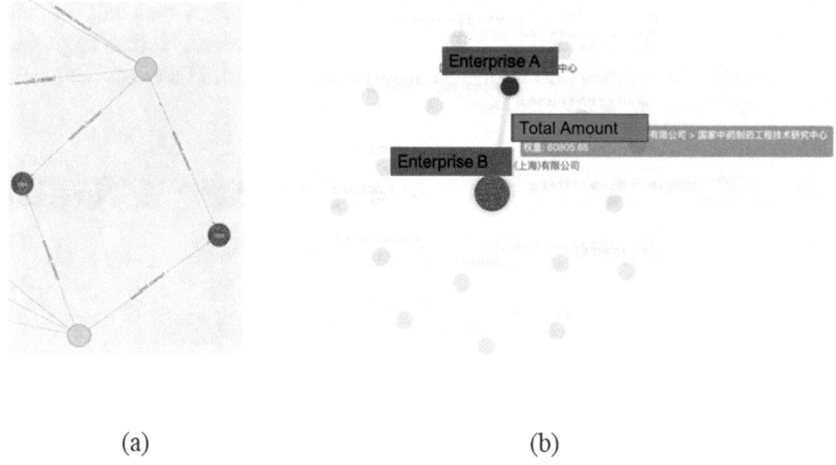

(a)   (b)

Fig. 6. Multiple contracts are signed between two enterprises, resulting in multiple red nodes in the original knowledge graph stored in Neo4J graph database. (b) Multiple contracts are aggregated into one trade relation to facilitate upper level community mining service. Details of this relation can be retrieved as is shown in Fig. 7.

Details of aggregated nodes and relations in the dimension-reduced graph can be retrieved, As is shown in Fig. 7.

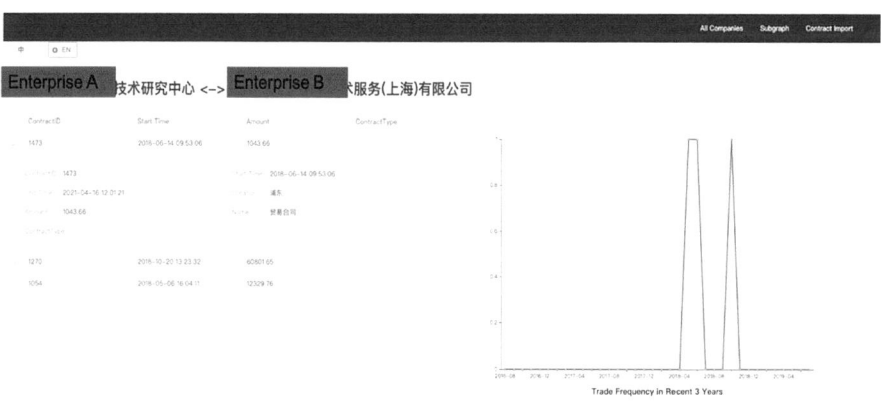

Fig. 7. Details of the aggregated relation can be retrieved.

There are 2 ways of generating enterprise correlation chains from the above-mentioned knowledge graph. The first one is through pattern matching. Enterprises are tagged by their functionality, and grouped by certain supply chain patterns. Data required by such service include tags indicating enterprises' functionality as well as their ids and names.

So when the pattern mining based graph service executes, it evokes data extracting services which read data payload from the related database tables, and data aggregation services organize the corresponding information to be the required format. An example of the pattern matching result is given in Fig. 8.

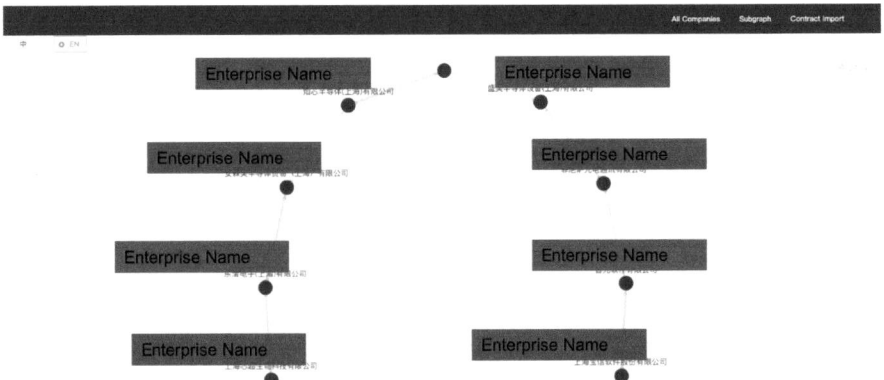

Fig. 8. Two supply chains following the pattern 'third-tier supplier –second-tier supplier – first-tier supplier – core enterprise – distributor' interlace with each other to form a long enterprise correlation chain. Edge width is associated with the amount of total transactions between two companies.

Another category of graph mining services utilize community detection mechanisms. In this paper we apply BMLPA with the registered capital of enterprises as criteria property for the initialization of cores is conducted on the enterprise linkage view, giving us communities of enterprises in the chain form. Considering the real-time requirement of industrial applications and the distribution of data that make the BMLPA execution process slow when converging, we set the iteration times to be 200 and threshold to be 0.75, giving us the resulting entity correlation chains, which are later stored in the graph database as supply-chain nodes, to fuel more business-related analytical tasks.

After dynamic transactional information is aggregated into entity correlation chains, more business-specific services consume the correlation and conduct tasks including (1) Analyzing the structural stability of supply chains; (2) Credit assessment of the companies involved in one supply chain; (3) Influence ranking of companies involved in identical supply chains; (4) Business partner recommendation for new entrant of market.

5.2   Discussion

Traditional methods that conduct supply chain risk analysis give credit only to large core enterprises in supply chains, with their non-current assets as guarantee of their credit. Small or medium sized companies rely on the strong credit guarantee of core enterprises. In the data processing phase, without proper fusion of transactional data and relatively static information, the transformation in form of current asset triggered by daily transactions cannot be captured. Conversely, with transparency and real-time guarantee, knowledge graph – based systems with transactional data is able to assess risk emerging and spreading in business processes. As is shown in Table 1.

Table 1. A comparison between utilizing knowledge graphs in addressing supply chain credit risks and other possible means.

	Knowledge graph - based system with transactional data	Other systems [16]
Information captured	Printed contracts recording financing and logistics information, as well as static information showing basic credit status of supply chain participants are captured, thus the transformation between different forms of assets can be accounted	Without proper fusion of multi-sourced data, assets cannot be accounted in time
Transparency	Partial information is exploded to clients, giving privacy assurance to clients' trade data. But there exists a data center	Attempting to utilize block chain and smart contracts to protect sensitive data
Real-time	Trading data are updated in a batched formatted. The frequency of updating can be determined by end-users	Vary between systems
Risk assessed	Risk emerging and propagating in business processes can be captured	Usually, only the credit risk of large core enterprises can be assessed

6   Conclusion and Future Work

This paper demonstrates the top-down construction process of domain knowledge graph, with ontology-assisted multi-source data integration and dynamic updating of daily transaction data. The knowledge graph is casted into lower dimensions according to the required inputs of upper-level graph mining and business analysis services. The dimension reduction process is done with the assistance of the resource modelling approach, so as to better adjust to the dynamic composition of services in the running time period. This work also shows the expediency of using the methods of community detection on graph-based knowledge base to evaluate credit of a given company in a specific trading activity. A comparison with other systems is listed in Table 1.

Other possible use of the domain knowledge graph is fraud detection and potential business partners recommendation. To accomplish such tasks, the data sources of the framework needs to be extended and more ontologies should be constructed to capture domain concepts to form a more comprehensive knowledge graph to support more complex semi-structure decision-making processes. We can also consider expanding knowledge graph construction phase by applying data-driven methods to update the KG schema as more concepts are generated from source data.

Acknowledgment. This research is supported by the Development of E-commerce Service Platform Architecture and Data Service Project under Grant 2017C02036.

References

1. Xu, B., Xie, C., Cai, H.: Application of domain-ontology method in meta-data management of data warehouse. Appl. Res. Comput. **11**(27), 4162–4164 (2010)
2. Sun, H., Ren, R., Cai, H., et al.: Topic model based knowledge graph for entity similarity measuring. In: 2018 IEEE 15th International Conference on e-Business Engineering (ICEBE), IEEE, pp. 94–101 (2018)
3. Gomez-Perez, J.M., Pan, J.Z., Vetere, G., Wu, H.: Enterprise knowledge graph: an introduction. In: Pan, J.Z., Vetere, G., Gomez-Perez, J.M., Wu, H. (eds.) Exploiting Linked Data and Knowledge Graphs in Large Organisations, pp. 1–14. Springer, Cham (2017)
4. Heflin, J., Song, D.: Ontology instance linking: towards interlinked knowledge graphs, p. 7 (2016)
5. Ruan, T., Xue, L., Wang, H., Hu, F., Zhao, L., Ding, J.: Building and exploring an enterprise knowledge graph for investment analysis. In: Groth, P., Simperl, E., Gray, A., Sabou, M., Krötzsch, M., Lecue, F., Flöck, F., Gil, Y. (eds.) The Semantic Web – ISWC 2016, vol. 9982, pp. 418–436. Springer, Cham (2016)
6. Moody, D.L., Kortink, M.A.R.: From enterprise models to dimensional models: a methodology for data warehouse and data mart design. In: DMDW, p. 5 (2000)
7. Bakalash, R., Shaked, G., Caspi, J.: Enterprise-wide data-warehouse with integrated data aggregation engine: U.S. Patent 7,315,849, 1 January 2008
8. Cai, H., et al.: IoT-based configurable information service platform for product lifecycle management. IEEE Trans. Ind. Inform. **10**(2), 1558–1567 (2014)
9. Tang, L., Liu, H.: Community Detection and Mining in Social Media, Synthesis Lectures on Data Mining and Knowledge Discovery. Morgan and Claypool, California (2010)
10. Bedi, P., Sharma, C.: Community detection in social networks. Wiley Interdisc. Rev. Data Min. Knowl. Discovery **6**(3), 115–135 (2016)
11. Zhu, X., Ghahramani, Z.: Learning from labeled and unlabeled data with label propagation. Technical Report CMU-CALD-02-107, Carnegie Mellon University (2002)
12. Xie, J., Szymanski, B.K., Liu, X.: Slpa: Uncovering overlapping communities in social networks via a speaker-listener interaction dynamic process. In: 2011 IEEE 11th International Conference on Data Mining Workshops, IEEE, pp. 344–349 (2011)
13. Orman, G.K., Labatut, V., Cherifi, H.: Comparative evaluation of community detection algorithms: a topological approach. J. Stat. Mech. Theory Exp. **2012**(08), P08001 (2012)
14. Wu, Z.H., Lin, Y.F., Gregory, S., et al.: Balanced multi-label propagation for overlapping community detection in social networks. J. Comput. Sci. Technol. **27**(3), 468–479 (2012)
15. Yu, H., Cai, H., Zhou, J., et al.: Data service generation framework from heterogeneous printed forms using semantic link discovery. Future Gener. Comput. Syst. **79**, 514–527 (2018)
16. Zhang, S., Miao, Q., et al.: A Management Method and Platform of Credit Exchange Based on Supply Chain Finance. CN:107767269 (2012)
17. The Service Ontology. https://dini-ag-kim.github.io/service-ontology/service.html

Towards Self-automatable and Unambiguous Smart Contracts: Machine Natural Language

Peng Qin[✉], Jingzhi Guo, Bingqing Shen, and Quanyi Hu

Faculty of Science and Technology, University of Macau, Macau, China
{yb77428,jzguo,daniel.shen,yb77476}@um.edu.mo

Abstract. A smart contract originally drafted by natural language is an essential task of many applications in blockchain technology. Firstly, natural language cannot be directly executed by computers, self-executing requires terms of the smart contract be computer-readable and executable. Secondly, in crossing environments or parties, contract translation needs the overall meaning of a sentence to have a meticulous precision, besides, low tolerance of mistakes for reducing a tedious process. Lastly, many kinds of templates of smart contracts need a common sense of agreement where each party agrees on the context of the contract. This paper explores the problems of the smart contract in natural language and self-executing to redefine the smart contract through an approach, which supports a human-readable, computer-understandable and self-executable contract representations with enabling semantic structural based on Machine Natural Language (MNL). Meanwhile, a common dictionary (CoDic) transfers natural languages into universal machine codes or languages without the ambiguity across parties.

Keywords: Machine natural language · Smart contracts · Semantic document · Universal grammar

1 Introduction

A smart contract is an automated legal contract with the expression using code to execute, verify, enforce terms of agreements in hardware and software. It prevents damage or control assets by digital manner [1], and can simplify contract processes, reduces costs by eliminating intermediaries, and law enforcement to avoid seeking the protection of legal establishments such as courts [2, 8]. However, terms and structures of contracts are more complicated in the real world. Contract documents are natural language written in ordinary sentences instead of in a constrained and defined mathematical form. Every social, legal, economic, and technical interaction has informal and formal protocols to achieve certainty of meaning. Natural Language Processing could be a branch of analyzing sentences enabling computers to understand natural language expressions (such as e-mail, newspaper articles, or legal texts). Complicated contracts have to be forced into natural language and code. Table 1 shows a simple template of a contract.

© Springer Nature Switzerland AG 2020
K.-M. Chao et al. (Eds.): ICEBE 2019, LNDECT 41, pp. 479–491, 2020.
https://doi.org/10.1007/978-3-030-34986-8_34

Table 1. A sample of contract

Date:	February 10, 2019
From:	Hello School

System must send record before 04/09/2019
Student contacts the professor by 20/10/2019
Student must pay $20 at 3:00 pm

Thus, three key issues are found in Table 1:

- *Term – Ambiguity*: Creating smart contracts doesn't need the written agreement obligations in an exceedingly strict sense. However, if a contract is expressed in code, then the understanding of the code between the parties has relevancy. This relevancy is a matter of regulation, like whether specific parties (such as consumers) are perceived by the terms, whether there was adequate sympathy of the terms to create the contract in the slightest degree. Meanwhile, it is not possible for legal contracts which are written by meticulous precision for a single word that may lead to expected financial consequences, and long-term disputes [3] such as "record" may have multiple meaning in Table 1. It is very unclear how a smart contract could be coded to give effect to such terms, and how is a computer to judge whether the contract in Table 1 has been violated. It is worth considering though whether the meaning of terms by itself is enough and would also indicate that participants have certainty of terms. It should not be forgotten that this process is not limited to ambiguity in words, but it may reveal the existence of ambiguity. Therefore, in the context of smart contracts, the problem is not that the parties disagree about the meaning of the text, but those participants who decide to translate obligations into code may make incorrect interpretations.
- *Multilingual contracts – Common sense*: There are two issues: (1) Natural language is a complex, novel, abstract, and unstructured language sentence. Smart contracts can only "understand" structured natural language in a relatively limited way. Contract parties, lawyers, or other document readers (including legal texts) rely on higher-level cognitive language processing. Legal language may be more challenging to translate into code than ordinary natural language. Therefore, it cannot approach the understanding ability of ordinary literate people. [9, 10] suggested to use smart contract template based on Ricardo contract concept, but it bridges the gap that cannot achieve communication of intermediate language which lawyers could not understand the smart contract code. (2) A contract containing terms expressed in code is like a contract expressed in different languages. If a contract expressed in multiple languages is invalid, then a contract expressed by code and natural language may not be valid either. Translation problems arise when computer scientists believe that there is no problem with the performance and semantics of legal and commercial prose. [11, 12] only consider the issue of creating smart contracts from the perspective of subject-matter expert's views. Among them, [11]

proposed a semi-automatic method translating human-readable contracts into smart contracts on Ethereum. However, this solution is fixed on the Ethereum and not clear how it can be extended or adapted in another environment. Besides, it does not contain the legal and business semantics that lawyers use to draft the denotational semantics. For the state-of-art machine translation tools, Google and Bing perform an excellent translation result using artificial intelligent approach. Approximation, where the overall meaning of the sentence should be allowed in automatic translation among natural languages. However, accuracy or exact meaning seems to be crucial when smart contracts are self-executing, it cannot be stopped or modified because the meaning of words always depends on the context. Therefore, the successful performance of a contract may depend on the meaning of the sentence to produce a universal smart contract for each party. If a part of a contract is to be carried out in a foreign country, it may be more efficient to express contracts in that foreign language. Thus, those contracts should perform these terms in their languages.

- *Code Self-executing*: Smart contracts can improve efficiency and performance as the automation of contract terms. This automation is achieved by controlling the computer code of automation performance. It raises a fundamental question - whether sentences or paragraphs expressed in computer code are meaningful and effective. [7] develops a code generator that converts the code generated by the Idris compiler into Serpent code and then compiles it into EVM bytecode. However, this method does not solve the problem of the terms' semantic. The structured approach allows computers to identify core contract information, but it does not state that the computer system understands the "meaning" of the contract terms inside the structure (i.e., what an "*expiration date*" means). The state-of-art smart contracts approaches are self-execution based on specific templates, which largely limit the content of contracts. Contracts are composed of many complex elements, as shown in Table 1. To achieve real self-execution, we should solve the contract itself, not the template, so that make computer understandable.

From the facing problems of the state-of-art approaches, it suggests that: (1) Smart contracts should be understood by different backgrounds and develop a common-sense contract to interact and communicate, such as lawyers and developers. (2) It should allow computers automatically to execute and understand the language of smart contracts. (3) It should have an explicit term definition across languages that can be mapped from terms to code, so that reduces the ambiguity from natural language to the smart contract.

Smart contracts originate from a document written by human natural language and then is converted to code, but it is not possible to automatically convert natural languages into code without ambiguity [4]. To avoid the term ambiguity of translating legal language into code, a common dictionary based on the techniques of collaborative conceptualization approach [5] supports the smart contract both human-readable and computer understandable and executable. Meanwhile, to leverage a broader understanding of natural languages and make the contract universal and free, we propose a novel smart contract framework by translating the sentences from human natural language to a semantic structure contract by Machine Natural Language [6, 14] for further self-execution. Section 2 briefly introduces the background and our related works in term and sentence handling. We explore a novel framework and redefine the smart contract through MNL discussed in Sect. 3. Finally, Sect. 4 gives a conclusion and outlook.

2   Background

2.1   Term - CoDic

The Conex Dictionary (CoDic) is a common vocabulary designed by us under the CONEX project [5]. Each concept of a dictionary term or phrase in CoDic is uniquely identified as the internal identifier $iid \in$ IID, which is neutral and independent of any natural language. In our lab, a sign is denoted by a set of grammatical features, which are:

$$Sign: = (Signifier,\ Signified)$$
$$:= (Structure,\ Concept)$$
$$:= (Form,\ Sense)$$

where:

*Form := (IID, **Term** | unique, literal);*
Sense := Meaning *definition of a sign, i.e., the form conveys the sense.*

In this definition, term T is the literal representational form of a sign. For example, the term "*bank*" with a meaning "*A business establishment in which money is kept for saving or commercial purposes or is invested, supplied for loans, or exchanged.*" has an identifier iid_1 = 0x5107df 00321f (see Fig. 1). It's another meaning "*A supply or stock for future or emergency use*" has another identifier iid_2 = 0x5107df003226 (see Fig. 2).

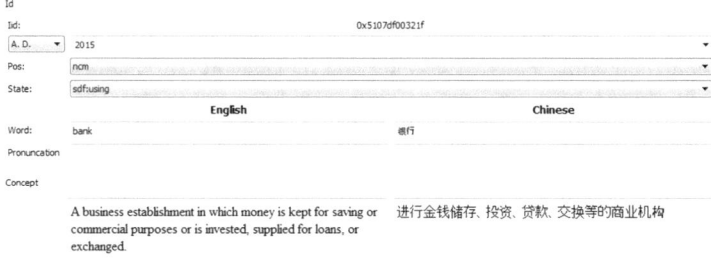

Fig. 1. "*Bank*" with the meaning of "*business establishment*" in CoDic

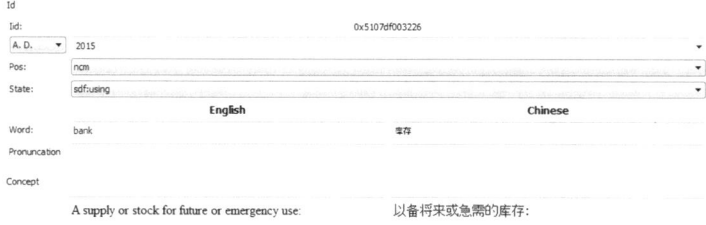

Fig. 2. *"Bank"* with the meaning of *"supply"* in CoDic

The form of a sign is again grammatically categorized into many sign classes based on predefined part of speech (PoS). Specifically, in CoDic, PoS is combined with the ID form of the sign, such that:

$$\text{IID} := \text{POS} + Y + \text{ID}$$

which is the universal sign representational form. For example, *iid* = 5107df022918, in which 1 after 5 refers to a common noun, 7df refers to the year 2015, and 022918 is ID. In CoDic, *Year* is originally designed for term deriving by time. The sign is the theoretical foundation of machine natural language described in this part. In the rest of this paper, the sign theory will be adopted as the atomic building terms for smart contract syntax and semantics.

2.2 Machine Natural Language

Machine Natural language (MNL) [6, 14] is the high-level sentence construction from terms designed by CoDic in part A. In the MNL, a sentence is a set of IID are presentational form, in which each *iid* ∈ IID has a sense and combined with other *iids* by means of attaching a grammatical case to each *iid*. Unlike many native human natural languages that construct a sentence by giving subject S, predicate V and objects O, MNL only has case concept for each *iid*. The case associated with an *iid* tells how an *iid* grammatically function in a sentence and combines with other *iids*. Thus, cases of signs are grammatical features of sentence construct, which are also called extrinsic properties of a sign. The structural combination of a sentence is theoretically determined by the connotation relationship of a sign in a bi-tree data model. Grammatical cases are the specific tool to implement the connotation relationship for a sentence, and a sentence is always confined in a meaning group (MG), a meaning group is a concrete and independent sense that can exist by its own. Most importantly, in the domain of computers, MNL makes the capable of representing any sentences readable and understandable by computers.

3   From MNL to Contract

3.1   A Novel Framework of Smart Contract

What is the mean of "computer-understandable" in a smart contract? Contracts are regarded as "natural language" texts. Therefore, a contract consisted of natural language should be understood and executed by computer without ambiguity. In our approach, the natural languages of smart contracts are automated execution with the use of terms based on CoDic to perform unambiguous terms and product universal contracts through MNL approach. It provides a translation between a given word and a set of computer instructions producing outputs that are consistent with what a person would understand the word to mean by CoDic. The novel smart contract framework based MNL approach is modeled and defined in Fig. 3, which comprises three processes: (1) *pre-defined contract* (2) *MNL universal representation* (3) *self-executing*. The first step, *Pre-defined contract* is a process of converting a contract which is only readable and understandable by the human being in a human natural language to a sequence of *iid* by CoDic, which conveys the "meaning" of contractual terms to computer systems and tells a computer what a word means is. The essence of this process is to transform a sequence of terms only understandable by the human in the contract to a set of symbolic signs (or symbols) readable and understandable by a computing program. The second step, *MNL universal representation* is the step of designing a universal sentence-based grammar to regulate how a list of concepts in terms of signs should be aligned [6]. It generates a universally readable and understandable sentence by a computer program in an MNL manner. In particular, this process sequentially experiences the activities of case generations (i.e., appending a grammatical case to each symbolic sign to indicate its grammatical functions and intrinsic property). The third step, *self-executing* is to decode and execute the sequences of *eiid* in the MNL sentence by computers through activities of contract rules (i.e., to extract corresponding terms from signs from CoDic). The goal of this paper is to build a novel MNL mechanism and accurately transform a simple sentence from a natural language contract to smart contract across parties, resulting in a semantic consistent sentence translation without semantic ambiguity in the smart contract.

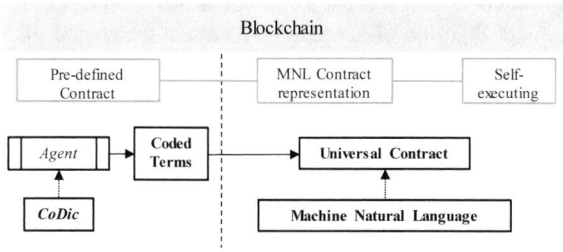

Fig. 3. A novel framework of smart contracts

In our approach, the characteristics of the proposed framework for a smart contract are:

- Making terms computer-readable and understandable through unique IID with unambiguity and accurate meaning,
- Designing a universal "template" which is semantic MNL sentence acquired from language grammar rather than trained machine data,
- Sharing a common sense of structure in mind for computer and human adapting to any crossing parties,
- Putting forward better self-execution by a sequence of meaningful *iids*.

3.2 Pre-defined Contract

The process of the pre-defined contract is conveying meaning terms from contract to the computer. Computers don't understand terminological meanings that correspond to the way people understand the meaning at a deeper level. As talked in Sect. 2, CoDic defines any objects as collaborative concepts to represent things as signs of semiotics. Thus, when any parties or agents from different contracts intend to communicate unambiguously, they must first build consistent signs collaboratively and semantically. A semantic input method called SIM [13] plays a role to allow the user to input data by selecting the exact meaning from the drop-down list when entering terms, users need to input any terms through SIM based on the CoDic [5].

Particularly, the smart contract is self-executing by codes, there exist some meaningful values to be solved, such as execution day and location of a contract. Thus, CoDic defines "noun of typed value" which is the *typed value* (*ntv*) specified by a data type for handling values of a smart contract. For example, *Decimal: 1234.66* denotes that the PoS is a decimal data type, and the run-time value of the data type decimal is 1234.66. Another example is *Expression: $e^{1/2}$*, which means that data type is "*expression*" and the expression is $e^{1/2}$. A typed value takes a TLV (tag, length, value) format, in which tag denotes data type, length tells the byte length of value followed. For example, *Int:1: EB* means that Int is a tag of PoS, *1* is the length for 1 byte of value, *EB* is hex value. Figure 4 shows that natural languages are generated into a machine-readable and understand the language in a common mind between humans and computers.

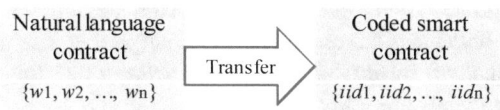

Fig. 4. A Code generation from natural language to machine

In summary, the *pre-defined contract* allows computers to automate and evaluate consistency with specific contract terms and is helpful for the computer to read and execute the value, such as time, location. It makes terms, all parties to convey the meaning with the common dictionary by computers. The computer system also "knows" the exact term meaning of contract but is automatable.

3.3 MNL Contract Representation

If a smart contract is efficient, it must be flexible enough to operate natural language, which is also a part of smart contracts following coding terms. MNL is fundamental in describing the grammar of a sentence, represented by a set of concepts and the relationships between them. Because the smart contract could not solve natural language contracts, we identified the concepts or entities of a term and defined them as elements of cases. Based on MNL, a sentence is to append a case to a term based on the analysis of a term on its PoS, its position in a sentence, and its structure. A case is a grammatical property of a term, which reflects the grammatical function performed by the term in a contract. The term relationships in a sentence are known through the defined term sequence and term position in a phrase, a clause and a sentence approach for any contracts. Available cases in MNL are Nominative (N), Accusative (A), Dative (D), Genitive (G), Predicative (P), Copulative (X), Adverbial (B), Complemental (C), Emphatic (E), Absolutive (L) and Fixed (F). Because smart contracts are much simpler than natural languages. Therefore, we defined a part of cases as the "template" of smart contracts. We defined the types of sentence and cases, and all sentence cases for the smart contract are designed as:

- Nominative (N)
  A nominative is consisting of words that function as the initiator of an action, an event or a state. It controls a verb that is followed. In the smart contract, it takes the function of a finite subject noun. It is critical that the code understands the context of the sentence to ensure that an appropriate semantic meaning is made.
  e.g. *White and black* cannot be told. (N)
- Accusative (A)
  An accusative is a function as the direct recipient of an action, an event or a state. In the smart contract, it takes the function of a finite noun for a direct object of a verb. The accusative describes the goal. The interpretation of the accusative will determine the object of the sentence. However, there is ambiguity about the meaning as applied to the sentence. For example, "*The task is finished by the student*", "*is finished*" may be coded as an implicit Predicative (P), in which case, the "*task*" would be appropriate. The Accusative is often equivalent to the direct object of the sentence, and it helps avoid ambiguity when interpreting a sentence to distinguish the actor (Nominative) from what the actor is acting upon (object).
  e.g. *Cat catches the mouse.* (A)
  e.g. *He writes 7659.* (A)
- Dative (D)
  A dative is denoting an indirect recipient of an action. In the smart contract, it takes the function of a finite noun for the indirect object of a ditransitive verb.
  e.g. *He gives me a book.* (D)
  e.g. *The program assigns X5T a value.* (D)

- Genitive (G)

  A genitive describes the attributes of a noun and takes the function of a finite adjective in the smart contract. It represents the part of the statement that modifies the N, A, and D. As such, the genitive can be thought of with the operators "*all*", "*three*" and "*red*". In other words, the genitive sets the prerequisites or restrictions on the N, A, and D.

  e.g. *Three men are walking along the street.* (G)

- Predicative (P)

  A predicative is denoting an action, an event or a state. In the smart contract, it takes the function of a finite verb. The predicative describes the action of the sentence. The N, A, and D typically consist of all non-verbs in the sentence. The interpretation of the predicative determines the N, A or D, and may also potentially modify the Adverbial (B). Figure 5 shows the rough steps based on the type of verb.

  e.g. *He finishes this task on time.* (P)

  e.g. *Alice cried and laughed for a movie.* (P)

- Copulative (X)

  A copulative a copulated component linked to copulative.

  e.g. *This task is difficult.* (X)

- Complemental (C)

  It is denoting additional attributes of an entity, an action, an event or a state. In the smart contract, it describes the operators "*when*", "*where*" and "*how*" that predicative is allowed, required, or forbidden. For example, TIME: 19/01/2019, which means that data type is "*time*" and the expression defines *Day/Month/Year*.

  e.g. *System must send prices before 04/09/2019.* (C)

- Adverbial (B)

  An adverbial modifies an action, an event or a state. In the smart contract, it takes the function of a finite adverb and prescriptive operator of a statement or sentence that describes what ideally is permitted, obliged, or forbidden. Adverbials may not always be written literally as "*permission*", "*obliged*" or "*prohibition*", but they can also be presented in other forms, such as "*should*", "*should not*", "must", "*must not*". For example, the verb "*vote*" suggests a "*must not*". Adverbial serves as useful markers for delineating sentence.

  e.g. *He must not vote this person.* (B)

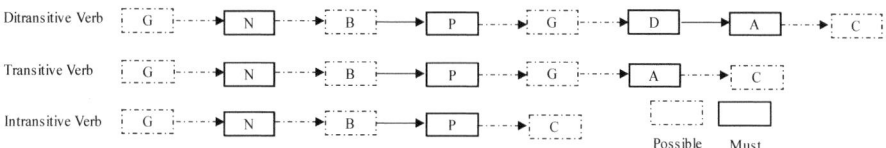

Fig. 5. Rules based on the type of verbs

Through this case analysis based on the MNL structure, we can express the most declarative sentence in the smart contract, and it is structured enough to be a machine-readable contract. Meanwhile, MNL allows the flexible arrangement of order without altering the meaning of the sentence. It is a key difference between order-dependent natural languages (e.g., SOV sentence pattern for English).

3.4 Self-executing

Once an MNL was created to sufficiently describe the content of a smart contract, we wrote parsing rules to specify the constraints. We define rules of the simple declarative sentence for English based on verbs as shown in Fig. 5 to express the given orders of the sentence. Meanwhile, there is default PoS for cases in MNL:

R1: n → (N | A | D) | v → P | adj → G | adv → B
R2: num, n → G, (N | A | D)
R3: adp, n → C,C

Figure 6 provides a baseline case example for three articles where there is a clear description and reflects the types of cases, based on principles we can now decompose complex structures into simple semantic that capture the concepts. Computers read the structured contracts, but it does not deal with the underlying meaning of words at a deep cognitive level, it requires a higher level of in the accuracy, depth and complexity of a contract. MNL universal contract representation shows a structured sentence by cases. A context-free grammar has been set up, which acts as a language device that has a case grammar to examine languages compatibility based on sentence rules. We also define the computer-understandable simple sentence, which is a set of EIID terms [14], such that:

$$\text{Sentence} = (eiid_1, \ldots, eiid_k, \ldots, eiid_n)$$

where an extended *iid* (*eiid*) = (*iid*.PoS.C) = ("*iid*" refers to a sense in CoDic "IID", "PoS" is the part of speech already defined in *iid*, "C" is the case based on MNL grammar. Thus, a sequence of terms in a sentence transfer to a sequence meaningful *eiid* that the computer understandable and executable. For everywhere, every party, the computer only reads and decodes the sequence of *eiid* based on local required the rule in the smart contract before taking further action.

We conclude that translating natural language into code does not constitute a direct process of translating legal prose into computer-readable instructions but requires prior interpretation of legal prose. MNL achieved a unique natural language structure to reduce the complexity of smart contract among crossing parties. Meanwhile, layers and programmer would share a common meaning or concept and would not require enough knowledge to reduce the cost.

	Statement	Contract structure in MNL	Coding
Article One	*System must send record before 04/09/2019.*	N = system B = must P = send A = record C = before [DATE=04/09/2019]	N.iid[1] B.iid[2] P.iid[3] A.iid[4] C.iid[5], iid[6]
Article Two	*Student contacts the professor by 20/10/2019.*	N = student P = contacts A = the professor C = by [DATE=20/10/2019]	N.iid[1] P.iid[2] A.iid[3], iid[4] C.iid[5], iid[6]
Article Three	*Student must pay $20 at 3:00 pm.*	N = student B = must P = pay A = [PRICE = $20] C = before [TIME=3:00pm]	N.iid[1] B.iid[2] P.iid[3] A.iid[4] C.iid[5], iid[6]

Fig. 6. Rules baseline case representation by MNL

4 Conclusion and Outlook

A smart contract is a complex syntactic and semantic phenomenon, which is affected by human language. Therefore, it needs to be decomposed into clear, simple and contextual types in grammar and semantics. When people read contract documents, they use high-level language processing capabilities to understand complex contracts, human-readable contracts need to be codified and validated to ensure that machine-readable representations conform to specified behavior. We think that the complexity of encoding smart contracts and the need to do correctly, computers can only understand every term in the fixed template, because the lack of advanced cognitive process, still need to be carefully designed and implemented, a freestyle structure, such as complex processing concepts - sentences cannot be solved in the right way. If the contract is not apparent or exploitable, its impact may be far-reaching. Because program code seems difficult to understand complex concepts.

In this paper, we outline the potential of smart contracts between independent entities (such as agents, humans, etc.). The core challenges are ambiguous, complex structures and the automatic execution of smart contracts. The main contribution of our work is to introduce a general dictionary to make human and computer-readable, understandable and automatable, and to translate human natural language into a coded machine-readable contract based on machine natural language. To achieve this goal, we have taken two steps. Firstly, we express human-readable terms as the unique IID that the computer understands. The most crucial task in this step is to eliminate ambiguity and execute the program in the translation of the smart contract. The second step is to construct these sentences based on high-level cases structure while maintaining conceptually and grammatically representations based on machine natural language. Meanwhile, MNL mechanism automatically generates meaningful *eiids*. Non-programmers are sufficient to express contract declarations and agreement on the contract's semantics. In fact, the proposed method can be coded by human beings, eventually by the machine itself.

Although this paper demonstrates methodology and proposes a novel the framework in the creation of a smart contract, we hope to automatically generate knowledge base by using the state-of-the-art natural language processing technology to analyze text-based constraints. In future work, we will discuss in-depth whether it is possible for human beings or autonomous entities, to verify the actual contract semantics and obligations and inference semantic based on machine code alone. We also consider security analysis and applications in blockchain [15, 16].

Acknowledgment. This research is partially supported by the University of Macau Research Grant No. MYRG2017-00091-FST.

References

1. Szabo, N.: Formalizing and securing relationships on public networks. First Monday, [S.l.] (1997)
2. Mik, E.: Smart contracts: terminology, technical limitations and real-world complexity. Law Innov. Technol. **9**(2), 269–300 (2017)
3. Dale, R.: Classical approaches to natural language processing. In: Indurkhya, N., Damerau, F.J. (eds.) Handbook of Natural Language Processing, 2nd edn. Taylor & Francis, Boca Raton (2010)
4. Katz, D.M.: Quantitative legal prediction–or–how i learned to stop worrying and start preparing for the data-driven future of the legal services industry. Emory Law J. **62**, 909–936 (2013)
5. Guo, J.: Collaborative conceptualisation: towards a conceptual foundation of interoperable electronic product catalogue system design. Enterp. Inf. Syst. **3**(1), 59–94 (2009)
6. Qin, P., Guo, J., Xu, Y., Wang, L.: Semantic document exchange through mediation of machine natural language. In: Proceeding of 15th IEEE International Conference on e-Business Engineering (ICEBE 2018), pp. 245–250. IEEE Computer Society (2018)
7. Pettersson, E.J., Edström, R.: Safer smart contracts through type-driven development. Ph.D. thesis, Master's thesis, Department of Computer Science & Engineering, Chalmers University of Technology & University of Gothenburg, Sweden (2015)
8. Idelberger, F., Governatori, G., Riveret, R., Sartor, G.: Evaluation of logic-based smart contracts for blockchain systems. In: 10th International Symposium, RuleML 2016, 6–9 July 2016, pp. 167–183 (2016)
9. Grigg, I.: The Ricardian contract. In: Proceedings of the First International Workshop on Electronic Contracting, pp. 25–31. IEEE (2004)
10. Grigg, I.: On the intersection of Ricardian and smart contracts (2017). http://iang.org/papers/intersection_ricardian_smart.html
11. Frantz, C.K., Nowostawski, M.: From institutions to code: towards automated generation of smart contracts. In: 1st International Workshops on Foundations and Applications of Self* Systems (FAS*W), pp. 210–215. IEEE (2016)

12. Clack, C.D., Bakshi, V.A., Braine, L.: Smart contract templates: essential requirements and design options, ArXiv eprints, December 2016
13. Xiao, G.: Semantic document exchange for electronic business through user-autonomous document sense-making. Doctoral thesis, University of Macau (2015)
14. Qin, P., Guo, J.: A novel machine natural language mediation for semantic document exchange in smart city. Future Gener. Comput. Syst. (2019). https://doi.org/10.1016/j.future.2019.07.028
15. Shen, B., Guo, J., Yang, Y.: MedChain: efficient healthcare data sharing via blockchain. Appl. Sci. **9**, 1207 (2019)
16. Almadhoun, R., Kadadha, M., Alhemeiri, M., Alshehhi, M., Salah, K.: A user authentication scheme of IoT devices using blockchain-enabled fog nodes. In: Proceedings of the IEEE/ACS 15th International Conference on Computer Systems and Applications (AICCSA), October–November 2018, pp. 1–8 (2018)

Application of Big Data Analytics
for Facilitation of e-Business

A Content Based e-Commerce Recommendation Approach Under the Veristic Framework

Imen Benzarti[1(✉)], Hafedh Mili[1], and Amandine Paillard[2]

[1] LATECE Laboratory, University of Quebec at Montreal, Montreal, QC, Canada
benzarti.imen@courrier.uqam.ca, mili.hafedh@uqam.ca
[2] University of Montpellier, Montpellier, France
amandine.paillard@etu.umontpellier.fr

Abstract. A recommendation system is an automated tool that suggests an ordered list of appropriate items to a user. In this paper, we propose a recommendation algorithm that takes into account the variable and complex semantics of multi-valued properties, and the level of uncertainty or fuzziness inherent in the representations of users and items. In particular, we will rely on the concept of veristic variables, proposed by Yager, and propose algorithms for building user profiles, and for matching product descriptions to those profiles. We tested our algorithms on the Movielens datasets. Our results show statistically significant improvements in the f1-score and the Root Mean Square Error (RMSE) compared to baselines like Bayesian and K-nearest neighbors (KNN) approaches.

Keywords: Recommendation algorithms · Veristic variable · Fuzzy sets · Content based recommendation

1 Introduction

Customer Experience Management (CEM) denotes a set of practices, processes, and tools that aim at personalizing a customer's interactions with a company around the customer's needs and desires [16]. *Recommendation* is one such personalization technique. Within the context of a broader research program that aims at developing a methodology and *software development framework for customer experience management applications* [1]. We aim to develop a generic recommendation algorithm that can handle the variety of product and service types and a variety of user profiles and that implement different recommendation strategies.

When we started working on our *recommendation framework*, we faced *conceptual difficulties* in three related areas: (1) handling multiple values for a given property, (2) handling multi-valued properties in concept matching, and (3) handling uncertainty. We will illustrate *part* of the first issue. Take the example of an on-demand video streaming service–call it Webflix–that offers movies and

© Springer Nature Switzerland AG 2020
K.-M. Chao et al. (Eds.): ICEBE 2019, LNDECT 41, pp. 495–514, 2020.
https://doi.org/10.1007/978-3-030-34986-8_35

that maintains profiles of its customers, including the properties *likesGenre* and *residesIn*. Jane, a given customer, likes several genres and hence *likesGenre(Jane)* = {Comedy, SciFi}. We may also have reason to believe that Jane's main residence is in either Canada, Mexico, or the US, but we are not sure which. Thus, we may write *residesIn(Jane)* = {Canada, Mexico, USA }. Clearly, the meaning of the property value *sets* are different. Jane likes *both* genres–*conjunctive property*–whereas Jane's country of main residence is *either* Canada, *or* Mexico, *or* the USA–a *disjunctive property*. The interpretation of multiple values is important for the concept matching–and recommendation.

In this paper, we propose a generic *content-based recommendation algorithm* that takes into account the variable and complex semantics of multi-valued properties, and the level of *uncertainty* or *fuzziness* inherent in the representations of consumers and products. Specifically, we use the concept of *veristic variables*, proposed by Yager in [22] and [21], to represent multi-valued properties. We propose an algorithm for content-based recommendations, which uses the theory of veristic and inferences from user ratings to derive user preferences to the movie genre. The algorithm was tested on MovieLens100K and MovieLens1M datasets. Our results show that our approach outperforms baselines like Naive Bayes, K-nearest neighbors (KNN) and fuzzy tree based approaches largely used to implement recommendation systems.

Section 2 frames the 'recommendation problem' in terms of *concept matching* between feature-based descriptions of consumers and products/services. In particular, we illustrate the different interpretations of multi-valued properties, and their impact on the concept matching. Section 3 introduces the concept of *veristic variables* and how they may be used to represent information about consumers and products/services. A *content-based* recommendation procedure that uses veristic variables is described in Sect. 4. We conducted an experiment to compare the results of our recommendation algorithm to those reported in the literature. The experiments and results are described in Sect. 5. We conclude in the Sect. 7.

2   The Recommendation Problem

In this section, we first frame the recommendation problem within the context of our CEM development framework. Next, we discuss recommendation within the context of feature-based representations to highlight the semantics of multi-valued properties and how they impact concept matching. Finally, we discuss the issue of fuzziness.

2.1   Framing the Recommendation Problem

Our goal is to build a recommendation framework to be used within the context of a *consumer experience management development framework* that in turn will be used with e-commerce applications. This means three things. *First*, our representation of consumers (users) and products needs to be generic. *Second*, our representation of *products* is *typically* or *potentially* fairly rich, corresponding to the contents of a *product catalog*. *Third*, the functionalities that we introduce need to be usable *from day one*. In other words, we need to be able to bootstrap the recommendation system with domain knowledge, for example, and then enhance its accuracy as a retail e-commerce system builds consumer data/transaction histories. For all practical purposes, these characteristics exclude a collaborative filtering approach [4] which does *not* require a rich representation of consumers and products, but *does require* a large transaction history.

For the purposes of this paper, we will assume a *feature-based* representation of consumers and products; the reader is referred to [1,9] for an actual *metamodel* of consumers and products/services. Recommendation can be seen as matching feature-based descriptions of consumers, on one hand, with feature-based descriptions of products, on the other. In the remainder of this section, we discuss in turn: (1) concept matching within the context of feature-based representations, (2) dealing with fuzziness, and (3) the problem with multi-valued fuzzy properties, requiring the theory of veristic variables.

2.2 Feature-Based Representations

We may compare two classes C_1 and C_2 along some property F_i, and the result of the comparison depends on the relationship between $F_i(C_1)$ and $F_i(C_2)$. We have three possible relationships, and corresponding interpretations:

$$\begin{cases} F_i(C_1) \subset F_i(C_2) & C_2 \text{ subsumes (super} \\ & \text{class of) } C_1 \\ F_i(C_1) \cap F_i(C_2) = \Phi & C_1 \text{ and } C_2 \text{ unrelated} \\ F_i(C_1) \cap F_i(C_2) \neq \Phi & C_1 \text{ and } C_2 \text{ somewhat similar} \end{cases} \quad (1)$$

Within the context of recommendation, we deal with classes that belong to different *domains*. Thus, the properties usually use different names, and we would compare $F_{customer}$ (<Customer or customer class>) to $P_{product}$ (<Product or product class>). In this case, the customer's property is taken to represent the *needs/requirements* whereas the product's property is interpreted as representing the product's features, i.e. what it has to *offer*. In this case, we recommend a product if it *subsumes* (covers) the needs of the customer.

However, we quickly run into trouble if we don't distinguish between single-valued versus multi-valued properties. Consider again the property *likesGenre*(.) that applies to consumers. An *individual* may like *several genres*, and thus *likesGenre(Jane)* = {*Comedy, SciFi*} has a different meaning from *residesIn*(< *North Americans* >) = {*Canada, Mexico, USA*}: the values are *not* considered as *alternatives* (*ored*) as with country of main residence; they are true/applicable simultaneously (*anded*). In this case, the property is *multi-valued*: it can take several values.

On the movie side, *hasGenre(m)* returns the set of genres embodied in m. For example, most action movies have a romance/love interest. Considering the meaning of the multiple values in this case, it would make sense to recommend a movie if it has *any* of the genres liked by Jane. Thus, we recommend a movie m to consumer Jane iff:

$$P_{consumer}(Jane) \ \cap \ P_{Product}(m) \ \neq \ \Phi$$

This raises the question of whether a movie that embodies *several* genres liked by a consumer is *preferred* to one that embodies a single genre. Generally speaking, it is an *optimization* problem–e.g. how much screen time to use for romantic scenes versus action scenes. We show in Sect. 4 two formulas for property matching, one for single-valued properties and one for multi-valued properties, and we discuss in Sect. 5 the effect of using one versus the other.

2.3 Dealing with Uncertainty

A lot of our knowledge about people and things is approximate and uncertain. This may be due to: (1) lack of knowledge, and (2) fuzziness. We illustrate both types below.

Assume that we did not know that Jane resides in the US, but that we met her at a shopping mall in Plattsburgh, in upstate New York, near the Canadian border. It is fairly safe to assume that she lives there, but she could also be visiting from Montréal, Canada, which is 65 miles north. In this case, even though *residesIn*(.) is single-valued, we might express the uncertainty with a *fuzzy set* of countries, where each country is associated with a *membership degree*. Thus: $residesIn(Jane) = \{0.85/USA, 0.15/Canada\}$.

Contrast that with presenting Jane with a list of genres, and asking her "how well do you like these genres, on a scale of 0 (a little) to 1 (a lot)", and she answered 1 for 'Comedy', and 0.6 for SciFi. We could use these numbers to represent the genres liked by Jane as *fuzzy set*, e.g. *likesGenre*(*Jane*) = { (Comedy, 1), (SciFi, 0.6) }. These weights have a different meaning from those in *residesIn*(*Jane*): Jane has a single (main) country of residence, but we don't know which one, and the weights represent the likelihood of any given value being the one; with *likesGenre*(*Jane*), *all* the values belong, but to different degrees. The 'fuzziness' of the set *likesGenre*(*Jane*) does not change the semantics of the concept matching: a movie m is *still* considered interesting for–should be recommended to–Jane iff $hasGenre(m) \ \cap \ likesGenre(Jane) \ \neq \ \Phi$.

The next section introduces the concept of *veristic variables*, which provides a mathematically sound way of handling *fuzzy, multi-valued properties*.

3 Veristic Variables

3.1 Veristic Versus Possibilistic Variables

In everyday "conversation" and reasoning, we tend to make *qualitative* statements about the world, as opposed to strictly *logical* or *quantitative* statements, either by choice, or because we lack precise knowledge. For example, a *quantitative* attribute such as *age* becomes a *qualitative* or *linguistic attribute*, whose values might be *Child, Teenager, Young [Adult], Middle-aged, Senior*. The theory of approximate reasoning (AR) deals with this kind of representation and reasoning in a formal way: it maps a statement such as "Jane is young" to a *fuzzy set* of ages, where the membership level of a given value (say 28) is no longer 0 or 1, but some value in the between indicating the compatibility of the value with the "quality" (youth). We *know* that Jane has a *single age*; we just don't know which one with certainty. Jane's age is called a *disjunctive variable*: it can take a *single value* among the elements of the fuzzy set.

Yager extended the theory of approximate reasoning to include *conjunction* of *veristic variables*, where the multiple values of a variable are all "true" [18–20]. *likesGenre(Jane)* = {Comedy, SciFi} means that Jane likes *both* comedy and SciFi movies/series. In this case, *likesGenre(Jane)* it is the *veristic variable*, and {Comedy, SciFi} is the value for that variable. We write V *isv* A to say that variable V is *veristic* and that its (veristic) value (*isv*) is the set A = {Comedy, SciFi}. Veristic variables and reasoning differ from *disjunctive* or *possibilistic variables* such as Jane's age. For example, Yager noted in [20] that with *possibilistic variables*, the *conjunction of knowledge statements* leads to a *conjunction* (intersection) of the corresponding *fuzzy sets*, whereas with *veristic variables*, the conjunction of knowledge statements leads to a *disjunction* or *union* of the sets of values. The converse is true for the *disjunction* of statements, which leads to the *union* of sets, for possibilistic variables, and the *intersection* of sets, for veristic variables.

To take advantage of the theoretical framework provided by *approximate reasoning* (AR), which deals with *conjunctive/possibilistic variables*, Yager proposed a way to map veristic variables–and veristic statements–to a corresponding *disjunctive/possibilistic* equivalent. Thus, instead of saying "the genre(s) liked by Jane", which can take several values, we say "the *set* of genres liked by Jane". Formally, assume that V ("genres liked by Jane") is a veristic variable whose values are taken from some set X–the set of *all* movie genres. We associate with V a *possibilistic/disjunctive* variable V^* that has, as a universe of discourse, the *powerset* of X, or 2^X–the set of *subsets* of X, i.e. the set of sets of movie genres [21].

Yager identified four different *canonical forms* for veristic variables, corresponding to different *modalities*, and for each canonical form, he associated the set W of possible values of V^* [21], as follows:

- *Open positive statements*, such as our example of "genre(s) liked by Jane" above, where V *isv* A means that the corresponding possibilistic variable V^* can take *any superset* of A as a value, i.e. $W = \{F \in 2^X | A \subseteq F\}$.
- *Closed positive statements*, which we note as V *isv(c)* A, where A is the *only* value for V^*. This corresponds to the variable V = "the *only* genres liked by Jane". Thus, if A is a value for V, then it is also the only value for V^*. In other words, $W = \{A\}$.
- *Open negative statements*, which we note as V *isv(n)* A as in "genre(s) liked by Jane" (V) "excludes" (*isv(n)*) {Crime, Horror} (A). Jane may "exclude" other genres too, e.g. Paranoid Fiction. Thus, if V *isv(n)* A, then "any set of genres that excludes the elements of A" is a potential solution for V^*. This means that the set of genres liked by Jane is a subset of (Y - A), or $\bar{A}$, i.e. $W = \{F \in 2^X | F \subseteq \overline{A}\}$.

– *Closed negative statements*, which we note $V\ isv(n,c)\ A$ as in "the genre(s) liked by Jane" (V) "excludes *only/exactly*" (isv(n,c)) {Crime, Horror} (A). This means that Jane likes every genre, except the ones listed. This is the *only value* for V^* and is $(Y - A)$, or $\bar{A}$, i.e. $W = \{\overline{A}\}$.

Given a veristic variable V, we will refer to the solution set W for its possibilistic equivalent (V^*), to assess the *degree of membership* of a given value (e.g. a particular genre) to the solution (fuzzy) set for V^*. This is discussed next.

3.2   Fuzzy Veristic Variables

One of the basic tenets of AR theory is what we might call the "open world assumption": what is *not* explicitly stated to be true, cannot be assumed to be false. *Possibility theory* defines the *probability of an event* as an interval, low-bounded by a *necessity measure N* and upper bounded by *possibility measure* Π, i.e. $N(A) \leq P(A) \leq \Pi(A)$ where A is an event defined in a set of reference X and $P(A)$ is the probability of the event A [15]. Along a similar vein, Yager defined $Ver(x)$, $Rebuff(x)$ as measures to define, respectively, how confident we are that an element $x \in X$ is a value for the veristic variable V, and how confident we are that x is *not* a value for V [19,21]. Given the veristic statement $V\ isv\ A$, and the corresponding possibilistic statement V^* is W, where W is a subset of 2^X consisting of all the subsets F of X that are *possible/alternative* solutions for V^*, Yager defines $ver(x)$ and $rebuff(x)$, for all $x \in X$, as follows

– $ver(x) = \min_{F \in W} \mu_F(x)$, where $\mu_F(x)$ is the *membership function* of F.
– $rebuff(x) = 1 - \max_{F \in W} \mu_F(x) = \min_{F \in W}(1 - \mu_F(x)) = \min_{F \in W} . \mu_{\overline{F}}(x)$

Each one of the four *canonical* veristic statements discussed above induces a different distribution of $ver(x)$ and $rebuff(x)$.

The reader is referred to [21] to establish the results shown in Table 1, where $Poss(x) = 1 - Rebuff(x)$; see [21].

Table 1. Ver, Rebuff, and Poss distributions for the primary veristic statements

Statement	Ver(x)	Rebuff(x)	Poss(x)
$V\ isv\ A$	$A(x)$	0	1
$V\ isv(c)\ A$	$A(x)$	$1 - A(x)$	$A(x)$
$V\ isv(n)\ A$	0	$A(x)$	$1 - A(x)$
$V\ isv(c,n)\ A$	$1 - A(x)$	$A(x)$	$1 - A(x)$

4 Content-Based Recommendation Algorithm Using Veristic Variables

Section 3 showed the theoretical basis for representing consumer preferences when those are multi-valued, as was the case for "genre(s) liked by Jane". As mentioned in Sect. 3.2, those preferences may *not* be known *firsthand*, but are typically inferred from past consumer behavior. In this section, we show how to determine a movie genre preferences from their past ratings of movies with known genres; we later (Sect. 5) test our profile construction method, and the corresponding concept matching algorithm, on the MovieLens public dataset. We start by explaining the general idea, and then give more details about the implementation.

4.1 Building User Profiles from Movie Ratings: Principles

Roughly speaking, the MovieLens dataset contains a bunch of movies, identified by sets (vectors) of genres, and a bunch of user ratings of those movies. We propose to use this information to build a profile of those users, consisting of a single property representing the genres liked or disliked by the user in question. For a user U, the general idea is as follows:

1. if user U likes a movie M with genres $\{g_1, ..., g_n\}$, treat it as a *positive, open veristic statement* V *isv* A, where V = "the genre(s) liked by U", and $A = \{g_1, ..., g_n\}$,
2. if user U does *not* like a movie M with genres $\{g_1, ..., g_n\}$, treat it as a *negative, open veristic statement* V *isv(n)* A, where V = "the genre(s) liked by U", and $A = \{g_1, ..., g_n\}$,
3. combine the open positive veristic statements corresponding to the various "liked" movies using the union operator, as explained in Sect. 3.1. In other words, if V *isv* A and V *isv* B, then V *isv* $A \cup B$,
4. combine the open negative veristic statements corresponding to the various movies not liked by the user, using the union operator, as explained in Sect. 3.1. In other words, if V *isv(n)* A and V *isv(n)* B, then V *isv(n)* $A \cup B$,
5. use the resulting aggregate positive open veristic statement, and negative open veristic statement, to compute $ver(genre)$, and $poss(genre)$, for all the genres referenced in the movie set, and
6. combine $ver(genre)$ and $poss(genre)$, for a given user, to produce a single score between -1 and 1, for each and all genres, to reflect the extent to which the user likes–or does not–a given genre. This will produce a weighted vector of genres, to be used as a user profile.

We now elaborate on: (1) building veristic statements from movie ratings, and (2) combining $ver(genre)$ and $poss(genre)$ to a single score.

4.2   Deriving Veristic Statements from Movie Ratings

There are two issues that we need to contend with: (1) distinguishing between "positive" and "negative" ratings, and (2) attributing membership degrees of genres.

First, the MovieLens database contains ratings of movies on a single Likert scale, reflecting how well moviegoers liked the movies. We could use a score of 3 as neutral, and treat anything above as positive rating and anything below as negative rating, but the reality is that some generous users will give consistently high scores [25] while hard to please moviegoers will give low scores to movies they liked. Hence, we used *mean-centering* and *normalization, per user*, to turn all movie scores into values in $[-1, 1]$, and then treat negative ratings as instances of *open negative*. Let $r(u, m)$ be the rating of movie m by user u, $\bar{r}(u)$ the average rating by user u, $r_{min}(u)$ the minimum rating given by u, $rmax(u)$ the maximum rating given by u, we define the normalized mean-centered rating $\hat{r}(u, m)$ [11] as:

$$\hat{r}(u,m) = \begin{cases} \frac{r(u,m)-\bar{r}(u)}{r_{max}(u)-\bar{r}(u)} & \text{if } r(u,m) > \bar{r}(u) \\ \\ \frac{r(u,m)-\bar{r}(u)}{\bar{r}(u)-r_{min}(u)} & \text{if } r(u,m) < \bar{r}(u) \end{cases} \tag{2}$$

Thus, if $\hat{r}(u, m) > 0$, we consider it a case of a positive open statement, and if $\hat{r}(u, m) < 0$, we consider it a case of negative open statement.

Consider movies m_1, m_2 and m_3 with genres {Action, Romance}, {Romance}, and {Action, Drama}, respectively, with the ratings $\hat{r}(u, m_1) = 0.8$, $\hat{r}(u, m_2) = 0.7$, and $\hat{r}(u, m_3) = -0.8$. The veristic statement corresponding to the various movies lead to the following fuzzy value sets

- $m_1 \to$ {(Action,0.4), (Romance, 0.4)}, where we split the overall movie score (0.8) equally between the various genres of m_1
- $m_2 \to$ {(Romance, 0.7)}, where the movie rating is credited to the single genre
- $m_3 \to$ {(Action,0.4), (Drama,0.4)}. Recall that m_3 induces an *open negative statement* because of the negative rating, and the *absolute value* of the rating (-0.8) is split equally between the two genres of m_3.

When the two veristic statements induced by the ratings of m_1 and m_2 are combined, we get the fuzzy value set {(Action,0.4), (Romance,0.7)}, where we took the union of the fuzzy sets {(Action,0.4), (Romance, 0.4)} and {(Romance, 0.7)}.

In practice, we chose not to include *all* the movie ratings, but to take the $k-highest$ positive ratings and the $k-lowest$ negative ratings, for some integer k–in part to even the 'playing field' between the different users in the MovieLens database.

4.3   Computing a Single Score from $Ver(x)$ and $Poss(x)$

Recall (Sect. 3) that Yager defined $ver(x)$ and $poss(x)$ as two "bounds" on the knowledge/certainty we have that some element x is a solution of a veristic statement, except that $poss(x)$ is not always greater than $ver(x)$. Recall that $ver(x)$ means the "minimum certainty" we have that x is a solution, whereas $poss(x)$ measures *the absence of proof that x is not a solution*. In particular, there are cases where $ver(x)$ is greater than $poss(x)$. In our three movie example above, using the Table 1: according to the open positive statement (the movie m_1) $ver(Action) = 0.4$ and $poss(Action) = 1$, whereas, according to the open negative statement $ver(Action) = 0$ and $poss(Action) = 1 - 0.4 = 0.6$. Based on [21] and on the definitions of $poss$ and ver, we conclude that $ver(Action) = 0.4$ and $poss(Action) = 0.6$.

Accordingly, rather than taking some arbitrary arithmetic combination of $ver(genre)$ and $poss(genre)$ as a unique $score(genre)$, we choose to do our own *qualitative* reasoning to figure out what to do about different values of $ver(x)$ and $poss(x)$. For a given genre, we distinguish between four borderline cases:

- $ver(genre) = 1$ (or "high"), and $poss(genre) = 1$ ("high"). In our context, this means that: (1) wherever the genre was present, the movie was *consistently* highly rated, and (2) the genre did not appear in *any* of the negatively rated movies. This is the case for *Romance* ($ver(Romance) = 0.7$ and $poss(genre) = 1$) in our example.
- $ver(genre) = 1$ ("high") but $poss(genre) = 0$ ("low"). Recall that $poss(x)$ measures the *absence of evidence that x is not solution*. In this case, genre was associated with *both highly positively* rated movies, and *highly negatively* rated movies. Yager says that the information is contradictory [21]. This is the case for *Action* in our sample.
- $ver(genre) = 0$ ("low") and $poss(genre) = 1$ ("high"). This means that: (1) the genre didn't appear in any positively rated movie, and (2) the genre didn't appear in any negatively rated movie *either*; i.e. we have no information about the genre in this case [21]. This would be the case for *any genre* other than Action, Romance, or Drama, with our three movie database.
- $ver(genre) = 0$ ("low"), and $poss(genre) = 0$ ("low"). This means that: (1) we have no evidence that the genre is liked, i.e. it didn't appear in any of the liked movies, and (2) we *do have* evidence that it is disliked, i.e. it appeared in negatively rated movies. Thus, we *definitely* know that *genre* is no liked [21]. This is the case for *Drama* in our three movie database: it appeared only in the negatively rated movie (m_3).

It is clear that genres with "high" for both metrics should be scored highest (Romance), and those with "low" for both metrics should be scored lowest. Remains the issue of ranking genres with contradictory information (high, low) and those with no information (low, high). We choose to rank *higher those with contradictory information* for two reasons. *First*, moviegoers tend to gravitate around the movie genres that *do* like, thus the very *presence* of a genre in the movies that a user watches may indicate that they like the genre. *Second*, the genre is *probably* not the reason for disliking a movie—for otherwise $ver(genre)$ would be 0. There may be other reasons—e.g. a particular *actor/actress* or director.

Accordingly, we chose to use the following *fuzzy rules* to assign a global score to genres:

- if $ver = high$ and $poss = high$ then $score = high$,
- if $ver = high$ and $poss = low$ then $score = med$,
- if $ver = low$ and $poss = high$ then $score = low$,
- if $ver = low$ and $poss = low$ then $score = very\_low$.

Accordingly, we chose to embody these rules in a *fuzzy inference engine* that, given numerical values for $ver(genre)$ and $poss(genre)$, applies the four rules above to produce an overall score between -1.0 and 1.0. Intuitively, the fuzzy inference engine applies the four rules on the input values and "combines their conclusions"; the "strength" of a given rule conclusion depends on the extent to which the inputs "satisfy" or "fit the description" of the rule "conditions" (see [12]). Space limitations do not allow us to go over the details of the fuzzy inference system, but suffice it to say that we used: (1) (semi)triangular membership functions for *high* and *low*, for both $ver(genre)$ and $poss(genre)$, (2) (semi)triangular functions for *high*, *medium*, *low*, and *verylow*, centered at 1, 0.33, -0.33, and -1.0, respectively, and (3) *centroid defuzzification method* for the overall score (see e.g. [12]. Figure 1 plots $score(genre)$, as a function of $ver(genre)$ and $poss(genre)$.

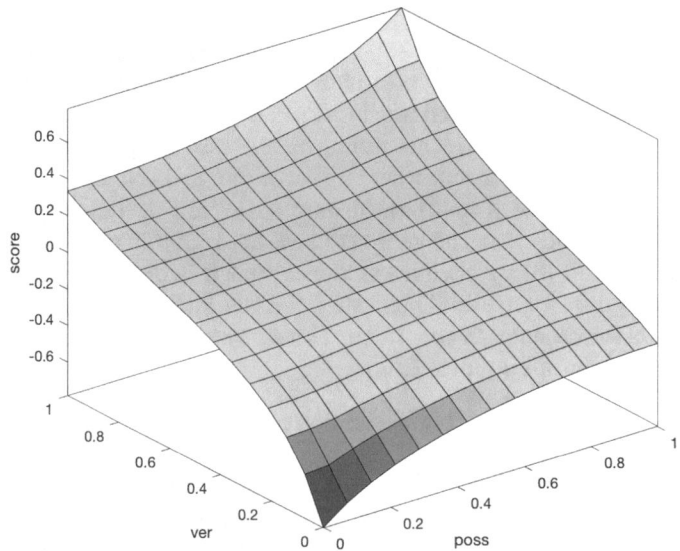

Fig. 1. A plot $score(genre)$ as a function of $ver(genre)$ and $poss(genre)$.

4.4   Recommending New Movies

The previous steps aim to build a predictive user profile that defines a preference score for each movie genre. This profile is used to recommend new movies to the user by predicting its matching degree with new movies to recommend.

We distinguish two types of matching: multi-valued matching and mono-valued matching. The choice of the matching method depends on the interpretation of the property.

In multi-valued interpretation, we query movies that match all user's genres preferences. For example, we want to recommend to a user u, having the following genres preferences: $u = (1/action, 0.75/comedy, 0.5/Romance)$, two movies $m_1 = (1, 1, 0)$ and $m_2 = (1, 0, 0)$. In a multi-valued interpretation the user u would prefer m_1 to m_2. We use the following similarity metric from [5]:

$$Sim_{multi}(user, movie) = 1 - \max_{g \in G} |user_g - movie_g| \qquad (3)$$

This measure is based on the maximum difference. To maximize the similarity degree, (and minimize distance) the movies genres have to match all genres in the user vector of genres. For instance, $Sim_{multi}(u, m_1) = 0.5$ and $Sim_{multi}(u, m_2) = 0.25$, thus, the movie m_1 gets a higher score because it matches more genres than m_2. Whereas, in a mono-valued match, we query movies that match at least a unique genre in the user genres preferences, in this case, the user u would prefer both movies. Thus, we use the following similarity metric from [2]:

$$Sim_{mono}(user, movie) = \max_{g \in G}(\min(user_g, movie_g)) \qquad (4)$$

This measure represents the guaranteed minimum similarity between sets, in other words, it will return the similarity degree between the most matching user and movie genre, for instance, $Sim_{mono}(u, m_1) = 1$ and $Sim_{mono}(u, m_2) = 1$.

5   Experimental Evaluation

5.1   Evaluation Methodology

To evaluate our approach we conducted experiments on MovieLens-100k and MovieLens1M which are publicly available datasets. The dataset MovieLens-100k contains 100,000 ratings given by 943 users on 1,682 movies and the dataset MovieLens-1M contains 100,000 ratings given by 6000 users on 4000 movies. The datasets are randomly partitioned into two overlapping sets for training and testing. For each user, 80% of his historical items are randomly chosen as the training data and the remaining 20% for testing. We randomly selected 5 random partitions of the training and testing from each dataset and reported average results. The final recommendation list is ranked according the predicted user preferences. To validate its consistency, this ranking order is compared to the actual ranking in the test set.

We measure the performance of different methods with the evaluation metrics: nDCG@N is a measure of ranking accuracy. Precision@N is percentage of the N top ranked relevant recommended items among recommended items. Recall@N is percentage of the N top ranked relevant recommended items among the set of relevant items. F-score@N is a single metric that combines precision and recall:

$$F1 - score@N = \frac{2 * Precision@N * Recall@N}{Precision@N + Recall@N}$$

To compute the differences between predicted ratings values and actual values in the test set we use root-mean-square error (RMSE) metric. To compute the average of all absolute errors between actual values in the test set and the predicted values we use MAE metric (Mean Absolute Error).

For each metric, we first computed the performance for each user on the testing set, and we reported the averaged measure over all the users.

5.2   Comparison with Other Approaches

The proposed approach of veristic content-based recommendation is compared with the following baseline approaches:

The Probabilistic approach [10]: this approach uses Naive Bayes classifier, a standard algorithm for text classification, as a profile model to predict the user ratings. This approach is a model-based approach were a predictive user profile is learned from past evaluations then matched with new items to recommend. We implemented this approach by predicting the movies genres in the user profile.

FuzzyTree approach [7]: is a fuzzy content matching-based recommendation approach, where user rating is used to derive user preferences to item features. User preference is then represented using fuzzy tree structure. This is a model-based approach.

K-nearest neighbors Approach [11, 17]: Contrary to the first two model-based approaches, this approach is memory-based (or similarity-based) that computes the users predicted rating of an item directly from his stored ratings. Thus, the predicted rating is the average rating of the evaluated items weighted by the similarity score between the previously rated item and the target item. To implement this approach, we selected the cosine similarity as similarity measure and weighted-sum as aggregation method.

For each approach we took the 20 best ranked items to build the user profile.

5.3   Comparison Results and Discussion

The results are summarized in the Table 2, we note that we used the datasets MovieLens100K and MovieLens1M as it stands without any modifications. It can be seen that veristic CB outperforms or at least compares favorably with all other compared approaches.

Veristic CB performs significantly well in f1-score, MAE and RMSE, for instance, veristic CB outperforms probabilistic approach by 10% on f1-score@5 and f1-score@10 and by 9,7% on NDCG@10 and by 0,16% on RMSE for the dataset MovieLens100k. The algorithm performs almost similarly for the dataset MovieLens1M compared to Bayesian approach.

Compared with KNN approach, veristic CB attains significant improvement by 2.9% on NDCG@5 and by 13.2% on f1-score and by 5% on RMSE. Compared with fuzzy tree approach, veristic outperforms by 10% on f1-score@10 and by 6% on RMSE. However, NDCG results are different according to the dataset: for the dataset MovieLens100k veristic CB outperforms on nDCG@10 by 6% and for the dataset MovieLens1M Fuzzy Tree approach outforms on nDCG@10 by 2%. This result needs more evaluation with more datasets.

These results show that the relevance of content-based recommendation can be improved by using veristic variables to build the predictive user profile. Veristic approach minimize the error of predicted values, compared to models based approach like the bayesian approach and similarity-based computationally expensive approach like KNN that uses the entire user-item database to calculate the similarity value between users or items [13].

Table 2. Comparison of f1-score, NDCG and RMSE on Movielens datasets

Dataset: MovieLens 100k						
Algorithm	F1-sc@5	F1-sc@10	nDCG@5	nDCG@10	RMSE	MAE
Veristic	**0,725**	**0,677**	**0,849**	**0,881**	**0,226**	**0,179**
Bayes	0,620	0,571	0,719	0,784	0,393	0,309
Fuzzy	0,626	0,580	0,763	0,815	0,318	0,267
KNN	0,617	0,573	0,802	0,846	0,277	0,228
Dataset: MovieLens 1M						
Algorithm	F1-sc@5	F1-sc@10	nDCG@5	nDCG@10	RMSE	MAE
Veristic	**0,743**	**0,700**	0,840	0,864	**0,248**	**0,201**
Bayes	0,636	0,595	0,705	0,760	0,373	0,291
Fuzzy	0,639	0,591	**0,857**	**0,889**	0,310	0,268
KNN	0,639	0,595	0,807	0,843	0,280	0,233

We now wish to check whether the ranking of algorithms in the Table 2 is statistically significant. That is, we want to know how sure we are, given the relatively small differences in RMSE, whether we can confidently claim that veristic CB achieves a better RMSE than the other algorithms in particular KNN. We therefore compute the sign test pairwise p-values as explained in [14]. We count for each pair of methods, A and B (for instance veristic and KNN). We have computed only results comparing an algorithm that achieved a better RMSE to an algorithm that achieved a worse RMSE. From the Table 3 we can see that veristic algorithm performs better than baseline methods and the p-value is above 0.05, therefore we can say that the results are statistically significant.

Table 3. Pair-wise p-values computed using the sign-test for the RMSE metric for the experiment reported in Table 2. Each method is compared to methods that achieved worse scores for the dataset MovieLens1M.

	KNN	Fuzzy	Bayes
Veristic	7.27E−44	1.93E−97	2.26E−196
KNN	X	1.69E−37	1.42–140
Fuzzy	X	X	1.35E−82

5.4   Configuration of the Parameter K

The parameter K is set to adjust the number of k-most and k-least preferred movies selected to compute users' profiles. To elaborate how it impacts the performance, we measure nDCG@10, f1-score@10 and MAE for both datasets. The results are reported in the Figs. 2 and 3. The best performance is reached at $k = 35$ for all metrics, we note that one user has rated in average 106 items and that our algorithm performs quite well starting from $k = 10$ and increasing the value of K doesn't improve significantly the performance. Thus, we need a small part of the train set to build the user profile. In the following K is fixed to 20.

5.5   Comparison of the Mono-valued and the Multi-valued Matching in Recommendation

The Fig. 4 shows the results of comparison between mono-valued matching and multi-valued matching as we mentioned in the Sect. 4.4. In the multi-valued matching we used the metric $Sim_{multi}(user, movie)$, defined in the formula 3 of the Sect. 4.4, which aims to match the exact user vector. In the mono-valued matching we used the metric $Sim_{mono}(user, movie)$, defined in the formula 4 of the Sect. 4.4, which aims to match the most preferred genre in the user profile. We compared both matching types with *cosine* similarity measure, where:

$$Cos_{Similarity}(x_1, x_2) = Cosine(F(x_1), F(x_2))$$
$$= \frac{\sum_{i=1,...,k} F_i(x_1) \otimes F_i(x_2)}{||F(x_1)|| + ||F(x_2)||}$$

Cosine similarity matches the properties of the user and the movie, thus we consider as a multi-valued matching that is less strict than the metric 3. The Fig. 4 shows that a multi-valued matching improves the recall by 15% and f1-score by 10%. However, cosine similarity performs the best MAE metric and the multi-valued matching with the metric $Sim_{multi}(user, movie)$ performs the worst compared to mono-valued matching with the metric $Sim_{mono}(user, movie)$.

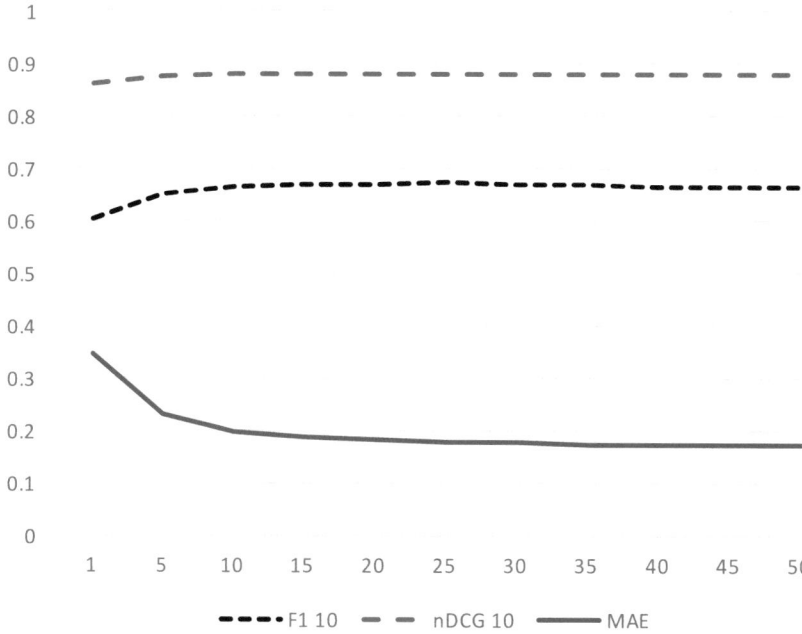

Fig. 2. F1@10, NDCG@10 and MAE for different values of K-most and k-least preferred movies with the dataset MovieLens100k

5.6   Assessing the Utility of Introducing k-Lowest Evaluated Movies in Building the User Profile

To assess what we gain when we consider negatively evaluated movie genres, we compared the original version of our algorithm with a set of modified versions that consider: (1) only top-k-positively evaluated items from the train set, (2) only least-k-negatively evaluated items from the train set, and (3) random evaluated items. Results in Fig. 5 show using both negative and positive statements improve the NDCG and the precision. And by using exclusively positive statements we improve recall and MAE. Using exclusively negative statements improves precision compared to random statements and performs the worst MAE. Then if we need a good precision with a better ranking of the recommended items the use of the mix of negative and positive statements, however, if we need a better recall of recommended items the use of the positive statement is recommended. The use of negative statements and random statement isn't reported to give good results according to this experience.

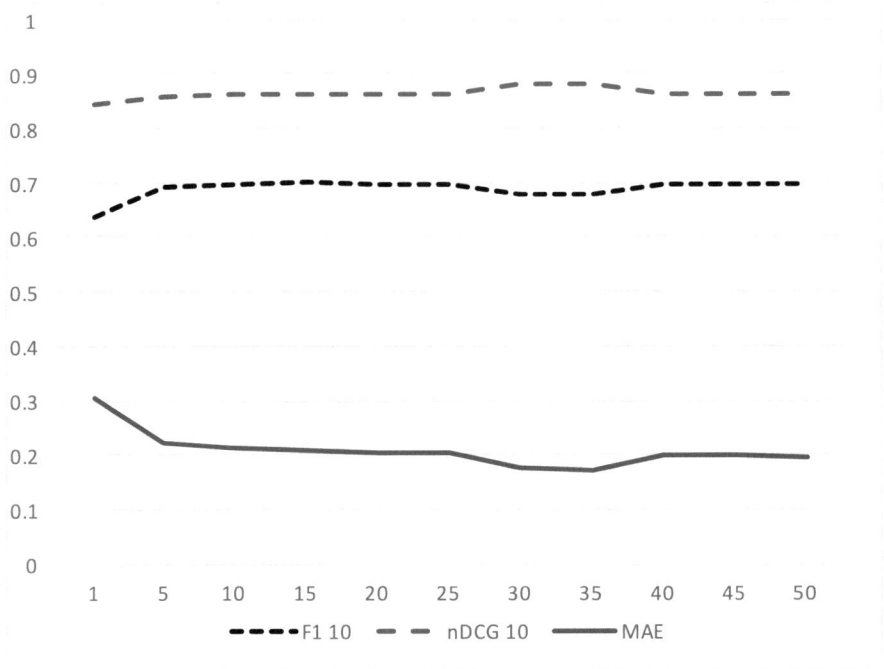

Fig. 3. F1@10, NDCG@10 and MAE for different values of K-most and k-least preferred movies with the dataset MovieLens1M

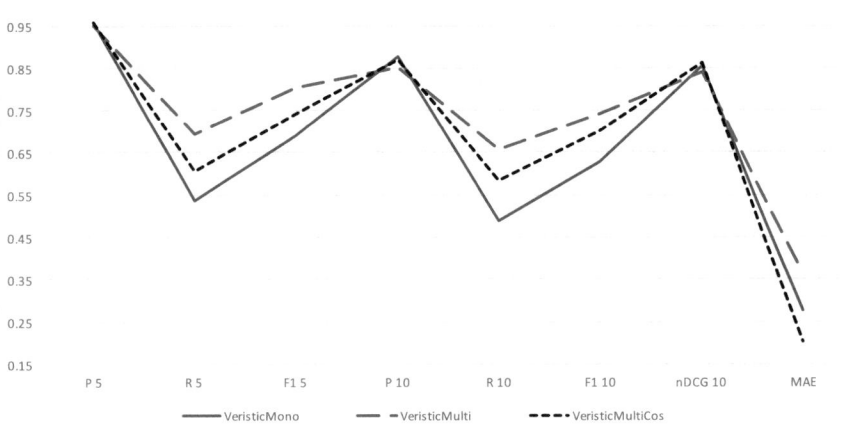

Fig. 4. Comparing multi-, mono-valued matching and cosine matching for Movielens1M

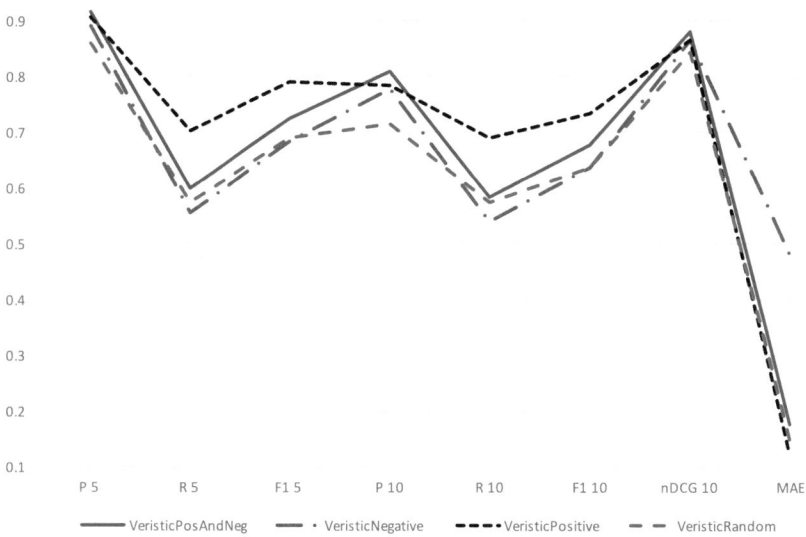

Fig. 5. Comparison between the combination of open positive and negative statements, open positive statements and random statements for Movielens1M

6   Related Work

The literature makes the distinction between two main categories of recommendation strategies: the collaborative filtering and the content-based. Hybrid recommendation systems combine both strategies. Content-based recommendation systems recommend items similar to those that the given user has liked in the past, they focus on items attributes to characterize users, by making such a proposal: *you may like this item which is similar to what you positively rated*, whereas, the collaborative recommendation systems identify users whose preferences are similar to the given user and recommend items they have liked, two users are similar if they rate the same items similarly [23] by making such a proposal: *Customers similar to you liked this item*. We also distinguish two predictions mechanisms in recommendation systems for each strategy: memory-based (or similarity-based) prediction and model based prediction. Contrary to, memory-based prediction which computes a user predicted rating of an item directly from his stored ratings, model-based methods use the user ratings to learn predictive a user profile with the most remarkable features of the items from train data. In this paper, we proposed a content-based recommendation system with a model-based prediction using veristic variable. Veristic Variables are multi-valued fuzzy variables, such variable takes many values simultaneously. Yager [22] was the first to propose the use of fuzzy logic in recommendation systems where he mentioned the problem of the different interpretations of multi-valued and mono-valued features. Later, authors in [25] proposed a first experimentation of Yager's recommendation system and showed that fuzzy method improved the recall. In his pioneer work [22], Yager posed the problem of multi-valued features in recommendation systems, without solving it but proposed the use of the veristic variables as a possible solution of this kind of variables. Later in [3], a preliminary approach for matching user profile and item descriptions which are described by fuzzy properties with veristic and possibilistic interpretations is proposed. This approach uses conditional possibility distribution [15] to match different interpretations of the user profile and the item description. Authors introduced an illustrative example with crisp values and they didn't conduct any experimental or formal validation of the approach. The veristic variable framework was used in [24] to propose a fuzzy k-nearest neighbor method for multi-label classification. Based on veristic statements combined conjunctively or disjunctively, the value taken by a variable is predicted, experimentation showed the competitiveness of this method with state-of-the-art methods. We were inspired by this method to build the predictive user profile for our recommendation algorithm. Multi-criteria recommendation approaches like Multi-Attribute Utility Theory [6] and Information Gain-based [26] do not distinguish multi-valued and mono-valued attributes of items and users. In [8] authors proposed an approach to learn user preferences over numeric and multi-valued linguistic attributes, however, this approach is based on the user selection of his preferences. To the best of our knowledge this is the first work that (1) proposes a method to build a recommendation system using veristic variables to derive a predictive user profile and (2) proposes mechanisms to match multi-valued and mono-valued properties of items and users with experimental evaluation.

7   Conclusion

We consider recommendation problem within the context of CEM as a concept matching problem between a feature-based description of customers and products. We raised the problems that come with such representation, especially, the problem of multi-valued fuzzy properties.

We proposed a content-based recommendation algorithm using varistic variable. Veristic Variable are conjunctive fuzzy variables that we used to represent the multi-valued attribute genre in the context of movie recommendation. The proposed recommendation algorithm consists on two steps (1) *building the predictive user profile* using the highest and lowest ratings of a user, and, open-positive and open-negative veristic statements to define a score that reflect to which extent the user likes or does not a given genre and (2) *Recommending new movies*: using the elaborated user profile, the matching degree of a user and a new movie is predicted, in this step, we defined multi-valued and mono-valued matching strategies using two different similarity measures. We evaluated our approach using MovieLens100k and MovieLens1M datasets. Results show that our approach outperforms baseline approaches in f1-score and RMSE and that a multi-valued recommendation matching improves significantly the f1-score compared to a mono-valued matching.

As a future work, we want to improve, user profiles building to support a mixture of features with different mano-valued and multi-valued interpretation and to improve similarity measure to support many features by associating different weights to them.

References

1. Benzarti, I., Mili, H.: A development framework for customer experience management applications: principles and case study. In: 2017 IEEE 14th International Conference on e-Business Engineering (ICEBE), pp. 118–125. IEEE (2017)
2. Costas, P., Nikos, I.K.: A comparative assessment of measures of similarity of fuzzy values. Fuzzy Sets Syst. **56**(2), 171–174 (1993). https://doi.org/10.1016/0165-0114(93)90141-4
3. Dell'Agnello, D., Mencar, C., Fanelli, A.M.: Item recommendation with veristic and possibilistic metadata: a preliminary approach. In: 2009 Ninth International Conference on Intelligent Systems Design and Applications, pp. 261–266. IEEE (2009)
4. Herlocker, J.L., Konstan, J.A., Borchers, A., Riedl, J.: An algorithmic framework for performing collaborative filtering. In: 22nd Annual International ACM SIGIR Conference on Research and Development in Information Retrieval, SIGIR 1999, pp. 230–237. Association for Computing Machinery, Inc. (1999)
5. Lee-Kwang, H., Song, Y.S., Lee, K.M.: Similarity measure between fuzzy sets and between elements. Fuzzy Sets Syst. **62**(3), 291–293 (1994). https://doi.org/10.1016/0165-0114(94)90113-9
6. Manouselis, N., Vuorikari, R., Van Assche, F.: Simulated analysis of maut collaborative filtering for learning object recommendation. In: Proceedings of the 1st Workshop on Social Information Retrieval for Technology Enhanced Learning, vol. 307, pp. 27–35 (2007)
7. Mao, M., Lu, J., Zhang, G., Zhang, J.: A fuzzy content matching-based e-commerce recommendation approach. In: 2015 IEEE International Conference on Fuzzy Systems (FUZZ-IEEE), pp. 1–8. IEEE (2015)
8. Marin, L., Moreno, A., Isern, D.: Automatic preference learning on numeric and multi-valued categorical attributes. Knowl.-Based Syst. **56**, 201–215 (2014)
9. Mili, H., Benzarti, I., Meurs, M.J., Obaid, A., Gonzalez-Huerta, J., Haj-Salem, N., Boubaker, A.: Context aware customer experience management: A development framework based on ontologies and computational intelligence. In: Sentiment Analysis and Ontology Engineering, pp. 273–311. Springer (2016)
10. Mooney, R.J., Roy, L.: Content-based book recommending using learning for text categorization. In: Proceedings of the Fifth ACM Conference on Digital Libraries, pp. 195–204. ACM (2000)
11. Ning, X., Desrosiers, C., Karypis, G.: A comprehensive survey of neighborhood-based recommendation methods. In: Recommender Systems Handbook, pp. 37–76. Springer (2015)
12. Sala, A., Guerra, T.M., Babuška, R.: Perspectives of fuzzy systems and control. Fuzzy Sets Syst. (2005). https://doi.org/10.1016/j.fss.2005.05.041
13. Shang, S., Kulkarni, S.R., Cuff, P.W., Hui, P.: A randomwalk based model incorporating social information for recommendations. In: 2012 IEEE International Workshop on Machine Learning for Signal Processing, pp. 1–6. IEEE (2012)

14. Shani, G., Gunawardana, A.: Tutorial on application-oriented evaluation of recommendation systems. AI Commun. **26**(2), 225–236 (2013)
15. Tzvieli, A.: Possibility theory: an approach to computerized processing of uncertainty. J. Am. Soc. Inform. Sci. **41**(2), 153–154 (1990)
16. Walker, B.K.: The emergence of customer experience management solutions (2011)
17. Wu, I.C., Hwang, W.H.: A genre-based fuzzy inference approach for effective filtering of movies. Intell. Data Anal. **17**(6), 1093–1113 (2013). https://doi.org/10.3233/IDA-130622
18. Yager, R.R.: Set-based representations of conjunctive and disjunctive knowledge. Inf. Sci. **41**(1), 1–22 (1987)
19. Yager, R.R.: Toward a theory of conjunctive variables. Int. J. Gen. Syst. **13**(3), 203–227 (1987)
20. Yager, R.R.: Reasoning with conjunctive knowledge. Fuzzy Sets Syst. **28**(1), 69–83 (1988)
21. Yager, R.R.: Veristic variables. IEEE Trans. Syst. Man Cybern. Part B (Cybernetics) **30**(1), 71–84 (2000)
22. Yager, R.R.: Fuzzy logic methods in recommender systems. Fuzzy Sets Syst. **136**(2), 133–149 (2003). https://doi.org/10.1016/S0165-0114(02)00223-3
23. Yera, R., Martinez, L.: Fuzzy tools in recommender systems: a survey. Int. J. Comput. Intell. Syst. **10**(1), 776–803 (2017)
24. Younes, Z., Abdallah, F., Denœux, T.: Fuzzy multi-label learning under veristic variables. In: International Conference on Fuzzy Systems, pp. 1–8. IEEE (2010)
25. Zenebe, A., Norcio, A.F.: Representation, similarity measures and aggregation methods using fuzzy sets for content-based recommender systems. Fuzzy Sets Syst. **160**(1), 76–94 (2009). https://doi.org/10.1016/j.fss.2008.03.017
26. Zhang, R., Tran, T.: An information gain-based approach for recommending useful product reviews. Knowl. Inf. Syst. **26**(3), 419–434 (2011)

Knowledge Graph Construction for Intelligent Maintenance of Power Plants

Yangkai Du, Jiayuan Huang, Shuting Tao, and Hongwei Wang[✉]

ZJU-UIUC Institute, Zhejiang University, Haining 314400, China
hongweiwang@intl.zju.edu.cn

Abstract. The intelligent maintenance of power plants greatly relies upon previous experiences and historical information to make informed decisions. Previous research is predominantly focused on collecting data from the maintenance process while little work has been done on the automatic capture and mining of knowledge resources from the data accumulated. Focusing on the experience feedback issue in power plants maintenance, this work proposes a novel process of automatic construction and reasoning of knowledge graphs to support the intelligent maintenance of complex power equipment. In this process, the Bi-LSTM-CRF model and the attention-based Bi-LSTM are specifically used to identify and extract entities and relations from unstructured status reports. On this basis, the knowledge graph construction method based on the neo4j graph database is developed. This paper details the preliminary work towards implementing the proposed process.

Keywords: Intelligent maintenance · Knowledge graph construction · Knowledge extraction · Knowledge organization · Bi-LSTM

1 Introduction

The high reliability of power equipment plays a vital role in ensuring the efficiency and safety of power plants. Corrective disposal and predictive maintenance methods have increasingly become important measures in improving the stability of equipment. These methods highly require useful information from all aspects of the equipment to support its operation such as the experience of handling fault in historical cases, the knowledge about critical properties and conditions of equipment, and the skill of quickly establishing a solution scheme. This feedback information is often stored in the form of unstructured text, which is termed status reports in the existing empirical feedback systems deployed in power plants. This kind of system generally accumulates millions of status reports in the current relational database and each report contains plenty of empirical feedback information and useful knowledge. However, massive and unstructured data make an acquisition of empirical maintenance information difficult and inefficient. Motivated by these issues, this research aims to propose a novel process of mining and managing useful knowledge from the massive data

© Springer Nature Switzerland AG 2020
K.-M. Chao et al. (Eds.): ICEBE 2019, LNDECT 41, pp. 515–526, 2020.
https://doi.org/10.1007/978-3-030-34986-8_36

through the identification and capture of maintenance knowledge using knowledge graphs. This is a key step towards implementing intelligent maintenance systems for power plants, which can effectively extract entities and relations from unstructured textual information. This process is also supported by the methods developed to attain quick-access of information and effective reuse of knowledge through advanced retrieval methods and information visualization.

As a new way of knowledge management and acquisition, knowledge graph has received extensive attention in recent years. Knowledge graph contains a large number of entities, concepts, properties, property values, and relations. The triples composed of entities, relations, entities and entities, properties, and property values are the basic structure of the graph. Many effective methods have been proposed to do entities extraction and classification of relations between entities from unstructured Chinese texts. Traditional named entities recognition (NER) methods are mostly rule-based and statistics-based. For example, Wang et al. designed a NER system based on structural features and contexture paradigm of texts which achieved 62.8% precision and 62.1% recall on the financial domain [14]. However, Rule-based methods require an expert to manually analyze the target text and set the rules – meaning that the performance largely depends on the quality of rules and its texts portability is limited. With the boosting of machine learning and increasing size of the corpus, the statistical model is widely used for NER. Zhou et al. [15] proposed the Hidden Markov Model (HMM) to classify the names, time and numerical quantities. A Maximum Entropy Model (ME) was described in Curran et al. [3]. Kazama et al. applied Support Vector Machine (SVM) in biomedical NE recognition [6]. McCallum et al. proposed a Conditional Random Fields (CRF) Model combined with domain knowledge in the form of the word list [9]. The aforementioned statistical-based NER models achieve good precision in the task of entity recognition, especially in English texts. However, these statistical-based methods require sophisticated feature engineering and semantic processing like dependency parsing, and the results tend to depend on the selection of features. In recent years, the rise of deep learning based natural language processing (NLP) which enables the end-to-end entities tagging and relation classification without sophisticated feature engineering get huge success, and the methods based on deep learning forms the basis of this research.

Industry-oriented knowledge graph (KG) is now widely adopted in the fields of medicine, finance, and other knowledge-intensive domains. KG can be an intelligent and effective tool to manage information generated during execution. However, the knowledge graph construction is still a challenging issue to overcome. On the one hand, for industry-oriented KG, knowledge extraction in the unstructured text requires high precision, unlike knowledge extraction in the open domain, which pays more attention to the breadth of knowledge. On the other hand, there is still no set of standards for the application and research of KG construction oriented to equipment and intelligent maintenance. This research aims to present an effective deep learning based frame for knowledge graph construction and realize the application of knowledge graph in the field of intelligent maintenance.

Table 1. Framework of entities.

ENTITY	MEANING	EXAMPLE
DEP (Department)	The name of Power Plant Department	维修处(**Maintenance Department**)
DEV (Device)	Equipment serial number and name	高压卡件(**High Voltage Device**)
STAT (Status)	An abnormal condition that occurs	正在施工(**Under Construction**)
OPER (Operation)	Power Plant operation	中断运行(**Interrupt Operation**)
TIME (Time)	Time	2018年1月31日(**January 31, 2018**)

The rest of paper is organized as follows. Section 2 will be an overview of the framework for knowledge graph construction and application for intelligent maintenance. Section 3 will introduce the details of the implementations of Information extraction, including entities extraction and relation classification. Section 4 will be an illustration of this research's knowledge organization. In the end, a discussion about this research and future work will be presented.

2   System Framework

The purpose of the framework is to automatically extract the structured knowledge from the massive unstructured plant status reports, and organize the knowledge to form a knowledge graph for intelligent maintenance of power plant. The premise of this proposed framework is the textual characteristics of the power plant status report (Fig. 1).

The status report is an abnormal-status report recorded by power plant maintenance staff during operation. It includes the status of the device at the time of the abnormal event, the operation to correct the abnormal state and the cause of the abnormal state. All of these are targets of knowledge extraction, which will serve as entities and relations in KG and have great value for making informed decision. The entities and relations are defined in Tables 1 and 2.

因超温超功保护双通道输入信号失效导致RPR保护动作
RPR protection action is caused by the failure of double
channel input signal of over temperature and over power protection
P4 信号触发，SB、SC、SD棒组由225步落入炉底
P4 signal triggered, rod groups of SB、SC、SD fell down to
bottom of the furnace on step 225

Fig. 1. Excerpt of status report.

A representative excerpt of status report is shown above. Characteristics of the status reports are listed below:

– There are a great quantity of terminologies and entities in the report. The terminology of power plant is quite complex, and a term is usually made up of several words, such as " 超温超功保护双通道".
– The status report contains numerous numbers and English letters like RPR and P4.
– The occurrence of some words is low, the problem of words sparsity and out of vocabulary (OOV) might occur if the model is word-based.

The aforementioned characteristics of status reports make it difficult to identify the boundaries of terms and entities from text. Besides, the sparsity of words and OOV problem may lead to the result that word vector of these words cannot be fully trained and represent complete semantic information.

To address these problems, we propose an automatic process to extract the knowledge and construct the knowledge graph effectively. The framework of the process is shown below.

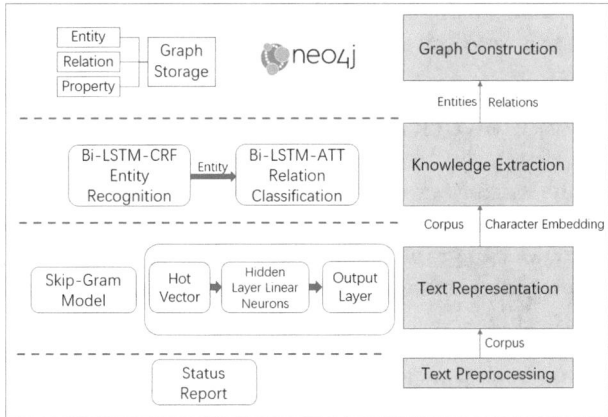

Fig. 2. The framework of the knowledge graph construction for intelligent maintenance.

Table 2. Framework of relations.

Relationship	Meaning
DEV-STAT	Device has abnormal condition
Cause-Effect	Causal Relationship between Operation and abnormal condition
OPER-STAT	Operation done to correct an abnormal condition
DEP-OPER	Operation done by Department
STAT-TIME	The time of abnormal condition
OPER-TIME	The time of Operation
DEP-STAT	Responsible Department for situation occurrence

As shown in Fig. 2, the process consists of four steps: text preprocessing, generation of character vectors, extraction of entities and relations, and the application of the graph database.

(1) Text Preprocessing

In this layer, stopping words and punctuations will be excluded from the status report through the stopping words vocabulary defined by users, and a corpus for character embedding training will be generated.

(2) Text Representation

Semantic features are captured through Distributed Representation of Character, Chinese characters will be mapped to vector space and represented by a low-dimension dense vector. The reason for using character vector instead of word vector is to tackle the problem of word sparsity and OOV, and specific analysis will be given in the next section. To achieve the goal, we use word2vec, which is a widely adopted distributed representation of the word and can represent the semantic information well by skip-gram or CBOW model. Finally, tuned character embeddings will be output to the next layer.

(3) Knowledge Extraction

Entities and Relations will be extracted in the knowledge Extraction layer. The input of the layer is the pre-trained character embeddings and sentence sequences; corpus will be divided into the training set and testing set. The corpus which is labeled with entities and properties will be passed to the Bi-LSTM-ATT model. In the last stage, relations between entities will be extracted and forms the triples with the labeled entities.

(4) Graph Construction

Use neo4j to store entities, relations and properties in the nodes and edges of the graph, and construct knowledge graph which supports high-performance accessing with the support of Cypher language provided by neo4j.

3   Knowledge Extraction

3.1   Text Representation

The proper representation of Chinese texts is the basis of natural language process (NLP). In many NLP tasks, words are seen as the basic units of input to models. Different from English, words in Chinese are not divided by space in written sentences. Therefore, splitting paragraphs into words is usually the first step of processing Chinese text. However, in this paper, instead of doing words segmentation, we choose to use characters as the primary input, the reasons are as follows.

In Chinese, characters are the most fundamental elements, which can be combined into words. Unlike alphabets in English, each Chinese character has specific meanings, which makes semantic analysis possible. In name entity recognition system, it has been found that the word-based model underperforms the character-based model in language modeling and sentence matching/paraphrase [10]. Two factors are leading to word-based models' disadvantages, as will be listed below.

Firstly, it has been confirmed that the data sparsity issue is possible to give rise to overfitting since more words need a larger number of parameters [10]. However, it is not practical to maintain a large word-vector table, so many words are treated as out-of-vocabulary word, which may have a negative impact on the model's learning ability. For a corpus, to learn the word semantics well, the model is supposed to have adequate exposure to each word. However, many Chinese words, especially the terminology we used in this paper, appear very infrequently, making the model impossible to fully learn their semantics.

Secondly, the current word segmentation technology is far from precise, and its error will be passed to the downstream NLP task. Taking Jieba, the most widely-used open-sourced Chinese word segmentation tool, as an example, when splitting texts of the news report, the rate of error is around 13% [7]. When Jieba processes the status report which contains digital data, the accuracy is lower. Since incorrect word segmentation can cause ambiguity, ambiguous word segmentation results in a nonnegligible interference effect in entity recognition.

Therefore, in this task, we use the character model instead of word model to represent the text, and we use model of Word2vec in Genism, which is an open-sourced word vector tool released by Google in 2013.

3.2   Named Entity Recognition

There are mainly two types of named entity recognition (NER). One is to classify each word on multiple tags, and the tag with the highest classification probability is selected. The other method is sequence labeling for different words at the same time, then the labeling sequence with the highest joint probability is selected. The sequence labeling method is favored in the research of named entity recognition, and the online sequence labeling method has been gradually developed in natural language processing research due to its scalability and adaptability [1].

This paper adopts the sequence labeling method. As mentioned in the text representation section, character vector based on Chinese character is chosen to be trained to avoid error from word segmentation, which requires a labeling set applied on Chinese characters. Considering the classification of entity in status report discussed previously, we use BIO labeling method to define the labeling set as L = {B_dep, I_dep, B_dev, I_dev, B_stat, I_stat, B_oper, I_oper, B_time, I_time, O} to represent the beginning, the interior and the ending of entity in different classifications.

For named entity recognition, as shown in Fig. 2, this paper proposes to use the bidirectional LSTM model with a CRF layer (Bi-LSTM-CRF) due to the following reasons. Firstly, Bi-LSTM shows superiority compared with other neural models like RNN and LSTM. On the one hand, it avoids gradient explosion and gradient vanishing of RNN. On the other hand, Bidirectional LSTM outperforms unidirectional LSTM due to importing the contextual information instead of only data previous to the sentence analyzed [4]. As shown in Fig. 3, the Bi-LSTM contains two directions of network structure: the forward layer is left-to-right propagation $\left(\overrightarrow{h}_1, \overrightarrow{h}_2, \overrightarrow{h}_3, \overrightarrow{h}_4\right)$; the backward layer is right-to-left propagation $\left(\overleftarrow{h}_1, \overleftarrow{h}_2, \overleftarrow{h}_3, \overleftarrow{h}_4\right)$. These two hidden state sequences are connected by vector splicing, and the output integrated hidden state sequence is then mapped to k-dimensional by the connected linear layer, where k is the number of label types in the training set. Thus, we obtain the extracted sentence features, recorded as (c_1, c_2, c_3, c_4) (Fig. 4).

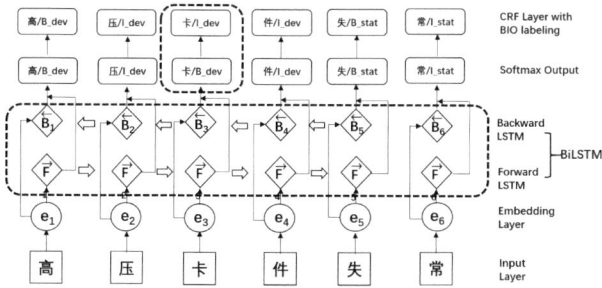

Fig. 3. Bi-LSTM-CRF model.

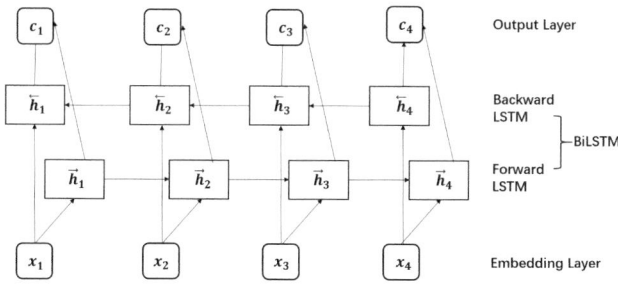

Fig. 4. Bi-LSTM model.

Although Bi-LSTM can perform sequence labeling tasks, label offset issues cannot be ignored. However, with the CRF model to complete the sequence labeling task, a global optimal output sequence can be obtained and label offset can also be avoided. Huang et al. [5] first apply a Bi-LSTM-CRF model to NLP benchwork sequence tagging data set, showing that this model could produce a state of the art accuracy on NER data sets. The reason is that the CRF model can adequately consider the dependence between the tags like B_dep and E_dep, so it can be correctly identified that the "power rail" is a complete term. On the contrary, the Bi-LSTM model would result in two entities "power" and "rail" regardless of the dependence between tags. Therefore, the addition of the CRF layer can effectively solve the problem of identification of domain terms, and mostly nested and composite structures. A similar example is also as shown in Fig. 3.

The specific method is to add the CRF layer after the softmax output layer of Bi-LSTM, and introduce the state transition matrix A as the parameter of the CRF layer. Matrix L represents the output of Bi-LSTM and $A_{i,j}$ represents probability of the chronological transfer from the i-th state to the j-th; $L_{i,j}$ represents the probability that the i-th word in the observation sequence is marked as the j-th labeling [8]. In this method, the maximum likelihood estimation is used as the cost function, and the Viterbi algorithm is used for decoding. $Y = (y_1, y_2, y_3, y_4, \ldots, y_n)$ is the label sequence to be predicted for observation sequence X. The formula to calculate the output of Y is as follows:

$$s\left(X, Y\right) = \sum\nolimits_{i=1}^{n} \left(A_{yi,yi+1} + L_{i,yi}\right)$$

$$\log L\left(y|x\right) = s\left(X, Y\right) - \log \sum\nolimits_{y'} \exp\left(s\left(X, Y'\right)\right)$$

Fang applies a similar model in Chinese electronic medical record field for NER and shows a good result: the five physical categories of body parts: symptoms and signs, examination and examination, treatment, disease, and diagnosis were identified, and the highest accuracy rate of 96.29%. The recall rate was 96.27% and the F value was 93.96 [13]. The good result of Bi-LSTM-CRF for NER in Chinese electronic medical record field indicates a considerable prospect of its utilization for status report for power plants.

3.3   Relation Classification

Neural network like Bi-LSTM can also be applied to relation classification. As mentioned in the introduction section, the traditional approach relies on existing vocabulary resources (such as WordNet), NLP systems, or some hand-extracted features [12]. Such an approach achieved high performance but may lead to an increase in computational complexity. Besides, the feature extraction work itself consumes a lot of time and effort. Another challenge is that critical information may be a word that can appear anywhere in a sentence, but neural models like Bi-LSTM may ignore this when receiving massive information.

To focus on the important information and data for relation classification, we intend to use an Attention-based Bi-LSTM model. The attention mechanism is a method that simulates the attention of the human brain. The core idea is to simulate the property that human brain pays attention to a certain key point while ignoring other non-key points at a certain moment [2]. Therefore, the main effect of this mechanism is to allocate more attention to keywords and less attention to other parts. The Attention mechanism was combined with the Bi-LSTM model, as shown in Fig. 5, to automatically discovers words that play a key role in classification, allowing the model to capture the most important semantic information from each sentence [16]. For instance, for the text "the batteries are nine years old and aging appears to be more rapid than expected", the keyword

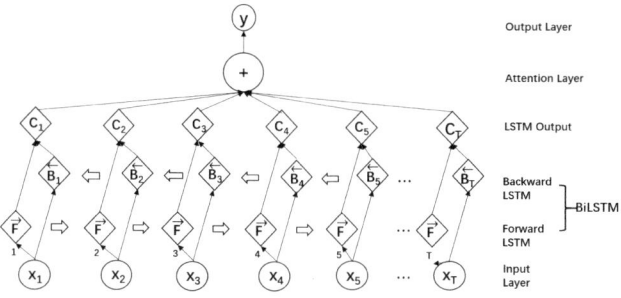

Fig. 5. Attention-based Bi-LSTM model.

is "batteries". However, without the attention mechanism, the Bi-LSTM model focus on the context information instead of the keyword "batteries". After adding the attention mechanism, the model can realize this function by calculating the related weight.

Zhou et al. designed an attention-based Bi-LSTM model to extract relation from text [16]. The Attention mechanism is realized as follows,

$$M = tanh\,(H)$$

$$\alpha = softmax\,(w^T M)$$

$$r = H\alpha^T$$

where H: $[h_1, h_2, h_3, h_4, \ldots, h_n]$ is the set of output vector produced by Bi-LSTM layer; T is the sentence length, and w^T is the transpose of the trained parameter vector; r is the representation of the sentence (Fig. 6).

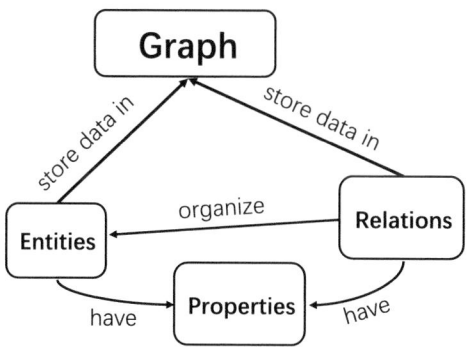

Fig. 6. neo4j data structure.

The final representation of sentence to be used for classification is obtained by:

$$h^* = \tanh\,(r)$$

4   Knowledge Organization

The entities(E), relation(R), properties(P) extracted by this model are highly connected data. It is necessary to store the relations and entities in one database which can maintain such connection between entities effectively. A graph database is an effective tool for data modeling when a focus on the relation between entities, and Neo4j is the most popular graph database for the knowledge graph construction. In Neo4j, a graph consists of three fundamental data structure, Nodes(N), Relations(R) and Properties(P). Nodes are interconnected by the relation defined by R set, and Properties can be attached to each node and Relations. The graph constructed by Neo4j can do the graph search effectively under the support of Cypher language provided by Neo4j, and this property provides the underlying database for high-performance KG based search and reasoning (Fig. 7).

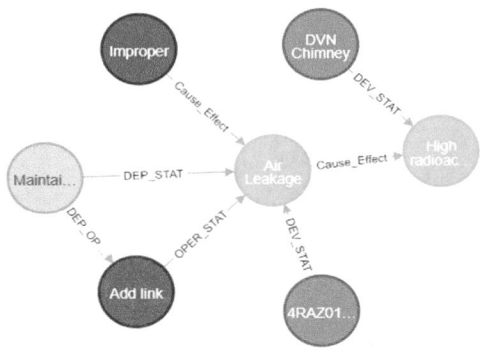

Fig. 7. Knowledge graph constructed by neo4j.

Neo4j enables the user to visualize the graph database with the support of D3 and web. A small part of the knowledge graph constructed by neo4j is shown above. The entities and relations in the graph are currently recognized manually from one of status report. The original reports recorded an abnormal condition (High redioation alarm), root analysis of the condition and the preventive operation. The red nodes represent Operation (OPER), orange nodes represent device (DEV), blue nodes represent the status (STAT) and yellow node represent the department name (DEP). The class of entities and relations follows the definition in Tables 1 and 2.

As can be seen from the figure, the root cause of the high radiation alarm was the improper size of gasket, which lead to the air leakage and finally trigger High radiation alarm. Other information such as responsible department, device name and preventive operation (Add link) is also organized clearly in the graph.

5   Conclusion

In order to solve the actual requirements of intelligent maintenance of power plants, based on the neural network model and the Neo4j database, we propose the idea of constructing a knowledge graph for the power field. Knowledge graphs provide the basis for quickly and accurately matching historically similar fault information because it can receive entities extracted from past maintenance reports and build relations as between them as received. In this paper, we propose a process for constructing a knowledge graph of the power plants. After choosing to train character vector as the pre-processing text, we envision training the Bi-LSTM-CRF model for named entity recognition and training the Attention-based Bi-LSTM model to extract the relations between the entities. Finally, the Neo4j database is used to receive the extracted structured data to construct a visual knowledge graph.

Models for extracting entities and relations are still being researched and their performance is gradually improving. Recently, Matthew et al. proposed a bidirectional language model trained by massive unlabeled corpus to obtain the language model vector of the current word to be annotated, which is then added as a feature to the original bidirectional RNN-CRF model [11]. The experimental results show that the addition of this language model vector can significantly improve the NER effect on a small amount of annotation data. This semi-supervised method may be applied to the process of constructing knowledge maps of professional fields in the future.

Acknowledgement. This work was supported by the Zhejiang University/University of Illinois at Urbana-Champaign Institute, and was led by Principal Supervisor Prof. Hongwei Wang.

References

1. Bu-Zhou, T., Xiao-Long, W., Xuan, W.: Confidenceweighted online sequence labeling algorithm. Acta Automatica Sinica **37**(2), 188–195 (2011)
2. Chong, Z.: Text classification based on attention-based LSTM model, vol. 10. Nanjing University (2016)
3. Curran, J., Clark, S.: Language independent NER using a maximum entropy tagger. In: Proceedings of the Seventh Conference on Natural Language Learning at HLT-NAACL 2003, pp. 164–167 (2003)
4. Graves, A., Schmidhuber, J.: Framewise phoneme classification with bidirectional lstm and other neural network architectures. Neural Networks **18**(5–6), 602–610 (2005)
5. Huang, Z., Xu, W., Yu, K.: Bidirectional LSTM-CRF models for sequence tagging. arXiv preprint arXiv:1508.01991 (2015)
6. Kazama, J., Makino, T., Ohta, Y., Tsujii, J.: Tuning support vector machines for biomedical named entity recognition. In: Proceedings of the ACL-02 Workshop on Natural Language Processing in the Biomedical Domain, vol. 3, pp. 1–8. Association for Computational Linguistics (2002)

7. Luo, R., Xu, J., Sun, X.: Peking university open source new Chinese word segmentation kit: accuracy far more than Thulac, Jieba word segmentation. https://github.com/lancopku/PKUSeg-python (19). Accessed 24 July 2019

8. Ma, J., Zhang, Y., Yao, S., Zhang, B., Guo, C.: Terminology extraction for new energy vehicle based on BLSTM_attention_CRF model. Application Research of Computers, pp. 1385–1389 (2019)

9. McCallum, A., Li, W.: Early results for named entity recognition with conditional random fields, feature induction and web-enhanced lexicons. In: Proceedings of the Seventh Conference on Natural Language Learning at HLT-NAACL 2003, vol. 4, pp. 188–191. Association for Computational Linguistics (2003)

10. Meng, Y., Li, X., Sun, X., Han, Q., Yuan, A., Li, J.: Is word segmentation necessary for deep learning of Chinese representations? arXiv preprint arXiv:1905.05526 (2019)

11. Peters, M.E., Ammar, W., Bhagavatula, C., Power, R., Peters, M.E., Ammar, W., Bhagavatula, C., Power, R.: arXiv (2017)

12. Rink, B., Harabagiu, S.: Classifying semantic relations by combining lexical and semantic resources. In: Proceedings of the 5th International Workshop on Semantic Evaluation, pp. 256–259. Association for Computational Linguistics (2010)

13. Tao, F.: Medical knowledge map construction based on Chinese language processing and deep learning Unpublished

14. Wang, N., Ge, R., Yuan, C., Wong, K., Li, W.: Company name identification in Chinese financial domain. J. Chinese Inf. Process. **16**, 1–6 (2002)

15. Zhou, G., Su, J.: Named entity recognition using an hmm-based chunk tagger. In: Proceedings of the 40th Annual Meeting on Association for Computational Linguistics, pp. 473–480. Association for Computational Linguistics (2002)

16. Zhou, P., Shi, W., Tian, J., Qi, Z., Li, B., Hao, H., Xu, B.: Attention-based bidirectional long short-term memory networks for relation classification. In: Proceedings of the 54th Annual Meeting of the Association for Computational Linguistics (Volume 2: Short Papers), pp. 207–212 (2016)

SPUNTB: A Stowage Planning via the Unified Neutral Theory of Biodiversity and Biogeography

Zongzhao Xie[1,2(✉)] and Wenbin Hu[1,2(✉)]

[1] School of Computer Science, Wuhan University, Wuhan, China
{xzz-1996,hwb}@whu.edu.cn
[2] Shenzhen Research Institute, Wuhan University, Wuhan, China

Abstract. Stowage planning, which raises when the ship industry determines the position of containers, is a key part in container terminal management. Literatures show that binary integer programming for that problem is impracticable because of large number of binary variables and constraints. To reduce the turnaround time and cost, this paper propose a algorithm for stowage planning based on the unified neutral theory of biodiversity and biogeography (SPUNTB). A greedy strategy is constructed to build the initial solution. Moreover, randomizing, migration strategy, unloading and reloading strategy, and filter are also introduced to make it more instructive and faster. The proposed algorithm, verified by extensive computational experiments, achieves a satisfying performance.

Keywords: Stowage planning · The neutral theory · Migration strategy · Filter

1 Introduction

Nowadays, containers, one of the greatest invention in 20th century, play a more and more important role in maritime transportation. Because of its wide application, ports, highways, bridges and airports are integrated into the logistics system. Over the past two decades, the world international maritime trades have increased continuously. The throughout of the world's busiest port, Shanghai port, has over 35 million TEUs (twenty-foot equivalent units) since 2014 and reached 40 million in 2017 [1]. The average carrying capacity of vessels grew from 3161 TEUs in January 2009 to 4449 TEUs in January 2014 [3].

Loading and unloading the same container in a port is called a shift. As for the containers can only be accessed from the top, shifts are common activities in container operation. But those activities are time-consuming and money-consuming, hence it is critical to reduce the number of shifts.

The task of determining where containers have to be placed in a port is called stowage planning. In addition to the access constraints mentioned above, many other constraints have impacts on the stowage planning, such as position

© Springer Nature Switzerland AG 2020
K.-M. Chao et al. (Eds.): ICEBE 2019, LNDECT 41, pp. 527–540, 2020.
https://doi.org/10.1007/978-3-030-34986-8_37

constraints that containers stand on the top of others in the stacks, rather than hung in the air, weight constraints when a container is stacked on others, stability constraints and so on.

It has been proven in [12] that the unified neutral theory of biodiversity and biogeography (UNTB) have a good performance on the block allocation problem (BAP) and this problem is similar to the stowage planning problem. Both are focused on the stack of containers, but the BAP doesn't have to consider the stability constraints. In this paper, we propose a algorithm based on UNTB to find out the optimal solution in stowage planning. The goal is to minimize the time and the cost a vessel spends in a port when keeping the stability of the vessel. The main research on stowage planning in this paper is as follows.

- We extend the stowage planning presented in [10] with the inclusion of weight of containers.
- We refer to the containers with the same destination and the same weight as a type, so we do not determine a special slot for a container, but a series of position for a container, which means the type of container matters.
- We propose a grouping strategy to reduce the solution set. It is used at the beginning of solving the stowage planning.
- Migration strategy, unloading and reloading strategy are used to improve the results.

The rest of paper is organized as follows. Section 2 summaries relevant literatures. Then, the problem description is defined in Sect. 3. In Sect. 4, the proposed algorithm is discussed. Section 5 analyzes the performance of algorithm based on numerical experiments and conclusions are discussed in Sect. 6.

2   Literature Review

A growing number of studies on stowage planning have been published in the past few years. Some of the studies formulate the problem as a 0–1 Integer Programming (IP). In order to minimize the number of shifts, Reference [7] presented a binary linear programming. They tried to get a optimal stowage planning for a single rectangular bay, but they only studied the accessibility constraints. Reference [5] and [18] also made use of binary linear programming to solve this problem. They thought about more constraints (like stability constraints, weight constraints, etc.) than the above one. However, Reference [9] and [8] showed that the stowage planning problem is NP-complete. The IP model is limited due to large numbers of binary variables.

Other papers proposed heuristic algorithms. Reference [11] adopted genetic algorithm to solve the stowage planning. They came out with a compact encoding which reduces the search space and could get a good result in reasonable time. Reference [21] also applied genetic algorithm to reduce the computational time and they developed different stack method for different types of containers. Branch and bound search was used in [17] to minimize the time the vessel spent on loading all containers. They took the accessibility constraints into account and the proposed algorithm had a better performance for big size ships. Reference [4,15,20] used tabu search, [14,16] used greedy algorithm, and [19] used tree search algorithm for this problem. However, the efficiencies and complexities of these algorithms still need to be verified.

3   Problem Description

A representative stowage planning was presented in [10]. We extend this definition by including weight restrictions.

Determining the position for all containers loaded at each port the containership calls during its journey is a very complicated task. Firstly, a full information of the cellular structure of the vessel is required. Next a complete view of containers (e.g. the number of containers, the weight, and the port of destination (POD)). When making the decision, all constraints should be met. Some of the important preliminaries about the proposed model are illustrated as below.

3.1   Vessel Model

A container vessel is a ship that transports containers from one port to another. The storage area of a vessel is divided into bays which consist of a number of cells. A vertical part of a bay is called a stack, and a cell is called a slot. The typical layout of a container vessel is shown in Fig. 1.

To simplify the vessel model, in this paper, a vessel has B bays. Each bay is a rectangular, with R rows labeled from bottom to top and C columns labeled from left to right. Slot in row r and column c is labeled (r, c). For irregular bays, they can be transformed into rectangular ones by adding some imaginary slots with additional constraints.

3.2   Container Model

In this paper, only one kind of container is discussed, which means all containers are of the same size (e.g. 20 TEU) and one container can be located in a slot. In spite of this, the weights of containers are different from others.

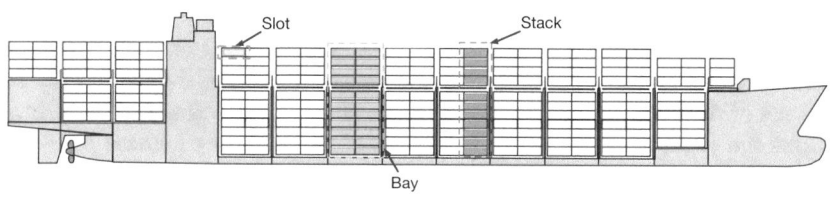

Fig. 1. A typical layout of a vessel

A container is called a j-container if and only if its destination is port j. Hence, all j-containers are loaded before the vessel visits port j and they should be unloaded at port j. A i,j-container is a j-container that loaded at port i.

A blocking happens when there exists a j_1-container in slot (r_1, c) where a j_2-container is in slot (r_2, c) with $j_1 > j_2$ and $r_1 > r_2$. In other words, the j_1-container is above the j_2-container. If we want to unload j_2-container, the j_1-container should be handled first. The blocking can occur at any time. The total number of blocking is given below.

$$Blocking = \sum_b \sum_{r \in R} \sum_{c \in C} bl_b^{rc} \tag{1}$$

where bl_b^{rc} is the number of blocking for the container in slot (r, c) of bay b.

3.3   Transportation Model

During a vessel's journey, it plans to visit N (≥ 2) ports sequentially. At port 1, containers destined for port $2, \ldots, N$ are loaded. At each port $i = 2, \ldots, N-1$, all containers destined for port i are unloaded and those destined for ports $i+1, \ldots, N$ are loaded. Shifts and blocking may occur in these ports. At port N, all containers are unloaded.

$T = [T_{ij}]$ is a $N \times N$ transportation matrix, where T_{ij} is the number of i, j-containers, for all $i, j \in \{1, 2, \ldots, N\}$. Note that $T_{ij} >= 0$ for all i, j and $T_{ij} = 0$ for all $i >= j$.

A transportation matrix is feasible if in each port the capacity and weight constraints can be satisfied. If T is not feasible, the decision about which containers should be loaded is beyond the scope of our considerations. Then, only feasible transportation matrices are considered throughout this paper. Reference [10] has proven that only stowage planning problems with $N >= 4$, $R >= 2$ and $C >> 1$ need to be pondered.

3.4   Stability Model

Before we discuss how to evaluate the stability of the containership, we introduce a reasonable assumption that each container has the same center of gravity, i.e. it is at the center of the container.

The stability of the vessel is assessed mainly from three aspect: GM, the distance between the vessel's center of gravity (G) and the metacenter (M); list, caused by containers being loaded on the vessel, and trim, defined as the total of change of drafts forward and aft. In all of them, GM is the most important.

Combine Fig. 2 and [13], we formulate GM, list, trim as follows:

$$GM = G_0M + \frac{\sum_i w_i * lh_i}{\Delta_T} \tag{2}$$

where G_0M is the distance between the center of gravity (G_0) and the metacenter (M) when the vessel is empty; w_i is the weight of container; lh_i is the vertical

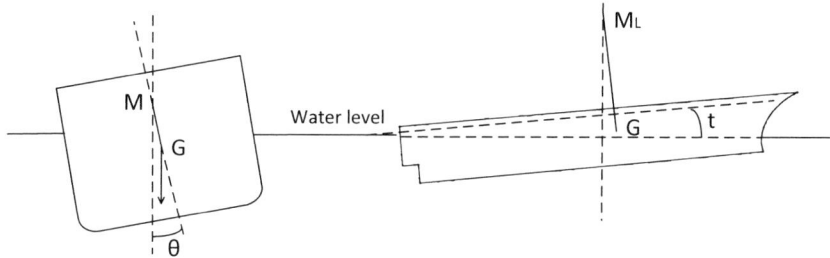

Fig. 2. The factors in evaluating the stability of a vessel.

distance between the vessel's center of gravity and the container's center of gravity; Δ_T is the displacement of the vessel, which is calculated as follow.

$$\Delta_T = \Delta_S + \sum_i w_i \tag{3}$$

where Δ_S is the displacement when the vessel is empty.

$$tan\theta = \frac{\sum_i w_i * lw_i}{\Delta_T * GM} \tag{4}$$

where lw_i is the horizontal distance between the vessel's center of gravity and the container's center of gravity;

$$t = \frac{\sum_i 12w_i * ll_i}{W * L^2} \tag{5}$$

where ll_i is the horizontal distance between the center of floatation and the container's center of gravity; W is the ship's width; L is the length of the vessel.

Since the transportation matrix is feasible, all containers can be loaded. So the displacement of the vessel Δ_T after all containers on board is constant. In (2), only $w_i * lh_i$ changes and it depends on where the container i is arranged. To ensure the stability of a vessel, we need to maximize GM, in other word, $max(\sum_i w_i * lh_i)$.

The list definition includes the GM. After maximize GM, to minimize list, we only need to minimize $\sum_i w_i * lw_i$. While the list value is either negative or positive, the exact objective is minimization of the absolute value of list ($GM \geq 0$). So is the trim. We need to minimize the absolute value of $\sum_i w_i * ll_i$.

Keeping in view of all the above discussed case, the objective function is illustrated as below.

$$minimizeZ = \alpha \sum_i (w_i * lh_i) + \beta | \sum_i w_i * lw_i |$$
$$+ \gamma | \sum_i w_i * ll_i | + \delta * Blocking \tag{6}$$

where, α, β, γ and δ are weights for the GM, list, trim and blocking, respectively. Note that α is a negative because of the maximization of the GM.

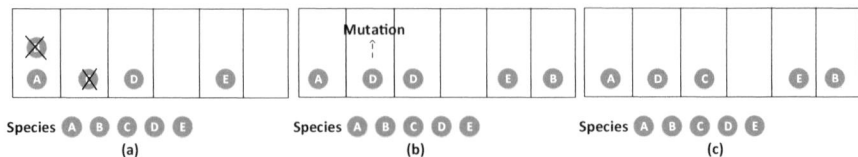

Fig. 3. The example for UNTB in the stowage planning. (a) Kill in random. (b) Generate new species (c) The final result

4   SPUNTB Model

It has been proven in [6] that containers with further destinations should be loaded before those with closer destinations. This strategy is also useful to reduce the number of blocking. So in the proposed algorithm, we deal with all i, j-containers $(j = N, \ldots, i+1)$ in sequence. But this may result in all i, j-container stacked together, which is not good for the stability of the vessel. Therefore, containers are divided into several groups. Each group has the same total weight and is stacked in different bay.

The UNTB is a hypothesis by ecologist Stephen Hubbell. In UNTB, each species follows a random walk and all individuals obey the same rules [2]. For stowage planning, a bay is referred to as an island and a container group is regarded as a species. Figure 3 illustrates the UNTB model for the stowage planning.

As shown in Fig. 3(a), five species is allocated to islands randomly. To get the next generation, kill a certain number of species in random, Fig. 3(a). Then in Fig. 3(b), give preference to the island with no species and randomly select a killed species to be born on it. During the process of generating, a mutation may happen and the new born species becomes another species that don't be killed. Repeat, and finally all species are in different islands 3(c).

4.1   The Modified UNTB Model

Because the number of containers in the stowage planning is large and the stowage planning is a NP-complete problem, when the original UNTB which is complicated and completely random is used to solve the problem, it has a slow speed and may fall into a local optimal solution. Thus, group strategy that dividing all islands into several equivalent groups is applied. This strategy can also reduce the search area. Figure 4 indicates how the conception of grouping strategy is introduced in the UNTB model. Meanwhile, some more modifications are made to improve the model, which will be described in detail in following paragraphs.

4.2   The Group Strategy

In order to reduce the range of solution set as well as satisfy multiple quay cranes work together, we group all bays into several groups according to the number

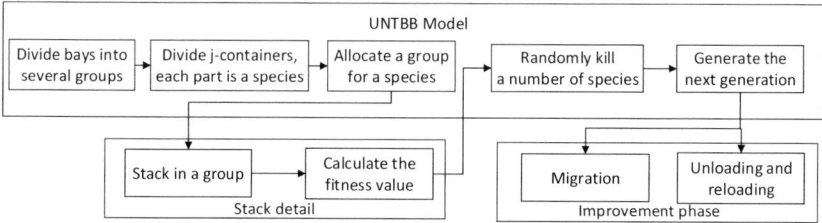

Fig. 4. The flow of SPUNTB model

of working quay cranes. Therefore, each group contains a successive subsection of bays. Due to fair division, the difference among groups is little and the free capacity of slots among groups is almost equal. To fit the group strategy, the number of groups for i, j-containers should not over this group.

Algorithm 1. Dividing the vessel

Input: b, the number of bays
    The free capacity of each bay
    The groups number, m
Output: the groups
 1: **for** $i = 0; i < b; i + +$ **do**
 2:   $totalNum \leftarrow$ free capacity of bay i
 3: **end for**
 4: $average \leftarrow totalNum/m$
 5: $minDiff \leftarrow -1$
 6: **for** $i = 0; i < b; i + +$ **do**
 7:   $groupNum \leftarrow groupNum +$ free capacity of bay i
 8:   $curDiff \leftarrow |groupNum - average|$
 9:   **if** $curDiff < minDiff || minDiff == -1$ **then**
10:     $minDiff \leftarrow curDiff$
11:   **else**
12:     save current position
13:     $groupNum \leftarrow$ free capacity of bay i
14:     $minDiff \leftarrow |groupNum - average|$
15:   **end if**
16: **end for**

To achieve this effect, we first calculate the average free capacity of the vessel (line 1–5). Add the free capacity of slots in a bay one by one (line 6). If the difference between the current sum and the average is smaller than last time, we find a break point (line 11–15). Repeat this process until we get all break points (line 6–16).

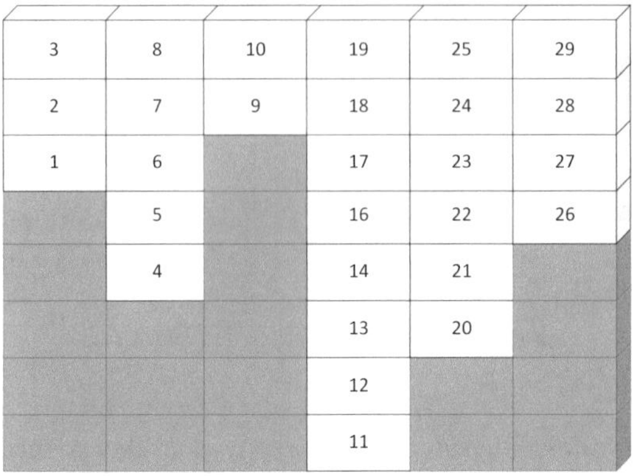

Fig. 5. A way to compare two scheme. The sequence in tier by tier scheme may be 1, 2, 3, 4, ... and the sequence in lowest scheme may be 11, 12, 13, 20, ...

4.3   The Stacking in a Group

After applying the modified UNTB model, we get the place of each container group. Furthermore, the way how containers stack in a group still needs to study.

There exist two schemes of stacking in a group. Both are greedy methods. The first one is tier by tier scheme. For the all containers to be stacked in bay A, If a stack is fully occupied or it reaches weight limit, another stack will be considered. In other word, if a stack cannot stack more container, then select another one. The other one is lowest scheme. In this way, the lowest slot for the containers is first chosen. If two or more slots are in the lowest position, randomly select one.

For a bay like Fig. 5, the sequence of loading containers may be 1, 2, 3, 4, 5, ... in tier bay tier scheme. And in lowest scheme, the sequence may be 11, 12, 13, 20 etc. The former is more useful for putting the further containers in the lower slot, but there may exist empty stacks while the latter is easier to satisfy the rule that the heavier containers are in the lower slot.

During this process, there may arise a problem that the number of containers in a group is greater than the free capacity of a bay group. To deal with this, a greedy methods is applied that selecting another group with the most free capacities.

4.4   Improvement Phase

Randomizing. Random processes are introduced into the proposed algorithm to increase the diversity of the result.

(a) During the process of dividing containers into several groups, all containers are sorted by weight. So the result is always in decreasing weight order. Instead, we do not sort the containers by its weight, but random order them decreasing or increasing.

(b) In tier by tier scheme, instead of putting containers from the left bay in a group, randomly select a bay from group as the beginning.

Migration, Unloading and Reloading. Extra steps are done to make the algorithm more useful.

At port 1, migration strategy is used. There have two strategy of migration: within-group and among-group. The within-group strategy means select two bay in random within a group and exchange their containers. The among-group strategy is that from all the groups, choose two bay from different group in random and then exchange all their containers.

For other ports, the vessel is not empty. If we still apply the migration strategy, the number of shifts may increase. So for port $i(i = 2, 3, \ldots, N - 1)$, The unloading and reloading method is employed. Before loading containers on board, first unload some containers from the vessel in order to reduce the number of shifts in the future. In detail, unload j-containers $(j \leq k)$ and those containers are loaded together with containers that should be loaded in this port.

Filter. Not all the solutions will be improved through above methods, so there is a filter to judge whether the solution has a high probability to become a better one. Because of less solutions are improved, the speed of iteration is speeded up.

In order to do this, 500 preliminary experiments are conducted to collect information about the fitness value Z_1 obtained before the improvement phase, and the fitness value Z_2 after the improvement phase. Calculate Z_1/Z_2 and find its median mid. Hence when new experiments are conducted, a solution with fitness value Z_1 only goes to the improvement phase if $Z_1 \leq mid * Z^*$, where Z^* is the value of the best known solution.

5 Experiments

The proposed algorithm was coded in Java and experiments are carried out under the configuration of Windows 10 Multiple Editions (Microsoft Corporation, WA, USA), and 8 GB RAM.

5.1 Data Set

The way we generate the data set is derived from the random integer partition (RIP) algorithm in [10].

Consider the problem of dividing a given number v into b parts, where $b < v$, and both v and b are positive number. If b equals 1, the result set contains only one number v (line 1–2). Otherwise, create a new array y and the value of each element equals to the index plus one (line 4–5). Exchange i-th element and j-th element, for $i = 0, 1, \ldots, b - 2$, $j = 0, \ldots, v + b - 2$ and $j \neq i$(line 7–12). Then sort the top $b - 2$ numbers (line 13). After that, any two adjacent number in top $b - 2$ is not equal and the latter one is always greater than the former. Let

---

Algorithm 2. Random integer partition, RIP

Input: The positive number,v
      The number of parts,b
Output: The array whose sum is b, x
 1: **if** $b == 1$ **then**
 2:     $x \leftarrow v$
 3: **else**
 4:     **for** $i = 0; i < v + b - 1; i + +$ **do**
 5:        $y[i] \leftarrow i + 1$
 6:     **end for**
 7:     **for** $i = 0; i < b - 2; i + +$ **do**
 8:        $j \leftarrow random(0, v + b - 2)$
 9:        **if** $i \neq j$ **then**
10:           exchange($y[i], y[j]$)
11:        **end if**
12:     **end for**
13:     sort $y[i]$ for all $i = 0, \ldots, b - 2$
14:     $x[b - 1] \leftarrow v + b - 1 - y[b - 2]$
15:     **for** $i = b - 2; i > 0; i - -$ **do**
16:        $x[i] \leftarrow y[i] - y[i - 1] + 1$
17:     **end for**
18:     $x[0] \leftarrow y[0] - 1$
19: **end if**

---

the first and the last element of result set equal to $y[0] - 1$, $v + b - 1 - y[b - 2]$ respectively and $x[i] = y[i] - y[i - 1] + 1$ for all $i = 1, \ldots, b - 2$.

As mentioned before, the transportation matrix T is used to store all the numbers of i, j-container. For port i, to generate the number of j-container $(j = i + 1, \ldots, N)$, let v equals the total number of container that should be loaded on board and $b = N - i - 1$, apply the RIP algorithm to obtain b non-negative integers, and assign them to T_{ij}.

After generating the transportation matrix T, we need to determine the weight of each container. The weight of all containers is between 15 tons and 25 tons. For port i, select k number from weight levels in random. Let $v = T_{ij}$ and $b = k$, and apply RIP algorithm to obtain b non-negative number for the container number of each weight level.

In order to extend the computational study, we create a set, composed of 30 cases. There are 3 groups in total: Weight, Capacity and Port. Each group is aimed at a specific aspect of the problem, e.g. in the weight group, all cases have a specific percentage of the maximal total weight allowed for the vessel.

- Weight cases (10), in which the weight is the most important factor. The total weight of the containers is set to 80%, 90% and 95% of the maximal total weight allowed for the vessel.
- Capacity cases (10), in which the number of containers stowed determines the problem. The percentage of stowed containers is set to 80%, 90% and 95%

– Port cases (10), in which the number of port visited during the journey matters. The value is between 5 and 15.

For Weight cases and Port cases, the number of stowed containers is between 75% and 100% of the maximal capacity; for Capacity cases and Port cases, the total weight of containers is set to between 75% and 100% of the maximal allowed weight; for Weight cases and Capacity cases, the number of visited port is 10.

5.2   Analysis of Parameters

The proposed algorithm is executed under the settings of several variables, such as the weight of each part in fitness function, iteration times and so on. These variables will affect the performance of the algorithm. For instance, different weight values result in different fitness value.

Table 1. Fitness function weights test result for different data set

Test case	Paraments				Result	
	a	b	c	d	Blocking ratio	The number of shifts
1	20	80	5	5	35.2%	237
1	20	80	10	10	36.8%	243
1	40	60	5	10	35.6%	250
1	40	60	20	15	34.3%	249
1	60	40	10	20	36.8%	246
1	60	40	15	5	30.9%	221
1	80	20	5	20	32.5%	227
1	80	20	15	10	36.2%	254
2	20	80	5	5	23.2%	190
2	20	80	10	10	25.6%	207
2	40	60	5	10	22.6%	198
2	40	60	20	15	24.6%	209
2	60	40	10	20	22.9%	200
2	60	40	15	5	19.8%	195
2	80	20	5	20	23.5%	205
2	80	20	15	10	21.1%	187
3	20	80	5	5	21.9%	62
3	20	80	10	10	20.8%	47
3	40	60	5	10	21.5%	49
3	40	60	20	15	23.0%	55
3	60	40	10	20	22.9%	56
3	60	40	15	5	18.5%	53
3	80	20	5	20	22.0%	46
3	80	20	15	10	21.3%	46

Weights for Fitness Function. To get the optimal fitness function that defined in (6), experiments are conducted to test the proper value of α, β, γ and δ. Like [13], all four weights α, β, γ and δ are tested by using the follow policy.

$$
\begin{aligned}
\alpha &= -a * 0.0008 \\
\delta &= b * 0.4 \\
\beta &= c * 0.002 \\
\gamma &= d * 0.003 \\
a + b &= 100 \\
c, d &\in [0, 20]
\end{aligned}
\tag{7}
$$

The result of the test is shown in Table 1. From it, we know that the algorithm owns a good performance for all kinds of data when $\alpha = -0.064$, $\beta = 0.03$, $\gamma = 0.015$ and $\delta = 1.6$. So, these parameters are used to implement stowage planning based on the SPUNTB.

The Iteration Times. To get the proper value of iteration time, we design this part experiments.

Set the iteration times to a range from 1 to 8 when using the within-group strategy, with the step value equals 1. The results are shown in Fig. 6.

This process is after the detailed process in a group. As shown in Fig. 6, when the iteration times get greater, the algorithm is almost random and the fitness value gets larger. Therefore, in later experiments, the iteration times is set to 1.

Set the among-group strategy iteration times range from 10 to 100 by 10. The results are shown in Fig. 7. Figure 7 illustrates that the optimal blocking ratio is got when the iteration times is 10.

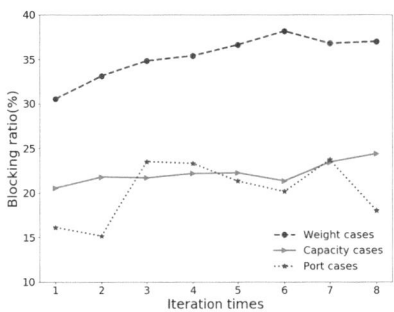

Fig. 6. The result of iteration times for within-group strategy

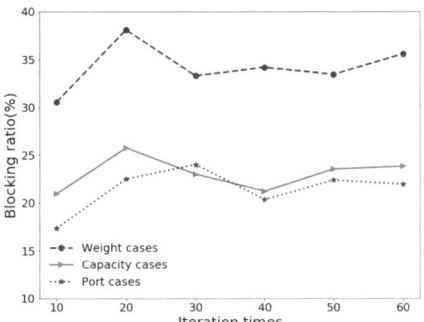

Fig. 7. The result of iteration times for within-group strategy

6   Conclusion

In this paper, we consider the stowage planning for a container ship and develop a algorithm with a reasonable blocking ratio for such problems. But, we do not determine a specific slot for a container, but take the type of containers into account. To keep the ship's stability, the computational formulas is introduced into the fitness function.

Several strategies are proposed, including group strategy, greedy strategy, random processes. All these strategies make the SPUNTB more instructive and faster to find an optimal solution. Experiments show that the proposed algorithm owns a good performance. It performs well on all types of cases.

Acknowledgment. This work was supported in part by the Key Projects of Guangdong Natural Science Foundation (No.2018B030311003).

References

1. List of busiest container ports. https://en.wikipedia.org/wiki/List_of_busiest_container_ports
2. Unified neutral theory of biodiversity. https://en.wikipedia.org/wiki/Unified_neutral_theory_of_biodiversity
3. Review of Maritime Transport. UNCTAD (2014)
4. Ambrosino, D., Anghinolfi, D., Paolucci, M., Sciomachen, A.: A new three-step heuristic for the master bay plan problem. Marit. Econ. Logistics **11**(1), 98–120 (2009)
5. Ambrosino, D., Sciomachen, A., Tanfani, E.: Stowing a containership: the master bay plan problem. Transp. Res. Part A Policy Pract. **38**(2), 81–99 (2004)
6. Aslidis, A.H.: Combinatorial algorithms for stacking problems. Ph.D. thesis, Massachusetts Institute of Technology (1989)
7. Avriel, M., Penn, M.: Exact and approximate solutions of the container ship stowage problem. Comput. Ind. Eng. **25**(1–4), 271–274 (1993)
8. Avriel, M., Penn, M., Shpirer, N.: Container ship stowage problem: complexity and connection to the coloring of circle graphs. Discrete Appl. Math. **103**(1–3), 271–279 (2000)
9. Botter, R.C., Brinati, M.A.: Stowage container planning: a model for getting an optimal solution (1991)
10. Ding, D., Chou, M.C.: Stowage planning for container ships: a heuristic algorithm to reduce the number of shifts. Eur. J. Oper. Res. **246**(1), 242–249 (2015)
11. Dubrovsky, O., Levitin, G., Penn, M.: A genetic algorithm with a compact solution encoding for the container ship stowage problem. J. Heuristics **8**(6), 585–599 (2002)
12. Hu, W., Min, Z., Du, B.: A novel algorithm based on the unified neutral theory of biodiversity and biogeography model for block allocation of outbound container. Int. J. Comput. Integr. Manuf. **27**(6), 529–546 (2014)
13. Imai, A., Sasaki, K., Nishimura, E., Papadimitriou, S.: Multi-objective simultaneous stowage and load planning for a container ship with container rehandle in yard stacks. Eur. J. Oper. Res. **171**(2), 373–389 (2006)
14. Kang, J.G., Kim, Y.D.: Stowage planning in maritime container transportation. J. Oper. Res. Soc. **53**(4), 415–426 (2002)

15. Monaco, M.F., Sammarra, M., Sorrentino, G.: The terminal-oriented ship stowage planning problem. Eur. J. Oper. Res. **239**(1), 256–265 (2014)
16. Parreño, F., Pacino, D., Alvarez-Valdes, R.: A grasp algorithm for the container stowage slot planning problem. Trans. Res. Part E Logistics Trans. Rev. **94**, 141–157 (2016)
17. Sciomachen, A., Tanfani, E.: The master bay plan problem: a solution method based on its connection to the three-dimensional bin packing problem. IMA J. Manag. Math. **14**(3), 251–269 (2003)
18. Sciomachen, A., Tanfani, E.: A 3D-BPP approach for optimising stowage plans and terminal productivity. Eur. J. Oper. Res. **183**(3), 1433–1446 (2007)
19. Wei-Ying, Z., Yan, L., Zhuo-Shang, J.: Model and algorithm for container ship stowage planning based on bin-packing problem. J. Mar. Sci. Appl. **4**(3), 30–36 (2005)
20. Wilson, I.D., Roach, P.A., Ware, J.A.: Container stowage pre-planning: using search to generate solutions, a case study. In: Research and Development in Intelligent Systems XVII, pp. 349–362. Springer (2001)
21. Yang, J.H., Kim, K.H.: A grouped storage method for minimizing relocations in block stacking systems. J. Intell. Manuf. **17**(4), 453–463 (2006)

A Big Data and Visual Analytics System for Port Risk Warning

Jie Song, Yinsheng Li$^{(\boxtimes)}$, and Xu Liang

Software School, Fudan University,
Shanghai 200433, People's Republic of China
18210240168@fudan.edu.cn

Abstract. Big data and visual analytics are critical for a product tracking and risk warning system to provide friendly GUIs, with which users can find the associated facts fast and precisely. In this work, the authors developed a big data and visual analytics tool for an underdeveloped product tracking and risk warning system for the smart port. The goal is to provide information statistics and further risk mining to reveal possible hidden facts. The proposed tool is to improve the data comprehensiveness, deep insight, and relevance. The Knowledge Graph is applied to organize and describe risk-related data, while data mining algorithms such as clustering and association analysis are used to tap the in-depth value of information. Based on Knowledge Graphs, the system solves the problem of searching related risk products with the same attributes, and analyzes the potential relationship between products of the same category through association analysis. The tool has been developed to support visual navigation in multi-dimensional, multi-faceted, multi-attribute, including association rules, risk warning and dynamic target.

Keywords: Smart port · Knowledge graph · Data mining · Big data visualization · Risk warning

1   Introduction

The port is an important part of guaranteeing national security. Inspection of illegal entry-exit items by the port can effectively protect the biological safety of the country and the health of the people.

Risk early warning, intelligent control and data visualization are the key supporting technologies and services to realize intelligent inspection and improve inspection efficiency and quality. At present, the existing big data and visual analytics on port risk early warning are inefficient in some key indicators. For example, the current risk warnings of port goods are predicted from historical declaration data, rather than based on real-time information such as epidemic information and consensus. Besides, the existing systems do not analyze the association rules of products of the same category in depth, or provide timely

© Springer Nature Switzerland AG 2020
K.-M. Chao et al. (Eds.): ICEBE 2019, LNDECT 41, pp. 541–558, 2020.
https://doi.org/10.1007/978-3-030-34986-8_38

warning to the risk-related products. During the on-site inspection, the officers found that the product was at risk, but the system would not issue warnings about other risk-related products with the same attributes as the product. The visualization of big data only based on statistical analysis, without in-depth analysis of data, and the existing systems are basically not involved in the visualization of product tracking and risk early warning, which can not allow users to realize further risk analysis and scientific inspection.

The emerging techniques of data mining and Knowledge Graphs have shot a light on the visual analytics of port systems. Data mining is aiming to mine hidden information from big data by using analysis algorithms, including classification, estimation, prediction, association analysis and clustering, and it has a very important application in the era of big data. As for the Knowledge Graph, since Google put forward the Knowledge Graph in 2012, the Knowledge Graph technology has developed rapidly. It changes data from meaningless character strings to hidden objects or things behind character strings, which can well reflect the attributes of various types of entities and the relationships between entities. Knowledge Graphs organize data in the form of the knowledge base, which is convenient for transforming unstructured data into structured data and establishing the correlation between data. At the same time, based on the advantages of Knowledge Graphs in relation analysis, product tracking can be easily visualized by using Knowledge Graphs. With various mining algorithms and Knowledge Graphs, data mining can analyze data in depth according to business needs. Also, with the development of visualization technologies such as Echarts, the development of big data analysis is increasingly moving towards multi-dimensional and multi-nature visualization of big data. People are easier and more willing to accept the results of data analysis visually. Big data visualization has become an important part of data analysis and information display.

Based on the lab's research and development experiences at port-related projects since 2013, the authors proposed a big data and visual analytics tool for an underdeveloped product tracking and risk warning system for the smart port. The tool is to provide information statistics and further risk mining to reveal possible hidden facts by the application of data mining, Knowledge Graphs and visualization technologies, and further support visual navigation in multi-dimensional, multi-faceted, multi-attribute. Based on Knowledge Graphs, the system solves the problem of searching related risk products with the same attributes, visualizes the process of product tracking, and analyzes the potential relationship between products of the same category through association analysis. The benefits of the tool are to provide decision support for risk early warning, dynamic control and product traceability of the port system, realize scientific automation of inspection, assist the port officers in customs clearance inspection and improve the work efficiency.

A comprehensive description of the big data and visual analytics tool, including its scenarios, principles, structure and evaluation, will be presented in the following sections. Section 2 will provide a literature review and previous work related to the lab. Section 3 will show the scenarios, problems, objective and principles of the visual analytics tool. Section 4 will present the risk warning and big data visualization system for the smart port. Section 5 will discuss the evaluation indicators of the tool. Section 6 will conclude the paper.

2 Literature Review and Previous Work

2.1 Literature Review on the Knowledge Graph

Knowledge Graphs provide a more effective way to express, organize, manage and utilize massive, heterogeneous and dynamic big data on the Internet, which is essentially the large-scale knowledge base of the Semantic Network, making the network more intelligent and closer to human thinking. Since the Knowledge Graph has been proposed, it has achieved great application effect in many scenarios. In addition to the application in classic scenes such as intelligent search, social networks and question answering systems, the Knowledge Graph has proved an effective solution to many vertical field problems. In the financial area, Li [1] applied Knowledge Graphs to identify fraud. In the medical area, Yale University has the world's largest neuroscience database Senselab [2]. In the e-commerce area, Tian and Ma improved user experience by applying Knowledge Graphs [3]. In the port, there is a lack of applications of Knowledge Graphs. Liao [4] studied the customs innovation system from the perspective of Knowledge Graphs.

2.2 Literature Review on Data Mining

Data mining refers to extracting people's interesting knowledge which is implicit, unknown in advance, potentially useful and easily understood, from large databases or data warehouses. Data mining techniques could be utilized to sift through the past data and develop predictive models for the examination of limited goods with a higher probability of fraud. Data mining technology has been widely used in the era of big data. Zhou and Zhang [5] used classification analysis in data mining to classify and predict customs risks. Yaqin and Yuming [6] put forward a classification model to predict the risk of the commodity through customs clearance by the use of data mining based on association rules. Watkins [7] creatively used data mining methods to explore the customs hidden rules and identify risks to prevent money laundering.

2.3 Literature Review on Big Data Visualization

In the era of big data, we can obtain deeper insight, better decision-making capability and stronger automatic processing capability by transforming big data into various forms of visualization. Data visualization has become an important trend in data analysis. However, the application of visualization in customs is still limited. Yang [8] proposed to link the advanced three-dimensional visual simulation technology with the single cargo supervision data to form the customs three-dimensional visual supervision system. Feng [9] made a useful exploration of the application of visual security intelligent lock supervision for container cargo.

2.4   Previous Work

The lab, to which the authors are affiliated, has been working on the risk evaluation and product tracking since 2013. Some risk-related concepts and models have been proposed and put into practice. In the product tracking area, the lab developed a tracing service based on traceable creditworthiness information, and the creditworthiness model and rating indicators in the service were developed to identify the recoverable products across China and Europe [10].

The lab also proposed to apply the blockchain into the creditworthiness systems to make an open, integrated, and scientific credit system [11]. Besides, the lab proposed a blockchain-based autonomous credit system to address separate brokerage systems, non-pertinence, centralized and static evaluation models, and insufficient supporting information in the traditional credit systems [12].

What's more, tens of large-scale creditworthiness and visualization services had been developed as public services for Jiangsu Province, Henan Province, and Beijing City. The lab also proposed a unique credit model in many education area [13].

3   Principles of Smart Port Data Visualization

3.1   Scenarios

The existing big data and visual analytics system can not provide powerful help in most port business scenarios, it can only serve as a data display screen. The proposed big data and visual analytics tool aims to solve the problems in the actual business scenarios of ports. Here are some typical scenarios.

Visual Analytics. Among the existing port systems, there is few visual analysis of port data for port officers to analyze the detail and relevance of the products risk. When there is a new batch of goods for customs declaration, the risk analysis model of the existing system can return the risk of the goods in the form of text. But owing to the lack of more risk-related information, the port officers can not further analyze and estimate risks of the goods. Based on Knowledge Graphs and big data visual technologies, the proposed big data and visual analytics tool displays the risks of goods, the process of risk transfer, and the related companies and consensus of the risks in multi-dimensional. With multi-faceted visualization, the port officers can easily find the associated facts of the risks and make risks tracking fast and precisely, then they can specify more precise solutions. Below are two specific scenarios of visual analytics.

Searching Related Risk Products with the Same Attributes. Among the products imported at ports, there are many products that have the same attributes, such as the manufacturer, origin or raw materials. In this circumstance, if the products have passed the assessment of the risk early warning system but one of the products is found to be defective during on-site inspection, other related products with the same attributes may also have problems. For example, as illustrated by Fig. 1, if there is a problem with the ingredients of fresh milk, other dairy products from the same batch of cows may also have problems because raw milk may be problematic. Traditional risk early warning systems only focus on the risk of a batch of products but do not correlate the risk of the product with other products. So even if the port officers found that there were problems in fresh milk, they would not notice other dairy products from the same batch of raw milk. But thanks to the fact that Knowledge Graphs can easily conduct in-depth searches of relationships, the proposed system can easily find out the risks of other products associated with fresh milk through relational networks, then realize the accurate risk early warning. Based on the visualization of Knowledge Graphs, the port officers can visually get the product tracking process and the correlation between products, and then implement precise control over the related risk products.

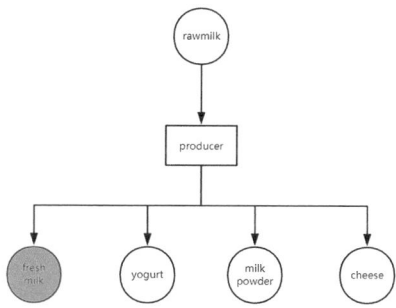

Fig. 1. Risk correlation

Risk Transmission. In the port system, there are credit scores for production enterprises, declaration enterprises, etc. When an enterprise has a consensus event, the existing system mostly analyzes the impact of consensus on the enterprise itself, without noticing how would it affect the enterprises associated with the enterprise.

As illustrated by Fig. 2, when enterprise 1 has a bad consensus event, there is no doubt that the credit value of enterprise 1 will be affected, and most existing systems do well in this respect. However, they do not analyze how this event would affect enterprise 2, enterprise 3, enterprise 4 and enterprise 5 which cooperate with enterprise 1. This also leads to a lack of intelligence and accuracy in data analysis. It is based on Knowledge Graphs which can well show the

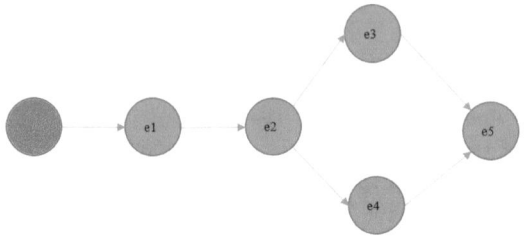

Fig. 2. Risk transmission

attributes of entities and their relationships that the proposed system addressed this problem. The system can easily analyze the impact of the enterprise's risk on the enterprise and the relevant enterprises by deep searching of relations on Knowledge Graphs. With the visual analytics, the port officers can trace the cooperative relationship between enterprises, get the impact of consensus among enterprises, and then make a scientific evaluation of credit value.

3.2 Problems and Objectives

As mentioned, the existing big data analysis and visualization in the port system suffer from simple statistical analysis, unscientific risk alert, or inadequate relevancy. They can not get information hidden in business data, or make an intelligent risk warning when the products, enterprises, or their related subjects may have the risk. When a batch of products is found to have problems, existing systems could not do warnings to the related risk products with the same attributes, which would enable these products to pass customs smoothly and may cause serious problems. The emerging Knowledge Graphs and Data Mining techniques are designed to organize and describe data, enable more detailed analysis and improve visualization. They make it more possible to get information hidden in the data.

   To check the performance improvement that Knowledge Graphs and Data Mining may bring on the big data analysis and visualization, a big data analysis and visualization system for the port has been proposed to reform the existing system. The objective is to solve the basic problems with the existing big data analysis and visualization system for the port, assist port officers to inspect, and realize the productivity increase.

   A visualization system becomes valuable and useful if only it can structure data in the right ways, analyze hidden and associated information in data, and visualize the data reasonably with various charts. Based on this assumption, there are three concerns to be addressed by a big data analysis and visualization system for the port, among which are (i) structuring data in a way that facilitates analysis and visualization, (ii) analyzing data in the port business scenario in-depth, and (iii) visualizing the data in multi-dimensional, multi-faceted, multi-attribute.

It is based on Knowledge Graphs, data mining and big data visualization technologies that the proposed system addresses the three basic concerns and has applied value for the port, as explained below.

(i) how to structure data in a way that facilitates analysis and visualization. Knowledge Graphs can well describe the logic contained in the business by using property graphs to present data, efficient query data and visualize the data. It's also suitable for the deep search of relationships, association analysis and other data analysis by using Knowledge Graphs to describe data.

(ii) how to analyze data in the port business scenario in depth. Data mining technologies such as association analysis, clustering and prediction are the most effective methods for big data analysis. Combined with port business scenarios and data mining technology, we can get a lot of potentially valuable information, such as products risk, clearance efficiency and so on. And with Knowledge Graphs, the system can make the risk analysis of related risk products with the same attributes.

(iii) how to visualize the data in multi-dimensional, multi-faceted, multi-attribute. On the basis of Knowledge Graphs, we use visualization technologies such as Echarts to ensure the scientificity and intelligence of visualization. On the one hand, Knowledge Graphs can intuitively show the relevance among the subjects. On the other, the various charts in Echarts can show data multi-dimensional and risk alert intelligently.

3.3   Principles and Features

As mentioned, the big data analysis and visualization system for the port are proposed to be a scientific and intelligent data analysis and visualization system. It is based on Knowledge Graphs, data mining and visualization technologies, the system realized in-depth data association analysis, risk early warning and dynamic control, and visualized the results obtained in a multi-dimensional and intuitive way. Basically, there are three principles with the proposed system, which make tremendous differences with the existing data analysis and visualization system for the port. The three principles are identified in the following.

(i) The data is structured to realize data analysis and visualization scientifically. The objective of the storage of data is service-oriented to provide a more efficient model for various data mining and multi-attribute data visualization. So We make full use of the characteristics of Knowledge Graphs and traditional databases, we storage the data commonly used and relevant to analysis in Knowledge Graphs, and storage the data that are not frequently used or important for the analysis in traditional relational databases. Comparatively, the existing systems only store data in traditional relational databases.

(ii) In data analysis, besides traditional statistical analysis, the proposed system attaches more importance to the analysis of risk correlation, risk warning and other aspects. In terms of statistical analysis, the system focuses on the important aspects of the port business system, including import and export efficiency, products flow and annual statistics information. In addition, the system conducts deeper data analysis in terms of risk transmission, risk correlation and so on.

(iii) Data visualization is multi-dimensional, multi-faceted, multi-attribute. The system improves the visualization from two aspects. Firstly, the system uses the characteristics of Knowledge Graphs to visualize the relationship between data and risk, so users can find the associated facts fast and precisely. Then the system uses visualization technology such as Echarts to visualize data from multiple dimensions, which can make risk analytics and warning more intelligent.

4   A Big Data Analysis and Visualization System for Port Risk Warning

4.1   Architecture of the System

The architecture of the system is described as Fig. 3. The big data analysis and visualization system for port risk early warning consist of three parts: data organization, data analysis and data visualization. The system uses Knowledge Graphs and traditional relational database to store data, so that we can analyze and visualize the data scientifically; then the system uses association analysis and prediction algorithms in data mining to analyze port data for the purposes of risk early warning; finally, based on big data visualization technology, we visualize the result of data analysis multi-dimensionally and intuitively, so as to realize risk early warning and assistant inspection.

4.2   Knowledge Graphs and Data Processing

Data processing module includes the following three aspects: data collection, data processing and data storage. Besides, the data mainly contain the historical data of the port, declaration data, basic information about declaration products and consensuses. Specifically, the historical data of the port cover historical data of import and export products, historical inspection data, and historical risks and solutions; declaration data refer to a new set of declaration commodity data that have not yet been cleared; basic information about declaration products consist of the name and brand of the product, declaration unit, transport unit, and country of origin, etc. In addition, consensuses include the latest policy information and laws, and epidemic information of relevant sites.

Data Collection. There are several data collection methods for different data types, as follows.

(1) For the historical data and declaration data of the port, we get it from the port database by means of interfaces.

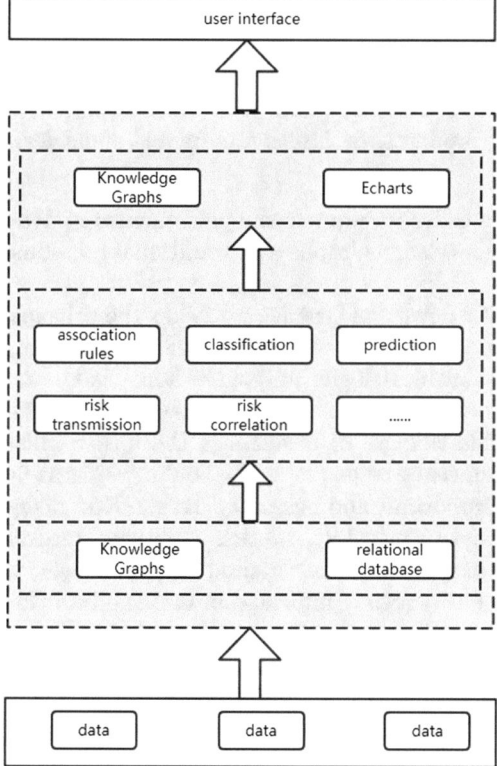

Fig. 3. The architecture of the system

(2) We extract basic information about declaration products from the block-chain of the smart port, which is also in the form of interfaces.

(3) For consensus and other information, it is gained from the smart port server by means of interfaces.

Data Processing. There are three data methods to regularize the data, the following are specific regularization methods.

(1) The missing statistics are made for the data, the data with fewer missing items are supplemented, and the data with more missing items are deleted. Specifically, when there are no more than three missing items, the system automatically calculates and completes them according to the logic of business or the similar data nearby (For example, if there is no date of declaration, the date in the previous declaration record is used as the date of the declaration); when there are more than three missing items, the record is considered invalid and deleted;

(2) Preprocessing data, modifying or deleting erroneous data manually, then the system will add some related information to the data, such as adding country latitude and longitude information according to the country name;

(3) The newly added consensus and other information will be related to the corresponding products, declaration units and countries.

Data Storage. In order to improve the performance of data visualization, the system combines Knowledge Graphs and traditional relational database storage to store data.

In the design of Knowledge Graphs, we follow the principles of business, efficiency, analytics and redundancy. Based on business principle, we design Knowledge Graphs from business logic, and make sure that it is easy to infer the business logic by observing the designed Knowledge Graphs. Besides, we also consider the possible changes of business in the future when we are designing graphs. Based on efficiency principle, we make the graphs as light as possible and only place part of important and necessary data in Knowledge Graphs. In addition, based on analytics principle, we delete entities unrelated to relationship analysis from the graphs. Based on redundancy principle, we place repetitive information and high-frequency information into the traditional database.

Therefore, before we place the data into Knowledge Graphs, we need to have an in-depth comprehension of the port business, and then design the Knowledge Graph based on the business. Taking milk commodities as an example, based on the customs declaration process of milk, we analyze the enterprises, personnel and commodities involved in each process one by one, determine the entities, entities attributes and the relationships between entities involved in the whole process, and then build the graphs. The relevant entities involved in the whole customs declaration process of milk are shown in Fig. 4.

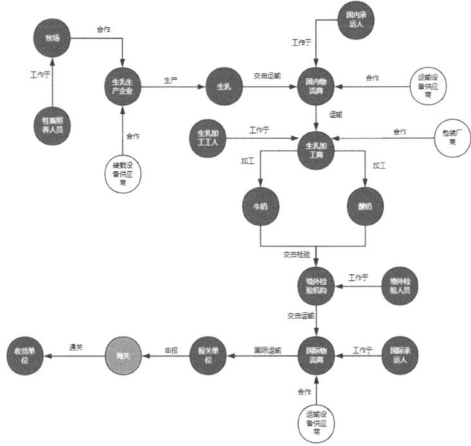

Fig. 4. The architecture of the system

Table 1. The attributes of milk

Name	Description
id	The id of the milk
name	The name of the milk
rawmilkId	The id of the raw of the milk
converterId	The id of the converter unit of the milk
hsCode	The HSCode of the milk

Once we have identified the entities, we need to identify the attributes of the entities. Based on the design principles, we only store the attributes commonly used and related to relational analysis in Knowledge Graphs. As described as Table 1, we only store some necessary information about milk in the knowledge map.

Finally, we have to face the choice of the storage system in the storage of Knowledge Graphs. Because the graphs we designed have attributes, the map database can be the first choice. And based on the principles of efficiency and redundancy, we placed some information in the traditional database to reduce the amount of information stored by Knowledge Graphs. So Neo4j is a good storage choice. We store the entities, the attributes of entities and the relationships between entities in the basis of the declaration process into Neo4j.

In addition, we place the data not involved in the relational analysis or infrequently used in the traditional relational database Oracle. The design of database tables is a relatively basic work and will not be explained in depth here.

4.3 Data Mining of the Port Business

Data analysis module includes analyzing data in depth by using traditional statistics, association analysis and prediction algorithms in data mining, and rule-based application of Knowledge Graphs. The system solves the problem of risk warning of related risk products with the same attributes through the rule-based application of Knowledge Graphs.

(1) Based on statistical analysis methods, we can obtain a general customs clearance situation such as changes in the volume of imports and exports, commodity flow information and customs clearance efficiency by analyzing the historical inspection data of the port. Taking the clothing customs declaration records as an example, through the comparative analysis of the clothing inspection records of each month, we can get the changes of clothing declaration every month. For the month with a low qualified rate of clothing customs declaration, we can improve the proportion of drawing appropriately.

(2) We apply the association algorithm in data mining to analyze the risk of customs declaration data, and cluster analysis is used to analyze the historical inspection data of applicants for inspection, consignees, transportation

units, and brands of the commodities for inspection to determine whether they belong to high-risk units. For example, we make a correlation analysis of the data in 2018 through the classical Apriori correlation analysis algorithm. Specifically, the input of the algorithm is the country of origin, product name, brand, customs declaration unit, transport unit, consignee and whether the sampling inspection is qualified or not, and the association rules are obtained as output. Furthermore, we select the association rules whose results are unqualified from the output to get the related items which lead to the customs declaration unqualified. Hence, if the attributes of a new set of customs declaration commodities include the related items above, then we can consider that the commodity has certain risks and make early warning for it.

(3) It is advisable to use the rule-based application in Knowledge Graphs to analyze the associated risk of commodities because the graphs make it easier to search for relationships in-depth. Take two typical scenarios mentioned in Sect. 3.1 as examples.

If an enterprise has bad public opinion, then the credit value of it will be affected, and for those enterprises that cooperate with it, it is obvious that their credit value should also be affected to some extent. The traditional credit models only consider the impact of consensus on the enterprise itself but ignore the impact on the credit of cooperative enterprises. To meet the real business scenario, we specify that the credit value of enterprises in the three-degree cooperative relationship of the enterprise with bad consensus will be affected by the decline. Moreover, traditional relational databases do not perform well in deep relational search. Instead, the proposed system solves the problem of credit influence in the enterprise relationship chain through the deep relationship-based search of Knowledge Graphs, therefore we can search enterprises directly based on depth. As mentioned in Sect. 3.1 above, if enterprise 1 has a bad consensus event, the system would correspondingly reduce the credit value of enterprises. Then the system will search for the cooperative relationship of enterprise 1, find all the cooperative enterprises in the three-degree relationship, and reduce the credit value of these enterprises to a certain degree according to the weight attenuation.

The related risk products with the same attributes refer to products that share some of the same attributes such as the manufacturer, origin or raw materials as the product which was found to be problematic. Risk warning of related risk products with the same attributes is an important and urgent problem to be solved. In the scene of port dairy product inspection, if the dairy products have passed the risk assessment but a batch of fresh milk is found to have problems with its ingredients, other dairy products from the same batch of raw milk may also have problems. But the existing risk early warning model cannot give risk warning to goods that have passed their assessment, thus the same problem probably exists in customs-cleared dairy products. However, relational searches based on Knowledge Graphs can solve this problem well. By defining the rules

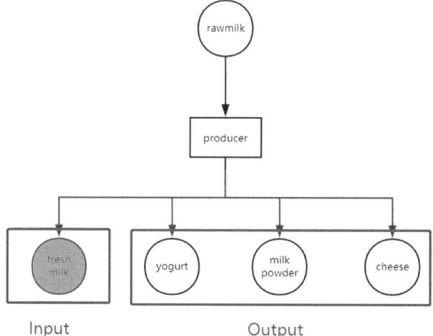

Fig. 5. Searching related dairy products processed from the same raw milk

in advance, the system can search the products declaration network according to the attributes, and return the risk-related products with some identical attributes (Fig. 5).

As shown in the figure above, if the fresh milk is found to have problems in the spot check, the system can take it as input, search along the customs declaration network of the fresh milk, and give the other customs declaration dairy products which are from the same batch of raw milk a certain degree of risk warning. By defining the rules in advance, the port officers also can search the related risk products with other attributes, such as the pasture, manufacturer or raw-processing company, then take further measures to analyze and inspect the products. The system returns the related risk products with the same attributes in the form of interface, which facilitates the visualization and further risk analysis for port personnel.

4.4 Visualization of Big Data

Data visualization is the most important part of the system. The system can provide multidimensional data analysis, risk early warning and product tracking by the visual interface, and the port officers can make scientific inspection

Fig. 6. The visualization of flow direction

decisions through visual analysis. The visualization of big data can present multidimensional data reports and charts by the big data visualization technologies such as Echarts, and the main visual screens are as follows.

Statistical Visualization. Statistical visualization mainly displays the results of statistical analysis, including the display of flow direction, the display of annual statistical, the display of customs clearance efficiency and the visualization of customs clearance units.

As illustrated by Fig. 6, the visualization of flow direction makes a statistical analysis of the flow of commodities, and then shows the main import and export countries of commodities. Above, the main import and export countries of the port are shown by the pie chart, bar chart and rotation column chart. In addition, the flow of commodities is shown by maps and flying lines, which makes it easy for the port officers to find the region of key import and export countries through maps, and then take different measures according to different national policies.

Risk Warning. The visualization of risk early warning mainly displays the results of risk analysis and realizes risk early warning and auxiliary inspection by showing the relevant information and the corresponding control strategy of risk.

Fig. 7. The visualization of risk warning

As illustrated by Fig. 7, based on the multidimensional visualization of the risk early warning, the port officers can get the relevant information and the corresponding control strategy of the risk. The system first presents the declaration data with risk on a round-the-clock basis. When clicking on the record in the round-the-clock list, the top side will show the risk score and risk grade of the declaration commodity. On the left side, it will show the high-risk information related to the declaration product, such as the qualified rate of the declaration unit, country of origin, and on the right side, it will show he corresponding control strategy of risk. The detail of the risk can help the port officers to analyze the risk, and the dynamic control strategies recommended can be used by port staff for auxiliary inspection. With the visualization of the risk early warning, the port officers can realize further risk analysis and scientific inspection.

Visualization of the Knowledge Graphs. The visualization of the Knowledge Graphs mainly displays the relational network of customs declaration commodities, including the entities, their relationships in the customs declaration process, and related risk products with the same attributes. The visualization facilitates the port staff to search for the relationships of the products and view the traceability process.

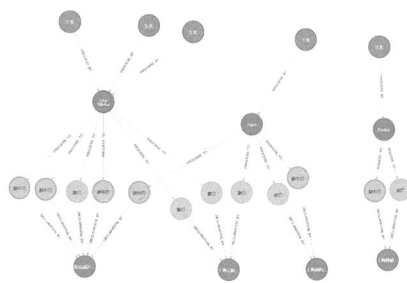

Fig. 8. The visualization of the knowledge graphs

The simplified visual display of the dairy products relationship network is shown in Fig. 8. The port officers can easily obtain the traceability of dairy products, dairy products from the same batch of raw materials, and the corresponding risks of the fresh milk by observing the visual network, and then deal with the risk of products accurately. If one of the dairy products has problems, the port officers can find the associated facts which may also have problems fast and precisely. Users also can get more attribute information by clicking the entities. Based on the visualization, the port officers can clearly and intuitively realize product tracking, and have a clear understanding of each risk involved in the traceability process.

5 Evaluation Indicators of the System

In this section, we'll evaluate our system by indicators. The evaluation methods will be discussed at first. Then we'll test the performance of the system.

5.1 Design of the Evaluation Methods

To evaluate the system functions and user experiences, a list of related indicators and evaluation methods are proposed which are shown in Table 2.

Table 2. Indicators and evaluations methods for system functions and user experience

Indicators	Description	Methods
Practicality	The practicality of the system refers to whether the information related to the actual business of ports	Give a score for each visual interface, then calculate the proportion of scores:
		0-Worthless information
		1-Part of valuable information
		2-Valuable information
Integrity	The integrity refers to whether the system incorporates all the information that can help users understand the data	Give a score for each visual interface, then calculate the proportion of scores:
		0-No information that can help users understand the data
		1-Part of information that can help users understand the data
		2-Has all the information that can help users understand the data
Acceptance	The acceptance refers to whether users can accept the system or are satisfied with it	Give a score for each visual interface, then calculate the proportion of scores:
		0-Its difficult for users to accept the interface
		1-Its a little difficult for users to accept the interface
		2-Its easy for users to accept the interface
Authenticity	Whether the information are convincing and accurate	Give a score for each visual interface, then calculate the proportion of scores:
		0-There are errors in the information
		1-There are few errors in the information
		2-There is no error in the information
Visuality	Whether the charts are easy to understand	Give a score for each visual interface, then calculate the proportion of scores:
		0-Its difficult for users to understand the charts
		1-Its a little difficult for users to understand the charts
		2-Its easy for users to understand the charts
Beauty	Whether there is a great creative design, and the system can help users to extract information	Give a score for each visual interface, then calculate the proportion of scores:
		0-The interface is unpretty
		1-The interface is pretty but unhelpful
		2-The interface is pretty and helpful
User participation	Whether the system have certain interactivity	Give a score for each visual interface, then calculate the proportion of score:
		0-The interface is not interactive
		1-The interface is little interactive
		2-The interface have great interactivity

5.2   Results and Analysis of the Evaluation

Follow the indicators we mention above, we test the system completely, and then get the results as Table 3.

Table 3. Results of system functions and user experience evaluation

Indicators	Results
Practicality	0.94
Integrity	0.85
Acceptance	1.06
Authenticity	0.87
Visuality	1.02
Beauty	0.76
User participation	0.79

We choose 5 main big data visualization interfaces of our system for the test. Based on the results of the test, we found that Since our system includes visual analysis in multi-dimensional, multi-faceted, multi-attribute, the system has great performance in practicality and visuality, which can realize scientific automation of inspection, assist the port officers in clearance inspection and improve their work efficiency.

For beauty and user participation, we found that the system did not perform well enough since there are some problems in the design of interfaces. Thus we need to further develop the interfaces.

6   Conclusion

The existing big data and visual analytics are not well recognized because of less in-depth data analysis, non-intelligence and insufficient visualization. A big data analysis and visualization system for port risk warning is proposed to address the issues of the existing systems based on Knowledge Graphs and data mining. Via Knowledge Graphs and data mining, the system solves the problem of searching related risk products with the same attributes and further risk mining to reveal possible hidden facts. Also, the proposed system realizes the visualization of product tracking by Knowledge Graphs in the Neo4j database through basic nodes and relationships, and the highly intelligent risk warning through visualization technologies.

The visual analytics tool is ambitious to be an important part of the product tracking and risk warning system in the port. Compared with the existing statistic-based visual analytic tool, the proposed tool dynamically visualizes product tracking, risk warning, and the relationship networks between products. The authors believe it is effective and helpful for the port users to find the associated facts precisely and realize intelligent inspection.

Despite the advanced techniques and visualization methods, the proposed visual analytic tool is still underdeveloped. The application of Knowledge Graphs and data mining algorithms need further research, and the methods of visualization need further development.

Acknowledgment. The authors would give our special thanks to all the colleagues from the Institute of E-business, the implementation of the system owes to their instructive suggestions and technical support. Special thanks are also given to the Shanghai ICIQ, for their help on the business.

References

1. Li, W.: Interner finance, how to identify fraud conduct using knowledge graph, 09 January 2016
2. Senselab. Center for medical informatics at Yale university school of medicine Yale university school of medicine, 08 January 2016
3. Tian, L., Ma, L.: Research on comprehensive evaluation of website information services based on user experience. Ecol. Econ. **10**, 160–162 (2013)
4. Liao, R., Kuang, Z.: The evolution and enlightenment of the research in free trade zone customs system innovation: perspective of knowledge map. Reform Econ. Syst. **05**, 41–48 (2017)
5. Zhou, X., Zhang, C.: Customs risk classification and forecasting model based on data mining. J. Customs Trade **38**(02), 22–31 (2017)
6. Yaqin, W., Yuming, S.: Classification model based on association rules in customs risk management application. In: International Conference on Intelligent Systems Design & Engineering Applications, pp. 436–439. IEEE (2010)
7. Watkins, R.C., Reynolds, K.M., Demara, R., et al.: Tracking dirty proceeds: exploring data mining technologies as tools to investigate money laundering. Police Pract. Res. **4**(2), 163–178 (2003)
8. Yang, B.: The application of visualization technologies in customs supervision. Wirel. Internet Technol. **06**, 114 (2012)
9. Feng, F., Sun, Z.: Research and application of safety intelligent lock for visual supervision of customs container goods. Internet Things Technol. **5**(06), 49–52 (2015)
10. Li, Y.: Design of RFID-based tracing service of WEEE. [Project Documentation] Project title: Globally Recoverable and Eco-friendly Eequipment Network with Distributed Information Service Management, Project No: 269122, European Commission within the Seventh Framework Programme, 31 July 2015
11. Li, Y., Xue, S., Liang, X., et al.: I2I: a balanced ecommerce model with creditworthiness cloud. In: IEEE, International Conference on EBusiness Engineering, pp. 150–158. IEEE Computer Society (2017)
12. Li, Y., Liang, X., Zhu, X., Wu, B.: A blockchain-based autonomous credit system. In: 2018 IEEE 15th International Conference on e-Business Engineering (ICEBE), Xi'an, pp. 178–186 (2018)
13. Wu, B., Li, Y.: Design of evaluation system for digital education operational skill competition based on blockchain. In: 2018 IEEE 15th International Conference on e-Business Engineering (ICEBE), Xi'an, pp. 102–109 (2018)

Analytics as a Service for e-Business

Privacy as a Service: Anonymisation of NetFlow Traces

Ashref Aloui[1]([⊠]), Mounira Msahli[2], Talel Abdessalem[2], Sihem Mesnager[1], and Stéphane Bressan[3]

[1] Université Paris 8, Saint-Denis, France
ashrefalouip8@gmail.com
[2] Télécom ParisTech, Institut Mines-Telecom, Paris, France
[3] National University of Singapore, Singapore, Singapore

Abstract. Effective data anonymisation is the key to unleashing the full potential of big data analytics while preserving privacy. An organization needs to be able to share and consolidate the data it collects across its departments and in its network of collaborating organizations. Some of the data collected and the cross-references made in its aggregation is private. Effective data anonymisation attempts to maintain the confidentiality and privacy of the data while maintaining its utility for the purpose of analytics. Preventing re-identification is also of particular importance. The main purpose of this paper is to provide a definition of an original data anonymisation paradigm in order to render the re-identification of related users impossible. Here, we consider the case of a NetFlow Log. The solution includes a privacy risk analysis process to classify the data based on its privacy level. We use a dynamic K-anonymity paradigm while taking into consideration the privacy risk assessment output. Finally, we empirically evaluate the performance and data partition of the proposed solution.

Keywords: Privacy · Anonymisation · Risk analysis · NetFlow

1 Introduction

NetFlow is a network protocol which monitors network traffic. It was initially developed by Cisco around the turn of the century and implemented in the company's routers. NetFlow began as a cache to improve the performance of IP lookups and later evolved into a widely used flow measurement tool. Routers running NetFlow maintain a "flow cache" including flow records that gathers details about the traffic forwarded by the router. These flow records can be collected, analysed and archived by a computer. This flow is exported using the User

This work has been supported, in part, by the IDOLE ANR project in France as well as the National Research Foundation in Singapore including the Prime Minister's Office, Singapore under its Corporate Laboratory@University Scheme, the National University of Singapore and Singapore Telecommunications Ltd.

K.-M. Chao et al. (Eds.): ICEBE 2019, LNDECT 41, pp. 561–571, 2020.
https://doi.org/10.1007/978-3-030-34986-8_39

Datagram Protocol. It is the primary technology for building core traffic matrices. NetFlow is also used as an intrusion detection system to enforce the security policy specially for identifying the denial of service attacks. Unfortunately, this traffic analyser is a double-edged sword. Often, in captured packets the header reveals information about the identity of the user, for example encrypted e-mails with the subject, sender and recipient, in clear text. This privacy violation can be more subtle by giving technical details about the used environment. For example, a full packet dump can provide malicious users the disk space requirements and performances linked to the network and its equipment.

Our paper provides the secure architecture of shared data with respect to privacy constraints. We choose NetFlow Logs as a case study and propose a new service called "Privacy as a Service" which sets out to ensure privacy in the flow network process. Finally, we employ a dynamic privacy model that intends to satisfy privacy user requirements and security challenges while sharing data.

In order to monitor security violations, it is crucial that the flow can be traced back to the source in a way that leaves no room for doubt. In this work, we consider two objectives: trust and data process privacy. High trust can be achieved by using a trusted element to securely manage (generate, store and use) users' credentials. The second requirement is the anonymity of users. Our research will answer questions such as how cryptographic parameters can be used to secure data sharing with respect to privacy among anonymous users and how effectively can an anonymous user of anonymous communication sessions communicate without compromising the anonymity of the two sides.

Previous works have proposed anonymisation approaches and are based on randomisation or pseudonymisation techniques. Here, we use the succession of two main processes. Firstly, data is classified based on its privacy level; a dynamic qualitative privacy risk assessment is used. For example, we employ the dynamic K-anonymity for the anonymisation process. K-anonymity [1] is achieved on different steps such as the generalisation and quasi-identifiers definition. We use this concept to add the anonymisation parameters. In this paper, we thus explore a new anonymous and secure data process scheme. Based on the privacy levels of each group of data, we define the K for K-anonymity and the pseudonymisation scheme used for generalisation.

The organisation of this paper is as follows. We begin with a brief background of the related work describing all anonymisation schemes. In Sect. 3, we present the threat model of this work. Then, the proposed solution with its related anonymity paradigm is provided in Sects. 4 and 5 followed by some experimental results. In Sect. 6, we use NetFlow to provide a privacy analysis and based on our findings we also discuss future work. Finally, in Sect. 7, we summarise our conclusions.

2   Background

Recent works have shown that there are two main approaches to ensure anonymity of data. The first is randomisation, which alters the veracity of the data in order to weaken the link between the data and the individual and to render the data sufficiently uncertain so that it can no longer be traced back to a particular individual. We can use several methods such as noise addition or permutation. The second is anonymity, which dilutes the attributes of those involved by changing their respective scale or order of magnitude (from a communal to a regional scale, for example). There are three main paradigms of anonymisation: K-anonymity, L-diversity and differential privacy. K-anonymity responds to the risk of revealing identity; it aims to prevent a person associated with an identification key from being isolated by grouping him with, at least, individuals. To do this, attributes are generalised to such an extent that all individuals share the same identification key. To generalise actually means "to remove some degree of precision" to certain fields. For example, lowering the geographical granularity of a city to a region includes a larger number of people involved. The generalisation can be global (replacement of all the cities by the corresponding region) or local (we only replace small towns by their region, we keep the big cities where there are at least k lines - capitals, metropolises etc.). There are several limitations that have been identified in this technique, mainly attacks such as unsorted matching, complementary release, minimality and temporal attacks [7–9].

L-diversity extends k-anonymity to ensure that it is no longer possible to achieve certain results by means of inference attacks, ensuring that in every equivalence class, each attribute has minus l different values. While this technique provides optimal defense against inferential revelation attempts when attribute values are properly distributed, it should be emphasised that it does not prevent information leaks if attributes within a segment are distributed over time. L-diversity can be redundant and laborious to achieve and is open to attribute-revelation attacks (risk of correlation). It's also susceptible to skewness attacks and similarity attacks and is inadequate at avoiding attribute exposure due to the semantic relationship between the sensitive attributes [8, 10].

Finally, with differential privacy, data sets are communicated to third parties in response to a specific request rather than being published as a single data set. Anonymised previews are produced by means of a subset of requests to a third party. The subset includes random noise deliberately added a posteriori. This technique does not modify the original data. Therefore, the data controller remains able to identify individuals in the results of differential privacy queries. Unfortunately, it is not possible to achieve maximum privacy and a high level of accuracy at the same time. Differential privacy is achieved by randomisation and therefore maximum privacy guarantees that are only possible by adding enough noise which reduces the accuracy.

3   Threat Model

In our approach, to ensure security and privacy, the proposed solution is based on an "honest-but-curious" model, which means that all entities of our system (i.e., service providers, and service users sharing data) will use our system honestly, but may be curious to collect and to analyse some information.

Here, we target two K-anonymity attacks. The first attack is the homogeneity attack, which leverages the case where all the values for a sensitive value within a set of K records are identical. In such cases, even though the data has been K-anonymised, the sensitive value for the set of K records may be exactly predicted.

The second attack is the background Knowledge Attack which leverages an association between one or more quasi-identifier attributes with the sensitive attribute to reduce the set of possible values for the sensitive attribute. For example, [11] showed that knowing that heart attacks occur at a reduced rate in Japanese patients could be used to narrow the range of values for a sensitive attribute of a patient's disease. In our proposal, we consider the use of a dynamic K-anonymity paradigm depending on the privacy of each group of data.

4   Context

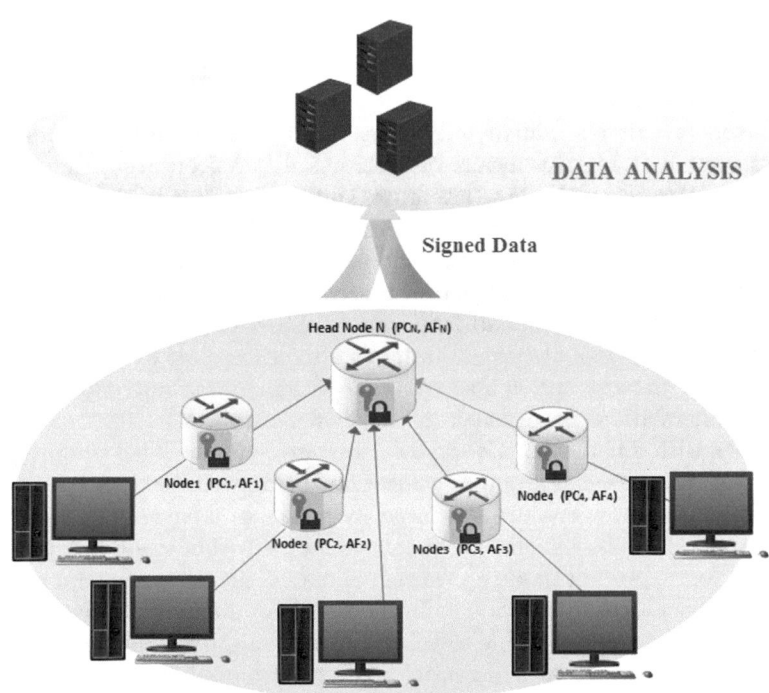

Fig. 1. Privacy as a service architecture

Network trace data provide valuable information which contributes to the modeling of network behaviours, defending against network attacks and developing new protocols. So releasing the data of a network trace is in high demand by researchers and organisations to promote the development of network technologies. However, due to the sensitive nature of network trace data, it is a potential risk for organisations to publish or to share the original data which may expose their commercial confidentiality and the customers' privacy within their networks. Several methods to defend the network trace attacks such as statistical fingerprinting and injection have been proposed, unfortunately, they are not enough to protect users' privacy because the correspondence between the source and destination IP addresses can also help the adversary to identify the target host.

We first, consider the decision version of the problem of achieving the optimal private K-anonymity by analysing the privacy risk level. In the following, we present our privacy modeling based on K-Anonymity and privacy assessment mechanism. Here, we consider different assumptions about the external knowledge available to an adversary as indicated in our threat model and the qualitative evaluation of privacy. Our methodology is fully adapted to the NetFlow data structure presented in Fig. 2.

First, we need to classify NetFlow data based on its privacy level. Here, we can qualitatively assess the privacy level by evaluating the relation between the data and the issuer identity. In our case, we define three privacy levels: very critical data, critical data and public data. Second, the output of privacy level assessment phase is the input of a dynamic K-anonymity process. Based on each data privacy level, we can define the groups of quasi-identifiers and their related values of K.

5   Proposal

In this section, we formalise the context and privacy risk approach. We start by identifying privacy and security requirements and consider the privacy risk analysis phase. First, we carry out a classification of data based on the privacy level of criticality. This corresponds to classifying IP addresses, ports and all logged data into groups in order to fix the respective K-Anonymity level. The next step is to use an anonymisation tool to apply the appropriate privacy function to the adequate field. Here, we consider that a data center collects all logged data of Netflow from different companies. The goal is to anonymously analyse logged data to retrieve useful information such as the percentage of connections to social media sites. Our work helps to generate high quality statistics through direct database queries without revealing sensitive personal information from the dataset.

5.1   Privacy Level Definition

Most existing anonymisation approaches focus on adding small amounts of statistical noise (e.g. Gaussian Noise). In our work, we aim to use random noise depending on the privacy level of grouped data. Each group has a defined level of privacy. We choose the amount of noise based on the level of privacy or sensitivity of data.

5.2   Modeling

Let $T(A_1, ...A_n)$ be the set of attributes and QI_T its associated quasi-identifiers. QI_T is a set of attributes $(A_i, ...A_j) \subseteq (A_1, ...A_n)$. Based on previous privacy assessment we can divide the set of quasi-identifiers to sub-groups. For example, we define:

– The QI_{TPr} as the set of quasi-identifiers associated to very critical data in terms of privacy.
– The QI_{TM} as the set of quasi-identifiers associated to medium critical data in terms of privacy.
– The QI_{TPu} as the set of quasi-identifiers associated to public data in terms of privacy.

This approach can be generalised as follows.

Definition 1. Let $QI_T(A_i, ..., A_k)$ the set of critical data to be anonymised and QI_{PL} the sub-group of this critical data with PL as privacy level. To each privacy level we assign the value K_i of dynamic K-Anonymity. The randomness of noise introduced by K-Anonymity is derived from the level of privacy sensitivity.

After defining different quasi-identifier groups, we focus on the domain generalisation hierarchy. We attribute to each QI_{PL} a domain hierarchy. We define our generalisation hierarchy based on the privacy level of anonymised data. A distinction is made between the anonymisation functions based on the privacy level of data. In case of IP addresses in NetFlow Logs, we can consider two main groups, the public and private group. Figure 1 illustrates the two domains D_{IP}: domain IP addresses and D_{Pr}: domain ports and value generalisation hierarchies for the two most important domains in case of NetFlow.

K has a dynamic value depending on the privacy level of the data. We consider our private table combining groups of attributes with their private level, K-anonymity value and corresponding anonymisation functions. As presented by Sweeney in [1], $D_i = dom(A_i, PT)$ denotes the domain associated with attribute A_i in the private table PT.

5.3   Dynamic Generalisation

The concept of generalising an attribute is a quite simple; a value is replaced by a less specific, more general value that is faithful to the original.

date	time	router_ip	sampling	src_ip	dst_ip	nexthop	input	output	pkts	bytes	first	last	prot	sport	dport	flags	tos	src_as	dst_as
17/08/201	0:00	68.148.39.	1	68.130.17	157.0.175.4	0.0.0.0	76	326	2	160	1502899218		1502899218	17	0	0	0	0	0	0
17/08/201	0:00	68.148.39.	1	68.131.16	208.230.71	0.0.0.0	36	43	2	50	1502899233		1502899233	6	33311	80	16	0	0	0
17/08/201	0:00	68.148.39.	1	196.166.21	240.172.7	0.0.0.0	322	326	2	181	1502899248		1502899248	6	60167	443	24	0	0	0
17/08/201	0:00	68.148.39.	1	2.41.205.1	68.129.53.	0.0.0.0	322	225	2	1133	1502899248		1502899248	6	443	63453	24	0	0	0
17/08/201	0:00	68.148.39.	1	68.129.93.	139.9.212.	0.0.0.0	217	393	2	162	1502899256		1502899256	50	0	0	0	0	0	0
17/08/201	0:01	68.148.39.	1	213.160.1	138.80.14	0.0.0.0	244	42	11	15246	1502899205		1502899262	6	4666	60000	24	0	0	0
17/08/201	0:01	68.148.39.	1	72.185.0.1	203.117.21	0.0.0.0	222	182	2	69	1502899272		1502899272	6	48278	8555	2	0	0	0

Fig. 2. NetFlow log

Given table PT, generalisation can be effective in producing a table RT based on PT. The number of distinct values associated with each attribute is non-increasing, so the substitution tends to map values to the same result, thereby possibly decreasing the number of distinct tuples in RT.

In order to distinguish between the privacy level of the pseudonymisation functions we define the generalisation function as $f_k i(t)$ on tuple t, $f_k i$:, where $f_k i : A_1, ..., A_n \rightarrow f_k i(A_1), ..., f_k i(A_n)$ and i is the variable privacy level. The classification of anonymisation functions depends on privacy level of data to be applied on.

$$
\begin{pmatrix}
A_1 \\
A_2 \\
A_3 \\
. \\
. \\
. \\
. \\
A_n
\end{pmatrix}
\xrightarrow{f_j(A_i)}
\begin{pmatrix}
f_0(A_1) \\
f_1(A_2) \\
f_2(A_3) \\
f_k(.) \\
f_k(.) \\
f_l(.) \\
f_l(.) \\
f_l(A_n)
\end{pmatrix}
$$

Definition 2. Let $Tk_i(A_1, ..., A_n)$ and $Tk_j(A_1, ..., A_n)$ be two tables defined on the same set of attributes or with the same set of attributes, k_i for dynamic privacy level. For every privacy group, we assign a group of data and consider the bijective mapping between two tables.

1. $|Tk_i| = |Tk_j|$, k_i is the privacy level
2. $\forall z = 1, ..., n : dom(Ak_z, Tk_i) \preceq_D dom(Ak_z, Tk_i)$
3. It is possible to define a bijective mapping between Tk_i and Tk_j that associates each tuples t_i and t_j such that $t_{i[Ak_z]} \preceq t_{j[A_z]}$.

Definition 2 states that a table $T_k j$ is a generalisation of a table $T_k i$, defined on the same attributes, if:

(1) $T_k i$ and $T_k j$ have the same number of tuples,
(2) the domain of each attribute in $T_k j$ is equal to or a generalisation of the domain of the attribute in $T_k i$, and
(3) each tuple $t_k i$ in $T_k i$ has a corresponding tuple $t_k j$ in $T_k j$ (and vice versa) such that the value for each attribute in $t_k j$ is equal to or a generalisation of the value of the corresponding attribute in $t_k i$.

We consider the bijective mapping between quasi-identifiers grouped by privacy level and generalisation elements. In static matching models, the match value function is exogenous. The value to any match depends on both the static production function and continuation values. In order to satisfy the K-anonymity condition, we associate significantly critical level privacy data to the highest value of K. In our case, we consider IP address classification as presented in Fig. 3 into public IP addresses and private IP addresses. For public IP addresses, we have three sub-groups: internal, special and external addresses. For the internal addresses, we are referring to the home routable IPs. The external public IPs are non-home routable IPs. Special public IP addresses are the well-known IP addresses such as DNS, Facebook, etc. Depending on the privacy criticality of data, we assign to each group a function of pseudonymisation to generate the attributes of generalisation (Fig. 4).

Table 1. Privacy as a service architecture

Fields	Privacy level	K value functions	Anonymisation
All addresses (IP, MAC, etc.)	Very sensitive	Highest value	Hash or encryption functions
Src and Dst ports	Sensitive	Mean value	Truncation or random permutation
IP source mask IP destination mask	Insensitive	1	Permutation
Other data	Public	0	No functions

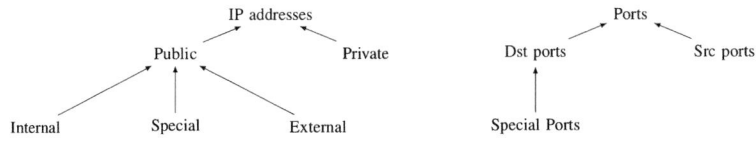

Fig. 3. Domain and value generalisation

6   Privacy Analysis

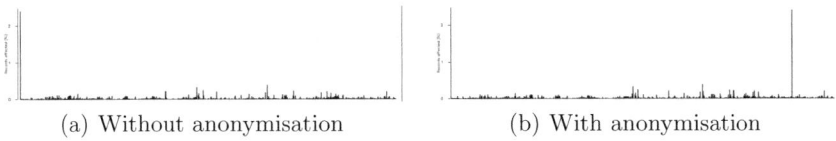

(a) Without anonymisation   (b) With anonymisation

Fig. 4. IP sources anonymisation

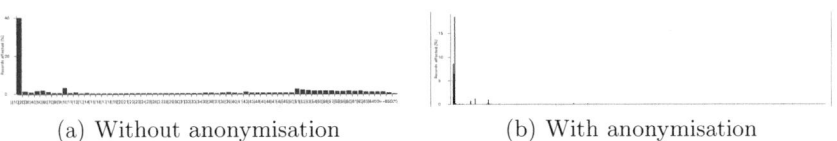

(a) Without anonymisation   (b) With anonymisation

Fig. 5. Source ports anonymisation

In order to satisfy K-anonymity, and in the case of NetFlow (see Fig. 2), we can distinguish four levels of privacy presented in Table 1. In this section, we will first give a brief introduction to Cisco NetFlow. Then, we will describe our experiments of applying the extended Crypto-PAn anonymisation method to sample Cisco NetFlow traces. Finally, we will analyse the results.

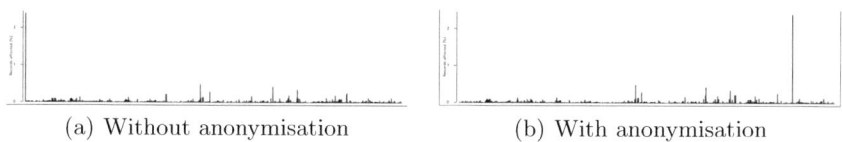

(a) Without anonymisation   (b) With anonymisation

Fig. 6. IP destination anonymisation

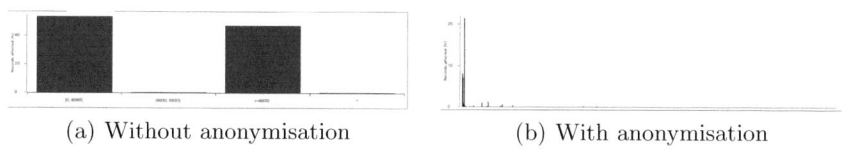

(a) Without anonymisation   (b) With anonymisation

Fig. 7. Destination ports anonymisation

Cisco NetFlow logs contain records of unidirectional flows between computer/port pairs across an instrumentation point (e.g. router) on a network. Ideally, there is an entry per socket. These records can be exported from routers or software such as ARGUS or NTOP. NetFlows are a rich source of information for traffic analysis consisting of some or all of the following fields depending on version and configuration: IP address pairs (source/destination), port pairs (source/destination), protocol (TCP/UDP), packets per second, time-stamps (start/end and/or duration) and byte counts. We setup the software to work on the binary data of the Cisco NetFlow logs directly. We could have simply compiled the code with a few changes to take a dotted decimal IP address through STDIN with the key as an argument. It would then send the anonymised IP address to STDOUT. A perl script could parse the log and call this C++ binary to do anonymisation on every IP address. Then, we apply the K-anonymity transformation.

After the classification of data depending on its privacy level and creating groups, we move on to the generalisation. Here, we consider that the group of quasi-identifiers has its own privacy level. Highly sensitive data, such as source and destination IP addresses, provide sensitive information on the inflows and outflows and can help a tracker trace the Internet browsing for examples for a network user. Source and destination ports are also in a group of very sensitive data. The second level of privacy is the medium sensitive data, which concerns data such as virtual private network IDs of source and destination. The IP source mask and IP destination mask could be in insensitive data and the rest of the data could be considered as a public information. As a consequence, we could assign the value 4 for the very sensitive group, the value 3 for sensitive data and the value 2 for insensitive data for the other data we do not need when using K-anonymity. The same analysis could be carried out when choosing the corresponding function for every group of data. For very sensitive data, we use the truncated hash function. For sensitive data, we use the permutation. The results of this simulation is illustrated in Figs. 5, 6 and 7. In every case, we can see the difference of data partition before and after implementing our proposal of anonymisation. In general, the ranges of data changed while applying the dynamic K-anonymity. More granular analysis could be done by refining the definition of more groups of data. For example for IP address, we can consider two groups for public and private addresses. This refining and more optimum K-anonymisation will be the subject of future work.

7   Conclusions

To avoid the hurdle of sharing logs, strong and efficient anonymisation techniques are needed. In this paper, we proposed a dynamic anonymisation paradigm by using privacy risk assessment. The scheme classifies data based on its privacy level and defines different levels of data privacy. Depending on this level, we fix the K for K-anonymity for every data group. The function of generalisation is also chosen based on the data privacy level. The implemented prototype was tested with NetFlow traces. We identify several levels of privacy criticality to classify the Log of NetFlow. The results of simulation shows that the data partition completely changes by applying an optimum and dynamic K-anonymity scheme.

References

1. Samarati, P., Sweeney, L.: Protecting privacy when disclosing information: k-anonymity and its enforcement through generalisation and suppression. Technical report, SRI International (1998)
2. Sweeney, L.: Computational disclosure control for medical microdata: the Datafly system. In: Record Linkage Techniques 1997: Proceedings of an International Workshop and Exposition (1997)
3. Slagell, A.J., Lakkaraju, K., Luo, K.: FLAIM: a multi-level anonymisation framework for computer and network logs. In: LISA, vol. 6 (2006)
4. Foukarakis, M., et al.: Flexible and high-performance anonymisation of NetFlow records using anontool. In: Third International Conference on Security and Privacy in Communications Networks and the Workshops, SecureComm 2007. IEEE (2007)
5. Farah, T., Trajković, L.: Anonym: a tool for anonymisation of the Internet traffic. In: 2013 IEEE International Conference on Cybernetics (CYBCONF). IEEE (2013)
6. Rajendran, K., Jayabalan, M., Rana, M.E.: A study on k-anonymity, l-diversity, and t-closeness techniques focusing medical data. IJCSNS Int. J. Comput. Sci. Netw. Secur. **17**(12), 172 (2017)

7. Hussien, A.A., Hamza, N., Hefny, H.A.: Attacks on anonymisation-based privacy-preserving: a survey for data mining and data publishing. J. Inf. Secur. **4**(2), 101–112 (2013). https://doi.org/10.4236/jis.2013.42012

8. Jain, P., Gyanchandani, M., Khare, N.: Big data privacy: a technological perspective and review. J. Big Data **3**(1), 25 (2016). https://doi.org/10.1186/s40537-016-0059-y

9. Sweeney, L.: Achieving k-anonymity privacy protection using generalization and suppression. Int. J. Uncertain. Fuzziness Knowl.-Based Syst. **10**(5), 571–588 (2002). https://doi.org/10.1142/s021848850200165x

10. Li, N., Li, T., Venkatasubramanian, S.: T-closeness: privacy beyond k-anonymity and l-diversity. In: ICDE 2007 IEEE 23rd International Conference on Data Engineering (2007). https://doi.org/10.1109/icde.2007.367856.

11. Machanavajjhala, A., Gehrke, J., Kifer, D., Venkitasubramaniam, M.: L-diversity: privacy beyond k-anonymity. In: 22nd IEEE International Conference on Data Engineering (ICDE 2006), Atlanta, Georgia, April 2006

12. Bild, R., Kuhn, K.A., Prasser, F.: Safepub: a truthful data anonymisation algorithm with strong privacy guarantees. Proc. Priv. Enhancing Technol. **1**, 67–87 (2018)

13. Slagell, A., Wang, J., Yurcik, W.: Network log anonymisation: application of crypto-pan to cisco netflows (2004)

Game Analytics Research: Status and Trends

Yanhui Su[(⊠)] 🆔

University of Skövde, 54128 Skövde, Sweden
yanhui.su@his.se

Abstract. This paper aims to perform a systematic literature review of the business intelligence used in the game industry which mainly focuses on the game analytics side. First, according to the game industry value chain, a review identifying and classifying the relevant papers which had been published, exploring them systematically to extract similarities and status. Results show how game analytics can be used in the game industry, with player analytics, game development analytics, game publishing analytics and also channel analytics. Second, considering the business intelligence problems or potential challenges in the game industry, how game analytics can help to solve that also be discussed. Third, as recent game analytics research is highly fragmented and the underexplored areas, especially for the potential research gaps and trends are also explored. The main contribution of this paper includes giving a clear and reasonable classification based on the game industry value chain about game analytics and making a detailed overview of current research status and also discussing the potential trends as the baseline for future research.

Keywords: Business intelligence · Game analytics · Game value chain · Game metrics · Retention · Prediction

1 Introduction

With the continuous development of the game industry, game research is also constantly improving. Many other interdisciplinary knowledge and theories have been used to promote the development of the game industry. As shown in Fig. 1, game analytics comes from Business Intelligence (BI) which reflects the combination of BI with the game research itself [1].

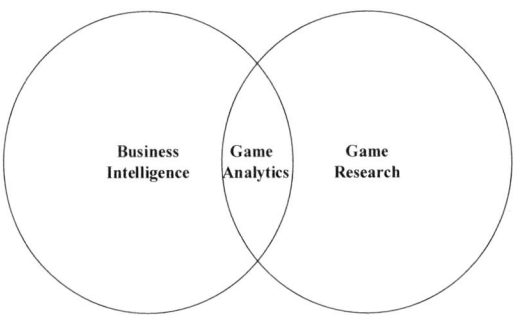

Fig. 1. Game analytics research.

© Springer Nature Switzerland AG 2020
K.-M. Chao et al. (Eds.): ICEBE 2019, LNDECT 41, pp. 572–589, 2020.
https://doi.org/10.1007/978-3-030-34986-8_40

Analytics is not just the querying and reporting of BI data but rests on actual analytics, such as statistical analytics, predictive modeling, optimization, forecasting [2]. The goal of game analytics is to support decision making. Analytics is the process of discovering and communicating patterns in data, towards solving problems in business or predictions for supporting the business decision, driving action and improving business performance [1]. The methodological foundations for analytics based on the mathematics combines with statistics, data mining, programming, operations research, and data visualization which is used for showing the analytics results to the relevant stakeholders. As shown in Fig. 2, there are several branches or domains of analytics, such as marketing analytics, risk analytics, web analytics, and game analytics [1]. According to the literature review, previously game analytics mainly focuses on game development and game research.

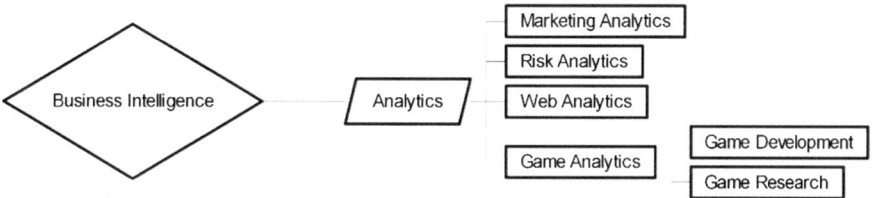

Fig. 2. Relationship between BI and game analytics.

Drachen et al. [1] point out that game analytic can be understood as the application of analytics to game development and research. In fact, game analytics not only can be used to identify in-game balancing issues [3], visualize players' movement on the game map [4], but also to save game development costs [5], and help with in-game bugs testing during the game development process [6]. However, by now, game analytics research is highly fragmented. There is no systematic research and reasonable classification about BI used in the game industry, especially for the game analytics.

Based on the literature review, according to the game value chain which is shown in Fig. 3, the traditional game industry value chain includes four parts [7]. The game developer develops games and then they find the game publisher to help with the game sales. The game publisher through the game distributor to connect the games with the potential game players together. As the traditional game value chain covers the whole game industry, so it is reasonable to use it as the base of the classification of game analytics.

Fig. 3. Traditional game value chain.

According to the development status of the game industry, followed the traditional game value chain, it's reasonable to divide the game analytics into four parts which include the game development analytics, game publishing analytics, game distribution analytics and also the game player analytics, as shown in Fig. 4.

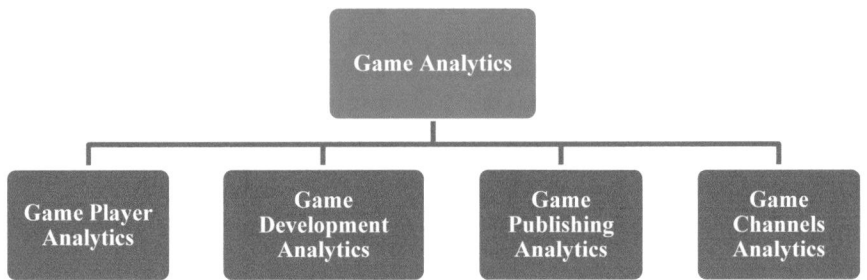

Fig. 4. Game analytics classification.

Game player analytics is vital for game analytics. The core of player analytics is to analyze the game behavior and specific preferences of the game player to determine the right direction of game development. The game development analytics include verification of the gameplay, interface analytics, system analytics, process analytics and also performance analytics. As for game publishing analytics mainly focuses on the retention analytics and revenue analytics. The channel analytics focuses on analyzing the channel's attributes and provides specific solutions for user acquisition.

As for game metrics, it can be defined as the behavioral data source which will be used for the game analytics [1]. They present the same potential advantages as other sources of BI which always be used for decision making in the game industry. Metrics can be variables or features, or complex aggregates and calculated values. The relationship between game metrics and game analytics is that game metrics are the numbers to track and game performance or development progress. Analytics can use metrics to find out the trends and help to make the right decision.

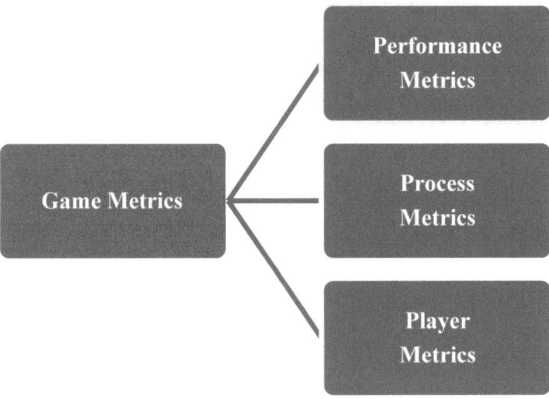

Fig. 5. Game metrics classification.

According to the classification [1], game metrics general can be divided into three categories as shown in Fig. 5, including the player metrics, process metric and also the performance metrics. Usually, player metrics focus on player's behavior and customer research. For the process metrics are used for the game development monitoring and management. For the performance metrics, they have a close relationship with the game technical monitoring such as the frame rate, the number of bugs and client executes.

2   Research Questions

In order to deeply understand the current literature about game analytics studies, the following three general research questions were formulated. The main target for the literature review is based on the reasonable classification to categorize and summarize all the work around game analytics. This research will help to identify research gaps in current research and to point to the main questions on the subject thus forming a baseline, for future research from the game industry and academic side. Therefore, the questions are summarized as follows:

- **RQ.1:** Regarding game analytics, what aspects have been explored so far in the available literature?
- **RQ.2:** What problems or challenges in the game area and whether game analytics can help to solve that?
- **RQ.3:** What topics have already been covered in literature but need further development and research?

3   Research Method

The literature review is a type of review article which includes the current knowledge classification, substantive findings, as well as theoretical and methodological contributions to a particular topic [8]. The main driver for conducting this review is to categorize and summarize all the work that has been done around game analytics for the game industry, to identify the potential gaps and commonalities in current research and to figure out the main questions on game analytics thus forming a baseline for future research.

The search strategy contained the following design decisions: Searched databases include Springer Link, Science Direct, IEEE Xplore, Web of Science, ACM and Google Scholar. Searched items include journal articles, conference papers, workshop papers, technical reports, and books. Search applied on full text and also including searched keywords in abstracts or titles, or those using a different variant of the terms which are relevant for the review. For the language, the search was only limited to papers written in English.

The reason why to choose these databases is that it covers most of the researches about BI used in the game industry, especially for game analytics. The search terms include the game analytics, game metrics, game analysis, game development and game distribution channel, game publisher, game publishing, and also the game data analysis with the year limitation. As some papers which indexed are not very relevant to the research areas, so this research only chooses the related papers for the research as shown in Table 1.

Table 1. Searching criteria for relevant research

Search criteria details	
Range	Peer-reviewed articles published in journals or full-length articles published in International conference/Workshop proceedings
Results	Presenting all related results
Year	Dated between 2005 and 2019
Language	English language studies

In fact, for the game analytics papers which got from the entire database, compared with the traditional game value chain, it is valuable to classify them into different subdivision topics. As it will give a clear map about how game analytics can be used in the game industry.

4   Research Details

Based on the literature review, according to the traditional game value chain [7], it is valuable and reasonable to divide the game analytic into four parts including the game development analytics, game publishing analytics, game distribution channel analytics and also the game player analytics. As for game player analytics is based on the segmentation analytics and also the in-game behavior analytics, which including the motivation of playing games, player game experience, etc. The game development analytics includes verification of the gameplay, interface analytics, system analytics, process analytics and also performance analytics. The game publishing analytics focuses on the fields of acquisition, retention and revenue analytics and the channel analytics focus on analyzing the channel's attribute and provides specific opinions for further game optimization and user acquisition.

4.1   Game Player Analytics

Game Player Analytics: Game player analytics focuses on the player itself. Traditionally, user research uses qualitative methods as part of practices, as they focus on subjective feedback from users, such as player experience, satisfaction and engagement survey. Therefore, most of these studies are conducted through interviews, in-depth questionnaires, and observations. Recently, a few quantitative and qualitative researches focus on successful cases, such as user researchers in determining the data analytics, subjective condition and depression research [9]. This research seeks to identify patterns of behavior that could point to potential frustration before players leaving a game.

Besides these, about the game remote sensing data, such as usability testing for game playability testing also provide insight into how people play these games [10, 11].

Player Segmentation: As for the game development, game designers not only need to focus on the gameplay but also need to know who will be the players and what will be their requirements. As game development and publishing process becomes closer, as discussed in this paper [12] the specific development needs to be carried out and meet the requirements from different game players. The recent trend is that in the early game design, more and more considerations will be given to the different requirements from player side, which will make the market promotion more effective [13, 14]. The segmentation can be traced back to the beginning of the 20th century which described the differentiation as meeting human requirements as accurately as possible. In practice, this means that in order to make sure the game is designed considering the needs of specific users, segmentation is the activity aimed at identifying different customer groups [15]. The goal of segmentation is to further classify the player's groups and provides game more in line with the needs of players.

In fact, players' needs for games are diverse, and the motivations for users to play games are diverse. These researches are based on the breakdown of user behavior. Player's classification has the great advantage which can be targeted according to the needs of different players during the game design. But sometimes there are some problems to divide into only two categories, hardcore games and non-core games. Because how to distinguish between core and non-core players is the vital indicators. As listed in these papers [16, 17], immersion is an important indicator to guide and evaluate the player's behavior and motivation factor. In order to conduct more effective segmentation of players, user research always takes these factors into account. Stanton et al. [18] present the first step towards self-refining games whose game systems continuously improve based on player analytics. They observe that game objectives cause players to explore only a small fraction of the entire state space. So they made data-driven simulation feasible even for complex dynamical game systems.

In-game Behaviors: In-game behaviors which include in-game actions and player's behavior, such as navigation, interaction with game assets like objects and other entities. Player's behavior research involves specific in-game behaviors throughout the game experiences. Darken and Anderegg [19] provide a new idea which regarded player's behavior as simulacra. They provide the candidate movement models which are based on the different types of players. Hamari and Lehdonvirta [12] compare the related game modes of players in different game level and find that the structure of the game is very similar to market considering the behavior from customers. Besides these, Thawonmas et al. [20] suggest detecting game bots based on their in-game behavior, especially those related to the designed purposes of bots. This approach has the potential to distinguish between human player's behaviors and automated programs. Nacke et al. [21] focus on a quantitative study of the player's behavior in a social health game called Health Seeker. Through analyzing, they conclude that having a well-connected in-game social network and also the in-game interaction can improve the player's behavior in solving game missions.

As a player, it's easy to generate thousands of behavioral measures over the game playing session. As every time a player inputs something to the game system, it always needs to react and respond. However, accurate measures of player activity include dozens of actions which need to be calculated for per second. For example, players in a famous game World of Warcraft (Blizzard Entertainment Games, 2004), how to measure the players' behavior could involve the position of the player's character, the health, mana, stamina, the character name, race, level, equipment and also the currency. Usually, this information collected from the installed game client and also the game servers. Drachen et al. [1] point out that analyzing behavioral data from games can be challenging for Massively Multiplayer Online Games. As each of these games has hundreds of thousands of simultaneously active users which means a huge data need to be collected. Drachen and Canossa [22] give the way of dealing with massive, high dimensional game data by clustering which can be used for improving the efficiency. Hadiji et al. [23] show that cluster analytics is widely used across disciplines and has been adopted in-game analytics to find patterns in the player's behavior. Drachen et al. [24] also compare the benchmark of games and improve player modeling using self-organization and provide an initial study on identifying different player types in a commercial game. Wallner et al. [25] use the lag sequential analytics (LSA) which makes use of statistical methods to determine the significance of sequential transitions to aid analytics of behavioral streams of players.

To sum up, player research is the foundation of game analytics. As it is not only track the game development on the demand of the player's requirement by player segmentation, but also discover some potential problems in the process of game development. Player research is also the vital safeguard for game publishing which can be collected by issuing process of players for different game levels through the analytics and gives the clear suggestions about the game optimizations which can be used to improve the game retention and extend the game life cycle.

4.2   Game Development Analytics

The initial game analytics has many applications in game development, mainly focus on monitoring the process of game development. It includes some technical performance and indicators of game development, such as the bug and crash monitors. Hullett et al. [26] explore how data can drive game design and production decisions in game development. They define a mixture of qualitative and quantitative data sources and present preliminary results of a case study about how data collected from the released game can guide game development. Game development analytics mainly focus on the analytics of core gameplay, interactive analytics, and in-game system analytics. However, for the game development analytics research which needs to cover the whole game development process. It not only includes the analytics of core gameplay, interface interaction analytics, but also includes in-game system analytics, process analytics, and performance analytics. As shown in Fig. 6, the game development analytics and classification.

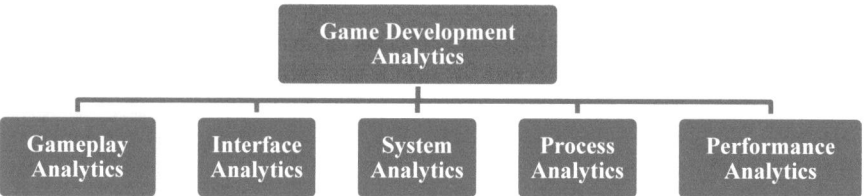

Fig. 6. Game development analytics classification.

Gameplay Analytics: Gameplay is the core of a game which is used for representing how this game is played. It relates to the real behavior of the user as a player such as in-game interaction, items trade, and navigation in the environment or game map. Gameplay analytics is significant to evaluate the game design and player experience. Gameplay analytics is used not only for guiding the game design, player research, and quality assurance but also for tracking any other position where the actual behaviors of the players interested in. This kind of game analytics can be recorded during all phases of game development [27, 28]. Besides these, Medler et al. [29] present how a visual game analytic tool can be developed to analyze the player's gameplay behavior. They develop and release the good analytic tools which can monitor millions of players after the game online. Mirza-Babaei et al. [30] provide new user research methods which have been applied to capture interactions and behaviors from players across the gameplay experience. Emmerich and Masuch [31] discuss the gameplay metrics used as one method to measure the player's behavior. They also present the conceptualization, application, and evaluation of three social gameplay metrics that aim at measuring the social presence, cooperation, and leadership respectively. Drachen and Canossa [32] point out that gameplay metrics are instrumentation data about user behavior and game interaction. Their research focuses on utilizing game metrics for informing the analytics of gameplay and guiding the development of commercial game products.

Interface Analytics: Interface analytics usually includes all interactions the player performs with the game interface and menus. This is usually be calculated by setting game variables, such as mouse sensitivity, monitor brightness. The data analytics of the interface part is based on the premise that all the menu and button settings can be recorded in advance. Only through the recorded data, the click volume of the interface icon and the validity of the design can be effectively analyzed. It also can be used for the players' game experience calculation and analytics. Interface analytics has a deep relationship with how players interact with the game UI and also the in-game interface. Xu et al. [33] analyze the bottlenecks in game designers' conventional practices and develop the single-match module of the visualization system to familiarize players with interactive analytics.

System Analytics: System analytics covers the actions game engines and also the sub-systems, such as AI system, automated events, Non-Player Character (NPC) actions and also the respond to player actions. System analytics can be used to measure the effectiveness of the system design. It also can give guidance about how to design the game system effectively. In fact, system analytics mainly targets to some game systems research and gives guidance about game development. Weber and Mateas [34] focus on the in-game system analytics and use the data mining techniques on large amounts of collected data to understand player strategies in the game StarCraft. In the research, more than 5000 replays of expert matches are used as training data for a machine-learning algorithm. This research becomes a component of an AI system that played StarCraft better than most other available techniques. It can help to improve the game AI system.

Process Analytics: Game process analytics focus on the game development process and give the full monitoring about the development process and provide the guidance about the game developing process, such as using the agile development method to measure the development process. Hullett et al. [26] focus on the collection and analytics of game data which can be used to inform the game development process. However, as games have gotten larger and more complex, the needs for different kinds of data to measure the game development process have increased as well.

Performance Analytics: Performance analytics relates to the performance of technical and software-based infrastructure behind a game itself. It includes the frame rate, the stable of client execute, bandwidth, game build quality and also the numbers of game bugs which had been found by QA testing. For example, Wang et al. [35] measure and analyze the performance of World of Warcraft (Blizzard Entertainment Games, 2004), as a representative of online games, over the mobile WiMAX networks which focus on the application level packet statistics such as the delay and bandwidth.

In short, from the game industry side, the effective data analytics during game development not only helps developers optimize games but also quickly verifies the game's core gameplay, interaction design and also user experience. It can help developers make the right decisions and improve the efficiency of game development and reduce the cost of game development. However, in the real data analytics, how to obtain the required data, how to effectively avoid mistakes in the direction of game development caused by analytics errors is still need to do further research.

4.3   Game Publishing Analytics

The initial game analytics focused on game development and game ontology research. However, the application of game analytics in game publishing is very limit which lack systematic studies. With the development of the game industry, especially the emergence of professional game publishers, game developers more and more realized the importance value from the game publisher side, especially for the data collection and analysis during game publishing. Effective game analytic can not only help the success of game promotion but also optimize the game in a targeted manner and extend the life cycle and increase the revenue. Moreira et al. [36] use the ARM (acquisition, retention, and monetization) funnel as the basic analysis for game publishing. According to the ARM funnel, the game publishing analytics can be divided into three parts. As shown in Fig. 7, game analytics in the whole game publishing process includes game acquisition analytics, retention analytics and also the game revenue analytics.

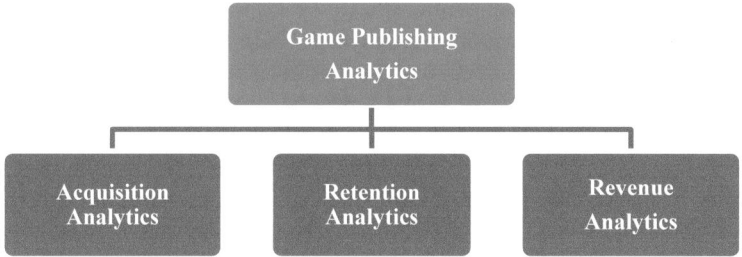

Fig. 7. Game publishing analytics classification.

Acquisition Analytics: Focuses on how to save the cost of attracting new users [1]. It also pays attention to how many new players enter the game, and how many players completed the tutorial. In order to acquire players, game developers must first invest in the development, testing, and then authorize their games for publishing on target platforms. The publisher often needs to get users by buying ads or by viral distribution on social networks. The aggregate of all these costs divided by the total number of users is called the User Acquisition Cost [1]. In fact, how to reduce the cost of user acquisition is a crucial part of game publishing, especially for global market promotion.

Retention Analytics: Retention rate is a vital indicator of measuring the stickiness of games. This index does not only measure how players are immersed in the game, but also can evaluate the game quality. The concept of retention rate comes from the marketing research side, which is provided by Hennig-Thurau and Klee [37]. They develop a conceptual foundation for investigating the customer retention process, with the concepts of customer satisfaction and relationship quality. At first, the retention rate is the factor in analyzing users' awareness of a brand. Then the concept of retention rate is applied to the game, especially in the analytics of player's retention in games. Debeauvaisand et al. [38] mainly analyze mechanisms of player retention in massively multiplayer online games and focus on how to improve the retention. Three key metrics are introduced which includes weekly playtime, stop rate, and how long respondents had been playing. The quantitative analytics shows how the game can efficiently wield retention system. Their research also utilizes several metrics to measure the players' retention, specifically: (1) hours of play per week, (2) stop rate and (3) the length of time. However, the metrics are only the first step towards a more comprehensive model of player retention in hardcore games. So for other kinds of games, it still needs different kinds of game metrics to measure the retention results such as the casual game which need to do the research as well.

Demediuk et al. [39] focus on the player retention research in League of Legends (Riot Games, 2009) by using survival analytics. The starting point of their research is to understand the influence of specific behavior characteristics on the possibility of player's retention. Based on the application of survival analytics, the churn rate of players is predicted by using cox-regression of mixed effect. Based on this research, an effective solution to the problem is proposed. Survival analytics is the practical approach as it provides the ratio and assesses the characteristics of the player who is at risk of leaving the game. The final results show that the duration between matches is a strong indicator of the potential churn rate. However, this article does not discuss the effects of other factors on retention rates. Because there are all kinds of reasons which may lead the players to leave the game. So for different games, the churn rate is also different, such as the churn rate of e-sports players also need to do further research.

Park et al. [40] focus on the key factors of player retention varies across player levels in online multiplayer games researches. They mainly discuss the online multiplayer game based on 51104 different individual log analytics, focus on exploration and analyze the key factors influencing the players' retention rate and also the key issue of retained throughout the entire game stage. They find out that the key indicators retained varied with the levels of the game. The achievements of players within the game features are significant to become senior players. However, once the players arrive at the highest game levels, social networking features are vital for the retention forecasts. This finding pointed out that social network positively affected retention when individuals form interactions with partners of appropriate levels. Yee [16] finds three motivational components which have a great relationship with retention: achievement, social, and immersion. This research is based on the factor analytics of survey results on MMORPG players. Hullett et al. [26] examine the impact of deviance on retention in the online multiplayer game League of Legends. In the research, they list two types of retention: short-term and long-term and by the deep analytics these retentions. They predict that players who frequently exhibit deviant behavior will have a negative impact on player retention. Andersen et al. [41] find out that music and sound effects have little effect on player retention in games, while animations can attract users to play more. They also discussed that minor gameplay modification will affect retention more than aesthetic variations. However, how they generalize to other games and genres requires further research.

In order to improve the retention rate of games in the virtual world, social factors can be recognized as an effective way to improve the retention rate. Krause et al. [42] in their paper investigate the potential of gamification with social game elements for increasing retention. Players in the experiment show a significant increase of 25% in retention and 23% higher average scores when the interface is gamified. Social game elements amplify this effect significantly. Besides these, Kayes et al. [43] analyze what factors cause a blogger to continue participating in the community by contributing content. The conclusion shows that the user who faces fewer constraints and has more opportunities in the community is more retained than others.

In short, the retention rate is a key indicator, especially for game publishing which is an effective way to measure the game quality. It is also the key benchmark to the game success in the market. However, for different types of games how to suggest a set of unified metrics for measuring the retention which needs to do further studies based on the player's behaviors analytics. Besides this, due to the analytics process, the acquired data will increase geometrically with the increasing of the player's information. How to effectively reduce dimensionality analytics is also an important research direction. Donoho and Tanner [44] and Thawonmas and Iizuka [45] discuss the visualization approach for analyzing players' action behaviors. The proposed approach consists of two visualization techniques: classical multidimensional scaling (CMDS) and Key Graph which is a visualization tool for the discovery of relations among text-based data. Wallner et al. [46] describe a case study regarding the use of telemetry data collection and data visualization during the development of a commercial mobile game. Stanton et al. [18] use survival analytics techniques for analyzing time-series data and drop-out rates during the game publishing process.

Revenue Analytics: With the rapid development of mobile games which take up the largest share in the game industry. As most of the mobile games are free, so players can download at any time. But for the freemium game, the revenue is mainly bought by players in the game from In-App Purchase (IAP) or advertising. Predicting the lifetime value of these players and their expenditure are the main challenges for game analytics. As only a small number of players make purchases. However, social factors are critical for the overall retention of games, the impact on revenue of mobile games has not been explored well. Drachen et al. [47] through the case study of more than 200,000 players, solve the problem about the relationship between the social features and the revenue in freemium casual mobile games. According to their research, classifier and regression models are used to evaluate the impact of social interaction in casual games for the player life-cycle value prediction. The final results show that social activities are not associated with the trend towards advanced players, but social activities will benefit for improving the game revenue.

As for freemium games, there is a big difference between payment players and also the No-payment players. According to the research from Drachen et al. [24], Non-payment players consist of the majority of freemium players, which in turn leads to highly uneven predictions in mobile games. For mobile game developers, the key challenge is to reduce the turnover rate and increase players, not only to improve the retention rate but also need to consider the changes from the junior players to senior players. A related goal is to increase the average life-cycle value (LTV) of players because of the significant increase in user acquisition costs (UAC) for mobile games in recent years. Considering these issues, the premium user acquisition costs and the market promotion fee continue to increase, the effective ability to predict the game revenue is important for the game publishing.

Alomari et al. [48] extract 31 features from the decision tree. Ten most important features of successful games are found which include the inviting friends' feature, skill tree, leaderboard, Facebook, time skips, request friend help, event offers, customizable, soft currency, unlock new content. The results of these studies are of great value to game developers in increasing their revenue. Besides this, the study also gives the conclusion that the related factor to revenue is the daily active user. In addition, other features will also play more significant roles in successful games, such as culture, lifestyle, and loyalty to game brand and promotion. Hsu et al. [49] present a novel and intuitive market concept called indicator products, for analyzing purchase decisions. Such kind of researches can help online game designers and researchers to observe the player's behavior more effectively.

4.4 Distribution Channel Analytics

In a broad sense, channels are the specific path to connect the games with players together. Most of the distribution channels aggregate a large number of users and form a platform for games. Krafft et al. [50] provide four insights into key domains in marketing channels research, including marketing channel relationships, channel structures, popular topics, and market strategies. Besides these, they also derive four trends that benefit for understanding the changes from channels side including the service economies, globalization, reliance on e-commerce technologies, and the role of big data for channel decisions. Similarly, Cramer et al. [51] feature the contributions from developers and researchers both from industry and academia side with experiences about the deployment and distribution. This research provides an overview of the challenges and methods to solve the potential issues for the distribution channels, such as the app stores, markets, new devices, and services. Latif et al. [52] visualize the In-App purchase rate of free and paid games from the channel Google Play. This visualization benefits for game developers in the development phases.

Channel distribution is vital for the success of the game. It is an important way to reach the user, such as the App Store and Google Play, which play an important role in the distribution of mobile games. Channel analytics is to combine channel data with game data together, to provide decision support for game development and publishing. It is possible to compare channels by player active rate, new player acquisition, retention rate, and also the payment rate, to find out the best channel for the game. Channels analytics can identify the potential target users, build a loyalty relationship with the games.

However, at present, the game channel research is listed as part of the market effect on statistics and analytics. As the game attributes, such as the size of the game packages which may affect the download, different game channels have different attributes, as well as the channels' benchmark have the impact on the distribution of games. So how to do the game analytics research combined with the attributes of channels is the potential research gap, requiring more research to focus on this area.

5   Problems and Challenges

According to the literature review, although the BI used in the game industry has attracted more and more attention, there are still potential problems in the research, especially for how to effectively use data analytics to drive the game publishing and how to set up the correct game metrics which can be used for the measuring the in-game activities and also the channels performance. In fact, considering the game industry status, these research areas are still full of the challenges.

Game Player Analytics Challenge: The game player analytics is based on the behavior analytics of the game users. However, the player's behavior changes with time, game preferences and also different game types. At present, the analytics results are based on enough data sampling, this will bring two problems. On one hand, how to obtain accurate results only by few data to save the cost is an issue. On the other hand, how to ensure that the collected data can represent the realistic game demands from the player side is also the challenge.

Game Development Analytics Challenges: At present, most of the analytics researches for the game development stage focus on ensuring that the gameplay is sufficient to meet the player's needs and also the game development process is controllable. But how to ensure that the in-game system design can adjust according to the player's feedbacks after game launch is almost ignored. Such as the design of the in-game economic system, how to verify the rationality of the numerical setting for the in-game system after launch should be the potential problem. As the design of the in-game economic system needs to be dynamically changed according to the player's feedback, rather than being unchanging. So game analytics used for driving game development still needs further research.

Game Publishing Analytics Challenge: So far, according to the literature review, game analytics related to game publishing research is not comprehensive. By now, most research mainly focuses on game retention and game revenue analytics side. There is no detailed analytics about the entire process of game publishing. For example, although the new content update performance is important for game publishing, few research focuses on this part. Such kind of game publishing analytics missing makes it hard to form an effective guideline for the game launch. It will also lead most of game publishing failures. So this part of the research is vital and valuable to do further research, especially from the industry side.

Game Channel Analytics Challenge: Channel analytics currently as an independent part, which is ignored by most game analytics researchers. The main reason is that game channel analytics is divided into market analytics. However, as for games, its distribution channels are different from other channels. The attributes of games also have a great influence on the distribution channels. In addition, from the game industry side, the channel attributes and benchmarks are also important factors that restrict game distribution. However, channel analytics is based on big data from different channels. As these data can be recognized as confidential information for channels. So how to get these data for the channel research is a potential challenge.

Besides these, on one hand, from the game industry side, by now, few of game analytics researches that can help to promote the BI knowledge sharing or standardization for games. As confidentiality data such as revenue and churn rate and retention make the knowledge sharing very difficult. That is also the potential reason why game analytics research is currently fragmented. In particular, there is limit research on game publishing from the channel side. This is to be expected in the explorative phase of a new domain being established, especially for the game publishing and distribution channels research. On the other hand, BI used in the game industry is ahead of the research from the academic side. The reason for the industry ahead of the academic is that for game analytics after the game launched, many data statistics are recognized as the secrets of companies. So the industry has concerns about collaboration with academia. However, the game industry mainly stays at the level of basic analytics to solve practical problems and it is difficult to rise to advanced academic research. Therefore, game analytics used in the game industry lacks a theoretical basis. That is also the main reason for the gap between the game industry and academic research.

6   Conclusion

In this paper, a systematic literature review and classification of game analytics is provided. First, as for Q1, this paper mainly discussed the benefits of game analytics used in the game industry and then based on the traditional game industry value chain, gave the reasonable classification and also the overview of related research areas. Second, as for Q2, according to the literature review and also combined with game industry requirements, this paper focused on the problems or challenges discussing especially for the player analytics, game development analytics, game publishing analytics and also distribution channel analytics. Third, as for Q3, this paper also gave the details about the potential research gaps which ignored by many researchers and also discussed the potential reasons from industry and academic side. Besides these, as discussed in this paper, game analytics can be used to predict trends, to understand the player's behavior, also to estimate game revenue. However, the current research still lacks comprehensive research especially for the analytics of the whole game publishing process and also the effectiveness of channel promotion. However, these researches are vital for the game industry, especially for the healthy development of game industry ecology. So based on the literature review, the game publishing and channels analytics need to do in-depth research in the future.

In summary, the main contribution of this paper is to provide a clear and reasonable classification about game analytics based on the game industry value chain and give the overview of current research status and also point out the potential research trends about how BI can be used in the game industry, especially for the game analytics. This research is valuable as a baseline for future research in this area.

References

1. El-Nasr, M.S., Drachen, A., Canossa, A.: Game analytics-the basics. In: Seif El-Nasr, M., Drachen, A., Canossa, A. (eds.) Game Analytics - Maximizing the Value of Player Data, chap. 1. Springer, London (2013)
2. Davenport, T.H., Harris, J.G.: Competing on Analytics: The New Science of Winning, 1st edn. Harvard Business Review Press, Brighton (2007)
3. Kim, H., Gunn, D.V., Schuh, E., Phillips, B., Pagulayan, R.J., Wixon, D.: Tracking real-time user experience (TRUE): a comprehensive instumentation solution for complex systems. In: Proceedings of SIGCHI Conference on Human Factors in Computing Systems, pp. 443–452 (2008)
4. Moura, D., Seif El-Nasr, M., Shaw, C.D.: Visualizing and understanding players' behavior in video games: discovering patterns and supporting aggregation and comparison. Game Pap, pp. 2–7 (2011)
5. Hullett, K., Nagappan, N., Schuh, E., Hopson, J.: Data analytics for game development, vol. 80, pp. 940–943 (2011)
6. Zoeller, G.: Game development telemetry in production. In: Seif El-Nasr, M., Drachen, A., Canossa, A. (eds.) Game Analytics - Maximizing the Value of Player Data, chap. 2. Springer, London (2013)

7. Kelly, C., Mishra, B., Jequinto, J.: The Pulse of Gaming. Accenture. Accenture (2015). https://www.accenture.com/t20150709T093434_w_/us-en/_acnmedia/Accenture/Conversion-Assets/LandingPage/Documents/3/. Accessed 23 Dec 2018

8. Creswell, J.: Research Design: Qualitative, Quantitative, and Mixed Method Approaches, 4th edn. SAGE Publications, California (2013)

9. Canossa, A., Sørensen, J.R.M., Drachen, A.: Arrrgghh: blending quantitative and qualitative methods to detect player. In: Proceedings of the FDG, pp. 61–68 (2011)

10. Thompson, C.: Halo 3: how microsoft labs invented a new science of play. Wired Magazin, vol. 15, no. 9 (2007)

11. Raharjo, K., Lawrence, R.: Using multi-armed bandits to optimize game play metrics and effective game design. Int. J. Comput. Inf. Eng. **10**, 1758–1761 (2016)

12. Hamari, J., Lehdonvirta, V.: Game design as marketing: how game mechanics create demand for virtual goods. Int. J. Bus. Sci. Appl. Manag. **5**(1), 14–29 (2010)

13. Huotari, K., Hamari, J.: Defining gamification - a service marketing perspective. In: Proceedings of the 16th International Academic MindTrek Conference. Presented at MindTrek 2012, pp. 17–22. ACM (2012)

14. Kotler, P., Keller, K.: Marketing Management, 12th edn. Pearson Prentice Hall, New Jersey (2006)

15. Hamari, J., Koivisto, J.: Social motivations to use gamification: an empirical study of gamifying exercise. In: Proceedings of the 21st European Conference on Information Systems, Utrecht, Netherlands (2013)

16. Yee, N.: Motivations of play in online games. J. Cyber Psychol. Behav. **9**, 772–775 (2007)

17. Kallio, K.P., Mäyrä, F., Kaipainen, K.: At least nine ways to play: approaching gamer mentalities. Games Cult. **6**(4), 327–353 (2011)

18. Stanton, M., Humberston, B., Kase, B., O'Brien, J.F., Fatahalian, K., Treuille, A.: Self-refining games using player analytics. ACM Trans. Graph. **33**(4), 1–9 (2014)

19. Darken, C., Anderegg, B.: Game AI Programming Wisdom 4. Charles River Media, pp. 419–427 (2008)

20. Thawonmas, R., Kashifuji, Y., Chen, K.T.: Detection of MMORPG bots based on behavior analysis. In: Proceedings of the International Conference on Advances in Computer Entertainment Technology, pp. 91–94 (2008)

21. Nacke, L.E., Klauser, M., Prescod, P.: Social player analytics in a Facebook health game. In: Proceedings of HCI Korea (HCIK 2015), pp. 180–187. Hanbit Media, Inc., Seoul (2015)

22. Drachen, A., Canossa, A.: Evaluating motion: spatial user behavior in virtual environments. Int. J. Arts Technol. **4**, 294–314 (2011)

23. Hadiji, F., Sifa, R., Drachen, A., Thurau, C., Kersting, K., Bauckhage, C.: Predicting player churn in the wild. In: Proceedings of IEEE CIG, pp. 1–8 (2014)

24. Drachen, A., Canossa, A., Yannakakis, G.: Player modeling using self-organization in tomb raider: underworld. In: Proceedings of IEEE Computational Intelligence in Games (CIG), pp. 1–8 (2009)

25. Wallner, G.: Sequential analysis of player behavior. In: Proceedings of the Annual Symposium on Computer-Human Interaction in Play, pp. 349–358 (2015)

26. Hullett, K., Nagappan, N., Schuh, E., Hopson, J.: Empirical analysis of user data in game software development. In: Proceedings of the ACM-IEEE International Symposium on Empirical Software Engineering and Measurement, pp. 89–98 (2012)

27. Isbister, K., Schaffer, N.: Game Usability: Advancing the Player Experience. Morgan Kaufman Publishers, Burlington (2008)

28. Andersen, E., Liu, Y.E., Apter, E., Boucher-Genesse, F., Popovic, Z.: Gameplay analysis through state projection. In: Proceedings of the Fifth International Conference on the Foundations of Digital Games, California, pp. 1–8. ACM (2010)

29. Medler, B., John, M., Lane, J.: Data cracker: developing a visual game analytic tool for analyzing online gameplay. In: CHI 2011: Proceedings of the SIGCHI Conference on Human Factors in Computing Systems, pp. 2365–2374 (2011)
30. Mirza-Babaei, P., Wallner, G., McAllister, G., Nacke, L.: Unified visualization of quantitative and qualitative playtesting data. In: CHI 14 Extended Abstracts on Human Factors in Computing Systems, pp. 1363–1368 (2014)
31. Emmerich, K., Masuch, M.: Game metrics for evaluating social in-game behavior and interaction in multiplayer games. In: Proceedings of the 13th International Conference on Advances in Computer Entertainment Technology, pp. 1–8 (2016)
32. Drachen, A., Canossa, A.: Towards gameplay analysis via gameplay metrics. In: Proceedings of the 13th MindTrek, pp. 202–209 (2009)
33. Xu, P., Ma, X., Qu, H., Li, Q., Wu, Z.: A multi-phased co-design of an interactive analytics system for MOBA game occurrences. In: Proceedings of the Designing Interactive Systems Conference, pp. 1321–1332 (2018)
34. Weber, B., Mateas, M.: A data mining approach to strategy prediction. In: Proceedings of IEEE Symposium on Computational Intelligence and Games, pp. 140–147 (2009)
35. Wang, X., Kim, H., Vasilakos, A.V., Kwon, T.T., Choi, Y., Choi, S., Jang, H.: Measurement and analysis of World of Warcraft in mobile WiMAX networks. In: Proceedings of the 8th Workshop on Network and System Support for Games, Paris, France (2009)
36. Moreira, Á.V.M., Filho, V.V., Ramalho, G.L.: Understanding mobile game success: a study of features related to acquisition, retention and monetization. SBC J. Interact Syst. **5**, 2–13 (2014)
37. Hennig-Thurau, T., Klee, A.: The impact of customer satisfaction and relationship quality on customer retention: a critical reassessment and model development. Psychol. Mark. **14**, 737–765 (1997)
38. Debeauvais, T., Schiano, D.J., Yee, N.: If you build it they might stay: retention mechanisms in world of warcraft. In: Foundations of Digital Games Conference, Bordeaux, France, pp. 180–187 (2011)
39. Demediuk, S., Murrin, A., Bulger, D., Tamassia, M.: Player retention in league of legends: a study using survival analysis. In: Proceedings of the Australasian Computer Science Week Multiconference, pp. 431–439 (2018)
40. Park, K., Cha, M., Kwak, H., Chen, K.T.: Achievement and friends: key factors of player retention vary across player levels in online multiplayer games. In: Proceedings of the 26th International Conference on World Wide Web Companion, pp. 445–453 (2017)
41. Andersen, E., Liu, Y.E., Snider, R., Szeto, R., Popović, Z.: Placing a value on aesthetics in online casual games. In: Proceedings of the Annual Conference on Human Factors in Computing Systems, pp. 1275–1278 (2011)
42. Krause, M., Mogalle, M., Pohl, H., Williams, J.: A playful game changer: fostering student retention in online education with social gamification. In: Proceedings of the Learning Scale Conference, pp. 95–102 (2015)
43. Kayes, I., Zuo, X., Wang, D., Chakareski, J.: To blog or not to blog: characterizing and predicting retention in community blogs. In: Proceedings of the 14 International Conference on Social Computing, pp. 1–7 (2014)
44. Donoho, D., Tanner, J.: Neighborliness of randomly projected simplices in high dimensions. PNAS **102**(27), 9452–9457 (2005)
45. Thawonmas, R., Lizuka, K.: Visualization of online-game players based on their action behaviors. Int. J. Comput. Games Technol. **5**, 1–9 (2008)
46. Wallner, G., Kriglstein, S., Gnadlinger, F., Heiml, M., Kranzer, J.: Game user telemetry in practice: a case study. In: Proceedings of the 11th Conference on Advances in Computer Entertainment Technology, pp. 1–4. ACM (2014)

47. Drachen, A., Pastor, M., Liu, A., Fontaine, D.J., Chang, Y., Runge, J., Sifa, R.: To be or not to be…social: incorporating simple social features in mobile game customer lifetime value predictions. In: ACSW 2018 Proceedings of the Australasian Computer Science Week Multi Conference, Article No. 40, pp. 1–6 (2018)
48. Alomari, K.M., Soomro, T.R., Shaalan, K.: Mobile gaming trends and revenue models: trends in applied knowledge-based systems and data science. IEA/AIE, pp. 671–683(2016)
49. Hsu, S.Y., Hsu, C.L., Jung, S.Y., Sun, C.T.: Indicator products for observing market conditions and game trends in MMOG. In: Proceedings of the 12th International Conference on the Foundations of Digital Games, pp. 21–30 (2017)
50. Krafft, M., Goetz, O., Mantrala, M., Sotgiu, F., Tillmanns, S.: The evolution of marketing channel research domains and methodologies: an integrative review and future directions. J. Retail. **91**(4), 569–585 (2015)
51. Cramer, H., Rost, M., Belloni, N., Bentley, F., Chincholle, D.: Research in the large using app stores, markets, and other wide distribution channels in Ubicomp research. In: Proceedings of the 12th ACM International Conference Adjunct Papers on Ubiquitous Computing - Adjunct, pp. 511–514 (2010)
52. Latif, R.M.A., Abdullah, M.T., Shah, S.U.A., Farhan, M., ljaz, F., Karim, A.: Data scraping from Google play store and visualization of its content for analytics. In: Proceedings of the 2nd International Conference on Computing, Mathematics and Engineering Technologies, iCoMET, pp. 1–8 (2019)

PCA Based Energy Network Temporal and Spatial Data Analysis and Prediction

Yifan Yin[1], Yan Sun[1], Han Yu[1], Zhuming Bi[2], Boyi Xu[1(✉)], and Hongming Cai[1]

[1] Shanghai Jiao Tong University, Shanghai, China
{cn_yyf, sun_yan, sharonhanz, byxu, hmcai}@sjtu.edu.cn
[2] Purdue University Fort Wayne, Fort Wayne, USA
biz@pfw.edu

Abstract. Nowadays, intelligent energy networks are rising with the development of Internet of Things. Relying on massive quantity of sensors, digital twin technology makes it possible to monitor the condition of an energy network in real time. However, for decision makers, future condition is more valuable. Common types of energy like cooling, heating and electrical power, are difficult to be stored. So, the production and consumption of energy shall be in a balance. The prediction of energy consumption helps to reduce cost and waste. Furthermore, the prediction equipment condition can support scheduling and maintenance. As a result, accurate and efficient condition prediction is strongly required in novel energy network. To solve the problem, a principle components analysis (PCA) based temporal and spatial view analysis and prediction (PTSVP) model is proposed in this paper. Because of the considerable amount of data in energy network, data analysis is difficult and inefficient. Dimension reduction can help to reduce data model complexity and lead to efficient prediction. Thanks to the linear dependency among network record, we use feature exaction technology to reduce dimension of energy network data. Besides, traditional prediction model tends to ignore the spatial relations among data. In this paper, both temporal and spatial factors are considered to make this prediction model more accurate and explainable. We apply this model to a practical energy network in Hongqiao, Shanghai, and compare it with traditional statistical learning and machine learning models. The result shows that our model is accurate, stable and efficient.

Keywords: Internet of things · Energy network · Temporal and spatial modeling · Dimension reduction

1 Introduction

In recent years, more and more multiple energy networks have been constructed for urban energy supply. An energy network is a relatively independent system to solve energy supply in certain area [1]. The size of a network ranges from several blocks to the whole urban area. The process of energy supply is quite complex, including production, transportation, consumption and sometimes short-term storage of energy resources. Nowadays, various types of energy are required for a city to function.

© Springer Nature Switzerland AG 2020
K.-M. Chao et al. (Eds.): ICEBE 2019, LNDECT 41, pp. 590–605, 2020.
https://doi.org/10.1007/978-3-030-34986-8_41

In industrial and commercial activities and citizen's daily life, cooling, heating and electric power are all necessary energy types.

Along with the rapid development of Internet of Things (IOT), IOT technologies are applied in various scenarios. The applications of IOT technologies lead to revolutionary growth of multiple energy networks. With the use of digital twins technology, the fusion of energy flow and information flow is able to be realized. The conditions of an energy network can be detected by sensors, and the records from different sensors constitute the data model with different attributes. This fusion is the basic of real-time energy network conditions monitors and big data analysis. Further applications are based on the massive quantities of sensors and the considerable amount of data provided by the sensors.

However, because of the massive scale of energy network data, the dimension of the data model is usually high, so data analysis can be time-consuming. If the data dimension could be reduced, further data analysis will be more efficient. In energy network data model, there are linear relations among dimensions. That is to say, some records may be calculated from others. This phenomenon makes it possible to present the origin high dimension data with much lower dimension features. Principal components analysis (PCA) is an effective method for feature extraction [2].

In an energy network, adequate production is an important standard for qualified service. Meanwhile, inadequate consumption leads to energy waste. Although in design of some energy networks, the storage process has already been taken into consideration, the cost is still quite high. Most types of energy like cooling, heating and electric power cannot be stored efficiently. As a consequence, the balance of supplies and demands shall be one of the primary problems of energy networks. For electric power supply, the production planning is fairly real-time. Even in case of cooling and heating delivered through water pipe, the period of energy transportation will be less than one hour generally. Therefore, prediction of network conditions, especially short-term prediction, is vitally helpful for production planning.

In general, temporal series analysis is an effective method to predict the data variations [3]. There are many machine leaning or statistical learning approaches for temporal series regression. Regression algorithms are a sort of statistical learning approaches. This kind of approaches predict the data based on the periodicity of time series. Auto Regressive Integrated Moving Average (ARIMA) is a typical regression algorithm [4]. There are also many reasonable machine learning methods. Deep Neural Network (DNN) is a classical solution for regression problem. However, for temporal series, Recurrent Neural Network (RNN) will be more appropriate. Long-Short Term Memory (LSTM) successfully get rid of the deficiency of RNN [5].

All the approaches above focus on temporal factors. However, temporal methods are unilateral. Spatial factors also have impact on prediction result. For a massive energy network, many of the attributes are correlative and interdependent. Support Vector Regression (SVR) is a practicable method to explore the relation among attributes [6]. The temporal prediction result can be verified and corrected in spatial view.

The operation conditions of an energy network are influenced by many external factors as well. For accurate prediction, the external factors such as weather, climate and society should be considered while modeling. The Internet is a reliable data source for historical and real-time external data. By introducing Internet plus technology, the result will be more meaningful and accurate.

To satisfy the demands of intelligent energy network prediction, we proposed PCA based temporal and spatial view analysis and prediction (PTSVP) model. Our method is applied in a practical energy network in Hongqiao, Shanghai. To evaluate the performance of our method, PTSVP is compared with traditional machine learning and statistical learning method like LSTM and ARIMA. the result proves that PTSVP is accurate, stable and efficient.

The rest parts of this paper are structured as follows. Section 2 introduces related works of this study. In Sect. 3, the framework of PTSVP is illustrated briefly. And in Sect. 4, the methodology of our method will be introduced in detail. Section 5 and Sect. 6 present an application of our method in a practical scenario and the evaluation based on this application. Finally, the paper and our contribution will be summarized in Sect. 7.

2   Related Works

Energy IOT constructions are widely developed in recent years. Energy prediction is the research hotspot of energy IOT studies nowadays. Energy storage is still a major challenge in practical production and daily life. The construction of energy storage system is very expensive and the performance is still not satisfactory for many types of energy resources [7]. As a result, the storage of energy is never a cost-efficient approach. In order to reduce waste, the energy supply and demand balance is meaningful in practical terms. Data driven prediction is a proved approach towards energy conditions prediction [3]. There are many researches about data driven prediction of energy networks, and most of the works can be divided into statistical learning and machine learning methods.

Most statistical learning methods can effectively take advantage of statistic features of data. Gaussian process regression is an archetypal statistical learning method. This method is used in short-term solar power prediction and proved to be practical [8]. Besides, Bayesian based regression is another significant approach. A variation Bayesian based regression model is used for wind power forecasting [9]. Furthermore, ARIMA is also one of the most typical statistical learning methods for data driven prediction. An ARIMA based method is put into applications for hybrid power vehicle system [4]. In that research, the vehicle running conditions and road situations are predicted. Form the studies above, the availability of statistical learning in energy prediction is proved.

Neural network is a representative and effective machine learning method for energy prediction. Artificial neural network is designed based on the process of thinking in the human brain, and it is applied to energy forecasting in different scenarios [10]. Intelligent algorithm is also introduced into neural network. A method combining conventional neural network and particle swarm optimization is proposed to solve energy demand prediction problem [11]. Among the advanced neural network approaches, RNN shows potential to process data of time series. In RNN, the historical data in temporal series support the prediction result. But in practical applications, RNN always leads to gradients vanishing or exploding. So RNN cannot process the long-term factors well as it should be in theory. LSTM is an improved RNN model that has better capacity for long series based prediction [5].

Dimension reduction is another important part of work in this paper, and related techniques are of significance in big data process [12]. Many mainstream dimension reduction methods are surveyed, classified and evaluated in [12]. The evaluative criteria are inspiring to this paper. Feature extraction is the most vital measure of dimension reduction. Such techniques are widely used in image processing. The spatial features are also considered in image data dimension reduction, and a deep learning based approach is proposed to extract features from massive images [13]. Among various approaches to dimension reduction, PCA is proved effective, especially in extract linear features [2].

Over the years, temporal and spatial modeling is a newly research field. One of the application fields of this technology is missing value imputation. Reference [14] has proposed a multiple view learning framework, combining recurrent neural network technology, support vector regression and collaborative method, to recover missing data in the monitoring system of Interstate Highway 80 in California. The framework considers both temporal view and spatial view and catch both local variation and global variation. Besides the domain of traffic and transportation, there are many other scenarios for temporal and spatial modeling. Reference [15] proposes a kernel-based learning method to realize spatio-temporal imputation of infectious diseases. By combining multiple risk factors, the trend of disease spreading can be predicted based on incomplete data, which helps government to make decisions in disease control. Image process is another important scenario of temporal and spatial modeling. This idea is introduced to image inpainting and image style transforming [16]. From those related studies, temporal and spatial approaches is proved to be useful in data imputation, prediction and decision supporting. It is helpful to involve both temporal and spatial factors in prediction models to make the result accurate and explainable.

3   Framework

PCA based temporal and spatial view analysis and prediction (PTSVP) model is proposed in the paper. The architecture of our model is illustrated in Fig. 1. The framework consists of three main procedures. These are dimension reduction, temporal series analysis, and spatial view analysis.

Fusion data from the energy network IOT system are the foundation of our prediction process. As shown in Fig. 2, the data are formed as a matrix $X = [x_1, x_2, \ldots, x_m] \in R_{n \times m}$, consists of m temporal series of data. The series represent attributes in data model and describe the conditions recorded by sensors. Each temporal series $x_i = \left[x_i^1, x_i^2, \ldots, x_i^n \right]^T \in R_n$ denotes a series of data record of one certain attribute at n intervals. The data of different attributes should be aligned to the same time intervals series. The time intervals should be continuous and the space between

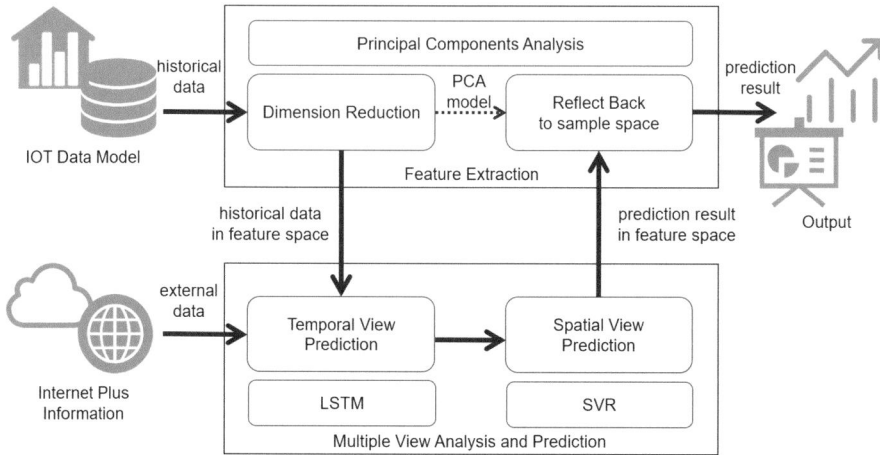

Fig. 1. PTSVP framework

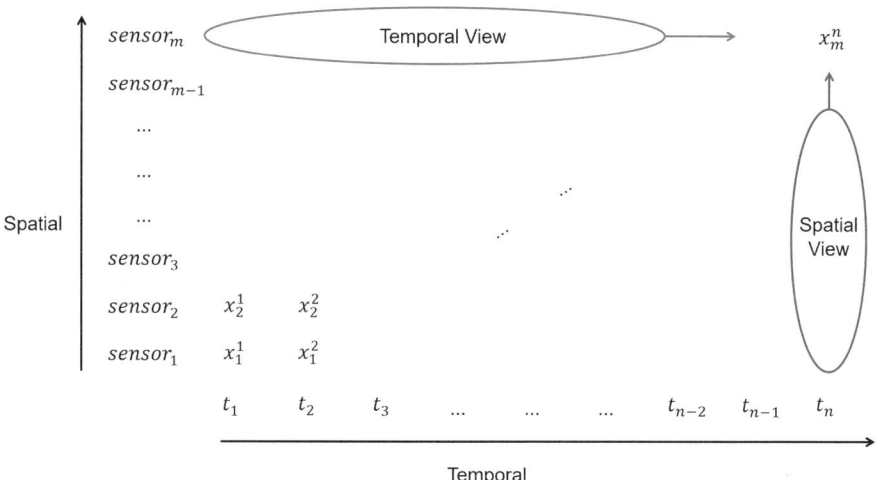

Fig. 2. Temporal and spatial prediction model

contiguous intervals should be fixed. The data matrix X contains historical records of m attributes at n intervals, and x_i^j means the value of the ith attribute at the jth time interval.

As is mentioned above, the data dimension tends to be high, making prediction process inefficient. So, the first procedure is dimension reduction. In the paper, PCA is applied to reduce the data dimension. All the historical data tuples are transformed from origin data space to a lower dimension feature space. The following prediction procedures are based on the dimension reduced data. The Next procedure is to predict tuples at the following intervals through temporal series method like LSTM. In this procedure, Internet plus data like weather information are considered to improve the accuracy of prediction result. Spatial view prediction is the third procedure of the prediction process. Spatial relations among attributes are analyzed based on SVR. The SVR model is constructed and applied to the result tuples' verification and correction. The prediction result in feature space is inversely transformed back to the origin high dimension data space as output of the proposed process.

4 Methodology

4.1 Dimension Reduction

Energy network data models can be considerably high dimensional. For the purpose of efficient data analysis, PCA are applied to dimension reduction of the data space. PCA is a sort of practical dimension reduce method. Its powerful capacity to detect the linear features has been widely proved. The PCA procedure are realized in several steps. These are preprocessing, principle components selecting and dimension transforming.

In this paper, a sample can be represented as $x \in R^m$, where m is the origin dimension of samples. Preprocessing step includes nondimensionalization and centralization. The attributes represent various physical conditions of different types and measurement units. For example, the unit of pipeline water flow is gallons per minute while that of motor power is kilowatt hour, which hinders comparison between these attributes. Nondimensionalization helps to unify the value of different dimensions. Then, for the flowing steps, the average value of each attribute in the samples should be zero, which is processed in centralization. Preprocessing is realized as (1) and (2), x' denotes the nondimensionalized sample and x'' means the centralized sample.

$$x' = \begin{bmatrix} x'_1 \\ x'_2 \\ \dots \\ x'_m \end{bmatrix} = \begin{bmatrix} x_1/a_1 \\ x_2/a_2 \\ \dots \\ x_m/a_m \end{bmatrix} = xA \tag{1}$$

$$x'' = x' - \overline{x'}, \ \overline{x'} = \frac{1}{m} \sum_{i=1}^{m} x' \tag{2}$$

Eigen is calculated based on sample covariance and singular value decomposition (SVD). Eigenvalues reflect the impact of features. Sort the eigenvalues in descending order, and select the tops that occupy most of the impact of all the eigenvalues. In general, it is good enough if more than 95% impact is selected.

4.2 Temporal Series Analysis

To assess the temporal changes of data series, LSTM is used in our process. LSTM is a specific kind of recurrent neural network. It has been proved effective in fields like natural language processing and automatic speech recognition [5]. Because of gradient exploding and vanishing, traditional RNN is not good at processing long series. LSTM is designed to solve this problem.

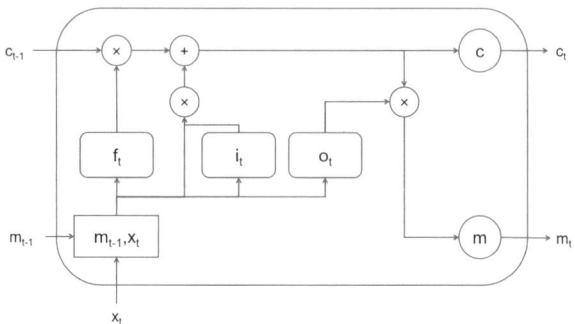

Fig. 3. LSTM architecture of recurrent layer

The architecture of LSTM is shown in Fig. 3, which is quite complex. In LSTM, cell state is introduced to store the long-term information. LSTM use three kinds of control switches to control the data flow between memory and cell states. The switches are named as input gate, forget gate, and output gate, which are represented by i_t, f_t and o_t respectively. The state of three gates are decided by current input x_t and the last memory state m_{t-1}.

In order to get accurate and meaningful prediction result, external data are considered in this procedure. We use Internet plus technology to support external input data like weather information. For the LSTM framework, the input are periods of dimension reduced historical data combined with the external data of the interval. And the output of the model are the corresponding tuples at the following intervals in the low dimension feature space.

4.3   Spatial View Analysis

The spatial factors should also be considered for a comprehensive prediction process. In an energy network, the sensors are placed in different locations. It is easy to see, there is spatial relevance among the attributes of the digital model. Regard the data as a series of time slices where each slice is a tuple of all the attributes at an interval. The spatial view analysis aims to discover the rules of tuple structure. The rules can be used to verify the validity of predicted tuples, so tuples can be corrected, which leads to appropriate prediction result.

As shown in Fig. 4, many of the covariances absolute values between data of different sensors is close to 1, which means the correlativity between those data. Subfigure (a) indicates the covariances of chilled water return temperature of chillers in Station 1. Most of the chillers have similar running schedule, so the covariances tend to be near 1 and the data are positively corelated. Meanwhile, Subfigure (b) indicates the covariances of several streams of data in different stations, and both positive correlation and negative correlation have been displayed. The meaning of each stream of sensor data is listed in Table 1. The results indicate the significance of spatial relations among data.

Table 1. Experimental statistics of PTSVP and baselines

Data name	Meaning
1LDSHS_T	Chilled water return temperature of Chiller 1 in Station 1
2LDSHS_T	Chilled water return temperature of Chiller 2 in Station 1
3LDSHS_T	Chilled water return temperature of Chiller 3 in Station 1
4LDSHS_T	Chilled water return temperature of Chiller 4 in Station 1
5LDSHS_T	Chilled water return temperature of Chiller 5 in Station 1
6LDSHS_T	Chilled water return temperature of Chiller 6 in Station 1
7LDSHS_T	Chilled water return temperature of Chiller 7 in Station 1
8LDSHS_T	Chilled water return temperature of Chiller 8 in Station 1
14BOILER_GAS_TOTAL	Gas consumption of Boiler 1 and 4 in Station 1
23BOILER_GAS_TOTAL	Gas consumption of Boiler 2 and 3 in Station 1
JZ1_Starttimes	Start-up time of Chiller 1 in Station 1
SLG_JZ3_BATTERY_IP	Battery voltage of Generator 3 in Station 1
S_5LDSHS_T	Chilled water return temperature of Chiller 5 in Station 2
S_7LDSCS_T	Chilled water support temperature of Chiller 7 in Station 2
S_SLG_JZ1_BATTERY_IP	Battery voltage of Generator 1 in Station 2
S_TT1012_T	Chilled water support temperature of main pipe in Station 2

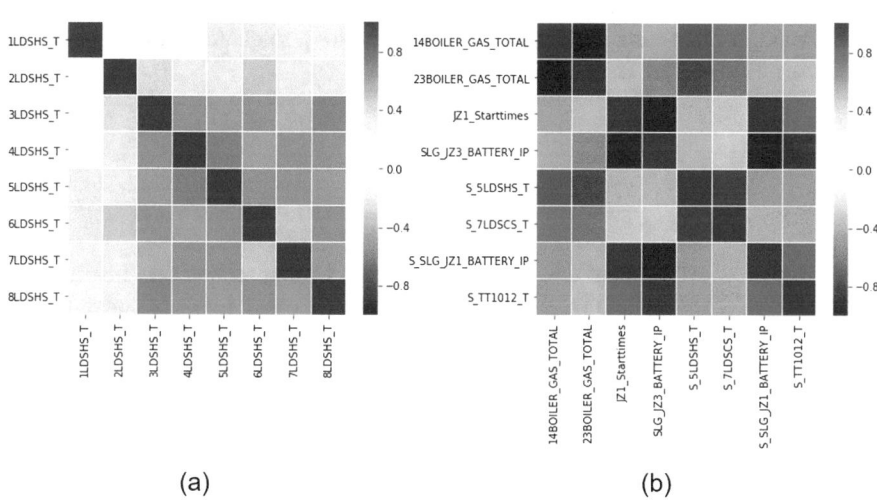

(a)   (b)

Fig. 4. Covariances between data of sensors

SVR is a helpful method to discover the spatial relevance. SVR is the regression version of support vector machine (SVM). In SVR, regression plane is represented as $y = f(x) = \omega x + b$, which is determined by the parameter ω and b. The most cost-efficient plane can be found by solving the following minimization problem.

$$min \frac{1}{2} \|\omega\|^2 + C \sum_{i=1}^{n} \left(\xi_i + \xi_i^* \right) \tag{3}$$

The cost or risk of a given regression plane can be calculated according to (3). The left part $\frac{1}{2}\|\omega\|^2$ represents structural risk, which reflects the complexity of the model. While the right part $C \sum_{i=1}^{n} \left(\xi_i + \xi_i^* \right)$ represents empirical risk, which reflects the distance between training points and the regression plane. Here ξ_i and ξ_i^* denotes slack variables.

Kernel function $K(x, y)$ is applied in order to reduce SVR computation. Feasible kernel functions are always specifically designed. Among them, radial basis function (RBF) has been proved simple and efficient, so we apply RBF as the kernel function to our process. Here $\gamma = -\frac{1}{2\sigma^2}$, and σ is a parameter to control the function.

$$K(x, y) = exp\left(\gamma \|x - y\|^2 \right) \tag{4}$$

The historical tuples are used to construct the SVR model. Substituting the prediction result into SVR model, if the predicted tuple is in the margin of regression plane, the prediction shall be proved valid. If the initial result is proved not valid enough, the spatial regression result will be calculated through SVR model for every attribute. The tuple will move gently towards the regression result. Until the tuple is proved valid or after certain times of iterations, the loop shall stop and the prediction result shall be corrected in spatial view.

5   Case Study

5.1   Data Description

In this study, the experiment data are obtained from Hongqiao business district intelligent energy system, which is a district scale multiple energy network constructed in Hongqiao business district, Shanghai, China. This network builds connections among energy companies, energy users and other entities on the energy supply chain. Involved energy types are diverse. Main sorts of energy resources supplied on the network are cooling and heating. And there is also electric power exchange between this network and power grid. Two power stations have been successively constructed. The energy customers cover various units in the business district such as hotels, shopping malls, office buildings and residential housings. Actually, the scale of this network is still expanding now. Each power station contains a combined cooling, heating and power (CCHP) system. There are several boilers, refrigerators and tri-generators supporting the energy production. Boilers consume natural gas and supply heating. Refrigerators consume electric power and supply cooling. Tri-generators consume gas and supply all three kind of energy resources. The electric power is used internally, while cooling and heating are exported. Both cooling and heating are stored in water and transported through pipelines. The pipeline network is a complex system, but it is well ordered. On the network, energy can be transported to no matter which position from power station in less than one hour. Through heat exchange equipment of user side, customers can enjoy steady, adequate, flexible and inexpensive cooling and heating energy service. To sum up, Hongqiao business district intelligent energy system is a well-constructed energy production, transportation, storage and consumption network. To monitor the conditions of the network, more than thousands of sensors are placed in production equipment, pipelines and consumption terminals. Sensors record conditions for every few milliseconds, so massive amount of historical data are stored in the intelligent system. The operation architecture of this energy network is shown in Fig. 5.

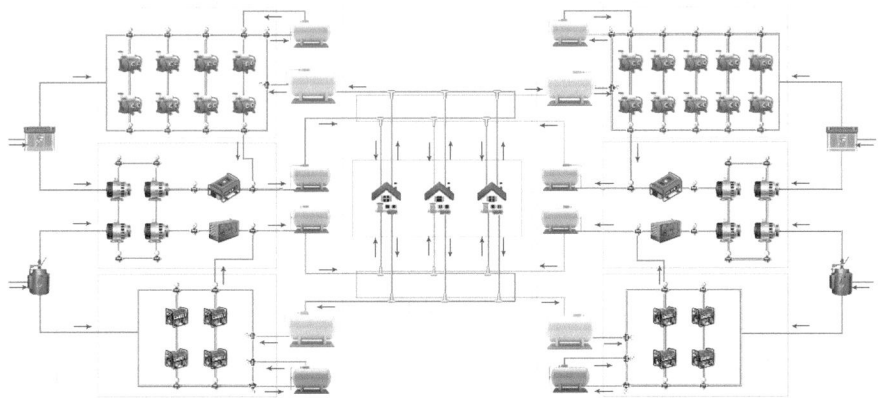

Fig. 5. Scenario: Hongqiao energy network

The energy network data used in this study are extracted from the intelligent system data storage. A valid group of data sections are provided. The dataset contains records of 302 sensors of five months in 2018. The period between neighboring records is 5 min, so there are 288 intervals in each day.

The external data used in this study are obtained from the Internet. The external weather data contain various dimensions of attributes, including temperature, humidity, air pressure, wind direction, wind speed, rainfall and snowfall probability. The weather information is updated every day. External social data include daily information like whether the date is a workday. Historical external data are obtained from several open dataset, and the real-time external data can be got from online API service like Moji Weather of aliyun cloud.

The origin dataset is divided into two parts. 80% of data are used as training dataset while the other 20% are used as testing dataset. The model is built based on the training dataset, and the quality and performance of our method is evaluated based on the testing dataset.

5.2 Baseline

PTSVP is be compared with ARIMA, LSTM and two variants of PTSVP.

PTSVP-T: This model uses the PCA based temporal view analysis and prediction model, but the Internet plus information is not employed.

PTSVP-TI: This model uses the PCA based temporal view analysis and prediction model with Internet plus information, but the spatial view is not considered.

5.3 Key Indicators

Suppose the size of testing dataset is s. For ith value, the ground truth is y_i, and its predicted value is y_i'. The performance of this regression problem is evaluated in the following key indicators including accuracy, stability and efficiency.

Accuracy: The accuracy is assessed by mean average percentage error (MAPE) and mean squared error (MSE). MAPE denotes the error percentage of the prediction based on ground truth, ranging from 0 to 100%. MSE is also named as 12-norm loss. Lower MAPE and MSE equate to better performance of accuracy.

$$MAPE = \frac{1}{s}\sum_{i=1}^{s}\left|\left(y_i - y_i'\right)/y_i\right| \times 100\% \tag{5}$$

$$MSE = \frac{1}{s}\sum_{i=1}^{s}\left|\left(y_i - y_i'\right)^2\right| \tag{6}$$

Stability: Determination coefficient and correlation coefficient are used to assess model stability. Coefficient of determination is denoted by R^2. It shows the dependent variable error explainable by independent variable error. The model is stable if R^2 is close to zero. Correlation coefficient between ground truth and prediction result can be represented by r. It is calculated by covariance and standard deviation, ranging from -1 to 1. Correlation coefficient of two series reflect the similarity between them in linear view. Higher correlation coefficient reflects better stability of the model.

$$R^2 = 1 - \frac{\sum_{i=1}^{s}\left(y_i' - \bar{y}\right)^2}{\sum_{i=1}^{s}(y_i - \bar{y})^2} \tag{7}$$

$$r = \frac{cov(y, y')}{\sqrt{var(y)var(y')}} \tag{8}$$

Efficiency: The efficiency of models is mainly evaluated based on the time efficiency of training process. All the methods are realized in the same computing environment (12 CPU processors, 1 GPU, 16 GB memory). The method with shorter training and processing time is more efficient.

6   Evaluation

6.1   Feature Space Determination

The dimension of feature space should be enough to reflect most of information in sample space. Capacity of a feature can be evaluated by its eigenvalue in PCA. In general, to ensure the quality of dimension reduction, eigenvalues summary proportion should be no less than 95%.

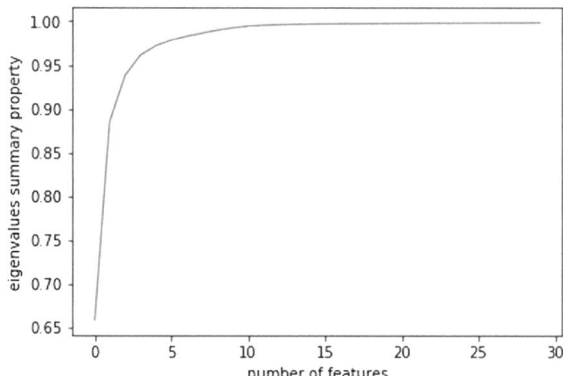

Fig. 6. Relation between eigenvalue summary and features number

As shown in Fig. 6, in this experiment, the first 14 features hold 99% of the eigenvalue summary of all 302 sample dimensions, and the first 29 features holds 99.9%. The eigenvalue percentage is quite high, which illustrates that most of dimensions in sample space is linear dependent. This result confirmed our opinion that the massive energy network data are correlated among dimensions and it can be simplified by dimension reduction. 99% eigenvalue summary is already high enough to realize effective dimension reduction. In order to analyze the relation between model performance and feature space dimension, we choose both 14 and 29 as the dimension of feature space for following experiments.

6.2   Performance

We applied our PCA based temporal and spatial view analysis and prediction model on the dataset. Compared with all the baselines introduced above, the accuracy, stability and efficiency of our method are shown in Table 2.

MAPE and MSE can represent the accuracy of a model. The accuracy of our method is close to that of LSTM and ARIMA. And in most indicators, our method is better. But if only temporal view is considered, the model will be much more inaccurate. Higher feature dimension leads to better performance in accuracy.

Determination coefficient and correlation coefficient can describe the stability of prediction result. ARIMA is the most stable model in our experiment, and then is

LSTM. If the real-time information from the Internet is not used, the model will be lack of stability. However, with the use of external data, the stability of our method can be similar to LSTM and ARIMA. As well as in accuracy, model with higher feature dimension is more stable.

Table 2. Experimental statistics of PTSVP and baselines

Method	MAPE	MSE (×106)	R2	r	Run time (s)
ARIMA	0.126	2.595	−1.759	0.889	416.050
LSTM	0.125	1.496	−2.451	0.855	328.930
PTSVP-T(14)	0.172	1.864	−7.986	0.040	19.957
PTSVP-TI(14)	0.169	1.842	−4.833	0.382	18.566
PTSVP(14)	0.126	1.426	−4.297	0.381	20.480
PTSVP-T(29)	0.171	1.867	−4.740	0.075	32.222
PTSVP-TI(29)	0.167	1.770	−3.215	0.810	32.194
PTSVP(29)	0.122	1.467	−3.008	0.810	33.941

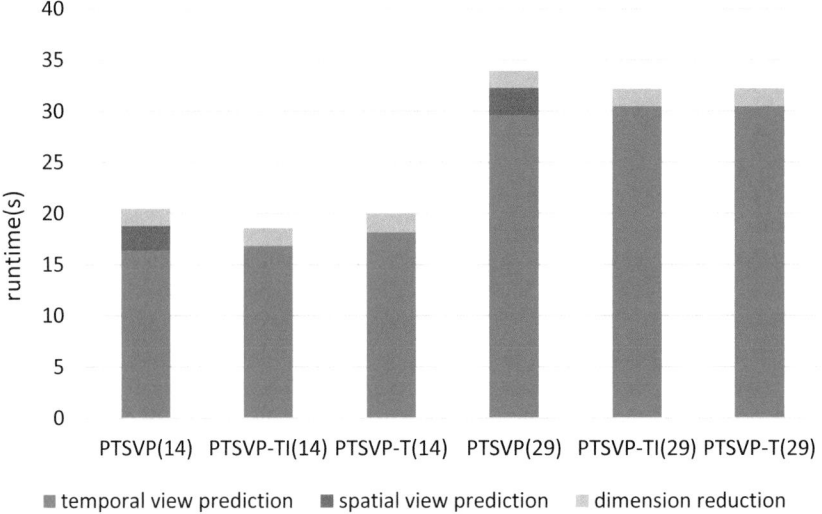

Fig. 7. Model training runtime of PTSVP

The efficiency is one of the prime advantages of our method. The time used to train our model is less than one tenth of both LSTM and ARIMA. From Fig. 7, we can find that temporal view prediction is the performance bottleneck of model training, so the efficiency of our model can be improved by optimize this part. Actually, temporal view prediction efficiency and feature dimension are positively correlated. And effective dimension reduction leads to time efficient model.

6.3   Analysis

Performance of the methods above can be summarized as Table 3. Dimension reduction significantly reduces the time to train temporal model, which is the most time-consuming part of our method. If only temporal view is considered, the prediction result will be inaccurate and unstable. The use of Internet plus data can help to enhance the stability of our method. And when spatial view is also considered, the accuracy of our model will be effectively improved.

Table 3.   Performance analysis of PTSVP and baselines

Method	Accuracy	Stability	Efficiency
ARIMA	+	+	−
LSTM	+	+	−
PTSVP-T	−	−	+
PTSVP-TI	−	+	+
PTSVP	+	+	+

In summary, our proposed method is much better than ARIMA and LSTM in efficiency, and the accuracy and stability of our method is similar to that of those traditional methods. If we increase the feature dimension, the model will be more accurate and stable. And if we decrease the feature dimension, the model will be more efficient.

7   Conclusion

Aiming to construct efficient and intelligent energy network, we focus on the analysis and forecast of massive amount of data in energy network IOT system. To increase the efficiency, accuracy and stability of data analysis and forecast, we have proposed a novel analysis and prediction model which considering both temporal and spatial view and combined with Internet plus data support. Our model is realized and applied on a practical energy network, and evaluated with several different forecasting models. The main contribution of our work is summarized as follows:

Firstly, we have proposed a PCA based temporal and spatial view analysis and prediction model. In our proposed method, we use PCA to realize dimension reduction, and significantly improve the efficiency of the prediction model. Both temporal and spatial view are considered in our model, which makes the prediction result more accurate. And we also have applied Internet plus technology to our model, using external information like real-time weather and holidays scheduling to support prediction, which makes the model more stable.

Secondly, our method is applied in an actual running intelligent energy network in Hongqiao, Shanghai. In order to address the practical demands of this energy network, we have built a IOT platform which supports intelligent data monitoring and analysis.

Furthermore, we have performed experiments to compare our method with traditional regression forecasting methods, such as LSTM and ARIMA. The experiment result is inspiring. Our method shows outstanding efficiency and satisfactory accuracy and stability.

In the future, our research will aim at much higher dimension samples of a bigger-scaled energy network or a coupled smart energy system with several energy networks. The capacity of our method to deal with greater orders of magnitude of data need to be evaluated. Choice of appropriate dimension reduction methods is another valuable research point in future.

Acknowledgement. This research is supported by the National Nature Science Foundation of China under Grant No. 61972243.

References

1. Yusheng, X.U.E.: Energy internet or comprehensive energy network? J. Mod. Power Syst. Clean Energy **3**(3), 297–301 (2015)
2. Chakravarthy, S.K., Sudhakar, N., Reddy, E.S., Subramanian, D.V., Shankar, P.: Dimension reduction and storage optimization techniques for distributed and big data cluster environment. In: Soft Computing and Medical Bioinformatics, pp. 47–54. Springer, Singapore (2019)
3. Amasyali, K., El-Gohary, N.M.: A review of data-driven building energy consumption prediction studies. Renew. Sustain. Energy Rev. **81**, 1192–1205 (2018)
4. Guo, J., He, H., Sun, C.: ARIMA-based road gradient and vehicle velocity prediction for hybrid electric vehicle energy management. IEEE Trans. Veh. Technol. (2019)
5. Greff, K., Srivastava, R.K., Koutník, J., Steunebrink, B.R., Schmidhuber, J.: LSTM: a search space odyssey. IEEE trans. Neural netw. Learn. Syst. **28**(10), 2222–2232 (2016)
6. Vrablecová, P., Ezzeddine, A.B., Rozinajová, V., Šárik, S., Sangaiah, A.K.: Smart grid load forecasting using online support vector regression. Comput. Electr. Eng. **65**, 102–117 (2018)
7. Parra, D., Swierczynski, M., Stroe, D.I., Norman, S.A., Abdon, A., Worlitschek, J., Bauer, C.: An interdisciplinary review of energy storage for communities: challenges and perspectives. Renew. Sustain. Energy Rev. **79**, 730–749 (2017)
8. Sheng, H., Xiao, J., Cheng, Y., Ni, Q., Wang, S.: Short-term solar power forecasting based on weighted Gaussian process regression. IEEE Trans. Industr. Electron. **65**(1), 300–308 (2017)
9. Wang, Y., Hu, Q., Meng, D., Zhu, P.: Deterministic and probabilistic wind power forecasting using a variational Bayesian-based adaptive robust multi-kernel regression model. Appl. Energy **208**, 1097–1112 (2017)
10. Sun, Y., Xu, L., Li, L., Xu, B., Yin, C., Cai, H.: Deep learning based image cognition platform for IoT applications. In: 2018 IEEE 15th International Conference on e-Business Engineering (ICEBE), Xi'an, pp. 9–16. IEEE (2018)
11. Muralitharan, K., Sakthivel, R., Vishnuvarthan, R.: Neural network based optimization approach for energy demand prediction in smart grid. Neurocomputing **273**, 199–208 (2018)
12. Meng, C., Zeleznik, O.A., Thallinger, G.G., Kuster, B., Gholami, A.M., Culhane, A.C.: Dimension reduction techniques for the integrative analysis of multi-omics data. Brief. Bioinform. **17**(4), 628–641 (2016)

13. Zhao, W., Du, S.: Spectral–spatial feature extraction for hyperspectral image classification: a dimension reduction and deep learning approach. IEEE Trans. Geosci. Remote Sens. **54**(8), 4544–4554 (2016)
14. Li, L., Zhang, J., Wang, Y., Ran, B.: Missing value imputation for traffic-related time series data based on a multi-view learning method. IEEE Trans. Intell. Transp. Syst. (2018)
15. Tan, Q., Liu, J., Shi, B., Liu, Y., Zhou, X.N.: Public health surveillance with incomplete data–spatio-temporal imputation for inferring infectious disease dynamics. In: 2018 IEEE International Conference on Healthcare Informatics (ICHI), New York, pp. 255–264. IEEE (2018)
16. Gatys, L.A., Ecker, A.S., Bethge, M.: Image style transfer using convolutional neural networks. In: Proceedings of the IEEE Conference on Computer Vision and Pattern Recognition, pp. 2414–2423. IEEE (2016)

A Behavior-Item Based Hybrid Intention-Aware Frame for Sequence Recommendation

Yan Chen, Jiangwei Zeng, Haiping Zhu$^{(\boxtimes)}$, Feng Tian, Yu Liu, Qidong Liu, and Qinghua Zheng

School of Electronic and Information Engineering, Xi'an Jiaotong University, Xi'an 710049, Shaanxi, China
chenyan@mail.xjtu.edu.cn, zxcvbn719270348@163.com, zhuhaiping@mail.xjtu.edu.cn, fengtian@mail.xjtu.edu.cn, 1347740318@qq.com, dong_liuqi@163.com, qhzheng@mail.xjtu.edu.cn

Abstract. Sequence recommendation is one of the hotspots of recommendation algorithm research. Most of the existing sequence recommendation methods focus on how to use the items' attributes to characterize the user's preferences, ignoring that the user behavior also can reflect the preference for items. However, user behavior often has problems of mis-interaction and random interaction, which leads to fully utilizing it difficultly. Therefore, this paper proposes a new Behavior-Item based Hybrid Intent-aware Framework (BIHIF). In this framework, the user's main intent is extracted based on user behaviors and interactive items, respectively, the two intent vectors are combined and extracted by the full connection layer to obtain the user's real intent. We use real intent and item vector to calculate the score of the candidate items and make Top-K recommendations. Based on the framework, we implement models respectively by MLP and GRU, which show good results in the experiments based on three real-world datasets.

Keywords: Sequence recommendation · Hybrid Intention-aware · User behavior · Attention mechanism

1 Introduction

The recommendation system, which can capture user's intention, explore user's focus of attention and recommend commodity or service to users, is an important component of e-business website [1]. In daily life, there is an obvious character of user's intention–variety. Users will have various interests on different commodity, or even on the same commodity under different environment at different time. Therefore, how to capture user's intention and how to accurately model user's intention become one of the most essential problems in recommendation system [2].

However, how to design the representation vector which can be characterized as user's intention is a real challenge. Traditional matrix decomposition methods usually map the users and items to low dimension fixed vectors in user-item combined space. But fixed user vectors largely limit the accuracy of modeling user's different interests

© Springer Nature Switzerland AG 2020
K.-M. Chao et al. (Eds.): ICEBE 2019, LNDECT 41, pp. 606–620, 2020.
https://doi.org/10.1007/978-3-030-34986-8_42

[2]. Similar items should have short distance in space. Besides, the distance between a user and an item represent the intensity of the user's interests of the item [3]. As shown in Fig. 1, user u_1's intent should be browsing or buying electronic products according to his or her click at commodity $\{i_1, i_2\}$ in an e-business website at some time, then the system should recommend i_3 to the user. But because the user representation vector is fixed, the system tends to recommend $\{i_4, i_5\}$ whose vector is nearer to user vector under this circumstances. This example vividly indicates that using low dimension fixed vector to represent user's intention may cause error result of recommendation. We think the reason why don't use fixed vector is that the user's intention is various and may change with time. So we should represent user's intention dynamically instead of fixed vector.

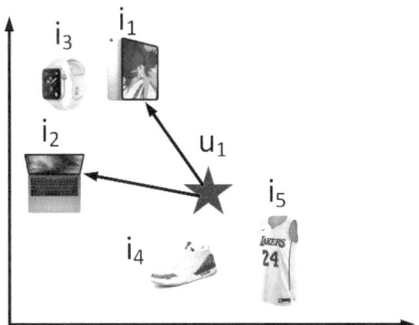

Fig. 1. Fixed vector to represent user's intention may cause error result.

Sequence recommendation is just to solve this problem, which extract user's intention from the current user behavior sequence (session) to recommend a correct item to user at next moment [4]. Recently, Recurrent Neural Network (RNN) is used to solve the recommendation problem based on session [5]. Compared to previous methods, RNN not only can make full use of all of click information in session, but also can model the click sequence. Therefore, it can promote performance remarkably. But in real life, the condition may be more complex. When users browse commodities, they are with some intent, so they would browse the commodities connected with their intention emphatically. Yet it may produce some irrelevant interaction records due to some mis-operations and some random interactions in a session. These irrelevant records can be seen as noise item in session data. For instance, a user wants a upper outer garment which means that his or her main intent is upper outer garment in this session. But in this session, he or she click a popular or discounted trousers (the user's intent is not trousers, only for curiosity). Such a noise item from random interaction misleads recommendation results easily. Therefore, how to extract user's real intention becomes an important factor of promoting recommendation performance.

Another problem is how to give a credible recommendation result from the session which includes noisy items. Although current research use attention, pooling etc. [6] methods to extract user's real intent, but most of research concentrate on how to use properties of items to represent user's preference, ignoring that user's behavior also can reflect preference for item. When we refer to user's shopping with intention, we often pay attention to the items related to user's intention in the session. When user pay attention to an item, he or she not only browse and click it, but also query it, put it into shopping cart or even buy it; while if user pay no attention to an item, user often just browse it. Therefore, extracting user's real intent according to user's different behaviors will be an effective approach. Due to this idea, this paper proposes a Behavior-Item based Hybrid Intention-aware Frame (BIHIF) for sequence recommendation to extract user's real intention. Major work of this paper can be concluded as follow:

- Propose a kind of neural network to extract user's real intention, and design a Behavior-Item based Hybrid Intention-aware Frame (BIHIF) for sequence recommendation.
- Use MLP and GRU to generate BIHIF examples respectively.
- The proposed models are evaluated on three real world datasets to prove that our BIHIF model is more effective than those models which extract user's intention only from items.

2   Relate Work and Problem Description

2.1   Relate Work

Session-based recommendation is a subtask of recommender system, in which the recommendations are made according to the implicit feedbacks within the user session. This is a challenging task because the users are usually assumed to be anonymous, and the user preferences (such as ratings) are not provided explicitly, instead, only some positive observations (e.g. purchases or clicks) are available to the decision makers [7]. The research on session-based recommendation system has attracted much attention since 1990s [8], and was labelled under different names: pattern-based recommendation, using patterns for recommendations, rule-based recommendation, sequence-based recommendation, sequential recommendation, transaction-based recommendation, session-aware recommendation, next-item recommendation, next-basket recommendation, next-song recommendation, next-movie recommendation, and next-POI recommendation, etc. [16].

With the development of statistics and machine learning techniques, especially some time-series-related models like Markov chain models, Recurrent Neural Network (RNN) models, etc. Thanks to the fast development of deep learning techniques in recent years, model-based recommendation system have reached its peak since 2017. Many researchers rushed into this area and developed various neural models for the next-item or next-basket recommendations in the past two years [4].

Inspired by recent advances in natural language processing area [9] some deep learning based solutions have been developed and some of which represent the state-of-the-art in SRS research field [10–14]. Hidasi et al. [12] use deep recurrent neural networks with a gated recurrent unit to model session data, which learns session representation directly from previous clicks in the given session and provides recommendations of the next action. This is the first attempt to apply RNN networks for solving the session-based recommendation problem, thanks to the sequential modeling capability provided by the RNNs, their model can take into account the users' historical behavior when making predictions of the next move. Tan et al. [14] propose a data augmentation technique to improve the performance of the RNNs for session-based recommendation. Yu et al. [15] propose a dynamic recurrent model, which applies RNN to learn dynamic representation for each basket for user general interests at different times and captures global sequential behavior among baskets [16].

2.2   Problem Description

Problem definition: Session-based recommendation, which can also be called sequence recommendation, next item recommendation, etc., is to recommend items of interest to the user at the next moment based on the user's current session.

This article uses $I = \{i_1, i_2, \ldots, i_{t-1}, i_t\}$ to denote a collection of user interaction items in a user session, where i_x is the index of a total of m items. And this paper uses $B = \{b_1, b_2, \ldots, b_{t-1}, b_t\}$ to denote a collection of user's behavior for each item in a user session, where b_x is the index of a total of n behaviors. This article proposed a model M, whose input is $I = \{i_1, i_2, \ldots, i_{t-1}, i_t\}$ and $B = \{b_1, b_2, \ldots, b_{t-1}, b_t\}$, and output is $y = M(I, B)$. As for the output, $y = \{y_1, y_2, \ldots y_m\}$, and y_x denote the rating of the item to be recommended. And then the model selects the top k highest-rated items from all the candidate items for recommendation.

3   BIHIF and Its Instances

In this part, we introduce the Behavior-Item based Hybrid Intention-aware Frame (BIHIF), and then introduce the models that we implement using MLP and GRU respectively.

3.1   BIHI Framework

This article proposes a session-based recommendation framework for extracting user's true intentions in combination with behaviors and items, which named Behavior-Item based Hybrid Intention-aware Frame. The basic idea of the framework is to generate an implicit vector that can represent the true intent of the user in the current session, and then calculate the score for each item in the candidate item set based on the implicit vector. As shown in Fig. 2, the entire framework is divided into an Input Layer, an Embedding Layer, Intent Abstract Layers, and a Score Layer.

In the model, each item in a session and the behavior that occur on each item are represented by a one hot vector, which is used as input to the model. Above the Input Layer is the Embedding Layer, which is a fully connected layer that maps the input sparse vector to a d-dimensional dense vector to obtain the hidden vector set of the items $I = \{i_1, i_2, \ldots, i_t\}$, $i_x \in \mathbb{R}^d$, Similarly, the implicit vector set of behaviors $B = \{b_1, b_2, \ldots, b_t\}$, $b_x \in \mathbb{R}^d$. The resulting dense vector is then input into the Intent Abstract Layers, and the Intent Abstract Layers are divided into three parts: the item layer, the behavior layer, and the hybrid layer. In item layer, the output is the user's intention v_i extracted from the items. In behavior layer, the output is the user's intention v_b extracted from the behaviors and the items, and the two vectors are concatenated and then passed to the hybrid layer through a fully connected layer to extract the user's true intention v_r, and finally pass v_r to the Score Layer to calculate the scores of all candidate items according to the user's intent. And then the model selects the top k highest-rated items for top-K recommendation.

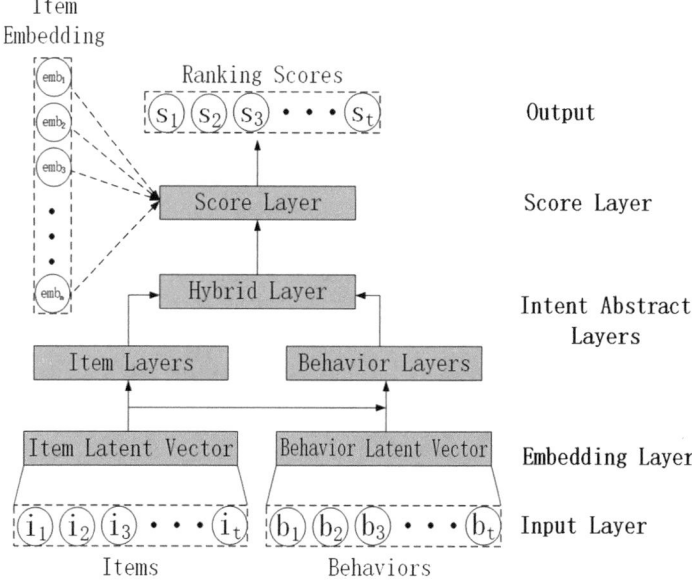

Fig. 2. Detail of BIHI framework.

3.2 Generate Models Using MLP

This section describes how to use MLP to extract the user's true intent base on the BIHI framework. As shown in Fig. 3, comparing the framework of Sect. 3.1, the input layer and the embedding layer of the model are unchanged, and the output of embedding layer are the set of items and behavior hidden vectors $I = \{i_1, i_2, \ldots, i_t\}$, $B = \{b_1, b_2, \ldots, b_t\}$. In the item layer, we process the hidden vector of the item, and get two vectors v_g and v_c. First, v_g denotes the general intent of the user in the current session, which is defined as the average of all the hidden vectors of the session:

$$v_g = \frac{1}{t} \sum_{x=1}^{t} i_x \tag{1}$$

Second, v_c denotes the current intent of the user in the current session. The last item of the session is used to represent the user's current intent: $v_c = i_t$.

The proposed attention net uses the attention mechanism to calculate the user's preference for the items, namely the attention weight. In this article, attention net uses a simple feedforward neural network to calculate the attention weight of each item in the current session. For a given item $i_x \in I$, the attention weight function is defined as:

$$a_x = W_0 \sigma \left(W_1 v_g + W_2 v_c + W_3 i_x + b \right) \tag{2}$$

which $W_1, W_2, W_3 \in \mathbb{R}^{d \times d}$ are weight matrices, $b \in \mathbb{R}^d$ is a bias vector, and $\sigma(\cdot)$ is a sigmoid function. $W_0 \in \mathbb{R}^{1 \times d}$ is the weight vector, a_x denotes the user's attention weight to item i_x.

After calculating the attention weights of all items in a session, all the item vectors are weighted and averaged and then input to a fully connected layer. The resulting output is the user intent extracted based on the items. The formula is as follows:

$$v_i = f \left(W_i \sum_{x=1}^{t} a_x i_x + b_i \right) \tag{3}$$

which $v_i \in \mathbb{R}^d$ denotes the user intent extracted based on items, $f(\cdot)$ is a non-linear activation function (we use tanh in this paper), $W_i \in \mathbb{R}^{d \times d}$ is a weight matrix, $b_i \in \mathbb{R}^d$ is a bias vector.

In the behavior layer, the extraction process of the user intent is similar to the extraction process of the item layer. The difference is that the user's behavior embedding vector $b_x \in B$ and the item embedding vector $i_x \in I$ are concatenate together to obtain a new embedding vector I_x, which is defined as follows:

$$I_x = concatenate(i_x, b_x) \tag{4}$$

The next process is still to process the embedding vector, and then get the vectors V_a and V_c that denote the user's general intention and current intention respectively. Input V_a and V_c into the attention net, calculate the attention weight like the Eq. 2, and calculate user intent based on behaviors and items like the Eq. 3.

Next, the user intent vectors v_i and v_b are concatenated and then passed to the hybrid layer. The output of hybrid layer is v_r, which denotes user's real intent, the function is defined as:

$$v_r = f(W_r(v_i, v_b) + b_r) \tag{5}$$

which $f(\cdot)$ is a non-linear function (we user tanh in this paper), $W_r \in \mathbb{R}^{d \times 2d}$ is a weight matrix, $b_r \in \mathbb{R}^d$ is a bias vector.

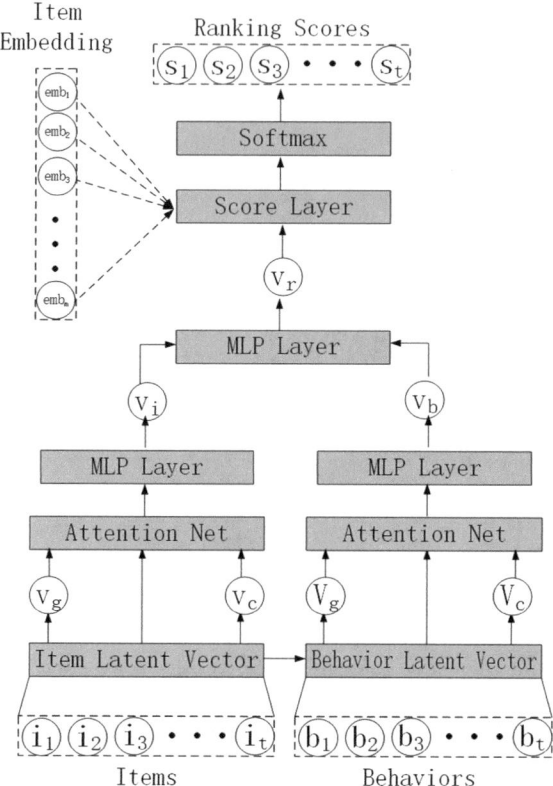

Fig. 3. Generate models using MLP

In the predicting phase, candidate item set $E = \{emb_1, emb_2, \ldots, emb_m\}$ are passed to the score layer, which $emb_i \in \mathbb{R}^d$. Given a candidate item $emb_i \in E$, use similarity score function to calculate score s_i. Past research usually directly uses the product of the user's real intent vector v_r and the candidate item embedding vector emb_i as the predicted score for the item. The function defined as:

$$s_i = v_r^T emb_i \tag{6}$$

However, considering the possible intent transfer [16], there is also a popular method to put the user's current intent into the score layer. Specifically, put the user's current intent v_c into a fully connected layer, resulting in a vector v_c'. Then, given a candidate item emb_i, similarity score function defined as:

$$s_i = \sigma\left(<v_r, v_c', emb_i > \right) \tag{7}$$

which $\sigma(\cdot)$ is a sigmoid function, $< \cdot >$ is the trilinear product of three vectors as:

$$<a,b,c> \; = \sum_{i=1}^{d} a_i b_i c_i = a^T (b \odot c) \qquad (8)$$

which $a, b, c \in \mathbb{R}^d$, and $\odot$ denotes the Hadamard product, i.e. the element-wise product between two vectors b and c.

We use two similarity score function to build models and compare the effects in the experiment. After calculate all candidate items' score, the score vector $S = \{s_1, s_2, \ldots, s_m\}$ is passed to a softmax layer, and the output denotes user's preference degree of each item in candidate set. The model will select top-K items as the recommended result.

3.3 Generate Models Using GRU

Section 3.2 uses MLP to extract the user's real intent, but the more popular method is to use RNN for sequence recommendation. Therefore, this section uses GRU to extract the user's real intent. Experiments show that the model using GRU is better than the model using MLP in some scenarios.

The architecture of the model is shown in Fig. 4. The specific user intent extraction process is similar to the process in Sect. 3.2.

In the intent abstract layers, we first extract the user's main intent v_i based on items. The GRU layer's input is embedding vectors $I = \{i_1, i_2, \ldots, i_t\}$, and the output is $h = \{h_1, h_2, \ldots, h_t\}$, which $h_x \in \mathbb{R}^h$. And then h_g and h_c are used to denote user's general intent and current intent respectively, which h_g is the average vector of all the vectors of h, and the last vector of h is used to denote the current intent: $h_c = h_t$. Pass h and h_g and h_c to the attention net, and the output is v_i, which denotes user intent. Extract the user intent v_b based on behaviors and items in a similar way. Next, the extracted user intent vectors v_i and v_b are passed to the hybrid layer, and the user's real intent vector v_r is extracted from the intent abstract layers. The calculation process is similarity to that in Sect. 3.2. In the predicting phase, we also use two different similarity score function to construct models. In this paper, loss function is defined as the cross-entropy, which can learn the parameters of the model using the standard small batch gradient descent method:

$$L(p, q) = -\sum_{i=1}^{m} p_i \log(q_i) \qquad (9)$$

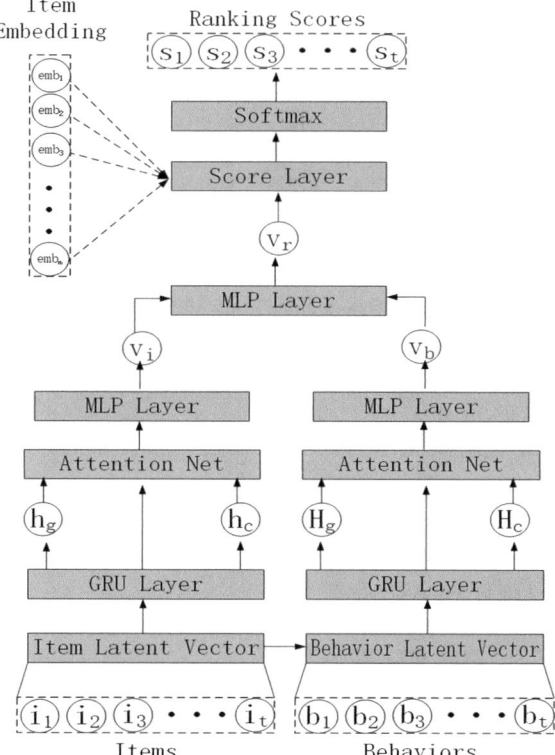

Fig. 4. Generate models using GRU.

4 Experiments

This section will first introduce datasets, comparing algorithm and evaluation metrics in experiments. Then we analyze performance of our model with baseline algorithms.

The experiment is to answer two questions:

- Are the two kinds of models generated based on BIHIF superior to the models with no use of user's behavior data?
- Do different similarity score function lead to distinction of model performance? And which function is better?

4.1 Datasets

This paper use three real world datasets: Steam dataset, Retailrocket dataset and Recsys challenge 2019 (test) dataset.

Steam dataset is a dataset that are players behavior dataset published on Kaggle website. It records about 200000 interaction, including 6508 clients and 5155 commodities. User's behavior consists of *purchase* and *play* in this dataset.

Ratailrocket dataset is a dataset about retail recommendation system that is published on Kaggle website. It includes 26987 commodities and 23073 clients. And records 2756101 interaction in four months. The user behavior in this dataset contains *view, add2cart* and *transaction*.

Recsys2019 dataset is the test dataset in challenge 2019 competition. It includes 60510 clients and 76273 commodities, and record 3872336 interaction. The user behavior in this dataset contains *interaction item image, clickout item, interaction item info, interaction item rating, interaction item deals* and *searching for item*.

To ensure the experimental results, we have preprocessed the data set. First, we filter out the users whose interaction records are less than 10 and the commodities that appear less than 5. Then we delete the sessions whose interaction items are less than 3. We take out 80% of all the sessions as training dataset and the rest as test dataset. The last item of each session becomes label. And the statistical information of preprocessed datasets is shown in Table 1.

Table 1. The statistical information of three datasets

Dataset	Steam	Retailrocket	Recsys2019
#users	6508	23073	60510
#items	5155	26987	76273
#behaviors	2	3	6
Avg. session length	28	9.9	27.9

4.2   Baseline Methods

Because the foothold of this paper is to use the user behavior data for sequence recommendation. The comparison algorithm selects three sequence recommendation neural network models that do not use user behavior data: GRU4Rec, NARM, and STAMP. Here is a brief introduction to the three models:

GRU4Rec [11]: An RNN based deep learning model for session based recommendation, which consists of GRU units, it utilizes session-parallel mini-batch training process and also employs ranking-based loss functions during the training.

NARM [13]: An RNN based state-of-the-art model which employs attention mechanism to capture main purpose from the hidden states and combines it with the sequential behavior as final representation to generate recommendations.

STAMP [16]: This is a model proposed in 2018. This model uses MLP and attention mechanism to extract the long-term and short-term intent of the user in the session for the next-item recommendation.

4.3   Evaluation Metrics

Because the recommendation system only recommends a limited number of items at a time, items that meet the user's needs should appear in the limited number of items listed in the system recommendation list [6]. So we use the following two evaluation indicators to evaluate the quality of the recommended list generated by the model.

P@20: The first evaluation metrics we used is P@20, which is the probability that the user who actually clicks on the test set will appear in the top 20 of the recommended list [16]. P@N does not consider the order in which the user actually clicks on the item in the recommendation list, as long as it appears in the top N position of the recommendation list. The formula is as follows:

$$P@K = \frac{n_{hit}}{N} \tag{10}$$

which N is the number of samples in the test set, and n_{hit} is the number of samples that the actual clicked item appears in our top-K recommendation list.

MRR@20: Another evaluation metrics used in the experiment is MRR@20, which is the reciprocal average of the serial number of the user's actual clicked item in the recommended list [13]. If the item does not appear in the top 20 of the recommended list, then the MRR is set to zero. MRR considers the location of the item within the quality of the evaluation recommendation list and is a very important indicator in some order-sensitive tasks. The formula is as follows:

$$MRR@K = \frac{1}{N} \sum_{t \in G} \frac{1}{Rank(t)} \tag{11}$$

4.4   Parameter Settings

In order to ensure the fairness of the experiment, the parameters of all experiments in this paper are set as follows: the dimension of the item and behavior embedding vector is 50, the hidden layer dimension of GRU is 100, the learning rate is 0.001, the optimizer selects Adam, batch size = 512, and the epoch selects 10. For the training set, we use popularity-based negative sampling for each session. The number of negative samples is 128 [6, 17]. For the test set, we randomly sample each session from the user's negative items. The number of negative samples is 128 [6]. Although this will increase the evaluation index of the model, we believe that such a comparative experiment setting is still fair [18, 19].

4.5   Experimental Results and Analysis

In this paper, we propose a neural network framework BIHI for sequence recommendation, and based on the framework, MLP and GRU are used to construct the model, and the user's real intent is extracted for recommendation. In the prediction phase of the model, we also used two different similarity score function, so we implemented four BIHI-based models, namely MLP_2, MLP_3, GRU_2 and GRU_3, which 2 and 3 represent the different similarity score function. Because one is the multiplication of two vectors and the other is the multiplication of three vectors.

The experimental results are shown in Table 2. The optimal values for each column are marked in red. Next, analyze the experimental results and answer the questions raised at the beginning of this chapter.

Table 2. The experimental results

Dataset	Steam		Retailrocket		Recsys2019	
Measure	P@20	MRR@20	P@20	MRR@20	P@20	MRR@20
GRU4Rec	18.2	5.6	14.6	2.9	53.3	20.5
NARM	25.1	12.5	46.8	30.6	52.1	38.9
STAMP	20.1	4.1	19.2	3.8	70.1	56.2
MLP_2	32.3	20.5	**51.5**	27.6	73.9	59.5
MLP_3	28.3	12.3	40.6	**32.7**	**81.7**	**78.9**
GRU_2	**40.3**	**23.3**	37.6	11.9	56.4	24.9
GRU_3	30.2	13.3	40.7	26.5	81.6	78.5

First, are the two kinds of models generated based on BIHIF superior to the models with no use of user's behavior data?

When GRU4Rec is not considered, NARM and STAMP represent models that use GRU and MLP to extract user intent for sequence recommendation. Then this experiment wants to verify that the model generated by MLP and GRU can be better than the corresponding two models in the performance based on BIHI framework. From Table 2 we can see that in the Steam dataset and RecSys2019 dataset, MLP_2 and MLP_3 are superior to STAMP in both evaluation indicators, and GRU_2 and GRU_3 are superior to NARM in both evaluation indicators. However, on the Retailrockt dataset, although MLP_2 and MLP_3 are still better than STAMP, GRU_2 and GRU_3 are not superior to NAMR. So we have conducted some analysis on three datasets. We count the number of times the three datasets correspond to different behaviors, as shown in Table 3.

Among them, the *purchase* of the Steam dataset means to buy the game, the *play* means to play the game; the *view* of the Retailrocket dataset means to browse the goods, the *add2cart* means to add the goods to the shopping cart, and the *transaction* means that the user purchases the goods; the *clickout* item of the Recsys dataset means user makes a click-out on the item and gets forwarded to a partner website, the *interaction item rating* means user interacts with a rating or review of an item, the *interaction item info* means user interacts with item information, the *interaction item image* means user interacts with an image of an item, the *interaction item deals* means user clicks on the view more deals button, and the *search for item* means user searches for an accommodation.

Table 3. The number of times the three data sets correspond to different behaviors

Dataset	Behavior type	Nums
Steam	Purchase	129511
	Play	70489
Retailrocket	View	2664312
	Add2cart	69332
	Transaction	22457
Recsys2019	Interaction item image	2485778
	Clickout item	172207
	Interaction item info	51801
	Interaction item rating	39655
	Interaction item deals	30947
	Search for item	21918

As can be seen from the Table 3, the ratio of the two behaviors in the Steam dataset is close to 1.8:1; for the Recsys dataset, the ratio of the six behaviors is about 113:8:2.4:1.8:1.4:1, the ratio of most behaviors to minority behaviors is about 8:1; for the Retailrocket dataset, the proportion of the three behaviors is about 118.6:3:1, the ratio of most behaviors to minority behaviors is about 30:1. This may be the reason why the effects of GRU_2 and GRU_3 are not good for NARM, and it will be our next direction.

Second, do different similarity score function lead to distinction of model performance? And which function is better?

In the Table 2, we can find that for the Steam dataset, MLP_2 and GRU_2 are better than MLP_3 and GRU_3, respectively. For the RecSys2019 dataset, this situation is reversed. Our analysis is as follows. For the dataset of the game field of Steam, the item, which is game, the user viscosity is very high. Even if the user has played a game, the probability that he will actually want to play again is very large. At this time, it is not very useful to consider the possible transfer of intent of the user. For RecSys dataset, a dataset that records the user's browsing of the hotel, the user is more likely to have a transfer of intent, so introducing the user's current intent into the prediction stage can make the model work better.

5   Conclusion

In this paper, we propose a new behavior-item based frame for sequence recommendation—BIHIF. BIHIF extracts user's intent based on items and behaviors respectively, and then use hybrid intention-awareness to extract user's real intent. Next, this paper use MLP and GRU to generate models based on this frame respectively. The experimental results on both two datasets exceeded the baseline method, but there is a problem with a dataset with a large proportion of user behavior types. In the future we plan to research the impact of user behavior ratio on the model, and for different user behavior we also consider adding some other information (e.g. dwell time, etc.) to improve the effect of the model.

Acknowledgments. This work was supported by National Key Research and Development Program of China (2018YFB1004500), National Natural Science Foundation of China (61877048), Innovative Research Group of the National Natural Science Foundation of China (61721002), Innovation Research Team of Ministry of Education (IRT_17R86), Project of China Knowledge Centre for Engineering Science and Technology, the Natural Science Basic Research Plan in Shaanxi Province of China under Grant No. 2019JM-458.

We thank the anonymous reviewers for taking time to read and make valuable comments on this paper.

References

1. Loyola, P., Liu, C., Hirate, Y.: Modeling user session and intent with an attention-based encoder-decoder architecture. In: Proceedings of the Eleventh ACM Conference on Recommender Systems, pp. 147–151. ACM (2017)
2. Zheng, L., Lu, C.T., He, L., et al.: Mars: memory attention-aware recommender system. arXiv preprint arXiv:1805.07037 (2018)
3. Bai, T., Du, P., Zhao, W.X., et al.: A long-short demands-aware model for next-item recommendation. arXiv preprint arXiv:1903.00066 (2019)
4. Wang, S., Cao, L., Wang, Y.: A survey on session-based recommender systems. arXiv preprint arXiv:1902.04864 (2019)
5. Zhang, S., Tay, Y., Yao, L., et al.: Next item recommendation with self-attentive metric learning. In: Thirty-Third AAAI Conference on Artificial Intelligence, November 2019
6. Fang, H., Guo, G., Zhang, D., et al.: Deep learning-based sequential recommender systems: concepts, algorithms, and evaluations. In: International Conference on Web Engineering, pp. 574–577. Springer, Cham (2019)
7. He, X., Zhang, H., Kan, M.-Y., Chua, T.-S.: Fastmatrix factorization for online recommendation with implicit feedback. In: Proceedings of ACM SIGIR 2016. ACM, Pisa, Italy, pp. 549–558 (2016)
8. Wasf, A.M.A.: Collecting user access patterns for building user profiles and collaborative filtering. In: Proceedings of the 4th International Conference on Intelligent User Interfaces, pp. 57–64. ACM (1998)
9. Sutskever, I., Vinyals, O., Le, Q.V.: Sequence to sequence learning with neural networks. In: Proceedings of NIPS 2014, 08–13 December, pp. 3104–3112. MIT Press, Montreal (2014)
10. Balázs, H., Quadrana, M., Karatzoglou, A., Tikk, D.: Parallel recurrent neural network architectures for feature-rich session-based recommendations. In: Proceedings of ACM RecSys 2016, pp. 241–248. ACM, Boston (2016)
11. Hidasi, B., Karatzoglou, A., Baltrunas, L., Tikk, D.: Session-based recommendations with recurrent neural networks. In: Proceedings of ICLR 2015, 2–4 May. CoRR, San Juan, Puerto Rico (2015)
12. Hu, L., Cao, L., Wang, S., Xu, G., Cao, J., Gu, Z.: Diversifying personalized recommendation with user-session context. In: Proceedings of IJCAI 2017, IJCAI, Melbourne, Australia, pp. 1858–1864 (2017)
13. Li, J., Ren, P., Chen, Z., Ren, Z., Ma, J.: Neural attentive session-based recommendation. In: Proceedings of ACM CIKM 2017, Singapore, Singapore, pp. 1419–1428 (2017)
14. Tan, Y.K., Xu, X., Liu, Y.: Improved recurrent neural networks for session-based recommendations. In: Proceedings of DLRS 2016, 15 September, pp. 17–22. ACM, Boston (2016)

15. Yu, F., Liu, Q., Wu, S., Wang, L., Tan, T.: A dynamic recurrent model for next basket recommendation. In: Proceedings of ACM SIGIR'16, 17–21 July, pp. 729–732. ACM, Pisa (2016)

16. Liu, Q., Zeng, Y., Mokhosi, R., et al.: STAMP: short-term attention/memory priority model for session-based recommendation. In: Proceedings of the 24th ACM SIGKDD International Conference on Knowledge Discovery & Data Mining, pp. 1831–1839. ACM (2018)

17. Kang, W.C., McAuley, J.: Self-attentive sequential recommendation. In: 2018 IEEE International Conference on Data Mining (ICDM), pp. 197–206. IEEE (2018)

18. He, X., Liao, L., Zhang, H., et al.: Neural collaborative filtering. In: Proceedings of the 26th International Conference on World Wide Web. International World Wide Web Conferences Steering Committee, pp. 173–182 (2017)

19. Hidasi, B., Karatzoglou, A.: Recurrent neural networks with top-k gains for session-based recommendations. In: Proceedings of the 27th ACM International Conference on Information and Knowledge Management, pp. 843–852. ACM (2018)

Author Index

© Springer Nature Switzerland AG 2020
K.-M. Chao et al. (Eds.): ICEBE 2019, LNDECT 41, pp. 621–622, 2020.
https://doi.org/10.1007/978-3-030-34986-8

Printed in the United States
By Bookmasters